Fishery Ecosystem Dynamics

Fishery Ecosystem Dynamics

Michael J. Fogarty
Northeast Fisheries Science Center,
National Oceanic and Atmospheric Administration, USA

Jeremy S. Collie
Graduate School of Oceanography,
University of Rhode Island, USA

OXFORD
UNIVERSITY PRESS

OXFORD
UNIVERSITY PRESS

Great Clarendon Street, Oxford, OX2 6DP,
United Kingdom

Oxford University Press is a department of the University of Oxford. It furthers the University's objective of excellence in research, scholarship, and education by publishing worldwide. Oxford is a registered trade mark of Oxford University Press in the UK and in certain other countries

First Edition published in 2020

Impression: 1

Published in the United States of America by Oxford University Press
198 Madison Avenue, New York, NY 10016, United States of America

British Library Cataloguing in Publication Data
Data available

Library of Congress Control Number: 2019948498

DOI: 10.1093/oso/9780198768937.001.0001

ISBN 978–0–19–876893–7 (hbk.)
ISBN 978–0–19–876894–4 (pbk.)

Printed and bound by
CPI Group (UK) Ltd, Croydon, CR0 4YY

Preface

Longstanding calls for the adoption of more holistic approaches to fisheries management incorporating broad ecosystem principles are now being translated into action on a global basis. The transition from concept to implementation is accompanied by the need to further establish and evaluate the conceptual and analytical framework for Ecosystem-Based Fishery Management. Training the next generation of scientists to take on this important challenge emerges as a critical need. Here, our objectives are to provide an introduction to this topic for the students who will carry on this work, to illuminate the deep and often underappreciated connections between basic ecology and fishery science as a whole, and to explore the implications of these linkages in crafting management strategies for the twenty-first century.

It is important to set out with some common understanding of the scope and motivation for this work. Here, we define a fishery ecosystem as one in which the direct and indirect effects of extraction of biomass as yield and other human interventions play an important role in shaping the overall dynamics of an aquatic ecosystem. We view the ecosystem through the lens of exploited species, their interactions with other species, and the interplay of environmental forcing and ecosystem dynamics. A fishery ecosystem is an important form of Coupled Social–Ecological System (CSES). Fishers provide an invaluable ecosystem service and contribute substantially to food security throughout the globe. The challenge is to define harvesting policies that maintain ecosystem structure and function to protect both the intrinsic value of these systems and the flow of vital ecosystem services.

Our objective is to show how the effects of harvesting are manifest through direct and indirect pathways affecting ecosystem structure. The scope of human intervention in fishery ecosystems is not limited to the direct effects of extraction of resources. In many instances, humans have actively engaged in attempts at ecosystem engineering with the objective of enhancing yields. For example, large-scale hatchery operations for salmonids and other species involve the addition of large numbers of juvenile fish into aquatic ecosystems in ways that can affect overall ecosystem structure and function. The deliberate introduction of non-native species in general has also been practiced, for example with addition of prey species to enhance production of economically important predator species. These actions have often been accompanied by unintended consequences. Accordingly, we will not restrict our attention just to extractive activities but will adopt a broader view encompassing a range of human interventions (and reciprocal interactions between human and natural communities) under our definition of an aquatic CSES.

We define Ecosystem-Based Fisheries Management (EBFM) as an integrated strategy for a defined fishery ecosystem that directly considers interrelationships among the parts of the system (including humans) and accounts for other environmental influences (including climate variability) and impacts on the ecosystem. It is an approach that must be adaptive in response to changing conditions and as scientific understanding accrues. It accounts for uncertainty and the mix of different (and potentially competing) societal goals and objectives. Resource management is at heart a social

Fishery Ecosystem Dynamics. Michael J. Fogarty and Jeremy S. Collie, Oxford University Press (2020).
DOI: 10.1093/oso/9780198768937.001.0001

and political process informed by science. This book is not about Ecosystem-Based Fisheries Management per se, with its deeply interwoven connections between the socio-political and ecological sciences. Rather, it concerns analytical frameworks, and their ecological bases, that can support EBFM. Consideration of estimation issues and model fitting are largely taken up in a companion website in connection with examples provided in the main text. Finally, we note that, although the effects of fisheries are broadly recognized as perhaps the dominant human impact in many aquatic ecosystems, it is also critically important to recognize that EBFM must be ultimately established within the broader context of aquatic Ecosystem-Based Management (EBM), which addresses the cumulative impacts of the full spectrum of human activities affecting marine and freshwater ecosystems.

About this Book

We have structured this book to meet several objectives. First, for the student, we wish to demystify the process of developing and applying mathematical models for management purposes. It is our experience that students come to this field with a broad range of backgrounds and prior training in mathematics. Accordingly, we have adopted a stepping stone approach in which we introduce simple models as foundations for subsequent further development and added complexity. Particularly in the earlier sections of this book, we have shown derivations in more detail than is perhaps customary. For the most part, we have put these more detailed derivations in text boxes that can be skipped by students with more experience. We introduce an array of models along a continuum of complexity. Our objective is not simply to provide a catalogue of models but rather to show how factors can be sequentially added to address specific questions by altering the dimensionality of simple underlying models. We view models as statements of hypotheses concerning the governing factors in aquatic ecosystems. They are of necessity abstractions of reality. The real world is often considerably messier than the analytical metaphors we construct and this must be kept firmly in mind. If we are successful, we can specify models that reduce a problem to its essentials. If the student emerges with the confidence to alter models to address specific needs or questions of interest by adding or modifying elements of the models, we will have met one of our important goals. Depending on the course level and background of the students in mathematics and ecology, instructors may choose to skip earlier sections and concentrate on later ones.

We particularly wish to encourage the exploration of the models described in this book through alteration and manipulation of their parameters and basic structural elements—in essence to play with the models to see what they can do. To this end, we have provided code in R, an extremely powerful programming language now widely used in many scientific fields including ecology and fisheries science, in a companion to this book in which data, models, and R code allow the student to replicate many of the examples developed throughout the book (http://www.Oxford.com/companion/Fogarty&Collie/). As with any programming language, there is a steep learning curve but a number of excellent of texts are now available that demonstrate the essentials of the R environment with respect to: data structures and manipulation; programming and analysis; and visualization. We strongly recommend *Ecological Models and Data in R* (Bolker 2008); *A Practical Guide to Ecological Modelling: Using R as a Simulation Platform* (Soetaert and Herman 2009); and *A Primer of Ecology with R* (Stevens 2009) as excellent introductions into the analysis of fundamental ecological models in R. Ogle (2016) provides a very informative basic entrée into fisheries applications in his *Introductory Fisheries Analyses with R*. In addition, there are now a number of R packages available for ecological and fisheries applications that we will draw on.

We have divided this book into three main sections. In the first, we provide an overview of classical ecological models and their properties. Our objective is to provide an overview of the ecological processes and mechanisms underlying these models. We hope to provide both a foundation for the broadly applicable ecological training required of the students and the point of departure for more specialized issues related to fisheries ecosystems.

In Part I we start with the simplest models of population dynamics. We progressively add complexity to the models to explore the ramifications of increasing dimensionality in the behavior of the models. This increased dimensionality encompasses consideration of demographic structure, spatial processes, multispecies interactions, and energy flow and utilization in aquatic ecosystems. We show how extremely complex behaviors can emerge even in relatively simple deterministic systems. Following an Introductory chapter in which we identify the key issues to be explored in the book and some important historical developments, we introduce in Chapter 2 single-species models in which the abundance of the population itself does not affect its rate of growth. We explore the properties of these models and describe the implications of adding factors such as age structure.

In Chapter 3 consideration of the effects of abundance of the population on its own growth rate is added. This simple modification radically changes our perception of the resilience of the population to external forcing factors (including harvesting). We next add explicit consideration of interactions among species into the mix in Chapter 4 (predation and parasitism) and Chapter 5 (competition and mutualism). We illustrate how predator–prey interactions, competition between pairs of species, and parasitism and disease can be modeled. Extension to broader multispecies systems is taken up Chapter 6 in an examination of community dynamics. Descriptors of spatial attributes of populations and ways in which introducing spatial considerations into the models alter their dynamical properties and response to external perturbations are explored in Chapter 7.

Part II takes up the theme of production processes at the individual, population, community, and ecosystem levels. The productivity of all ecosystems is ultimately set by the amount of energy fixed at the base of the food web. In Part I we followed the tradition established in the ecological literature for single- and multispecies systems using numerical abundance or density as the model currency. However, in a fisheries context, we are most often concerned with biomass and production rather than numbers because the weight of the catch and production is usually more meaningful than the number caught in this context. More importantly, we wish to make a connection with fundamental energetic principles. In Part II, we show how framing our models in terms of biomass and production allows us to make this translation. The interpretation of these models and their parameters accordingly takes on additional dimensions relative to the number- or density-based models in the opening chapters of this book.

In Chapter 8, the elements of production at the level of the individual are introduced. While traditional quantitative fisheries texts develop descriptive models of these processes (growth and reproduction) in detail, our account differs in approaching these issues from a bioenergetic perspective to allow a connection with broader ecosystem principles that dominate the latter part of the book. Explicit consideration of population production is next taken up in Chapter 9 with a focus on recruitment, growth, and mortality. Recruitment is typically the most variable component of production at the population level. Our treatment of mortality provides the linkage to broader ecosystem considerations through an emphasis on the components of natural mortality (particularly predation and disease). We then address ecosystem-level production processes in Chapter 10. Patterns of energy flow and utilization are central to the themes explored in this chapter.

In Parts I and II of this book we focus on the ecology of the non-human elements of fishery ecosystems. In Part III we shift our attention to the interplay between human intervention in aquatic ecosystems and the dynamics of these systems. Ultimately we do not manage aquatic ecosystems per se, but rather attempt to manage human behavior and impacts on these systems. Management is much more likely to be effective if we consider fishery ecosystems as coupled social–ecological systems. Viewed in this light, it is the two-way interaction of social and "natural" components of the system that is critical.

In this Part we again follow the template of introducing the set of simpler considerations that emerge from systems of low dimensionality to introduce basic concepts. We then progress to higher dimensional systems in which multispecies and ecosystem considerations become prominent. We explore the ways in which multispecies models can provide a bridge between single-species models and full ecosystem approaches. In particular, we seek to make the connection with current management practices by expanding the scope to encompass broader ecological considerations while retaining familiar concepts such as management reference points, but now translated to a multispecies setting.

In Chapter 11 we set the stage by considering the realm of traditional fisheries analysis and management focusing on single-species models. Entire books have of course been devoted to this topic and we cannot do this important area justice in a single chapter. We seek to complement existing treatments of fisheries models by exploring some topic areas that are often accorded less attention, including the possibility of alternative stable states in exploited populations and complex dynamical behavior. Our objective is to provide the background for moving toward the broader incorporation of ecological principles in assessment and management. Chapter 12 introduces the concept of harvesting at the community level. We describe models spanning a range of complexity, again progressively moving from models of lower to higher dimensionality. In Chapter 13, we next move from assemblages of species to whole ecosystems with a focus on energy flow as a fundamental consideration, again arrayed along a continuum of model complexity. We trace the dynamics of these systems from primary producers through exploited food webs and consider how fundamental energetic constraints play a central role in the observed dynamical behavior of these systems. Prediction is a critically important objective in support of fisheries management. Many of the models described in preceding chapters play an important role in this process. In Chapter 14 we take up the topic of Empirical Dynamic Modeling to complement the standard approaches to prediction in fisheries analysis. Here, we let "let the data speak" in the construction of nonlinear, nonparametric models that have recently shown considerable promise in fisheries forecasting. We draw the book to a close in Chapter 15 with a summary of some of the key issues in moving toward successful implementation of EBFM.

Acknowledgments

We are most grateful for the helpful reviews of drafts of chapters in this book by Richard Bell, Tim Essington, Henrik Gislason, Simon Jennings, Julie Kellner, Joseph Langan, Scott Large, Jason Link, Alec MacCall, Steve Munch, Paul Rago, Marie-Joëlle Rochet, Paul Spencer, and Mark Wuenschel. Many colleagues graciously shared data, analyses, and/or software: Andy Beet, Richard Bell, Don Bowen, David Chevrier, Karyl Brewster–Giesz, Erin Bohaboy, Steve Cadrin, Steve Campana, Kiersten Curti, Ethan Deyle, Sarah Gaichas, Rob Gamble, Jon Hare, Kimberly Hyde, Jim Ianelli, Raouf Kilada, Gordon Kruse, Joe Langan, Michael LaPointe, Sean Lucey, Marissa Litz, Alec MacCall, Ryan Morse, Stefan Neuenfeldt, Jan Ohlberger, Charles Perretti, Anna Rindorf, Charles Stock, Tadayasu Uchiyama, Morten Vinther, Ian Winfield, and Hao Ye. Andy Beet provided indispensable advice in development of the companion electronic supplement. Students in our classes provided extremely valuable feedback and suggestions for improvements during the development of this book.

The views expressed in this work are those of the authors and do not necessarily reflects those of the National Marine Fisheries Service.

This work would not have been possible without the support of our wives Anne and Elizabeth and our children and we are most appreciative for their patience while this book was in preparation.

We dedicate this book to the memory of Terry Quinn, Saul Saila, J. Stanley Cobb, and John Steele – colleagues, friends, and mentors.

About the Authors

Michael J. Fogarty is a senior scientist at the Northeast Fisheries Science Center (NOAA Fisheries) and a visiting scientist at the Woods Hole Oceanographic Institution. He holds adjunct professorships at the Graduate School of Oceanography, University of Rhode Island, and the School of Marine Science and Technology, University of Massachusetts.

Jeremy S. Collie is a Professor of Oceanography at the Graduate School of Oceanography, University of Rhode Island.

Contents

CHAPTER 1

Introduction

1.1 Overview

Fisheries supply a critically important ecosystem service—providing high quality food resources to a burgeoning human population. Fishery products are among the most widely traded commodities in global markets. Fish and shellfish provide 3.1 billion people with nearly 20 per cent of their annual consumption of animal protein (FAO 2018). Perhaps more significantly, food derived from aquatic sources is particularly high in both micro- and macro-nutrients that in the developing world may not be readily available from other sources. Global consumption of fishery products recently reached an average of 20.3 kg per-person-per-year in 2016, a record high (FAO 2018). Food from aquatic sources is unquestionably a critical element of food security in many parts of the world. Recent total production from inland and marine sources (capture fisheries and aquaculture) reached 171 million mt in 2016 (FAO 2018; Figure 1.1). Capture fisheries in freshwater and marine systems accounted for 91 million mt while aquaculture production contributed 80 million mt. Marine fisheries accounted for 87 percent of the total landings from capture fisheries (Figure 1.1) while aquaculture from freshwater systems was dominant (64 percent of the total aquaculture yield). Employment in fisheries and aquaculture is correspondingly important, with an estimated 40.3 million individuals engaged in capture fisheries and 19.3 million people involved in aquaculture enterprises in 2016 (FAO 2018).

It is clear that capture fisheries and aquaculture operations make a vital contribution to human health and well-being. It is equally clear that unless we provide the necessary safeguards to protect aquatic ecosystems both for their intrinsic value and their importance to society, the sustainable flow of ecosystem services derived from these systems will be imperiled. It follows that the livelihoods of fishers and the social and economic health of fishing communities ultimately cannot be protected without first protecting the ecosystems on which they depend.

In this book, we seek to describe the analytical framework for the development of holistic approaches to fishery management based on broad ecological principles. Production processes at the population, community, and ecosystem levels provide the foundation for understanding potential fisheries yield and the limits to sustainable exploitation at each level.

In the remainder of this chapter, we provide an historical perspective on the development of the field. We also provide examples of how processes at the population, community, and ecosystem levels translate into complex patterns reflected in catch histories. The importance of confronting complexity in the analysis and management of these systems is clear.

1.1.1 Historical perspectives

The essential role of aquatic food resources in the historical development of human societies is well-established (Fagan 2017). The longstanding importance of fishing to human cultures can be traced through the prevalence and diversity of harvesting implements found in archeological sites, artistic depictions throughout prehistory and antiquity, and the recorded history of many civilizations. It has recently been hypothesized that the incorporation

Fishery Ecosystem Dynamics. Michael J. Fogarty and Jeremy S. Collie, Oxford University Press (2020). © Michael J. Fogarty & Jeremy S. Collie 2020.
DOI: 10.1093/oso/9780198768937.003.0001

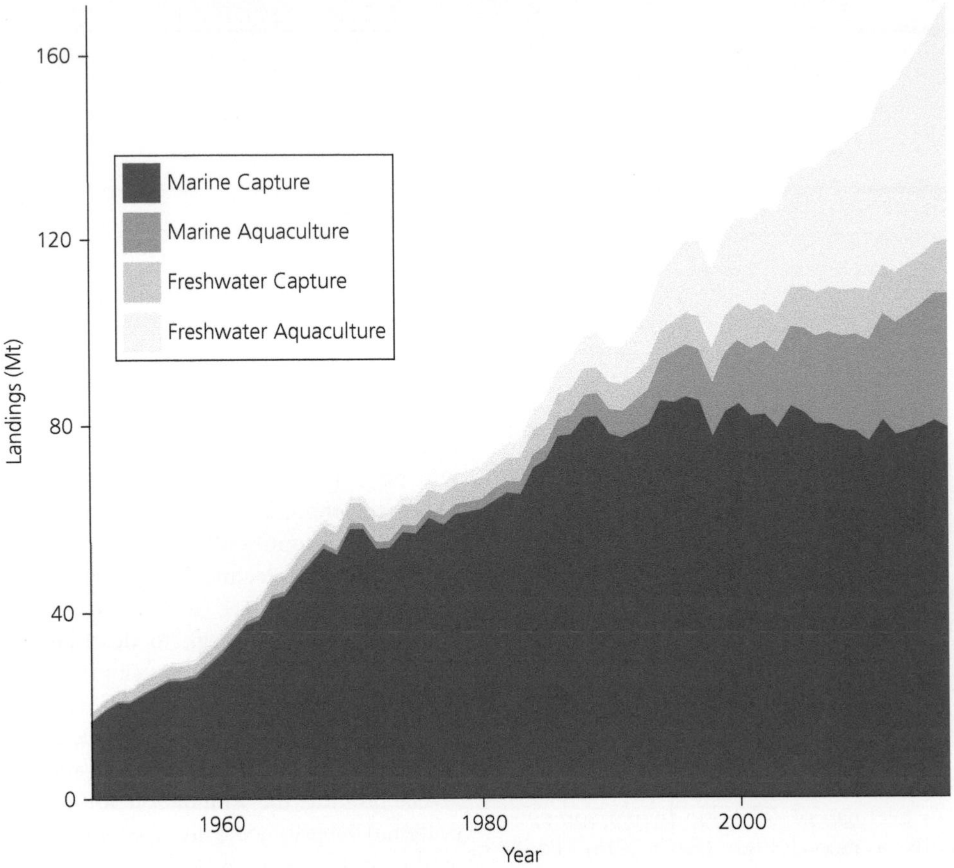

Figure 1.1 Global landings of marine and freshwater capture fisheries and aquaculture production (million t; FAO 2018).

of a protein and lipid-rich diet derived from aquatic sources in fact played a critical role in the evolutionary development of the human brain (Cunnane and Stewart 2010). A full appreciation of the deep connection between human societies and aquatic ecosystems is essential to understanding the past and the prospects for the future. Among the earliest fishing artifacts discovered are bone harpoons from sites dated to 90 000 yrs BP in Africa (Stringer and McKie 1998). The remains of fish and shellfish in extensive middens throughout the world attest to the prominence of aquatic resources in the diets of early coastal peoples. Human settlement sites were often placed in proximity to both water and food resources. As human populations grew, pressure on aquatic resources increased, ultimately expanding the range and scope of fishing activities to encompass both progressively deeper-water marine environments and enhanced culture in both freshwater and marine ponds and pools (Fagan 2017). In the west, demand for fish products increased dramatically under religious strictures against eating other forms of animal protein. In the Middle Ages, the number of days during which abstinence from meat was required approached half of the year (Fagan 2006), stimulating both increased fish culture (often associated with monasteries) and increased exploitation of marine resources.

Fishing has also served as a catalyst for exploration and trade for millennia (Kurlansky 1997; Fagan 2017) as fishing evolved from a subsistence level activity to a well-developed commercial enterprise. In northern Europe, the fortunes of the Hanseatic League during the medieval period were linked to the dominant role of fisheries in the trade of nations. Tithing records and other sources from the Middle Ages provide important indications of the importance of fishing in local and regional economies in Europe (Cushing 1988). Reports of cod harvest from Newfoundland extend back to the fourteenth century as Basque fishermen and others sought untapped fishing grounds in the new world (Kurlansky 1997). Although the importance of commercial fishing over the centuries has garnered much attention, we would be remiss in not mentioning the role of angling as a recreational and aesthetic pursuit, so wonderfully reflected by the enduring influence and popularity of Izaak Walton's *The Compleat Angler*, first published in 1653 and still in print.

Concerns about the broader ecological implications of fishing are not new. Interestingly, there is evidence that the direct and indirect effects of fishing on marine ecosystems elicited concerns as early as the fourteenth century (Anon. 1921; Sahrhage and Lundbeck 1992). In Britain, Acts passed in Parliament in 1350 and 1371 called for strict observance of "ancient statutes" directed to the preservation of the early life stages of fishes (Anon. 1921). In 1376, a petition was placed before the Commons to ban the Wondyrchoun, a type of dredge, because: "*... the great and long iron of the Wondrychoun presses so hard on the ground when fishing that it destroys the ... plants growing on the bottom under the water, and also the spat of oysters, mussels, and other fish, by which the large fish are accustomed to live and be nourished*" (Anon. 1921). In 1499, the use of trawls was prohibited in Flanders in response to observations that: "*... the trawl scraped and ripped up everything it passed over in such a way that it rooted up and swept away the seaweeds which served to shelter the fish; it robbed the beds of their spawn or fry ...*" (Anon. 1921).

Further evidence of longstanding concern over fishing practices deemed to be destructive to habitat includes bans imposed on the use of trawls in estuarine waters in the Netherlands in 1583 and in Britain in 1631 where the use of "*... traules [was] forbidden as well as of other nets which shall not have the meshes of the size fixed by law and orders ...*" and the establishment of trawling as a capital offense in France in 1584 (Anon. 1921).

1.1.2 Scientific developments

Fishery resources, particularly in the oceans, were long thought to be boundless. Thomas Henry Huxley, Darwin's champion, maintained a long-standing involvement in fisheries issues, serving on several commissions devoted to assessing the status of fisheries in Britain in the nineteenth century (Smith 1994). Huxley's interest appears to have been motivated, in part, by a deep seated commitment to the welfare of the working classes, including those in sea-going trades (Desmond 1994). He also maintained a life-long interest in the biology and evolutionary history of fishes. Huxley famously remarked in 1884 "*... the cod fishery, the herring fishery, the pilchard fishery, the mackerel fishery, and probably all the great sea-fisheries, are inexhaustible; that is to say that nothing we do seriously affects the number of fish ... given our present mode of fishing. And any attempt to regulate these fisheries consequently ... seems to be useless.*" Although Huxley had carefully qualified his remarks, the paradigm of inexhaustibility was broadly accepted, shaping the attitudes of many fishers, scientists, and politicians and complicating efforts to establish effective restrictions on fishing activities.

By the time of Huxley's remarks, however, declines in exploited fish populations had in fact been recognized. For example declines in coastal fishery resources in New England led to the establishment of the U.S. Fish Commission in 1871. Spencer Fullerton Baird, assistant secretary of the Smithsonian Institution and later the first Commissioner of US fisheries, was charged with leading the investigation into the potential causes of the decline. Baird identified five potential factors (none mutually exclusive) for the decline:

- *The decrease or disappearance of the food upon which the fish subsist, necessitating their departure to other localities.*
- *A change in location, either entirely capricious or induced by the necessity of looking for food elsewhere as just referred to.*
- *Epidemic diseases or peculiar atmospheric agencies such as heat, cold, etc.*
- *Destruction by other fishes.*
- *The agency of man: this being manifested either in the pollution of the water by the discharge into it of the*

refuse of manufactories, etc. or by excessive overfishing, or the use of improper apparatus.

Baird's prescient statement of the key issues and questions remains as valid today as it was in the late nineteenth century. His broad multidisciplinary perspective set an important precedent for future fisheries investigations, now reflected in the recognition of the importance of invoking ecosystem principles in fishery resource management. We will touch on many of these topics throughout this book. It has long been recognized that fishery resource species face multiple stressors and it is essential that we understand the cumulative effects of these factors on their productivity.

In contrast to the long history of human harvest of the oceans, large-scale scientific endeavors in support of resource management (and the understanding of underlying basic ecological principles) are comparatively recent. Indeed, the term ecology (*oecologie*) itself was not coined until the 1860s by Ernst Haeckle, while extensive written records of large-scale fisheries pre-date this landmark by several centuries. In many instances, fish and shellfish populations had already been substantially altered by fishing prior to the development of a scientific framework within which to evaluate these changes. True baseline conditions therefore are not known and can only be inferred.

Our focus is on quantitative methods; accordingly here we will briefly trace the early development of modeling approaches applied to fishery systems encompassing both single-species and broader multispecies/ecosystem contributions. We will emphasize historical developments from the turn of the twentieth century until the mid-1980s, leaving more recent developments over the last several decades to the remainder of the book.

The distinguished Russian scientist, Fedor I. Baranov, provided perhaps the earliest full quantitative treatment of fishery dynamics with a focus on single-species issues. Baranov's contributions, starting in 1914 with his treatise "On Overfishing" were unfortunately not widely known in the west for several decades and a full compilation of his translated works was not available until much later (Baranov 1976). His famous catch equation, later independently derived by Thompson and Bell (1934) remains a central contribution to understanding fishery dynamics.

Intriguing changes in relative species composition of the catch in the Adriatic Sea following the cessation of fishing during World War I were noted by the Italian fisheries ecologist, Humberto D'Acona. These observations prompted him to enlist the help of the noted mathematician Vito Volterra and his future father-in-law, in modeling the dynamics of this system. In describing the insights garnered from his now famous predator–prey equations in relation to closure of the Adriatic fishery during the war, Volterra (1926) noted that "*... a complete closure of the fishery is a form of 'protection' under which the voracious fishes were much the better and prospered accordingly but the ordinary food fishes on which these were accustomed to prey, were worse off than before.*" As we shall see, Volterra's work provided the basis for the development of early fishery harvesting models (humans as predators). The above quote also holds significance for understanding the potential direct and indirect effects of establishing aquatic protection areas when viewed in an ecosystem context.

Alfred Lotka (1925), had earlier derived a very similar system of predator–prey (consumer–resource) equations inspired by plant–herbivore interactions. The Lotka–Volterra equations became one pillar of the (first) golden age of ecology (Kingsland 1995). Interestingly, Lotka, also developed a food-web model for aquatic systems in his classic treatise and devoted a chapter to interspecies equilibria in aquatic ecosystems (Lotka 1925). Lotka perceptively noted a critical issue in mixed-species fisheries that remains important today—a rarer species caught in concert with a more abundant one can be endangered because harvesters will continue to pursue the more common species even as the rarer one declines to unprofitable levels (Lotka 1925; p.95).

Raymond Pearl's use of the Verhulst logistic equation to model human population growth in the early 1920s (Kingsland 1995) was rapidly followed by adaptations for fishery applications, including specification of target exploitation rates. Hjort et al. (1933) provided the earliest use of the logistic production model to assess the optimum catch in fishery ecosystems (Smith 1994). Hjort's work at the time centered on exploitation of marine mammal resources. Graham (1935) applied the logistic model to individual growth rather than population size for aggregate species groups. Graham's method was taken to remedy data limitations at the individual species level (Smith 1994), an approach that we shall revisit later in this book. G. Evelyn Hutchinson, who derived many important basic ecological insights from his interest in limnology, demonstrated that the introduction of time delays in the logistic model resulted in the potential for complex dynamical behavior (Hutchinson 1948), touching on another of the important themes in this book.

The 1950s ushered in an extremely productive era in modeling fish population dynamics with the publication of such seminal works as "Stock and Recruitment" by William Ricker (1954) and Raymond Beverton and Sidney Holt's (1957) classic monograph "On the Dynamics of Exploited Fish Populations." Ricker not only established an analytical framework for understanding the fundamental connection between generations and its implications for management, he was the first to demonstrate in an ecological context that very complicated dynamics can emerge from simple discrete-time population models. While Beverton and Holt principally concentrated on single-species dynamics, they touched on a broader array of topics including ecosystem considerations. They noted: "*This is a generalization of what is perhaps the central problem in fisheries research: the investigation not merely of the reactions of particular populations to fishing, but also of the interactions between them and the response of each marine community to man's activity*" (Beverton and Holt 1957, p. 24), clearly presaging the need to address the broader ecological dimensions of the harvesting issue. Also during this period, Milner B. Schaefer revived application of the logistic model in surplus production models in two influential papers (Schaefer 1954, 1957). Pella and Tomlinson (1969) followed with a generalization of the logistic model that allowed a wide range of functional forms and William Fox developed his asymmetrical production function based on the Gompertz growth model rather than the logistic (Fox 1970).

The development of a formal and well-established analytical framework for providing scientific advice in support of fishery management at the single-species level led to broad global application of these principles. The Food and Agriculture Organization of the United Nations played a key role in the world-wide dissemination of these methods with the establishment of courses in quantitative methods and the publication of a series of influential training manuals (e.g. Gulland 1968, 1969) and related subsequent publications (Gulland 1977, 1983, 1988). Again in the spirit of providing basic and practical training materials, Ricker (1975), whose direct experience encompassed both freshwater and marine systems, produced his "Handbook of Methods of Fisheries Science." These efforts collectively centered on formalizing the mathematical description of population dynamics to determine the condition of exploited aquatic populations (or stocks) in relation to objectively defined targets and limits to exploitation (now known as biological reference points).

While advances in population dynamics and stock assessment continued apace, approaches based on a broader ecosystem perspective were not forgotten. In the 1920s, Alister Hardy, the noted biological oceanographer (and inventor of the Continuous Plankton Recorder) constructed a detailed marine food web with Atlantic herring as a focal point (Hardy 1924), providing an important qualitative description that was amenable to quantitative treatment. Clarke (1946) developed a simple energy flow model for Georges Bank, a historically important fishing ground located off New England, that built on some of the concepts put forward by Lotka (1925). In a landmark study, Ryther (1969) provided a high-profile attempt to estimate global marine fishery production potential based on consideration of energy flow through successive trophic levels. Ryther's estimate of 100 million metric tons, although not uncontroversial at the time, now seems remarkably prescient (worldwide catches from marine capture fisheries are now quite close to this estimate).

Related attempts to estimate the determinants of fish production in freshwater systems were vigorously pursued including energy-flow models and predictive models using physiographic characteristics (e.g. McConnell et al. 1977). These efforts, including the development of the morphoedaphic index used to predict fish yield from fundamental limnological metrics, (Kerr 1974; Ryder et al. 1974) provided an early example of a macroecological (Brown 1995) approach in fisheries. The conceptual importance of these developments in freshwater ecosystem science cannot be overstated. The focus on development of simple predictive models, exemplified by the work of Rigler (1982a,b) and championed by Peters (1991) in a broader ecological context, foreshadowed the development of macroecology as an important subdiscipline of ecology.

Throughout the 1930s and culminating in work completed after World War II, Ludwig von Bertalanffy strove to develop a General Systems Theory. Although von Bertalanffy is well known to all students of fishery science for his eponymous model of individual growth, widely applied in single-species fishery models, his research into the dynamics of systems (including ecosystems) holds greater relevance as a framework for understanding coupled human–natural systems such as fishery ecosystems.

1.2 Process and pattern in fishery ecosystems

Examination of long-term records of abundance and catch of aquatic species immediately reveals a fascinating diversity of dynamical patterns. The most extensive record of historical change comes from paleoecological studies in both marine and freshwater ecosystems (Finney et al. 2010; Cohen et al. 2016). For example, reconstruction of Pacific sardine (*Sardinops sagax*), anchovy (*Engraulis mordax*), and hake (*Merluccius productus*) population levels in the Santa Barbara Basin off central California (Soutar and Isaacs 1969; Baumgartner et al. 1992) reveals dramatic change on multidecadal time scales over nearly two millennia (Figure 1.2). Anchovy and sardine are planktivores and play similar functional roles in the ecosystem. Hake are predators of both sardines and anchovies. These three species play a dominant role in the pelagic food web of the California Current ecosystem. The bottom waters and sediment of the basin are anoxic and subject to little biogenic perturbation. Varved core samples taken within the basin contain a history of scale deposition patterns of these species that can be used to construct indices of abundance. The broad temporal extent of these observations provides a window into population change prior to the twentieth century when human influence on these populations was minimal. It is clear that the potential for high levels of variability is an intrinsic property of these species. Finney et al. (2010) provide an intriguing synthesis of paleoecological studies in marine ecosystems documenting the critical role of climate variability and change in these patterns on multidecadal to centennial time scales. Cohen et al. (2016) show that climate effects in Lake Tanganyika over the past 1500 yrs have strongly affected fish and bivalve populations. The recent period of warming is linked to declines in overall ecosystem productivity in this ecosystem that pre-dated intensive fishing activity.

Although no single measure can reflect in full the dynamical behavior of coupled human–ecological systems and its underlying causes, documentation of harvest levels in fishery ecosystems on timescales from decades to centuries does provide important insights into the complexity of these systems and the interaction among the social, economic, and ecological factors at play. We see well defined cycles in the catch and sudden shifts in catch levels. A sampler of these patterns will suffice to highlight dramatic change in marine and freshwater fishery ecosystems (Figure 1.3).

The blue pike (or blue walleye, *Stizostedion vitreum glaucum*) in Lake Erie underwent pronounced cyclic fluctuations during the period 1915–1960 (Figure 1.3a). The hypothesized mechanism underlying this periodicity relates to a form of compensatory population regulation, a topic we will take up in Chapter 3. As we shall see, population mechanisms of this type can give rise to extremely complex dynamical behavior. Under the combined stress of harvesting pressure and eutrophication of the lake, this subspecies unfortunately appears to have gone extinct in the early 1960s.

Pink salmon (*Oncorhynchus gorbuscha*) catches from Bristol Bay Alaska exhibited large fluctuations in interannual catch levels during the period 1900–1920 (with peaks in 1906, 1912, and 1920) and then dropped to a much lower level and remained in this new state for approximately 30 years

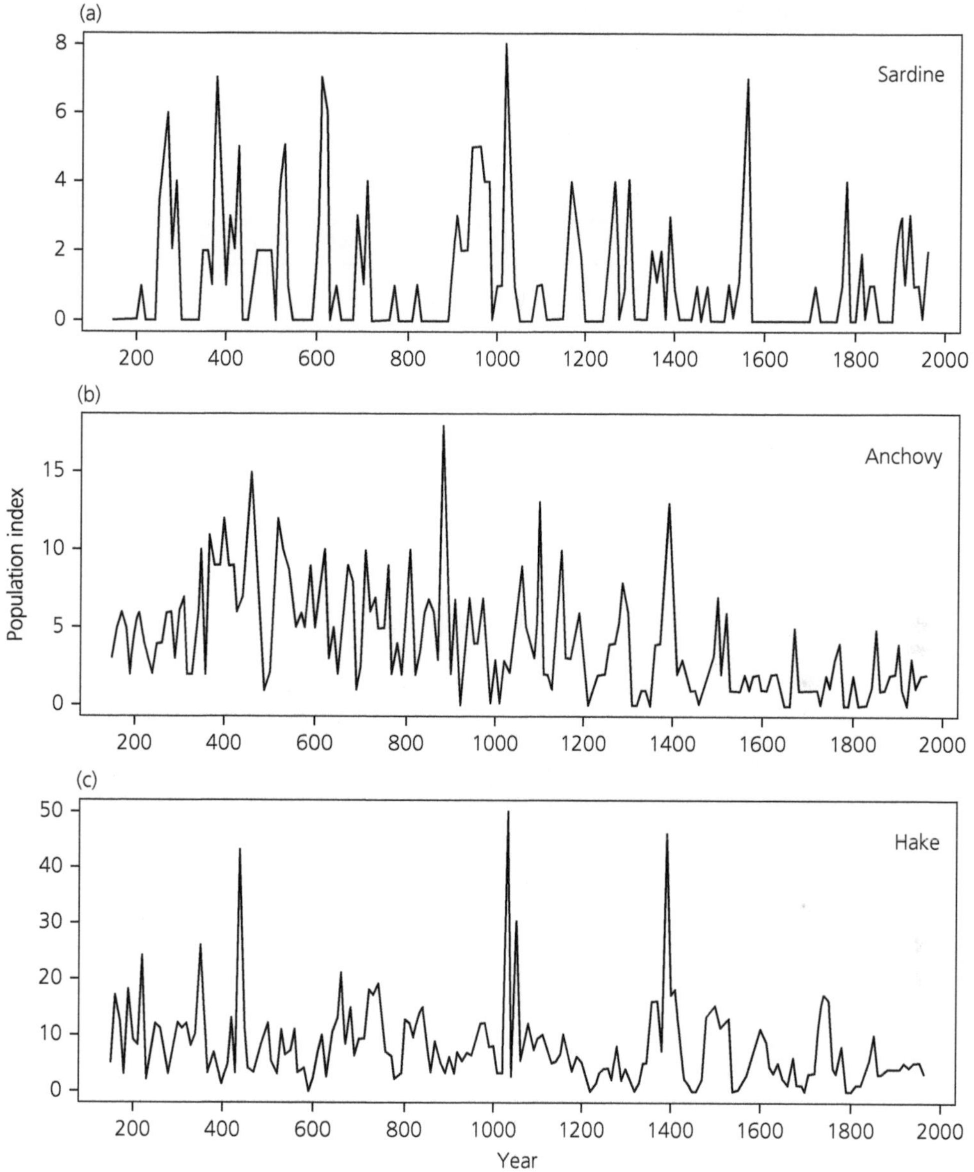

Figure 1.2 Indices of population size of sardine (*Sardinops sagax*), anchovy (*Engraulis mordax*), and hake (*Merluccius productus*) as measured by scale deposition rates over the Santa Barbara Basin off California (Soutar and Issacs Isaacs 1974; Baumgartner et al. 1992).

(Figure 1.3b). The average catch during 1900–1920 was approximately 500 000 fish with a coefficient of variation (a measure of relative variability) of approximately 110 percent. After 1920, the catch dropped to less than 20 per cent (on average) of its former level but actually had a higher coefficient of variation (~160 percent). Pink salmon, like all Pacific salmon of the genus *Oncorhynchus* are anadromous. The adults return (with some straying) to their natal streams and rivers to spawn once and die. A closer look at the pink salmon catch series reveals an additional interesting dimension. Pink salmon have a two-year life span. The young spend less than half a year in freshwater before going to sea

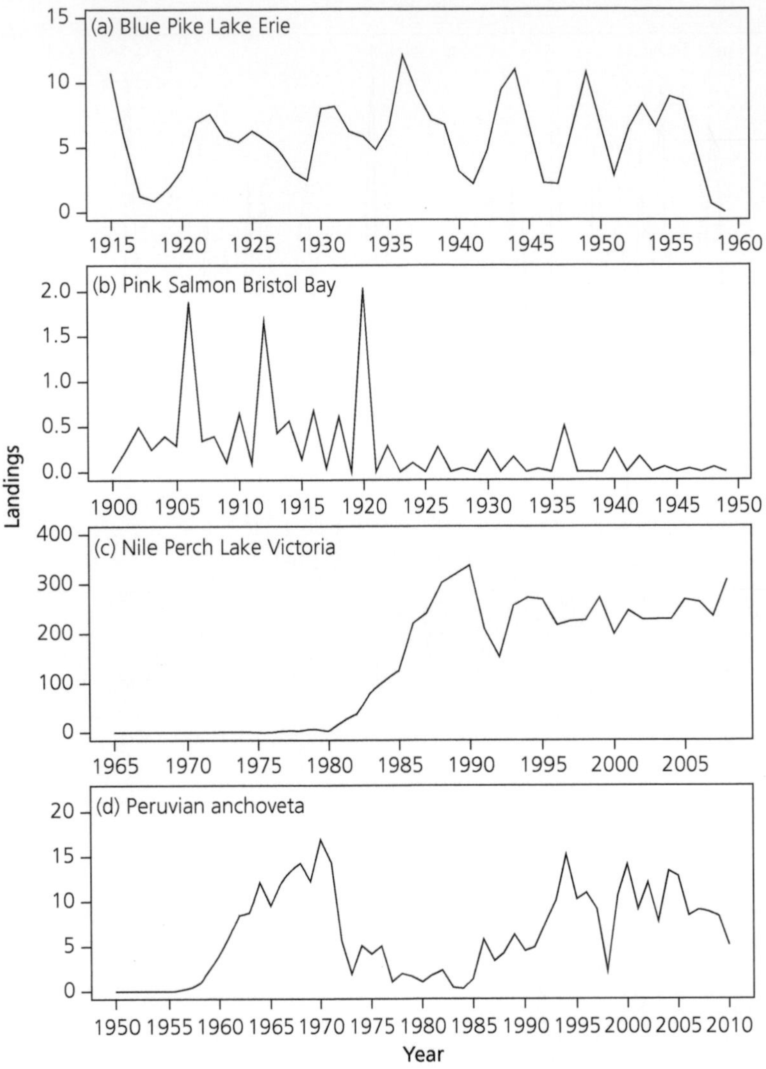

Figure 1.3 Landings trajectories over time of (a) blue pike (*Sander vitreus glaucus*) in Lake Erie (thousand t) (data from Parsons 1967), (b) Bristol Bay pink salmon (*Oncorhynchus gorbuscha*) (millions of fish) (data from https://www.fishbase.org/recruitment/), (c) Nile Perch (*Lates niloticus*) in Lake Victoria (thousand t) (data from Kolding et al. 2014), and (d) Peruvian anchoveta (*Engraulis ringens*)(million t) (data from /www.fao.org/fishery/statistics/)

to spend the remainder of their life cycle before returning to spawn. The fish returning to spawn in adjacent years can be effectively thought of as separate populations (called lines). Notice that all the major catches prior to 1921 were from even year lines. After the sudden drop following the large catches in 1920, the odd year lines essentially went at least economically extinct for the remainder of the period examined in Figure 1.3b.

As noted in the preface, the impact of fishery-related activities is not confined to the removal of biomass through catch but also encompasses deliberate attempts at ecosystem engineering through introduction of non-native species, hatchery operations for both native and non-native species, and other forms of ecosystem manipulation. The Nile Perch (*Lates niloticus*) was introduced into Lake Victoria in the 1950s to support the development of

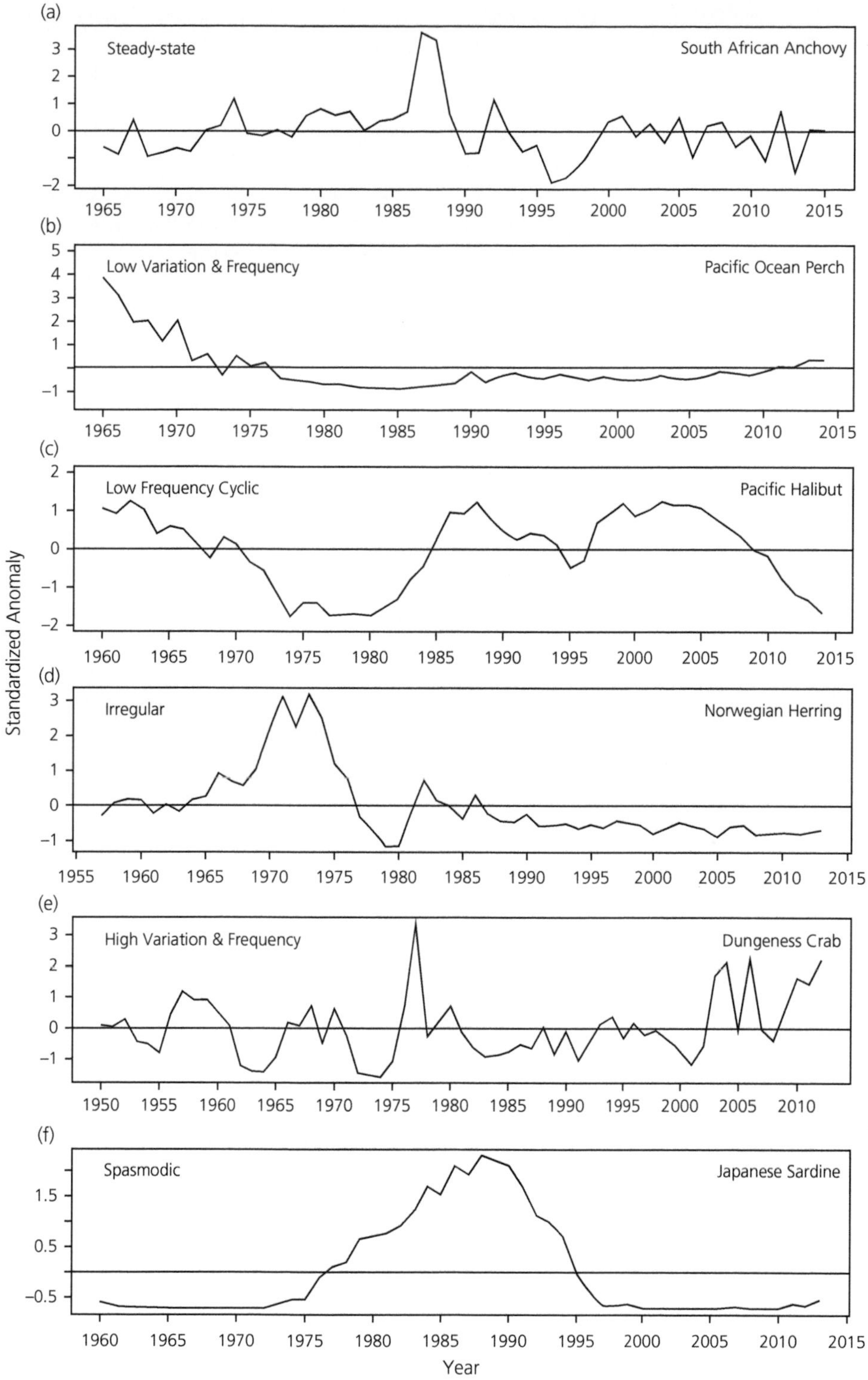

Figure 1.4 Examples of patterns of variability in marine populations following the taxonomy proposed by Spencer and Collie (1997a). Time series are expressed as standardized anomalies.

a new fishery (Pringle 2005). Nile perch is among the largest freshwater fish species, reaching a maximum length of nearly two meters and weights of up to 200 kg. Although warnings had been issued by the East African Fisheries Research Organization about the adverse ecological implications of any such introduction, this species was released into the lake, with large-scale consequences for the lake ecosystem. Following the initial introduction into the lake, catches of Nile perch remained relatively low until 1980 when a rapid expansion in the fishery developed for this species. Catches subsequently peaked in 1990 at approximately 340 thousand tons (Figure 1.3c); during the period 1990–2000, catches averaged approximately 240 thousand mt. It has been hypothesized that native fish species, including the astonishingly diverse cichlid community preyed on the early life stages of perch following their introduction, initially keeping them in check (Walters and Kitchell 2001) until the abundance of Nile perch reached a sufficient level to depress the native fishes, initiating a new feedback loop that further depressed the native fish assemblages. Fishing pressure and cultural eutrophication may also have resulted in declines of cichlids and a reduction in predation pressure on young perch. Kolding et al. (2008) provide an important review of the broad suite of observed changes in the lake and possible causes.

As a final example, we turn to the Peruvian anchoveta (*Engraulis ringens*) in the Humboldt Current system. The anchoveta supports the largest single-species fishery in the world during periods of high abundance. However, the anchoveta population has undergone large-scale changes in abundance on multidecadal timescales (Figure 1.3d). A sharp decline in anchoveta landings in the early 1970s has been attributed to the interplay of fishing pressure and a particularly strong El Niño event which resulted in sharply reduced productivity. The decline of the anchoveta population also held important ecosystem consequences, resulting in starvation of predators, particularly seabird populations strongly dependent on this species as a staple of the diet.

Caddy and Gulland (1983) offered an early attempt at constructing a taxonomy of variability in marine fishery resources. Four principal patterns of variability were recognized: Steady, Cyclical, Irregular, and Spasmodic. These evocative labels convey the diverse spectrum of observed population patterns in fish and shellfish stocks. Steady populations are those characterized by globally stable equilibria. Cyclic species undergo well-defined periodic changes in abundance. Irregular dynamics are characterized by high levels of variability and are often associated with stochastic environmental forcing. Spasmodic populations are ones distinguished by alternating periods of high and low levels of abundance.

The initial classification scheme proposed by Caddy and Gulland (1983) was expanded by Spencer and Collie (1997a) to encompass Steady State; Low-variation Low-Frequency; Cyclic; Irregular; High-variation High-Frequency, and Spasmodic dynamics. Spencer and Collie (1997a) further provided an objective protocol for classifying stocks into these categories. In Figure 1.4 we show representatives of each of the categories identified by Spencer and Collie. The main message that emerges from these analyses is that quite complex patterns in population levels and yield can emerge over time in fishery ecosystems.

The noted theoretical ecologist Ramon Margalef (1960) observed that:

Ecosystems result from the integration of populations of different species in a common environment. They rarely remain steady for long, and fluctuations lie in the very essence of the ecosystems and every one of the ... populations. (Cited in Smith 1994.)

Yet the concept of steady-state dynamics remains at the heart of most fishery management precepts as embodied in concepts such as Maximum Sustainable Yield. We cannot ignore fundamental changes in the environment or in food web structure, whether the latter reflect natural changes, the indirect effects of fishing due to species-selective harvesting patterns, or deliberate attempts at ecosystem engineering by humans.

1.3 Confronting complexity

Whether we can deal with the daunting complexity of ecosystems and the associated management challenges as we move toward Ecosystem-Based Fisheries Management (EBFM) is a frequently voiced, and legitimate, concern. Models of exploited aquatic

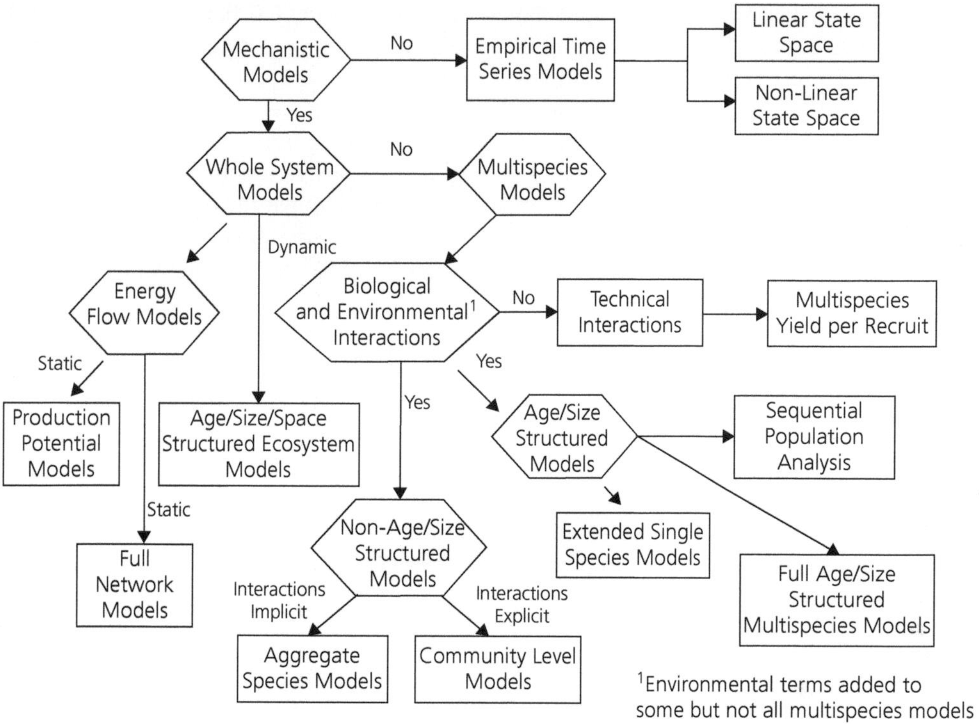

Figure 1.5 A dichotomous key showing the classes of multispecies and ecosystem models to be considered in this book.

ecosystems provide the principal tools for synthesis and integration of information on the structure and function in fishery ecosystems, the role of environmental change, and the effects of human interventions of various types (including the efficacy of management actions). They further provide vehicles for prediction. In Figure 1.5, we provide a roadmap to the classes of multispecies and ecosystem models to be covered in this book. The types of models shown in Figure 1.5 encompass a broad spectrum of possibilities, ranging from simple empirical models to full ecosystem models.

We believe that a deliberate strategy of managing complexity in models of fishery ecosystems is essential. In constructing a model, a clear understanding of the role it is to play is essential. We view models as embodying hypotheses about states of nature. They are abstractions of reality. The development of models in support of EBFM can be arrayed along a continuum of complexity involving tradeoffs in realism, mechanistic detail, and parameter and/or model uncertainty. By adding complexity to models, we

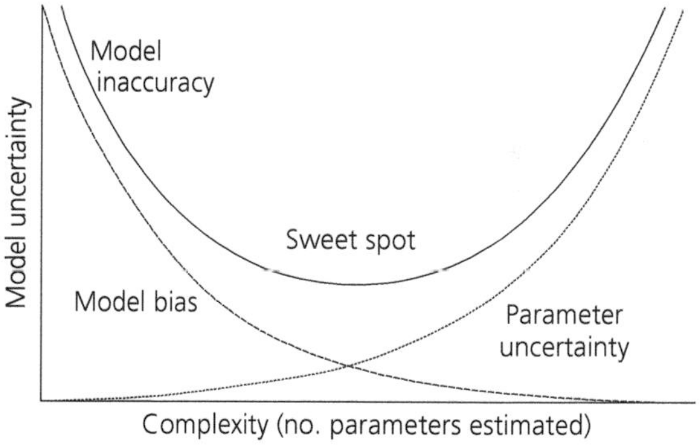

Figure 1.6 Tradeoff between model bias and parameter uncertainty as a function of model complexity (number of parameters; adapted from Collie et al. 2014).

hope to provide a more realistic portrayal of the system. However, this enhanced realism comes at a price. Very often we find that overall model uncertainty increases with complexity because of the increased demands on data and information, and difficulties in estimation (Figure 1.6). Collie et al. (2014) note that a "sweet spot" in model performance occurs at intermediate levels of model complexity. We are further keenly aware that the availability of ecosystem level information differs substantially in different parts of the globe. If we are to provide usable information for ecosystem-based management, we must consider options for data-limited situations. We have attempted to build this perspective into our approach to describing practical models to support operational management strategies.

1.4 Summary

In this chapter, we have sought to convey why fisheries constitute such an important element of global food security and why they are so important in the fabric of human societies throughout the world. We have traced the importance of these considerations over time and recounted key developments in the scientific understanding and approaches so critical to effective fisheries management. We have shown how processes at the population, community, and ecosystem levels are manifest in complex ways in simple descriptors such as catch histories. We describe the importance of confronting complexity in the analysis and management of these systems if we are to move toward operational EBFM and approaches to defining aquatic ecoregions as a precursor to specifying spatial management domains.

Additional reading

Fascinating accounts of the history of fisheries can be found in Sahrhage and Lundbeck (1992), Kurlansky (1997), Fagan (2006, 2017). Kingsland (1995) trace the early development and evolution of ecological models. For a comprehensive treatment of the history of fisheries science see Smith (1994).

PART I

Ecological Models: An Overview

CHAPTER 2

Density-Independent Population Growth

2.1 Introduction

Our exploration of basic population growth begins with consideration of elemental rates of birth and death. We would expect that following the initial establishment of a population in a new environment where resources are abundant (and particularly if the risk of predation is also low), it would initially experience relatively unconstrained growth. In other instances, we may be interested in the rate of recovery of a population after it has been reduced to low levels by natural or anthropogenic disturbances. Although resource limitation will ultimately slow the rate of increase as the population grows, it is worth considering how we might construct a model of the initial phases of population growth or recovery.

Consider the case of the grey seal (*Halichoerus grypus*) in eastern Canada. By the early 1960s, hunting and other pressures had reduced the grey seal population to a few thousand individuals (Bowen et al. 2003). In general, the analysis of aquatic populations entails special challenges because births and deaths in the population often cannot be directly observed and enumerated. However, many seals and other pinnipeds have discrete breeding colonies on land where it is possible to count newborns (pups) and thereby directly census at least segments of the population. Direct observation of deaths at sea of course remains problematic. On Sable Island, off Nova Scotia, researchers have counted pups since 1963 using tagging and photographic methods. The estimated number of pups on the island increased at a sustained exponential rate during the period 1963–1997 (Bowen et al. 2003; Figure 2.1). Subsequent estimates of the seal population on Sable Island (Hammill et al. 2014), show a decrease in the growth rate consistent with the emergence of density-dependent population growth as the population has recovered.

In the following, we show how to construct simple models building on important clues such as an apparent constancy in the rate of change in population size during an initial recovery period, as in the grey seal example. Extension to the case of age- and stage-structured models is then presented and we demonstrate the connection with simpler (non-structured) models.

2.2 Simple population models

We begin with models that ignore factors such as the age, size, and genetic structure of the population and its spatial configuration. We will further assume that the population is closed (no immigration or emigration), or alternatively, that net migration is zero. Ecological models can be framed in either continuous or discrete time. Populations characterized by birth and death rates occurring throughout the year (for example in tropical and semi-tropical environments) are often best treated with models framed in continuous time using differential equations. In contrast, those in temperate and arcto-boreal systems typically have a

Fishery Ecosystem Dynamics. Michael J. Fogarty and Jeremy S. Collie, Oxford University Press (2020). © Michael J. Fogarty & Jeremy S. Collie 2020.
DOI: 10.1093/oso/9780198768937.003.0002

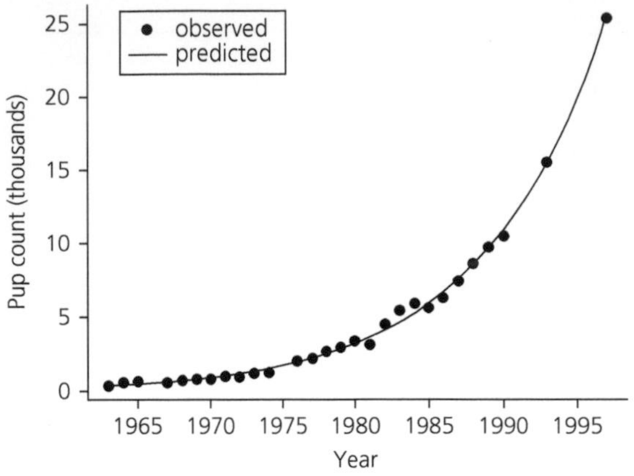

Figure 2.1 Abundance of grey seal (*Halichoerus grypus*) pups on Sable Island (Bowen et al. 2003).

well-defined seasonal reproductive period and the application of discrete-time models (or difference equations) may be more appropriate.

2.2.1 Continuous-time model

If the birth and death rates in the population are constant over all ranges of population density, then the rate of change of population size can be expressed as:

$$\frac{dN}{dt} = (b - d)\,N \tag{2.1}$$

where N is population size, b and d are the instantaneous rates of births and deaths. Note that the per capita or relative growth rate $[(1/N)\mathrm{d}N/\mathrm{d}t]$ is a constant equal to the birth rate minus the death rate. This constant is often designated as the intrinsic rate of increase [$r = b - d$; (Wilson and Bossert 1971; Vandermeer and Goldberg 2003; Gotelli 2008)]. The rate of change of the population is zero if the birth and death rates balance, positive if $b > d$ and negative if $b < d$. The solution to this differential equation is:

$$N_t = N_0 e^{rt} \tag{2.2}$$

(see Box 2.1) and a sketch of this equation for several values of r is provided in Figure 2.2. The population will either increase or decrease exponentially, unless $b = d$ ($r = 0$). We can readily linearize this model by taking natural logarithms and estimate the parameters r and N_0 using simple linear regression (see Box 2.2).

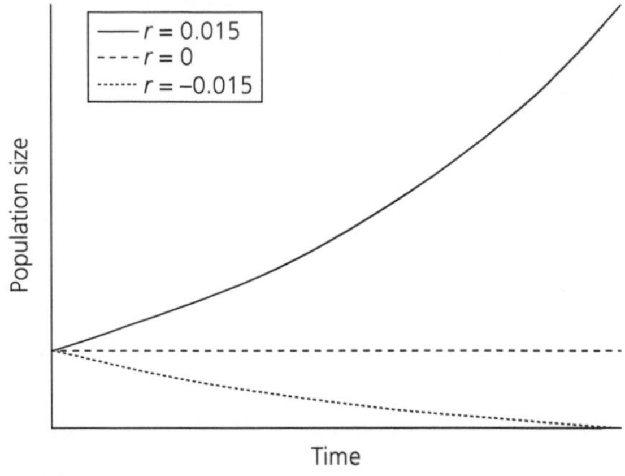

Figure 2.2 Population trajectories for the exponential growth model for three levels of the intrinsic rate of increase.

Box 2.1 Solving the Density-Independent Growth Model

The steps used in solving the density-independent growth model are shown below. Our model is a first-order separable differential equation where $r = b - d$:

$$\frac{dN}{dt} = (b - d)\,N = rN$$

Separating variables and integrating from time t_0 to t we have:

$$\int_{N_0}^{N_t} \frac{dN}{N} = (b - d)\int_{t_0}^{t} dt$$

Recall that by the chain rule:

$$\int \frac{dN}{N} = \log_e N$$

The solution is then:

$$\log_e N_t = \log_e N_0 + r(t - t_0)$$

Taking antilogs, and setting $t_0 = 0$, we obtain the solution:

$$N_t = N_0 e^{rt}$$

Box 2.2 Applying Models to Data

The solution to the density-independent population growth model is a simple non-linear equation with an exponential form. How can we relate the model to real observations described at the start of this chapter? Well-established regression techniques are available for fitting equations to data for both linear and non-linear models. If we return to the penultimate step in Box 2.1, we see that the model on logarithmic scale is linear. Expressing this in the form of a simple linear regression we have:

$$\log_e N_t = \log_e N_o + r \cdot t + \varepsilon_t$$

where ε_t is an error term representing the difference between the observed and expected value of $\log_e N_t$. To apply standard linear regression techniques, we assume that the error term is normally distributed with mean zero and constant variance. Further, it will be assumed that successive values of ε_t are independent (i.e. are uncorrelated). Notice that the error term is additive on the logarithmic scale.

Here, we are interested in finding the relationship between N_t and t. We can find a solution to estimating N_0 and r on the right hand side of the model by choosing values of these parameters that minimize the sum of the squared difference between the observed and predicted values of N_t:

$$\min \sum \left(\log_e N_t - \log_e \hat{N}_t\right)^2$$

This is the approach used to fit the population growth curve in Figure 2.1.

2.2.2 Discrete-time model

We next consider how to construct a model for populations with seasonal spawning periods and non-overlapping generations. For example, Pacific salmon (*Oncorhynchus* spp.) breed in fresh-water streams and rivers and, after an initial developmental phase, enter the marine environment where they remain until returning with high fidelity to their natal river to spawn. The adults die following reproduction and hence the generations are non-overlapping; this life-history pattern is called semelparous. The number of years at sea before returning for mating differs markedly among salmon species, but is relatively fixed within species (Burgner 1991). The return of adult salmon to streams and rivers and their subsequent death and decomposition plays an important role in the productivity of these ecosystems with respect to nutrient transfer. Viewed from this perspective, extracting too many salmon before they return to their natal spawning areas can hold adverse consequences for nutrient dynamics and system-wide production, an unintended consequence of exploitation of Pacific salmon at the ecosystem level.

In the following, we present the case of sockeye salmon (*Oncorhynchus nerka*) in the Quesnel River, a tributary of the Fraser River, British Columbia, in greater detail. This population was at extremely low levels during the period 1945–75 as a result of a sequence of disease outbreaks causing high pre-spawning mortality, apparently exacerbated by high water temperatures (Roos 1991). Sockeye salmon in this watershed primarily return to spawn at age four. Notice that there is great disparity in the population sizes from year to year during the recovery phase (Figure 2.3) with one "line" clearly dominant and the others at much lower levels of abundance (although a subdominant line is clearly discernible as the population increases). These lines effectively represent different subpopulations appearing at four-year intervals. Although not universal, this pattern is common in a number of sockeye salmon stocks and is referred to as cyclic dominance (Ricker 1975). The possible mechanisms underlying cyclic dominance have been the subject of a series of intriguing analyses (Ricker 1975; Collie and Walters 1987; Myers et al. 1998; Guill et al. 2014). We will focus on the development of a simple model for the dominant line of sockeye in this system. While the abundance of this line increased dramatically, the ratio of the number of sockeye in successive generations was relatively stable, implying a constant rate of increase over the time period considered in Figure 2.3. Because of the strong fidelity to spawning location, we will, for our present purposes, assume that the population is closed (i.e. straying to other spawning

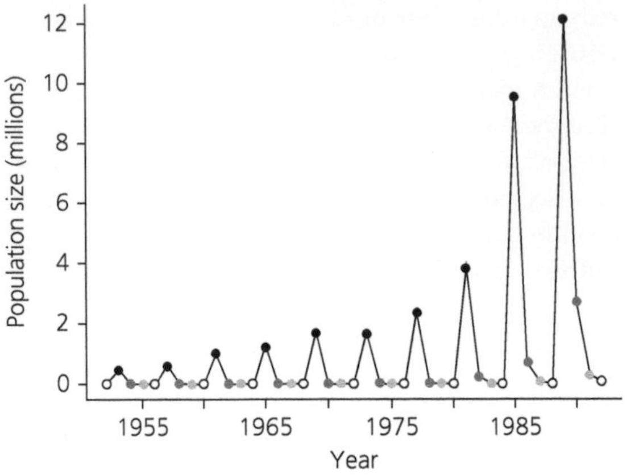

Figure 2.3 Population trajectory of sockeye salmon (*Oncorhynchus nerka*) from the Quesnel River, British Columbia. The four lines are indicated by different shading patterns. (Data courtesy of Michael Lapointe, Pacific Salmon Commission.)

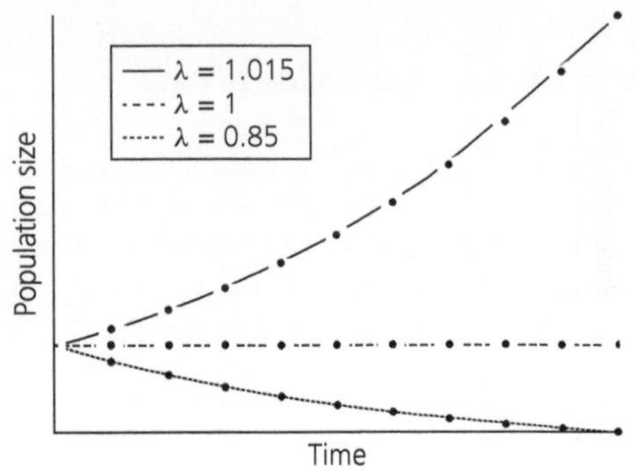

Figure 2.4 Trajectory of a population undergoing geometric growth for three levels of the finite rate of increase. Because the population is assessed at discrete time intervals we represent the growth trajectories as points at each census interval.

sites or inputs of strays from other rivers will be considered negligible). Again for convenience, we will assume a fixed age at reproduction although adjustments to account for individuals returning on different schedules can be readily accommodated.

The change in the number of individuals in the population at successive generations for a population with a constant rate of increase is given by:

$$\Delta N = N_{T+1} - N_T = r_d N_T \tag{2.3}$$

where N_T is the number in the population at generation T, and r_d is the discrete rate of population growth. The model can be expressed more simply as:

$$N_{T+1} = \lambda N_T \tag{2.4}$$

where $\lambda = 1 + r_d$ (the finite rate of increase). The population is stable ($N^* = N_{T+1} = N_T$) only when $\lambda = 1$.

This model describes a population at fixed points in time. We can write the model as a simple recursion formula as a function of the population size at time 0:

$$N_T = \lambda^T N_o \tag{2.5}$$

(see Box 2.3) and plot the population trajectory as illustrated in Figure 2.4.

Box 2.3 Recursion Formula for Geometric Population Growth

The development of a recursion formula for our finite-difference model is particularly simple. We begin by showing the population sizes at successive generations. The first step is:

$$N_1 = \lambda N_0$$

where N_0 is the initial generation size and λ the finite rate of increase. Next we provide the populaton size at generation 2:

$$N_2 = \lambda N_1 = \lambda(\lambda N_0) = \lambda^2 N_0$$

after substituting in the results from our first iteration. The number at $T = 3$ is then:

$$N_3 = \lambda N_2 = \lambda\left(\lambda^2 N_0\right) = \lambda^3 N_0$$

and it is clear that a simple rule has emerged as our recursion formula:

$$N_T = \lambda^T N_0$$

The dynamical properties of difference equations can be readily illustrated in graphical form (Figure 2.5). A plot of N_{T+1} against N_T is provided in Figure 2.5a for the case in which the finite growth rate is greater than 1 compared with the equilibrium case—the dashed line in Figure 2.5a shows a 1:1 relationship between the population size at successive points in time. This type of graph,

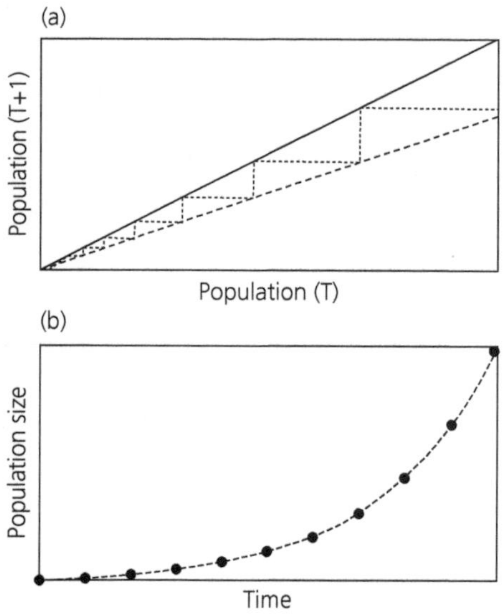

Figure 2.5 Number in the population at generation T+1 as a function of the population size at generation T (top panel). The dashed line represents the 1:1 replacement line when the number at successive time steps remains unchanged. The solid line represents population size at successive time steps. In this example, the population growth rate exceeds replacement and the population grows geometrically (bottom panel).

often called a cobweb plot, provides a handy visual representation of the way in which the recursion formula in Box 2.3 unfolds with successive steps in the process. Because we will use this type of graph again, it is worth understanding how to read and interpret cobweb plots in general. From any starting value on the x-axis, simply read up to the intersection point with the population line (solid line in Figure 2.5a). If we then read over to the y-axis from that point we obtain the number in the population at generation T+1. We could then, somewhat laboriously, transfer that number to the x-axis and repeat the process to get the population number at generation T+2, etc. We can, however, streamline the process by simply following the equivalent sequence of steps shown by the dotted line in Figure 2.5a. By drawing a horizontal line from the point of intersection with the population line to the 1:1 line and then vertically to again intersect the population function, we trace the population trajectory in a series of steps. Notice the increasing heights of the dotted lines as we iterate between the dashed and thick solid lines,

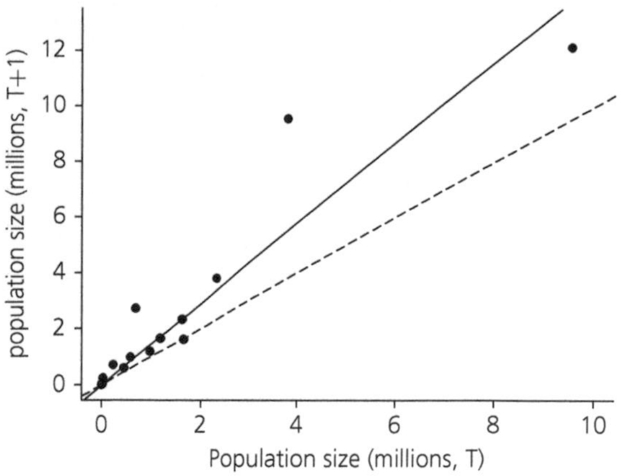

Figure 2.6 Relationship between population sizes at successive generations for Quesnel River sockeye salmon (*Oncorhynchus nerka*) for the dominant (black symbols) population lines. The solid line is a linear regression fit to the population data.

showing a geometric increase in the population. The bottom panel of Figure 2.5 shows the change in population size at each time step (each closed circle represents an event in which the dotted line hits the solid population line). As we view cobweb plots in subsequent sections of this book, note particularly that when the population line or curve is above the 1:1 replacement line, the population will grow; the population will decrease when it is below this line.

If we examine the Quesnel River sockeye salmon data, we see a clear relationship between the population size at time t and time t+4 (i.e. between generation T and generation T+1, Figure 2.6). The population line lies above the 1:1 replacement line and we therefore predict geometric growth within the range of observed population sizes. In Chapter 3 we will explore the full implications of changing population growth rates as population size increases. Again, the principal lesson is that we should not assume that an apparent exponential or geometric increase in population size will continue unabated.

2.3 Age- and stage-structured models

In the simple models described above, we considered the birth and death rates of populations in which no distinction was made among individuals according to age, life stage, or size. Here we extend this approach to consider the influence of the demographic structure of the population. This structure can be thought of as comprising different stages, which for our purposes can represent ages or age groups, size classes, or life-history stages (e.g. eggs, larvae, juveniles, adults). Our interest in the demographic composition of the population centers on the fact that aquatic organisms of different stages produce different number of offspring and have different survival rates. In our earlier models of exponential and geometric growth, we implicitly assumed average values for birth and death rates over all demographic stages.

In the following, we will focus on discrete-time matrix models (see Caswell 2001). Although continuous-time models are available for age-structured populations, we find that the concepts underlying age- or stage-structured models in discrete time are more intuitive and easily understood than the corresponding continuous-time versions, which require more advanced mathematical treatment.

2.3.1 Age-structured model

To motivate our treatment of this topic, we begin by showing the age composition of the haddock (*Melanogrammus aeglefinus*) population on Georges Bank during 1931–40 (the first decade for which estimates are available; Figure 2.7). These estimates, taken during a period of apparent relative stability in the haddock population, illustrate several important features. The estimated overall age composition remained relatively constant over this period as indicated by the roughly similar sized circles (indicating abundance) in each of the rows representing different ages. However, the effect of variation in survival of young haddock during the first year of life is evident and we begin to get a small hint of the population variability that will figure prominently elsewhere in the book (manifest in the larger shaded circles in Figure 2.7). We can trace the trajectory of stronger than average abundance for haddock of the 1929 year class (starting with its first appearance in these estimates in 1931 as two year olds; a year class comprises individuals born in a given year). Similarly, relatively strong 1936 and 1939 year classes are also evident (shaded circles in Figure 2.7) and we can trace their numbers along the diagonals

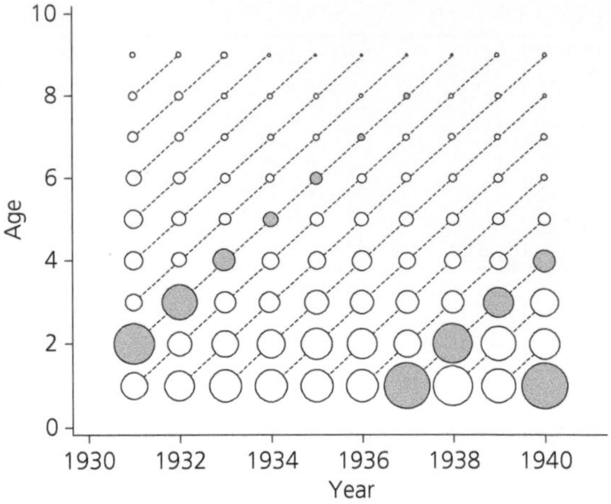

Figure 2.7 Estimated number-at-age of haddock (*Melanogrammus aeglefinus*) on Georges Bank during 1931–40. The size of each circle is proportional to numerical abundance. Dotted lines indicate the progression of cohorts. The shaded circles indicate the 1929, 1936, and 1939 year classes. Data from Clark et al. (1982).

connecting time and age. We will treat the issue of the characteristically high variability in year class strength in fish populations in further detail when we take up the topic of recruitment in Section 9.3. The issue of whether the age distribution is stable is of immediate importance as will be shown in the next section. Understanding the age composition of a population (or related demographic structure) is also important because anthropogenic stressors often have a differential effect on different ages and demographic attributes.

To construct population models that account for age or size structure, we will require information on age- or size-specific survival rates, maturation schedules (proportion mature at each age or size), and fecundity (the mean number of viable eggs produced by a female at each age or size). Fecundity is the maximum reproductive output thought to be physiologically possible for each age class. We will also be interested in fertility—the realized reproductive output-at-age. In aquatic ecosystems, we find a broad diversity of life-history patterns reflecting different reproductive strategies. For example, many invertebrates and teleost fishes have both very high fecundity and very low rates of survival during the early life stages. Exceptions include certain species that invest substantial parental care and energy in their offspring. In general, most elasmobranch fishes exhibit ovoviviparous or viviparous reproductive patterns and relatively low fecundity. Aquatic reptiles are ovoviviparous while mammals are viviparous. Survival during the first year of life for these species is very much higher than in most fish and invertebrates. Again, for now we will consider only the case in which these vital rates do not vary with density.

In age-structured population models, it is common to consider only females, based on the assumption that reproductive output is limited by the number of females and that a sufficient number of males are available to fertilize the eggs. By convention the number of offspring produced is designated age 0 (often called the 0-group or young-of-year in fisheries analyses); subsequent ages are numbered accordingly. The individuals born in a given time period will also be referred to as a cohort. Individuals between age $a-1$ and age a constitute age-class a. For example, individuals between ages 0 and 1 are designated age-class 1. The fundamental model is based on the survivorship between successive age classes and on the reproductive output of the population. In the following, we will assume a birth-pulse formulation in which mature females give birth upon entering a new age class. We will also assume that the population is enumerated immediately prior to the breeding season. The formulae provided below are predicated on these assumptions. It is possible to specify models that consider a birth-flow formulation in which reproduction occurs throughout an age class, although this approach is more complicated and involves approximations to permit the use of a continuous reproductive process within a discrete-time model. If population counts are taken following the breeding season, other adjustments are necessary for correct specification of the reproductive and survival terms (Kendall et al. (2019)[1]. We begin with the observation that the number of individuals in age class a at time $t+1$ is simply the number of survivors from that cohort from the previous time step. If the proportion surviving from age class a to age class $a+1$ is denoted s_a we then have:

$$N_{a+1,t+1} = s_a N_{a,t} \tag{2.6}$$

Notice that if we can estimate the number in the population at successive ages, we can estimate the survival rate. When population estimates are made during a pre-breeding census, the estimated survival rate is given by the ratio $N_{a+1,t+1}/N_{a,t}$. To complete our description of the population dynamics, we need to specify the initial number in the cohort. This number is a function of the contribution of each age class to reproduction and is obtained by simply summing the product of fertility and the number of females over all age classes:

$$N_1 = \sum_{a=1}^{a_{\max}} \mathcal{F}_a N_a \tag{2.7}$$

where $\mathcal{F}$ is fertility. In the following we will define fertility at age a as:

$$\mathcal{F}_a = s_0 \upsilon_a p_a f_a \tag{2.8}$$

where υ_a is a measure of age-specific egg viability (the fraction of eggs that can be successfully fertilized and develop into normal embryos), s_0 is the probability of survival of an age 0 individual, p_a is the proportion of females mature at age a, and f_a is the fecundity (considering only female eggs) for each age.

When the population is estimated using a pre-breeding census, the newborns must survive for nearly a year to be available to be counted in the next census and they are assigned an age of 1 in the matrix model (see Kendall et al. 2019). Equation 2.8 accordingly must account for survivorship during the first year of life.

There is emerging evidence that the viability of eggs produced by females of different ages may differ for some species, with younger females

[1] For a post-breeding census, the youngest age class comprises age 0 individuals and the index of age used for this group reflects this feature. The parents must survive through the age-class to breed and contribute to the young-of year population. It is therefore necessary to account for the survival of the parents. We therefore apply a survival coefficient of s_a rather than s_0 in Equation 2.8. Further, when a post-breeding census is employed, accounting for reproductive output can be counter-intuitive. Some juveniles who survive may become mature by the end time step of the time step, necessitating their inclusion as adults (see Kendall et al. 2019 for the appropriate formulae). Kendall et al. (2019) note that an alternative solution involves assigning this first age class an age of 1 corresponding to the end rather than that start of the time step.

producing less viable eggs (Hixon et al. 2014; Marshall 2016). This issue is a critical consideration in evaluating age- or size-specific harvesting strategies and the relative impact of preferentially harvesting larger (and older) individuals as is common in traditional fisheries management. Age- or size-selective harvesting can have unintended consequences, ranging from genetic alteration of the population through artificial selection to disproportionate negative effects on the reproductive output of the population if larger, older females are preferentially removed.

Obtaining estimates of age-specific (or more generally, size- or stage-specific) survival rates over time is a critical endeavor in fisheries analysis. In particular, we ultimately wish to be able to attribute total mortality to different sources, including fishing and various sources of natural mortality (see Section 9.2.2) . Typically, we estimate instantaneous rates of fishing and "other" mortality. These instantaneous rates are additive, a convenient feature that facilitates further analysis. The instantaneous rate of total mortality for each age class (Z) can, however, be readily converted to a survival rate:

$$s_a = \exp(-Z_a). \tag{2.9}$$

This system of difference equations can be used to determine the population size at successive points in time for each age class. Given a set of initial population estimates for each age, and estimates of survival rates (assumed to be constant in time) and fertility, we can make a *projection* of the numbers in the population over time. In this context, a projection is an assessment of what *would* happen given a set of assumptions and model structures (Caswell 2001). In contrast, a *prediction* is an attempt to specify what *will* happen, usually couched in probabilistic terms, and accommodating possible changes in system state that go beyond the set of assumptions used in the projections considered here.

We can conveniently write the age-structured model in matrix form:

$$\begin{pmatrix} N_{1,t+1} \\ N_{2,t+1} \\ N_{3,t+1} \\ \cdot \\ \cdot \\ \cdot \\ N_{n,t+1} \end{pmatrix} = \begin{pmatrix} \mathcal{F}_1 & \mathcal{F}_2 & \mathcal{F}_3 & \cdot & \cdot & \cdot & \mathcal{F}_n \\ s_1 & 0 & 0 & \cdot & \cdot & \cdot & 0 \\ 0 & s_2 & 0 & \cdot & \cdot & \cdot & 0 \\ \cdot & \cdot & \cdot & \cdot & \cdot & \cdot & \cdot \\ \cdot & \cdot & \cdot & \cdot & \cdot & \cdot & \cdot \\ \cdot & \cdot & \cdot & \cdot & \cdot & \cdot & \cdot \\ 0 & 0 & 0 & \cdot & \cdot & s_{n-1} & 0 \end{pmatrix} \cdot \begin{pmatrix} N_{1,t} \\ N_{2,t} \\ N_{3,t} \\ \cdot \\ \cdot \\ \cdot \\ N_{n,t} \end{pmatrix} \tag{2.10}$$

with the matrix elements representing the transitions from the columns to the rows. For a review of the mechanics of matrix algebra, including the rules for matrix multiplication needed here, see Hastings (1997; Chapter 2). In Box 2.4, we provide a simple numerical example.

This model can be compactly expressed in matrix notation as:

$$\mathbf{N}_{t+1} = \mathbf{L} \cdot \mathbf{N}_t \tag{2.11}$$

where $\mathbf{N}_{t+1}$ is the vector of numbers at age for time $t+1$ and $\mathbf{L}$ is the matrix of age-specific fertilities and survival rates, commonly known as the Leslie matrix (note that bold type is used to distinguish matrices and vectors). It can be readily verified that multiplying the vector $\mathbf{N}_t$ by the matrix

Box 2.4 Leslie Matrix Projection

An example of projecting with a Leslie matrix is provided below for several time steps for a starting population of 10 individuals in the 0-group and 2 individuals at age 1. The fertility at age 0 and 1 is 5 and 10 respectively and the survival probability for age 0 is 0.5 in this example.

$$\begin{pmatrix} 70 \\ 5 \end{pmatrix} = \begin{pmatrix} 5 & 10 \\ .5 & 0 \end{pmatrix} \begin{pmatrix} 10 \\ 2 \end{pmatrix} \quad \text{Time 1}$$

$$\begin{pmatrix} 400 \\ 35 \end{pmatrix} = \begin{pmatrix} 5 & 10 \\ .5 & 0 \end{pmatrix} \begin{pmatrix} 70 \\ 5 \end{pmatrix} \quad \text{Time 2}$$

$$\begin{pmatrix} 2350 \\ 200 \end{pmatrix} = \begin{pmatrix} 5 & 10 \\ .5 & 0 \end{pmatrix} \begin{pmatrix} 400 \\ 35 \end{pmatrix} \quad \text{Time 3}$$

$$\begin{pmatrix} 13750 \\ 1175 \end{pmatrix} = \begin{pmatrix} 5 & 10 \\ .5 & 0 \end{pmatrix} \begin{pmatrix} 2350 \\ 200 \end{pmatrix} \quad \text{Time 4}$$

$$\begin{pmatrix} 80500 \\ 6875 \end{pmatrix} = \begin{pmatrix} 5 & 10 \\ .5 & 0 \end{pmatrix} \begin{pmatrix} 13750 \\ 1175 \end{pmatrix} \quad \text{Time 5}$$

This repeated matrix multiplication is an example of a simple Markov chain, in which population numbers at time $t+1$ depend only on the population numbers at time t. The recursive equation can be expressed as:

$$N_t = L^t N_0$$

L results in the system of difference equations in Equation 2.6. We can handily use the matrix model to make projections about the growth of the population and the changes in the age composition over time.

To demonstrate this approach, we show an interesting application of the Leslie matrix approach to modeling an invasive species, the rusty crayfish (*Orconectes rusticus*) in Sparkling Lake in northern Wisconsin (Hein et al. 2006). Leslie matrices were constructed to examine possible methods of population control for this species, which has caused the extinction of native crayfish species and other forms of ecosystem disruption in North American lakes. Hein et al. (2006) specified a birth-pulse model with a pre-breeding census as in Equation 2.8. The Leslie matrix for a 4 age-class crayfish model in the absence of deliberate control measures is:

$$\begin{pmatrix} 0 & 1.41 & 1.98 & 2.58 \\ 0.652 & 0 & 0 & 0 \\ 0 & 0.363 & 0 & 0 \\ 0 & 0 & 0.128 & 0 \end{pmatrix} \qquad (2.12)$$

(Hein et al. 2006). We used estimates of age-specific survival rates in their Table 2 to specify the entries on the subdiagonal and discounted their estimates of reproductive output by the young-of-year survival rate to specify the first line of the matrix. Here we see that fertility increases with age (top row) while the survival rate decreases. The projected population abundance for the first 15 time steps for each age class for each time step is provided in Figure 2.8a and the ratio of total population numbers at successive time steps is shown in Figure 2.8b. Notice that the ratio of the total number in the population at successive time steps has quickly attained a constant level, indicating that the population is growing at a constant rate at this point. As we show below, the relative proportion of each age class also remains constant at this point—this is referred to as the stable age distribution (SAD).

When the SAD has been attained, the population model can be written:

$$\mathbf{N}_{t+1} = \lambda \mathbf{N}_t \qquad (2.13)$$

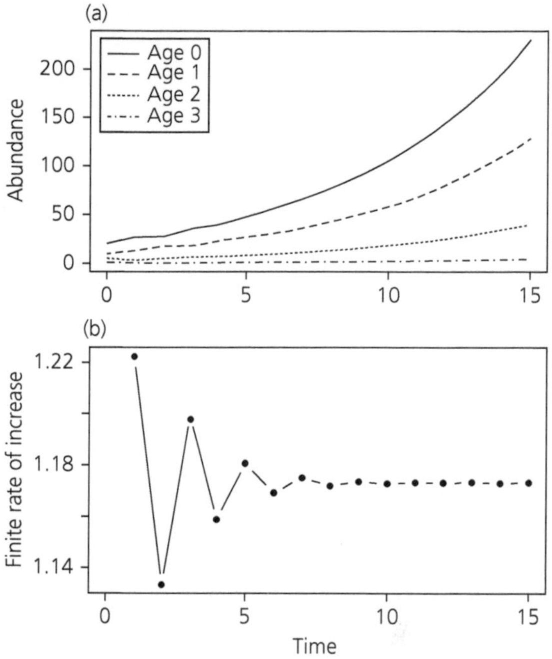

Figure 2.8 (a) Population trajectories by age class for invasive rusty crayfish (*Orconectes rusticus*) in the absence of control measures (Hein et al. 2006). (b) Ratio of the total number of individuals in the population at successive time steps. We have retained the age number conventions of the authors.

where λ is again the finite rate of increase. It is now clear that the general behavior of the age-structured model is identical to that for the non-age structured case in discrete time once the SAD is attained. Recall that the basic equation including the full projection matrix is: $\mathbf{N}_{t+1} = \mathbf{L}{\cdot}\mathbf{N}_t$ and therefore when the *SAD* has been reached, we can write:

$$\lambda \mathbf{N}_t = \mathbf{L} \cdot \mathbf{N}_t \qquad (2.14)$$

We next multiply the left-hand side by the identity matrix (this is a simply square matrix comprising 1s on the main diagonal and zeros elsewhere as described in Hastings (1997; Chapter 2)). We can then write:

$$\lambda \mathbf{I} \cdot \mathbf{N}_t = \mathbf{L} \cdot \mathbf{N}_t \qquad (2.15)$$

or, bringing all terms on one side of the equation we have

$$(\lambda \mathbf{I} \cdot \mathbf{N}_t - \mathbf{L} \cdot \mathbf{N}_t) = (\lambda \mathbf{I} - \mathbf{L})\ \mathbf{N}_t = 0 \qquad (2.16)$$

and we can solve this equation directly for λ. Note that one trivial solution to the equation is $\mathbf{N}_t = 0$. The non-trivial solutions to this equation require that:

$$|\mathbf{L} - \lambda \mathbf{I}| = 0 \quad (2.17)$$

where the vertical lines enclosing the expression on the left-hand side indicate that we are taking the determinant of this expression. This is called the characteristic equation. If $\mathbf{L}$ has dimension $n \times n$, its characteristic equation has n roots (solutions), which may be either real or complex numbers. The roots to this equation (or the characteristic values) are of particular importance; they are also commonly called eigenvalues. A key result is that the largest real root of the equation (λ) gives the finite rate of increase once the SAD has been attained. The subdominant eigenvalues determine the degree of oscillation (if present) as the population approaches the stable-age distribution. If the largest eigenvalue is a complex conjugate, the population will cycle and a stable age distribution will not be attained (Caswell 2001).

For our crayfish example, the estimated finite rate of increase is $\lambda = 1.17$, indicating positive population growth in the absence of control measures (Figure 2.8). Hein et al. (2006) demonstrated that by reducing fishing pressure on crayfish predators and by selective trapping of larger crayfish, the population growth rate could in fact be reduced below 1.0, thus suggesting that controlling the invasion would be possible using a combination of methods. The stable age distribution for the crayfish example in the absence of any control measures is depicted in Figure 2.9a.

Each eigenvalue also has an associated characteristic vector (or eigenvector) with special properties. Basically, the eigenvectors of a square matrix such as the Leslie matrix are non-zero vectors whose elements retain the same relative proportions when multiplied by the matrix (although their absolute magnitudes do change). Here we can see a connection to the idea of the SAD, which is in fact given by the eigenvector corresponding to the largest eigenvalue of the Leslie matrix.

Once we have determined the magnitude of the dominant eigenvalue, we can solve for the

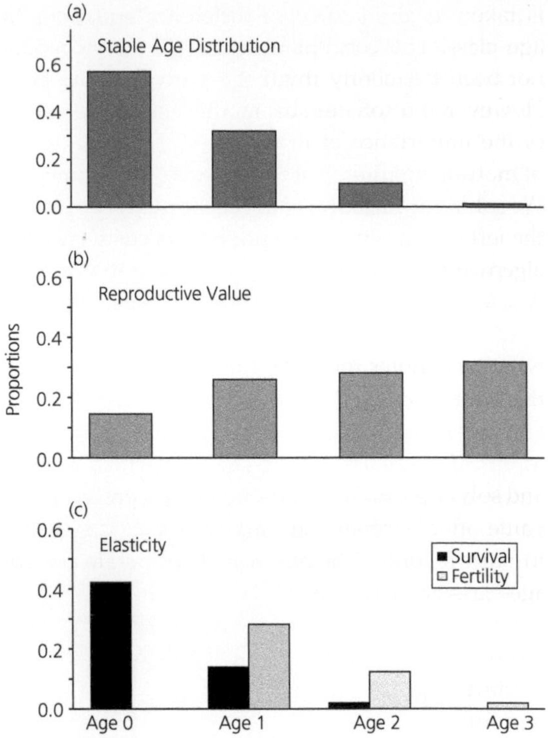

Figure 2.9 (a) Stable age distribution, (b) reproductive value, and (c) elasticities of invasive rusty crayfish (*Orconectes rusticus*) in the absence of control measures (matrix inputs from Hein et al. 2006).

corresponding eigenvectors. We begin with the vector equation:

$$\mathbf{Lw} = \lambda \mathbf{w} \quad (2.18)$$

and note that we can write this equation as:

$$(\mathbf{L} - \lambda \mathbf{I})\,\mathbf{w} = 0 \quad (2.19)$$

where $\mathbf{w}$ is the eigenvector corresponding to the dominant eigenvalue, λ. This is often referred as the right eigenvector. The system of simultaneous equations summarized in (2.19) can be solved using standard methods (for a 2×2 matrix this entails the use of the quadratic formula). The solutions provide the estimates of the fraction of each age class in the population once the stable age distribution has been attained.

We are also interested in the contribution of each individual to current and future reproduction. This is the reproductive value of an individual and it is taken to be identical for each individual in an age class. The concept of reproductive value has not been frequently invoked in fisheries analysis. However, it resonates strongly with the recognition of the importance of maintaining sufficient stocks of mature individuals to ensure viable populations. We solve for the reproductive value by determining the left eigenvector corresponding to the dominant eigenvalue. We then have:

$$\mathbf{vL} = \lambda \mathbf{v} \tag{2.20}$$

where $\mathbf{v}$ denotes the left eigenvector. We can write this equation as:

$$\mathbf{v}\,(\mathbf{L} - \lambda\,\mathbf{I}) = 0 \tag{2.21}$$

and solving, we obtain the estimates of reproductive value after normalizing all elements of the vector to sum to one. The reproductive value for each age-class of rusty crayfish is shown in Figure 2.9b. We see that individuals that have survived to reproductive age and that have the highest fertility dominate the contributions to current and future reproduction.

2.3.1.1 Sensitivities and elasticities of matrix models

In many instances, we wish to know how a change in any one of the matrix elements, a_{ij}, affects the finite rate of increase (λ_1) of the population. The sensitivity of the dominant eigenvalue to small changes in a given matrix element is given by:

$$\frac{\partial \lambda_1}{\partial a_{ij}} = \frac{v_i w_j}{\langle w, v \rangle} \tag{2.22}$$

where the denominator of the right-hand side of Equation 2.22 denotes the scalar product of the two vectors. This equation gives the rate of change of the dominant eigenvalue as a function of each element of the matrix **L**. Elasticities scale the sensitivities to account for the differing magnitudes of the matrix elements:

$$e_{ij} = \frac{a_{ij}}{\lambda_1} \cdot \frac{\partial \lambda_1}{\partial a_{ij}} \tag{2.23}$$

The elasticities sum to one over all elements of the matrix **L** (Caswell 2001). In our crayfish example, the elasticity in age-0 survival and age-1 fertility is highest (Fig. 2.9c). This normalized representation of the sensitivities permits a more direct interpretation of the relative importance of an individual within each class with respect to both fertility and survival.

We have seen that in many instances, a constant finite rate of increase and stable age distribution is attained for populations following the simple density-independent construct of the Leslie matrix. It turns out, however, that more complex patterns can emerge, notably in the special case of semelparous species such as Pacific salmon (*Oncorhynchus* spp.). Seven species of Pacific salmon exist along the North Pacific Rim, supporting extremely important commercial and recreational fisheries and holding particular cultural significance for native peoples throughout the range. An illustration of a four age-class model with all reproduction concentrated in the oldest age class is:

$$\mathbf{L} = \begin{pmatrix} 0 & 0 & 0 & \mathcal{F}_4 \\ s_1 & 0 & 0 & 0 \\ 0 & s_2 & 0 & 0 \\ 0 & 0 & s_3 & 0 \end{pmatrix} \tag{2.24}$$

An illustration of possible population trajectories for a model of this type is provided in Figure 2.10. Here we see that population levels at successive generations do not necessarily settle down to a constant rate of increase (or may be attained only slowly). Instead, an oscillatory population trajectory is possible. In the example shown in Figure 2.10, the dominant eigenvalue is a complex number, indicating that we should in fact expect oscillatory behavior.

We have seen that the age-structured models share with the non-structured population models a basic agreement in the dynamical behaviors possible. Age-structured models with no density dependence can only exhibit geometric increase, stasis, or geometric decline once the stable age distribution is attained. We can provide a graphical analysis for an age-structured population to illustrate some basic properties. The relationship between successive ages for a four age-class model with no density dependence between successive stages is shown in Figure 2.11. This representation is referred to as a Paulik diagram in the fisheries literature in recognition of the contributions of

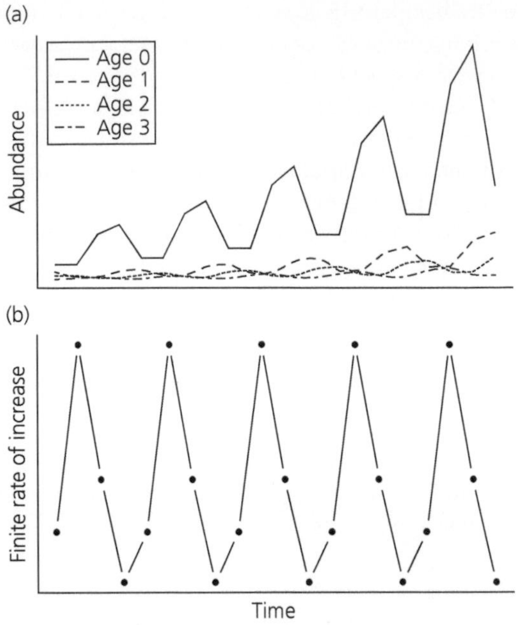

Figure 2.10 (a) Population trajectories (generation time) for a semelparous species and (b) estimates of the finite rate of population increase.

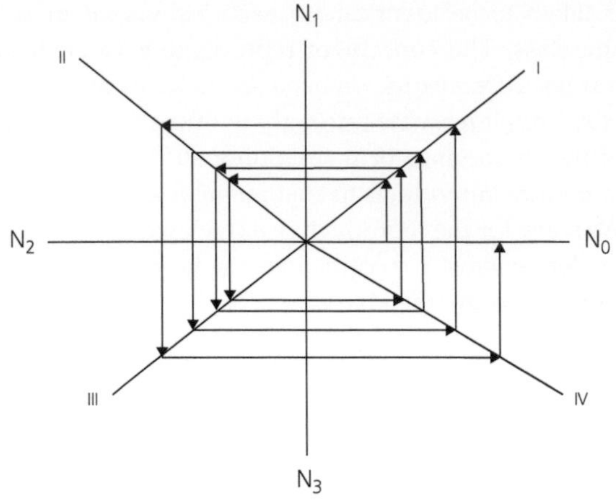

Figure 2.11 Paulik diagram for a semelparous population with four age classes. Roman numerals designate quadrants for the relationships between successive age groups.

Gerald Paulik to fish population dynamics and graphical depictions of this type in particular (Paulik 1973). This diagram is essentially an extension of the cobweb approach we used earlier. Note that in this particular visual representation, we can depict no more than four age classes (as in our example for sockeye salmon). From the starting point on the N_0 axis in Figure 2.11, trace the arrows through each quadrant showing the relationship between the numbers in successive ages. In this case, the relationships are all linear because we are only considering density-independent processes. Notice that the slopes of the diagonal lines represent stage-specific survival and that this population is growing.

2.3.1.2 Stage-structured models

It is sometimes more convenient (and/or more realistic) to deal with populations that are classified by size or life stage rather than age. Age determination is difficult for some species and even for taxa that are readily aged, many demographic processes (e.g. maturity, fecundity, mortality) depend more on size or life stage than age *per se*. We can specify matrix representations of these life stage-classified models in a way that is analogous to age-structured matrix models (see Caswell 2001). However, stage-structured models present certain challenges for analysis due to the fact that stages can represent a mix of several age-classes and estimates of some measures that are readily computed in age-stuctured models can be difficult or ambiguous in stage-structured models (Kendall et al. 2019). Botsford et al. (2019) evaluate these issues and examine alternatives.

The stage-structured projection matrix can be written:

$$\begin{pmatrix} p_1 & \mathcal{F}_2 & \mathcal{F}_3 & \cdot & \cdot & \cdot & \mathcal{F}_n \\ g_1 & p_2 & 0 & \cdot & \cdot & \cdot & 0 \\ 0 & g_2 & p_3 & \cdot & \cdot & \cdot & 0 \\ \cdot & \cdot & \cdot & \cdot & \cdot & \cdot & \cdot \\ \cdot & \cdot & \cdot & \cdot & \cdot & \cdot & \cdot \\ \cdot & \cdot & \cdot & \cdot & \cdot & \cdot & \cdot \\ 0 & 0 & 0 & \cdot & \cdot & g_{n-1} & p_n \end{pmatrix} \tag{2.25}$$

where the p_i represent the probabilities of surviving but remaining in the same size or stage class in the time interval and the g_i represent the probability of surviving and growing into the next largest category. Here $p_i = s_i(1-\gamma_i)$ where s_i is the survival probability and γ_i is probability of growing into the next stage class and $g_i = s_i\gamma_i$.

For simplicity, in Equation 2.25 we have shown the case in which growth only occurs into the next largest category within each time step of the analysis. These restrictions can be removed so that it is possible to define the projection matrix in such a way that any element of the matrix is potentially non-zero.

As noted earlier, estimates of survival rates are available for a broad spectrum of harvested species, often over many years. It then remains to estimate the transition probabilities. Determination of individual growth rates is critically important in generating estimates of productivity and yield (see Chapters 9 and 11). Accordingly, we also have a rich resource to draw on in estimating the transition probabilities. In the simplest case, under the assumption of a constant stage duration, a crude estimate of the stage-specific transition probability is given by the inverse of the mean time spent in each stage (Caswell 2001, p. 160). A more refined estimate for the case of a fixed stage duration is given by:

$$\gamma_i = \frac{\left(s_i/\tilde{\lambda}\right)^{T_i} - \left(s_i/\tilde{\lambda}\right)^{T_i - 1}}{\left(s_i/\tilde{\lambda}\right)^{T_i} - 1} \tag{2.26}$$

where s_i is the survival rate for a unit time step (typically years) for the ith stage, $\tilde{\lambda}$ is an initial estimate of the finite rate of increase, and T_i is the overall stage duration (Caswell 2001). An initial estimate of γ_i can be made by setting $\lambda = 1$. This estimate can then be refined by taking the resulting starting values of g_i and p_i, solving for the dominant eigenvalue of the matrix, substituting this new estimate into the matrix, and successively applying this process until convergence in the values of λ is attained.

The flexibility of stage-structured models is nicely illustrated by Brewster-Giesz and Miller (2000), who developed a life-stage model for sandbar shark (*Carcharhinus plumbeus*) in the northwest Atlantic. Five life stages were represented: neonates (N), juveniles (J), subadults (SA), adults (A), and "resting" adults (R). Neonates make the transition to juveniles within the one-year time step of the model. Juveniles and subadults can either survive and grow into the next class within one year or remain within the same category. Adults reproduce

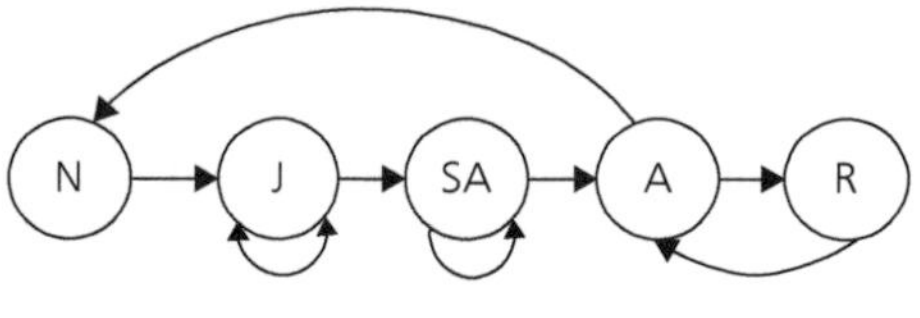

Figure 2.12 Life-cycle model of a sandbar shark (*Carcharhinus plumbeus*) population comprising neonates (N), juveniles (J), sub-adults (SA), adults (A), and resting adults (R).

and then make the transition to the resting stage. Following reproduction, adult females enter a resting stage of approximately one year. Thus the interval between parturition events is a minimum of two years. This life-cycle graph for this particular model is shown in Figure 2.12. To model this life-history pattern, Brewster-Giesz and Miller (2000) specified a projection matrix of the form:

$$\begin{pmatrix} 0 & 0 & 0 & g_A \mathcal{F}_A & 0 \\ g_N & p_J & 0 & 0 & 0 \\ 0\ 0 & g_J & p_J & 0 & 0 \\ 0 & 0 & g_{SA} & 0 & g_R \\ 0 & 0 & 0 & g_A & 0 \end{pmatrix} \tag{2.27}$$

A birth-pulse, post-breeding census structure was assumed. As with the age-structured population model, the dominant eigenvalue, λ_1, summarizes the demographic information contained in the matrix $\boldsymbol{L}$, once the stable-stage distribution is reached. For the case where no fishing is allowed, $\lambda_1 > 1$ and the population is predicted to grow given the specification of the parameters used in this projection. In contrast, the population is predicted to decline at the fishing rates estimated at the time of this analysis (Brewster-Geisz and Miller 2000). Estimates of the stable age distribution reproductive value, and elasticity for each stage in the absence of fishing are shown are shown in Figure 2.13.

2.4 Summary

Population growth rate depends on the difference between births and deaths. In continuous time, a population without density dependence will increase or decrease exponentially; a population with discrete generations will change geometrically, with finite rate of increase λ. Aquatic organisms

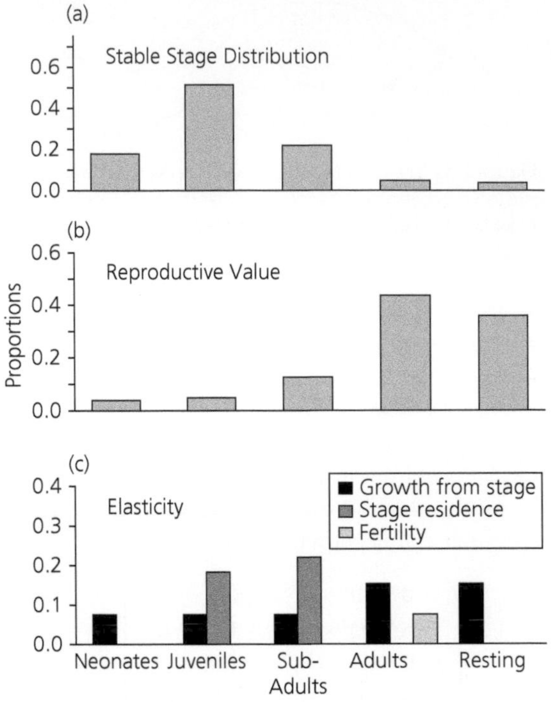

Figure 2.13 (a) Stable age distribution, (b) reproductive value and (c) elasticities of sandbar sharks (*Carcharhinus plumbeus*). Matrix data inputs courtesy of Karyl Brewster-Giesz.

of different life stages have different survival rates and produce different numbers of offspring; therefore, depending on the objectives of the analysis, it may be necessary to structure population models by age, size, or life-history stage. Structured populations will converge to a constant growth rate once a stable age/stage distribution is attained. Structured population models can be written in matrix form. Then, matrix algebra can be used to calculate λ, the stable distributions of age or stage, reproductive values, sensitivities, and elasticities. An additional source of anthropogenic mortality can shift the balance of births and deaths from population growth to population decline. In structured populations, anthropogenic impacts could reduce the survival or fertility of particular life stages. Using matrix methods, we can calculate the effect of these individual impacts on the population growth rate.

We note that although the classical size- and age-structured fishery models to be examined in Chapter 11 are not typically represented in matrix form, in virtually all cases they can be readily cast in this way (see Getz and Haight 1989). Matrix models should accordingly not be considered as somehow separate from other types of age- or stage-structured models. Rather they simply package the models in a different form and take advantage of certain features of matrix algebra.

Additional reading

A number of truly excellent introductory texts surveying the field of quantitative ecology have been published over the last several decades. Wilson and Bossert (1971) provided one of the earliest primers covering the topic. This student-friendly format has been updated over the years by Roughgarden (1998) and by Gotelli (2008) with the fourth edition of his primer. Other important introductory texts include Hastings (1997), Case (2000), and Vandermeer and Goldberg (2003). Botsford et al. (2019) offer more advanced treatments of these topics. More advanced treatments of theoretical ecology can be found in Kot (2001). Caswell (2001) remains the standard for matrix population models.

CHAPTER 3

Density-Dependent Population Growth

3.1 Introduction

No population continues to grow indefinitely and, in the absence of major perturbations, most populations persist over ecological timescales. We therefore need to consider extensions to the density-independent models described in the last chapter to account for these observations. The general concept of negative feedback processes provides a framework for understanding the modulation of unconstrained population growth. These feedback processes can confer a certain resilience to natural and anthropogenic disturbances. Density-dependent processes are one class of feedback mechanisms that are important in this way. These include processes such as intraspecific competition for food resources, space, and other commodities; cannibalism; density-related vulnerability to predation; and other factors. These mechanisms alter the types of trajectories we observed in the previous chapter where, in the absence of negative feedback effects, the populations either grow or decline exponentially (geometrically in discrete time), or remain in a precarious neutral equilibrium when births balance deaths. As resources become limiting or selective predation intensifies with increasing density, however, the rate of change of the population is reduced, leading to different outcomes including a possible stabilization of population size. The concept of sustainable harvesting depends critically on compensatory processes of this type.

3.2 Compensation in simple population models

We now show how simple modifications of the exponential and geometric growth models described in Chapter 2 can be developed to incorporate feedback processes. We shall consider situations in which both positive and negative feedback processes can be represented in the model. Casting birth and death rates as functions of population size provides the vehicle for incorporating these considerations into simple population models. The consequences of this modification are profound. Unlike the density-independent case, both stable and unstable equilibrium points are now possible for density-dependent population models. In turn, these characteristics have important implications for the resilience of the population to perturbations.

3.2.1 Continuous-time models

Controlled experiments have played a central role in testing hypotheses in ecology. However, for larger aquatic organisms, this not generally feasible and we are left to discern underlying mechanisms using a combination of observation and modeling to examine alternative hypotheses. However, important experiments have been performed on small-bodied fish in aquaria, providing a unique opportunity to see population processes unfold in a way not possible in natural settings. To introduce density-dependent processes we turn

Fishery Ecosystem Dynamics. Michael J. Fogarty and Jeremy S. Collie, Oxford University Press (2020). © Michael J. Fogarty & Jeremy S. Collie 2020.
DOI: 10.1093/oso/9780198768937.003.0003

to experiments of the response of guppy (*Poecilia reticulata*) populations to exploitation (Silliman and Gutsell 1958). In Figure 3.1 we depict the full population trajectory for an unexploited control group showing the number of adults and the number of immature individuals at each weekly census starting at week 15 of the experiment. The adult population appeared to increase exponentially during the initial phase of the experiment following a delay related to the generation time. Over the remainder of the experiment the number of adult individuals stabilized. Note, however, that the number of immature fish declined sharply as the adult population increased (Figure 3.1). In this experimental design, no refuge for immature guppies was made available. Cannibalism by adults on immature guppies, a phenomenon well known to aquarists, is the regulatory mechanism in this case. Intraspecific predation is one of several possible compensatory mechanisms that can result in a decrease in the rate of population increase at high population levels. We will explore models of cannibalism in greater detail in our examination of recruitment processes in Section 9.3. In the following sections, we will describe approaches to incorporate more general density-dependent processes into our population model.

In the exponential model described in Chapter 2, the per capita birth and death rates were independent of population size (i.e. these rates were constant for all population levels). We can however, make the birth and death rates density-dependent by expressing them as functions of population density (or size). The simplest form for these relationships is linear:

$$b = b_0 - b_1 N \tag{3.1}$$

and

$$d = d_0 + d_1 N$$

where we have now stated that the birth rate declines linearly as N increases and the death rate increases as the population increases (Figure 3.2a; see Wilson and Bossert 1971, Vandermeer and Goldberg 2003, and Gotelli 2008). This is the simplest form of density dependence because the addition of each individual to the population changes the per capita growth rate (b–d) by the

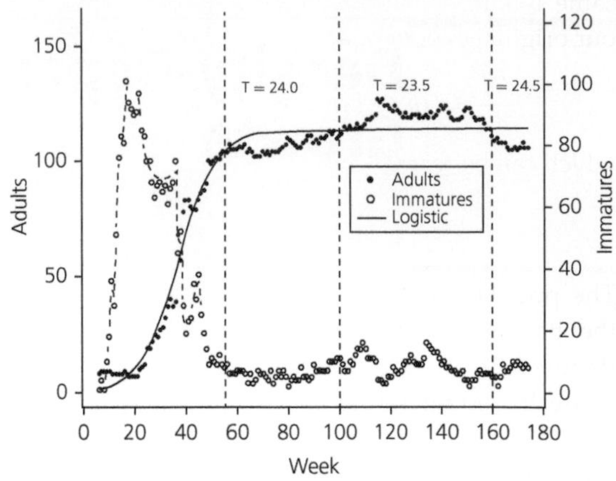

Figure 3.1 Observed abundance of immature and adult guppies (*Poecilia reticulata*) in one control tank (D) in the experiment conducted by Silliman and Gutsell (1958). The predicted population level based in the logistic model is also shown. The count of immature guppies is indicative of the birth rate. The rapid decline in immatures in the population is due to cannibalism by adults. The average temperature (T, °C) for three stanzas of time differed and affected population levels of both immatures and adults.

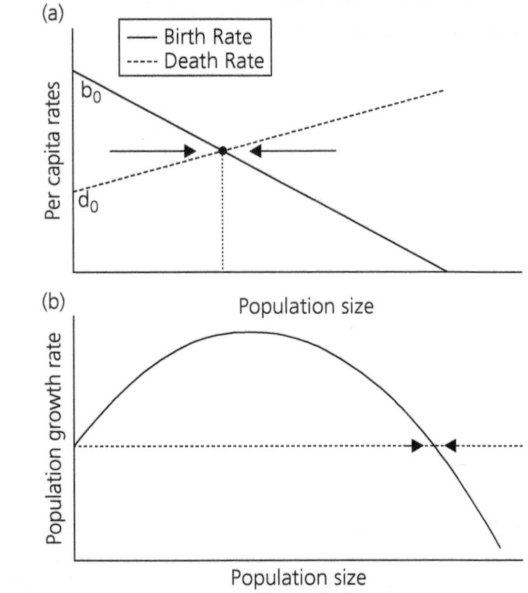

Figure 3.2 (a) Linear models for per capita birth and death rates. The intersection of the two lines marked by a closed circle represents a globally stable equilibrium point and (b) the rate of change of population size as a function of population size (N). The arrows converge at the equilibrium population size. The dotted line indicates zero population growth.

same amount. Substituting these new terms into our original exponential model we have:

$$\frac{dN}{dt} = [(b_0 - b_1 N) - (d_0 + d_1 N)]\ N \qquad (3.2)$$

which can be rearranged to give:

$$\frac{dN}{dt} = [(b_0 - d_0) - (b_1 + d_1)\,N]\ N \qquad (3.3)$$

The population equilibrium, N^*, is reached when the per capita birth and death rates are equal (Figure 3.3). We find the equilibrium point for this model by setting the derivative equal to zero so that:

$$(b_0 - d_0) = (b_1 + d_1)\,N \qquad (3.4)$$

and the equilibrium population size is therefore:

$$N^* = \frac{b_0 - d_0}{b_1 + d_1} \qquad (3.5)$$

Insights into the stability of this equilibrium can be inferred by inspection. For $N < N^*$ the birth rate exceeds the death rate and the population will grow. Conversely, for $N > N^*$ deaths exceed births and the population will decline. We refer to the negative feedback mechanism underlying this model as compensatory because the per capita population growth rate declines with increasing population size—changes in birth and/or death rates compensate for increases in the population density, stabilizing population growth. Depensation (defined below) occurs when there is positive feedback between the growth rate and population size over some range of abundance levels.

It is not necessary for both the birth and death rates to change as a function of abundance for compensation to exist. For example, if the birth rate remains constant but the death rate increases linearly with abundance, we have:

$$\frac{dN}{dt} = [b_0 - (d_0 + d_1 N)]N \qquad (3.6)$$

and the equilibrium point is now:

$$N^* = \frac{b_0 - d_0}{d_1} \qquad (3.7)$$

In the following, it will be convenient to simplify the model by letting $\alpha = (b_0 - d_0)$ and $\beta = (b_1 + d_1)$ giving:

$$\frac{dN}{dt} = (\alpha - \beta N)N \qquad (3.8)$$

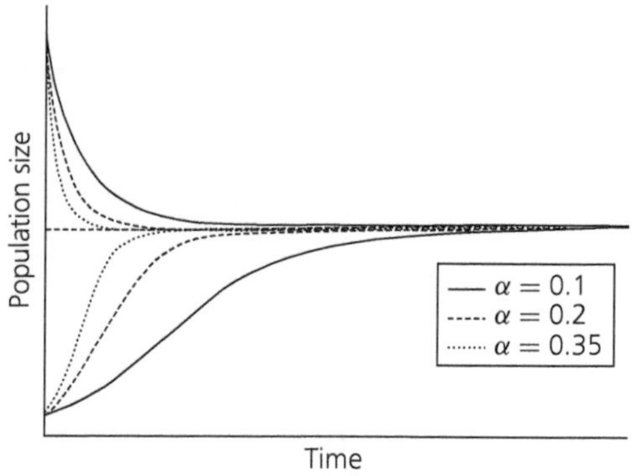

Figure 3.3 Population trajectories as a function of time for the logistic growth model in continuous time for three levels of the intrinsic rate of increase and two levels of the initial population size (N_0) starting above and below the equilibrium level.

(Remember that we can always convert back to birth and death rates as in our original formulation and we need not stipulate that births and deaths are potentially both functions of abundance). The rate of change of population size is a quadratic function (Figure 3.2b). The equilibrium population size is found by setting $dN/dt = 0$ and solving to give:

$$N^* = \frac{\alpha}{\beta}. \qquad (3.9)$$

By taking the derivative of Equation 3.8 with respect to N, setting it equal to zero, and solving we find that the maximum rate of population change occurs when:

$$N = \frac{\alpha}{2\beta}. \qquad (3.10)$$

For this symmetrical model, the population is growing at its maximum rate when it is one half the equilibrium population size. This calculation essentially involves finding the point at which a line with zero slope is tangent to the curve. We have simply taken the derivative of our function to determine the point at which the population is growing at its maximum rate. This point becomes critical in considering harvest strategies addressed in Chapter 11. In Box 3.1 we examine the local stability properties of this equilibrium and the predicted time required to recover from a (minor) perturbation.

Integrating Equation 3.8 by parts, we find that the solution can be written:

Box 3.1 Qualitative Stability Analysis and the Characteristic Return Time for the Logistic Model

The differential equation for the case of a linear per capita change in population growth as a function of abundance is:

$$\frac{dN}{dt} = \dot{N} = (\alpha - \beta N)N$$

and taking its derivative with respect to N we have:

$$\frac{\partial \dot{N}}{\partial N} = \alpha - 2\beta N$$

Evaluating at the equilibrium ($N^* = \alpha/\beta$) we have:

$$\frac{\partial \dot{N}}{\partial N} = -\alpha$$

The Schur–Cohen stability criterion requires that the partial derivative be negative and we conclude that the equilibrium point is stable.

To estimate the time required to return to the equilibrium following a small perturbation we begin by assuming that the perturbation (x) from equilibrium will decay exponentially:

$$\frac{dx}{dt} = -\alpha x$$

and its solution is:

$$x_t = x_0 e^{-\alpha t}$$

The time required to decay to approximately 37% (~1/e) of the initial perturbation is found by solving:

$$\frac{x_0}{e} = x_0 e^{-\alpha t}$$

Dividing through by x_0, taking logs of both sides and solving we have the characteristic return time

$$t_c = \frac{1}{\alpha}$$

Notice that anything that reduces the intrinsic rate of increase will increase the characteristic return time. For example if harvesting is introduced as a density-independent factor it will affect the intrinsic rate of increase and we can calculate its effect on the recovery time.

$$N_t = \frac{N^*}{\left(1 + \gamma \cdot e^{-\alpha t}\right)} \tag{3.11}$$

where $\gamma = (\alpha - \beta N_0)/(\beta N_0)$ and N_0 is the population at time 0 (see Box 3.2). Possible trajectories for a population growing according to this model are provided in Figure 3.3. Notice that for initial population size below the equilibrium, the population follows a sigmoidal trajectory. If the initial population is above the equilibrium, it decays exponentially to N^*.

Applying the logistic model to the guppy population numbers depicted in Figure 3.1 we see that it provides a reasonable representation of the dynamics of the population (for more information on estimation procedures, see Box 3.3). However, the observed population size exhibited persistent fluctuations above and below the predicted equilibrium level. In fact, water temperatures in the experiment could not always be precisely controlled due to equipment malfunctions (Silliman and Gutsell 1958), resulting in variation in the population around the equilibrium level (see Figure 3.1). Silliman and Gutsell note that the guppies were quite sensitive to temperature, and survival rates declined when temperature levels increased. This result serves as an important reminder that environmental factors cannot be ignored as we seek to understand the determinants of population change, even in laboratory settings.

Because Equation 3.8 retains a fundamental connection with elemental birth and death rates, we prefer this particular representation of the logistic growth model. It is, however, quite common in both ecology and fisheries texts to use an alternative form framed in terms of the intrinsic rate of increase of the population (r) and the carrying capacity (or equilibrium level, K):

$$\frac{dN}{dt} = r\left(1 - \frac{N}{K}\right)N \tag{3.12}$$

Note that when $N = K$, the expression in parenthesis is zero and the rate of change of population size is therefore zero (i.e. the system is in equilibrium). If we make the substitutions $\alpha = r$ and $\beta = r/K$ (and therefore $\alpha/\beta = K$), we recover our Equation 3.8. Note that expressed in terms of birth and death rates, K is given by Equation 3.5. The solution to the logistic model expressed in terms of r and K is:

Box 3.2 Solution to the Logistic Growth Model

The differential equation for the case of a linear per capita change in population growth as a function of abundance is:

$$\frac{dN}{dt} = (\alpha - \beta N)N$$

Separating variables and integrating, we have:

$$\int_{N_o}^{N_t} \frac{dN}{(\alpha - \beta N)N} = \int_{t_o}^{t} dt$$

and noting that

$$\int \frac{dN}{(\alpha - \beta N)\,N} = -\frac{1}{\alpha}\log_e\left[\frac{(\alpha - \beta N)}{N}\right]$$

we have:

$$\log_e\left[\frac{(\alpha - \beta N_t)\,N_0}{(\alpha - \beta N_0)\,N_t}\right] = -\alpha\,(t - t_0)$$

letting $t_0 = 0$, multiplying by -1 and taking antilogarithms gives:

$$\left[\frac{(\alpha - \beta N_0)\,N_t}{(\alpha - \beta N_t)\,N_0}\right] = e^{\alpha t}$$

Next, multiply both sides by N_0 and cancel N_t on the left hand side to give:

$$\left[\frac{(\alpha - \beta N_0)}{(\alpha/N_t - \beta)}\right] = N_0 e^{\alpha t}$$

Now isolate N_t to give:

$$\frac{1}{N_t} = \left[\frac{(\alpha - \beta N_0 + \beta N_0 e^{\alpha t})}{\alpha N_0 e^{\alpha t}}\right]$$

which, after inverting and dividing the numerator and denominator by $N_0 e^{\alpha t}$ and rearranging results in:

$$N_t = \frac{\alpha}{\frac{\alpha - \beta N_o}{N_0 e^{\alpha t}} + \beta}$$

Finally, if we divide the numerator and denominator on the right hand side by β and again rearranging we have an expression in which the equilibrium abundance ($N^* = \alpha/\beta$) appears in the numerator:

$$N_t = \frac{N^*}{1 + \frac{\alpha - \beta N_0}{\beta N_0} e^{-\alpha t}}.$$

$$N_t = \frac{N_0 e^{rt} K}{K - N_0 + N_0 e^{rt}} \tag{3.13}$$

3.2.1.1 The generalized logistic model

There is no a priori reason to suppose that populations are regulated solely by linear declines in the per capita rate of population growth. We can write a generalized form of the logistic model to permit a wider array of shapes for the population production function. These shapes in turn can be related to different life history strategies and characteristics (Gilpin and Ayala 1973, Fowler 1981). We now introduce an additional "shape" parameter, θ:

$$\frac{dN}{dt} = \left(\alpha + \beta N^{\theta - 1}\right) N \tag{3.14}$$

and note that the signs of the parameters α and β differ depending on whether θ is greater than or less than 1. In this representation, for $\theta < 1$, $\alpha < 0$ and $\beta > 0$; the signs are reversed when $\theta > 1$. Equation 3.13 is sometimes called the theta-logistic model (we have written it in a slightly different form than originally proposed by Gilpin and Ayala (1973) to comport with similar models in the fisheries literature). The logistic model is a special case of this more general model with $\theta = 2$. We will encounter a variant of this form in Pella and Tomlinson (1969) fishery production model (see Section 11.3). The population growth rate can now take on a range of shapes as a function of density or abundance (Figure 3.4). We can find the equilibrium point for this model by again setting the derivative equal to zero. The equilibrium point is now:

Box 3.3 Fitting the Logistic Growth Model to Data

As with the density-independent population model, we can linearize the logistic model to provide initial parameter estimates. In turn, we can then use non-linear regression to obtain final parameter estimates. Recall our solution for the logistic equation (Equation 3.11):

$$N_t = \frac{N^*}{(1 + \gamma \cdot e^{-\alpha t})}$$

Dividing both sides by N^* and inverting we have:

$$\frac{N^*}{N_t} - 1 = \gamma \cdot e^{-\alpha t}$$

Taking natural logarithms gives:

$$\log_e\left[\frac{N^*}{N_t} - 1\right] = \log_e\gamma - \alpha t$$

Given a visual estimate of N^*, linear regression can be used to estimate $\log_e\gamma$ and α.

We next use the resulting parameters of the linearized model as starting values in a non-linear least squares regression. Starting with initial values of the parameters, Equation 3.11 is used to predict $\hat{N}_t$ at each point in time.

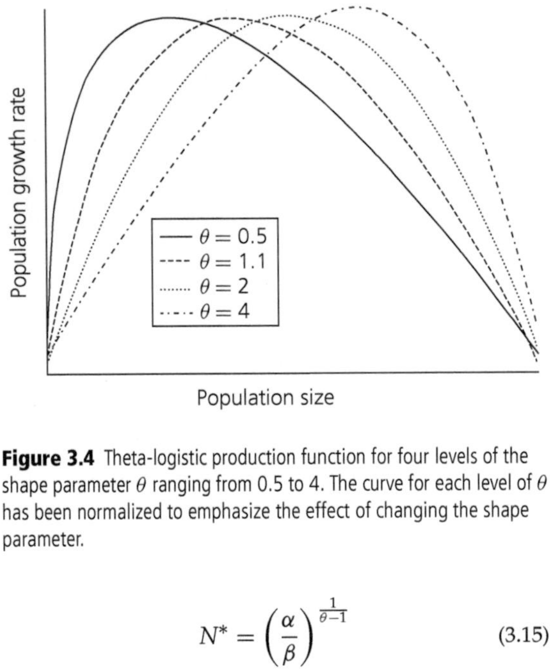

Figure 3.4 Theta-logistic production function for four levels of the shape parameter θ ranging from 0.5 to 4. The curve for each level of θ has been normalized to emphasize the effect of changing the shape parameter.

$$N^* = \left(\frac{\alpha}{\beta}\right)^{\frac{1}{\theta-1}} \tag{3.15}$$

Note that θ cannot equal 1 in this formulation (there is no stable equilibrium point for $\theta = 1$). As $\theta \rightarrow 1$, the model converges to a logarithmic form and we can write:

$$\frac{dN}{dt} = (\alpha - \beta \log_e N)\, N \tag{3.16}$$

This is one way of expressing the well-known Gompertz model. It is the basis for the Fox (1970) fishery production model referenced in Chapter 11 Box 11.2. Now the equilibrium is given by:

$$N^* = e^{\alpha/\beta} \tag{3.17}$$

and the maximum rate of change occurs at:

$$N = \frac{\alpha}{\beta e} \tag{3.18}$$

where e is the base of the natural logarithms. In contrast to the logistic model where the maximum rate of change occurs at one half the equilibrium level, the maximum occurs at approximately 37 per cent ($1/e$) of the equilibrium for the Gompertz model.

3.2.2 Multiple equilibria

The logistic model is one of the most renowned equations in the canon of population biology (Kingsland 1995). It has also played a central role in the development of harvesting models in fisheries science and resource economics. The heuristic value of the logistic model is undeniable—it shows in a simple but elegant fashion why negative feedback processes are so critical to the stability and resilience of populations. Linear per capita rates of change in birth and/or death rates are sufficient to confer the fundamental attributes so essential to population regulation. The continuous time logistic equation also has "well-behaved" dynamics with a globally stable equilibrium point.

However, some populations undergo very rapid changes in state from one level to another. For example, the Northern Cod population off

Newfoundland supported a sustainable fishery for nearly five centuries before suddenly collapsing in the early 1990s under heavy exploitation (Rose 2004). A closer examination of the population levels starting in 1984 show a decline culminating in a collapse by 1995 where it remained before showing some indication of a nascent recovery (Figure 3.5). The population remained at this lower level despite a partial moratorium on harvest implemented in 1992. Potential reasons invoked for the decline and slow recovery include continued exploitation, climate change, increased predation by seals, depensatory population dynamics, or some combination of these factors (Rose 2004).

In the following section, we will explore how non-linear changes in per capita birth and/or death rates as a function of population size can result in rapid shifts in population levels. As noted above, there is no reason that we should expect vital rates to conform to linear per capita rates of change. Perhaps more significantly, it can be difficult to know how these rates might actually vary in most aquatic populations where birth and death rates are generally impossible to observe and fully enumerate in any direct way. This highlights an important consideration throughout this book. Ecological (and fisheries) models are caricatures of reality—they serve as useful metaphors but

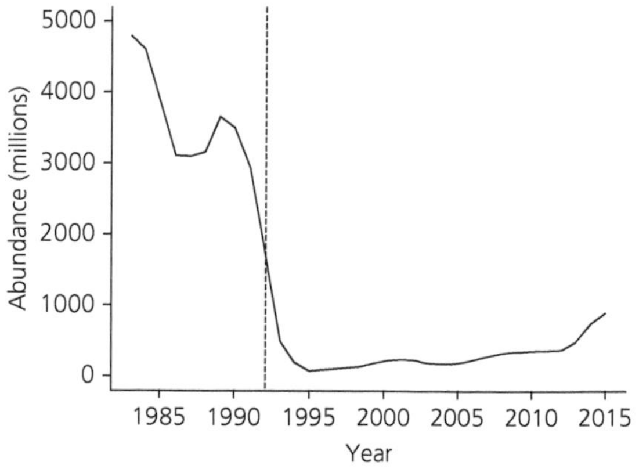

Figure 3.5 Estimated population abundance (age 2+) of Northern Cod (*Gadus morhua*) off Labrador-Newfoundland 1984–2015 (DFO 2016). Broken vertical line shows when the fishery was first closed.

necessarily entail varying degrees of uncertainty concerning the array of mechanisms affecting actual populations, communities, and ecosystems. Model uncertainty is an important and pervasive problem. We will take up this issue further in Chapter 14 where we describe the use of non-linear, non-parametric models that remove constraints in specifying particular structural forms in modeling population and community dynamics.

Interestingly, Alfred Lotka (1925) neatly sidestepped this issue in his derivation of logistic-type population models. Lotka (1925) began with an unspecified population function:

$$\frac{dN}{dt} = f(N) \tag{3.19}$$

and noted that by taking advantage of the immensely useful properties of the well-known Taylor series expansion, the function can be approximated by a polynomial expression giving:

$$\frac{dN}{dt} \approx a_1 N + a_2 N^2 + a_3 N^3 + a_4 N^4 \ldots a_n N^n \tag{3.20}$$

where the a_i are model coefficients (see Box 3.4 for details). Notice that if we truncate the expansion after the second term, we recover the logistic model (with $a_2 > 0$):

$$\frac{dN}{dt} \approx (a_1 - a_2 N)\, N \tag{3.21}$$

(cf. Equation 3.8). Lotka developed his approximation to the continuous-time model in the vicinity of the point $N = 0$—the only "known" point in the unspecified function. The inclusion of higher-order terms allows greater flexibility in approximating Equation 3.18 in the vicinity of the selected point. In the example shown in Box 3.3, a 4th-order polynomial provides a much closer fit to the function near the origin than does a 2nd-order (logistic) form. It also provides a much closer approximation to the entire curve in this case.

3.2.2.1 Depensation

To connect these ideas with our development of the logistic model in Section 3.2.1 we extend our earlier treatment of linear *per capita* rates of change. Suppose we now have a non-linear functional relationship for the per capita rate of change in births, which

Box 3.4 Taylor series approximation

We begin with the general model:

$$\frac{dN}{dt} = f(N)$$

and proceed to expand $f(N)$ as a Taylor series near $N = 0$. We know that a population of zero will have zero growth and therefore $f(0) = 0$ and $N = 0$ is one solution (root) of the unspecified function (Lotka 1925). Expanding the right hand side in a Taylor (Maclaurin) series we have:

$$f(N) = \frac{df(0)}{dN}N + \frac{d^2f(0)}{dN}\frac{N^2}{2} + \frac{d^3f(0)}{dN}\frac{N^3}{3} + \ldots + \frac{d^nf(0)}{dN}\frac{N^n}{n}$$

where we are simply taking a sequence of higher order derivatives of the function $f(0)$. If we actually knew the functional form we could specify these derivatives (assuming it is fully differentiable) and they would provide the coefficients for the polynomial approximation to the function. Because we have an unknown function we cannot take this step. Nonetheless, we can assign the coefficients

$$a_1 = \frac{df(0)}{dN}; a_2 = \frac{d^2f(0)}{dN}; a_3 = \frac{d^3f(0)}{dN} \text{ etc.}$$

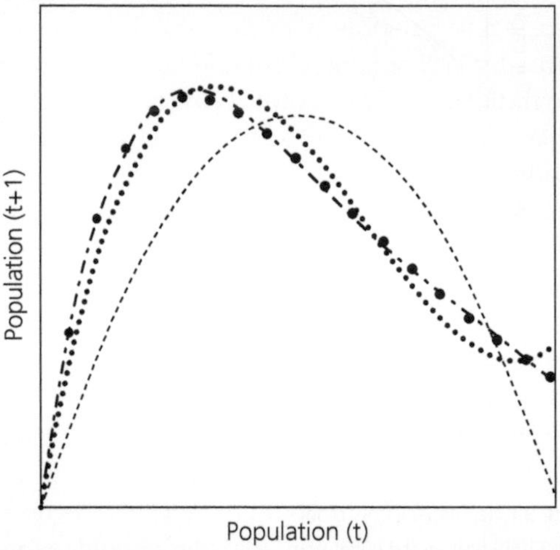

and our approximation is:

$$\frac{dN}{dt} \approx a_1N + a_2N^2 + a_3N^3 + a_4N^4 \ldots a_nN^n$$

In the figure above, we show approximations to a non-linear function (closed circles) for second-order (dashed line), third-order (dotted line) and fourth-order (dot-dashed) polynomials using this approach. The approximation converges provided that the higher-order derivatives become small.

peaks at an intermediate level of population size. We might encounter such a situation when the chance of successfully finding mates at low population levels is reduced and when intraspecific competition for food resources (and thus energy for reproduction) intensifies with increases in population size. For example, we might specify a quadratic function to represent the *per capita* birth rate:

$$b = b_0 + b_1N - b_2N^2 \tag{3.22}$$

If we assume that the per capita death rate remains a linear function of population size as in Equation 3.1, we now see that the line representing the death rate intersects the curve for the birth rate at two points (Figure 3.6a). This holds important implications for the stability properties of this model. When the birth curve lies above the death rate line, the population will increase; conversely, it will decrease when the birth rate curve lies below the death rate line. Our population model then becomes:

$$\frac{dN}{dt} = \left[(b_0 - d_0) + (b_1 - d_1)N - bN^2\right]N, \tag{3.23}$$

and we now have a cubic polynomial with two equilibria (in addition to a stable equilibrium point at $N = 0$)—an upper stable equilibrium and a lower unstable point (see Figure 3.6b). This situation, in which the population growth rate is negative at low abundance levels, is termed critical depensation. We again find the equilibrium points for this model by setting the derivative equal to zero and solving. In this case, using the quadratic equation gives the two roots to this equation:

$$\frac{-(b_1 - d_1) \pm \sqrt{(b_1 - d_1) + 4b_1(b_0 - d_0)}}{-2b_1} \tag{3.24}$$

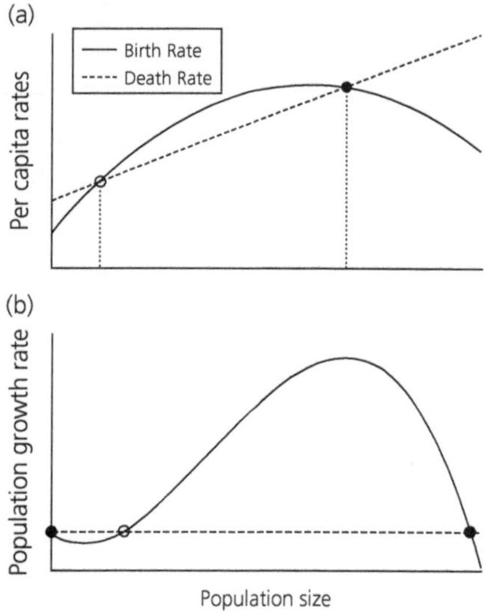

Figure 3.6 (a) Depensatory model resulting from a nonlinear per capita rate of birth and a linear death rate. The closed circle marks a stable equilibrium point. The open circle denotes an unstable equilibrium. (b) The population growth rate has an unstable point at low population size. The dashed line indicates the zero population growth rate.

The key issue is that we have now moved from the logistic case which admits only one globally stable equilibrium for the case $N > 0$ to a situation that entails two globally stable equilibria (one at $N = 0$) with an intermediate non-stable equilibrium point (Figure 3.6b). In this case, if the population falls below the threshold level at the non-stable equilibrium point and continues to decline, the population will collapse completely. The population is predicted to decline to extinction and can only be "rescued" by a subsidy from another population. If we had assumed that the logistic model (with its "well-behaved" dynamics) holds, this crash would be perceived to be sudden and unexpected.

3.2.2.2 Alternative stable states

We have seen that populations can change from one apparently stable state to another. More generally, such populations are said to have alternative basins of attraction. To generate patterns of this type, we require somewhat more complex models. Steele and Henderson (1984) described a population model that combines a logistic growth curve with an S-shaped death rate function. Although the model is motivated by consideration of predation processes affecting the death rate of a prey, the Steele and Henderson (1984) model does not explicitly include a population term for the predator—it is framed as a single-species model. In Chapter 4, we will take up the role of predators explicitly. The model can be written:

$$\frac{dN}{dt} = (\alpha - \beta N)\, N - \frac{cN^2}{D^2 + N^2}, \qquad (3.25)$$

where c is the maximum consumption rate per predator and D is the prey abundance at which consumption is one half its maximum and all other parameters are defined as before. Ludwig et al. (1978) had previously used a model of this type to represent the dynamics of a forest pest, the spruce budworm, in coniferous forests of North America. A graphical illustration of this model and its equilibrium points is provided in Figure 3.7a. Depending on the level of consumption, c, the predation curve can intersect the logistic curve one or more times, giving rise to multiple equilibria. As in the simple model with linear density dependence, some insights into the stability of these equilibria can be inferred by inspection. When production exceeds predation mortality, the population will increase; conversely the population will decline when predation exceeds production (Figure 3.7a). For low c there is a single equilibrium at high population sizes; for intermediate levels of c there are two stable equilibria separated by an unstable equilibrium; finally for high c there is a single low stable equilibrium. In this situation, a population can shift rapidly from one domain of attraction to another—the population levels can flip between alternative stable states (Steele and Henderson 1984). Spencer and Collie (1997b) invoked this mechanism in their analysis of the sudden decline of the Georges Bank haddock population.

We can represent these points of rapid change, and their attendant equilibria, in a bifurcation diagram. In Figure 3.7b we portray the population size as a function of the maximum consumption rate per predator. Notice that as the consumption rate increases from a low level (with high population size and a stable equilibrium point on the upper

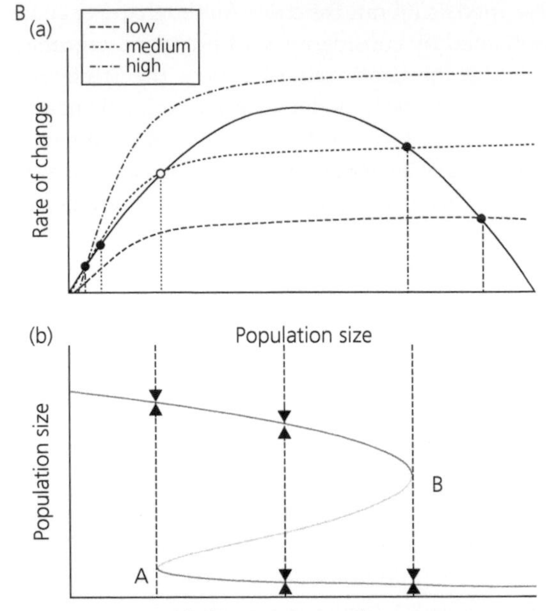

Figure 3.7 (a) Production model with predation. This model combines the logistic curve with a Type-III predator functional response (S-shaped curves). Predation curves are shown for three levels of consumption. Intersections between the lines and the production function define equilibrium levels of population size; filled circles indicate stable equilibria and unfilled circles unstable equilibria (after Steele and Henderson 1984). (b) Bifurcation diagram for Steele–Henderson model with an unstable region bounded between points A and B.

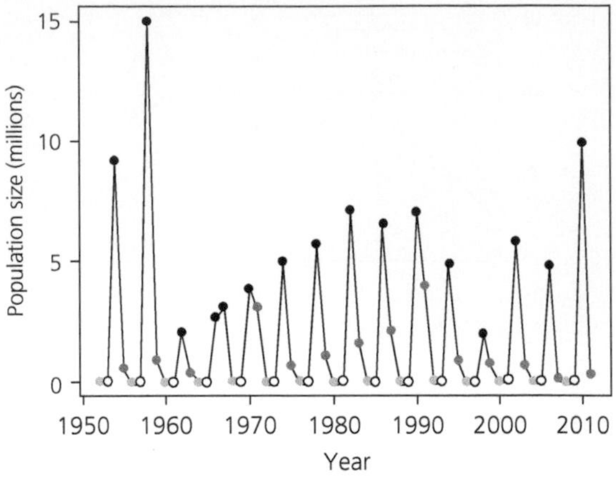

Figure 3.8 Abundance of sockeye salmon (*Oncorhynchus nerka*) in the Adams River, British Columbia. Shaded points indicate the four lines in this cyclic population. Data courtesy of Michael Lapoint, Pacific Salmon Commission.

branch of Figure 3.7b), we pass through an unstable region with two possible population levels bounded by points A and B in Figure 3.7b (light gray line). Beyond point B, a sudden collapse to a much lower stable level will occur. In this case, recovery to the upper stable population cannot be assured until the consumption rate drops to point A, well below the level that initially caused the sudden drop. This phenomenon is termed hysteresis. Viewing humans as predators of aquatic populations, it is clear that we need to understand the interface between fundamental fishing processes and fisher behavior affecting the human harvest rate. As we will see in Chapter 11, traditional fishery models almost invariably assume that there is no possibility of alternative stable states. It is implicitly assumed that if overfishing occurs and remedial reductions in fishing pressure are implemented that the population will return to the equilibrium point, reversing the path followed during the decline.

3.2.3 Discrete-time models

To motivate our development of a density-dependent model in discrete time (applicable to species with seasonal breeding period and non-overlapping generations), we again turn to sockeye salmon. In the following, we consider the population in the Adams River, British Columbia. Cyclic dominance again emerges as a dominant feature of the dynamics of this population (Figure 3.8) and there are clearly defined dominant and subdominant "lines." Note too that the "lines" do not show unconstrained population growth as was the case for our example of the recovery period of the Quesnel River sockeye population in Chapter 2.

The development of non-linear, discrete-time models has involved two principal pathways. The first entails direct specification of difference equations developed specifically for this purpose. The second, and very common, approach is to translate a differential equation to a discrete form using a simple approximation of the time derivative. Note that if we use increasingly small time steps in the approximation, the approximation converges to the analytical solution (if one exists) to the differential equation. In the following we will demonstrate both approaches and will return to this issue in subsequent chapters. Consideration of non-linear difference equations holds an unexpected surprise. Some single-species models in discrete time, unlike their continuous-time counterparts, can exhibit very complex dynamical behavior for certain regions of the parameter space (Ricker 1954, May 1972, 1974, 1976).

Perhaps the simplest direct modification of our geometric growth model to represent non-linear processes can be written:

$$N_{T+1} = \frac{\lambda N_T}{N_T + \psi} \tag{3.26}$$

where λ is defined as before, Ψ is a constant, and T is the generation time for a semelparous population (Hoppensteadt 1982). Notice that Equation 3.26 is of the same general form as Equation 3.13. Here, as N_T increases, N_{T+1} approaches an asymptote. Hoppensteadt's compensatory model was developed directly in difference equation form. We will encounter a model of this general type again in our study of recruitment processes in Chapter 9. The equilibrium point for this model is found by setting $N_{T+1} = N_T = N^*$ and solving to give:

$$N^* = \lambda - \Psi \tag{3.27}$$

for $\lambda < \Psi$; the population declines to zero if $\lambda > \Psi$.

A discrete-time version of the logistic model as a modification of the differential form can be derived in several ways. One approach is to convert the continuous form model to a discrete-time analog by the Euler method. The approximation is:

$$\frac{dN}{dt} \approx \frac{\Delta N}{\Delta t} \tag{3.28}$$

and typically Δt is set to a time step of one time interval (e.g. year or one generation). In this case we then have:

$$N_{T+1} - N_T = (\alpha - \beta N_T)N_T \tag{3.29}$$

or

$$N_{T+1} = (1 + \alpha)N_T - \beta N_T^2 \tag{3.30}$$

The equilibrium point of this model is:

$$N^* = \frac{\alpha}{\beta} \tag{3.31}$$

as in the continuous-time version. This version of the discrete-time logistic has the disadvantage that it can predict negative population sizes. For moderate levels of α the population reaches a stable equilibrium level as in its continuous time counterpart. In the following section, we show however, that as α increases, a very different picture emerges.

An alternative approach to converting the continuous-time logistic model to a discrete-time formulation can be used that avoids the problem of negative population sizes. Here, we employ a piece-wise approximation to the original differential equation (see Box 3.5). The resulting model can be written:

$$N_{t+1} = N_t e^{\alpha - \beta N_t} \tag{3.32}$$

where the equilibrium point is again found to be $N^* = \alpha/\beta$. This model is sometimes referred to as the Ricker-logistic model. This model, unlike the Euler approximation of the logistic model is asymmetric. To see the differences, refer to Box 3.4 where we used a Ricker-logistic model to generate the points and show the fits of polynomial approximations to these data (including quadratic or logistic-type). The fit of the discrete logistic and Ricker-logistic to the Adams River sockeye population is shown in Figure 3.9.

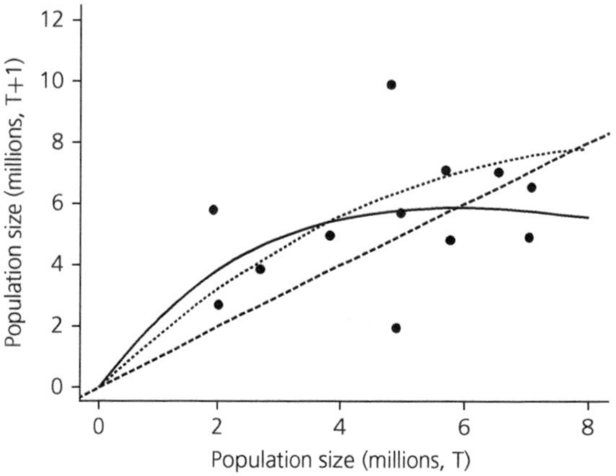

Figure 3.9 Discrete-time logistic model applied to the dominant line of Adams River (B.C) sockeye salmon (*Oncorhynchus nerka*). The dotted curve is the discrete logistic model (Equation 3.26), the solid curve is the Ricker-logistic model (Equation 3.32), and the dashed line is the replacement line.

Box 3.5 From Continuous to Discrete Time

The simplest and perhaps most common approach to converting a differential equation to a difference equation is a variant on the Euler approximation

$$\frac{dN}{dt} \approx \frac{\Delta N}{\Delta t}$$

For the logistic model we have:

$$N_{T+1} - N_T = (\alpha - \beta N_T)\ N_T$$

For the case $\Delta t = 1$.

This approximation to the logistic model has the disadvantage that it can predict negative population numbers. An alternative approach that does not have this problem can be developed as follows. The per capita rate of change for the logistic is:

$$\frac{1}{N}\frac{dN}{dt} = (\alpha - \beta N)$$

and by the chain rule, the relative derivative term can be written:

$$\frac{1}{N}\frac{dN}{dt} = \frac{d\,\log_e N}{dt}$$

Integrating over the interval k to t we have:

$$\int_k^t d\,\log_e N = \int_k^t (\alpha - \beta N) dt$$

or

$$\log_e\left(\frac{N_t}{N_k}\right) = (\alpha - \beta N)\,(t - k)$$

Letting t approach $k+1$ and taking antilogs we have:

$$N_{k+1} = N_k \exp\,(\alpha - \beta N_k)\,.$$

3.2.3.1 Complex dynamics

In Chapter 2 we saw that the general lessons concerning the trajectory of population size in density-independent population models were consonant in continuous and discrete time. In these density-independent models, we find only three types of behavior: monotonic increase, decrease, or stasis. Earlier in this chapter, we saw a much broader array of outcomes including the possibility of multiple stable states in non-linear models. For discrete non-linear population models, even more complex dynamical behavior is possible depending on the intrinsic rate of increase of the population.

The discovery that simple, first-order difference equation population models can exhibit extraordinarily complex behavior can be attributed to William Ricker (Ricker 1954). Moran (1950) had earlier noted the potential emergence of oscillatory behavior in models of this type. Ricker went on to show not only the possibility of cyclical behavior but also the potential for aperiodic behavior that we would now label as chaos.

It turns out that as the density-independent term in the discrete logistic model increases, complex behaviors emerge in these simple difference equations. We can further simplify these models to focus on this key issue (the carrying capacity term affects the scaling of the outputs but not the dynamical patterns in population trajectories). To emphasize the key role of the intrinsic rate of increase, we can write the discrete logistic model as:

$$N_{T+1} = \alpha\,(1 - N_T)\,N_T \tag{3.33}$$

and the Ricker-logistic function as

$$N_{T+1} = N_T e^{\alpha(1-N_T)} \tag{3.34}$$

These two domed-shaped models are often referred to as over-compensatory. When the functions are highly convex with a steep slope at the origin, very complex dynamics are possible. We can again use cobweb diagrams to illustrate these features. In the following we will illustrate the basic concepts using the Ricker-logistic model although complex dynamics readily emerge in the logistic model as well. We have plotted two cases differing in the intrinsic rate of increase (Figure 3.10). On the left side of Figure 3.10, we see that by tracing the trajectory of change for a Ricker-logistic curve with a relatively modest rate of change at the origin, the population ultimately settles down to a fixed point equilibrium (the bottom images in each panel show the time trajectory for each point in the corresponding cobweb diagram). Recall that the straight line through the origin is the one-to-one replacement line and the point of intersection between the production function and the replacement gives the equilibrium point. The stability of the equilibrium depends on the slope of the production function where it intersects the replacement line; the equilibrium is unstable if the slope < -1. It is this issue that controls the highly complex dynamics that can emerge in

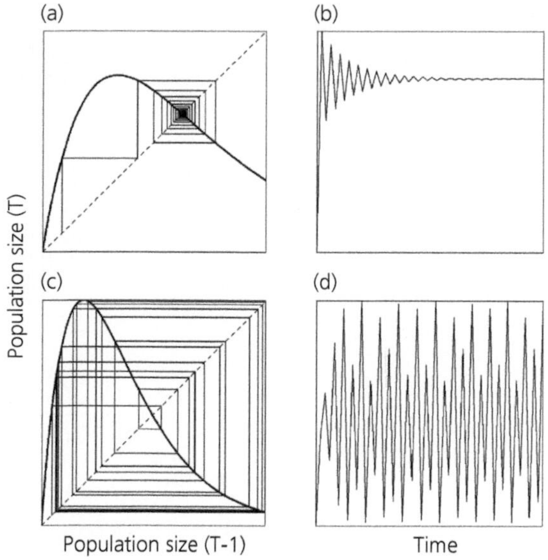

Figure 3.10 Cobweb diagram for (a) the discrete Ricker-logistic model for $\alpha = 6$ and (b) the corresponding time series trajectory showing the stable equilibrium point, (c) the discrete Ricker-logistic model for $\alpha = 20$ and (d) the corresponding time series trajectory.

these models. On the right side, we see that a more highly convex production function with a steeper slope at the intersection never settles down to a stable equilibrium point. Instead, the highly nonlinear production function essentially "kicks" the population to different levels at each time step.

The magnitude of the parameter α solely controls the dynamical behavior in these simple models. For both the Ricker-logistic and the logistic model, at lower values of α we see stable dynamics (Figure 3.11). However, at higher levels of α we observe a sequence of period doubling behavior with cycles of period 2, 4, 8 etc. and ultimately enter a realm of aperiodic behavior (chaos). The values of α at which this sequence of events unfolds differ in the Ricker-logistic and the logistic models but both exhibit period doubling behavior.

A principal hallmark of chaotic dynamics is an extreme dependence on initial conditions; arbitrarily close starting conditions diverge exponentially over time. The time-evolution of nearby points in phase space emerges as a critical consideration in understanding and classifying complex dynamical behavior. We will return to this topic in Chapter 14.

3.3 Time-delay models

In the continuous-time models described earlier in this chapter, individuals were implicitly assumed to mature instantaneously into adults with no time delays. The simple difference equations above have a time delay of one time unit (generations, years, etc.). We now introduce delay-differential equations and higher-order difference equations. We do so principally to represent reproductive delays reflecting the time it takes to make the transition from the egg stage to adulthood to complete the life cycle. The introduction of time delays can also provide an entrée into models with a simple age or stage structure. For discrete-time models, it also allows us to represent populations with overlapping generations. We will also see that incorporation of time delays also holds important consequences for the dynamics of the population. In particular, oscillatory behavior can emerge in continuous-time models when lags are introduced. In this case, the larger the time delay, the more pronounced the periodic behavior observed. For discrete-time models, we find that higher-order time delays can also introduce important changes in dynamical behavior.

3.3.1 Continuous-time models

The renowned ecologist G. Evelyn Hutchinson explored the consequences of time-delayed responses to population change in the continuous-time logistic model. Recall that the standard representation assumes an instantaneous response to changes in population density. In effect, it implicitly assumes that individuals age instantaneously to adulthood and contribute to the reproductive population. In reality of course, there is a time delay between birth and attainment of adulthood. Hutchinson (1948) examined a model of the form:

$$\frac{dN_t}{dt} = (\alpha - \beta N_{t-\tau}) N_t \qquad (3.35)$$

where τ is the time delay in the population response. Here, the per capita rate of change is a function of the population density or abundance τ time units earlier. A closed-form solution to this equation does not exist and we must apply numerical methods to solve Equation 3.35. The addition of the time delay can result in complex behavior including damped oscillations in the approach to equilibrium or sustained oscillations about the equilibrium point (a stable limit cycle). The amplitude of the oscillation essentially depends on the response time (which is inversely proportional to the density-independent coefficient) and the length of the time delay. The higher the magnitude of the density-independent coefficient and the longer the time delay, the higher the amplitude (Gotelli 2008). The period of the oscillation is approximately four times the length of the time delay. The standard form of the logistic model in continuous time cannot exhibit these types of dynamical response. An example of a population trajectory for a delay-differential model is provided in Figure 3.12 for lags of 1 and 2 time periods. For the parameters chosen, damped oscillations are observed for $\tau = 1$; for $\tau = 2$, however, sustained oscillations are observed.

We can also consider a variant on this theme in which we separate reproductive processes and adult mortality:

$$\frac{dN_t}{dt} = (\alpha - \beta N_{t-\tau})\ N_{t-\tau} - d_m N_t \qquad (3.36)$$

where we posit a delay of τ time units between birth and entry (or recruitment) of new individuals into the adult population (N_t) In this model, the time-delayed component represents a density-dependent recruitment process and the last term (d_m) is a density-independent mortality term (Fogarty and Murawski 1996). Notice that this structure is fundamentally different from Equation 3.35—the time-delay component does not enter as a product of current and past population levels to affect the rate of change but rather as a time delayed quadratic term reflecting density-dependence in recruitment.

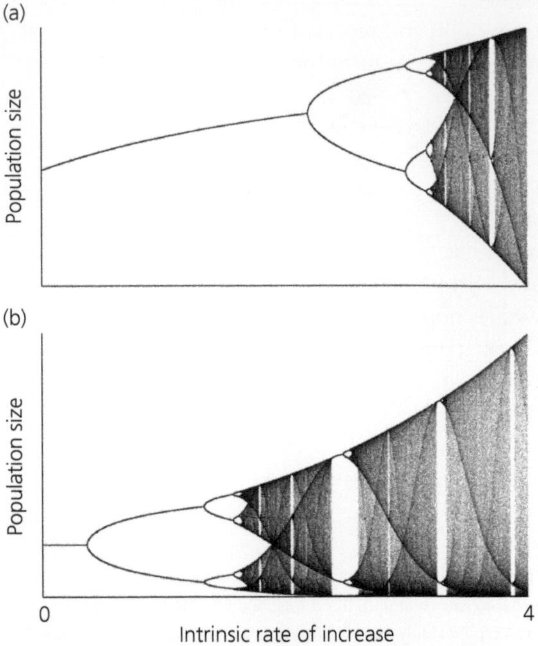

Figure 3.11 Bifurcation diagrams for (a) the discrete logistic and (b) Ricker-logistic population models.

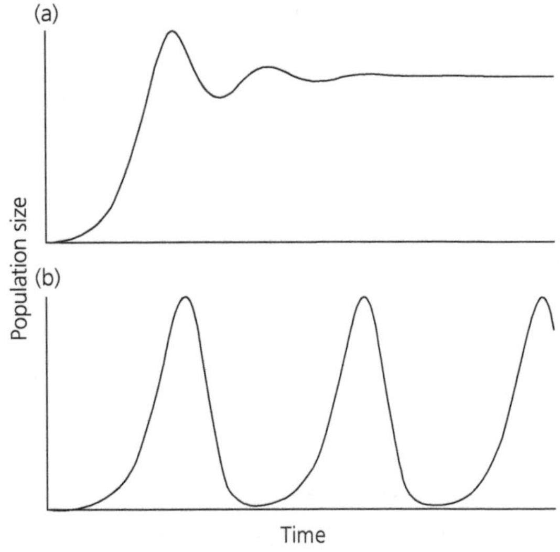

Figure 3.12 Example of population trajectories for a delay–differential model with time delays of (a) one time unit and (b) two time units. The delay–differential model has been solved using a numerical integration scheme.

3.3.2 Discrete-time models

We next consider higher-order time delays in the development of difference equation models. If we assume that the density dependence occurs during the pre-adult phase and that the density dependence is Ricker-type, we have:

$$N_{t+1} = sN_t + N_{t+1-\tau} e^{\alpha - \beta N_{t+1-\tau}} \qquad (3.37)$$

where s is the fraction surviving to time $t+1$ from time t and the second term on the right-hand side represents the introduction of new adults into the population based on reproduction r time units earlier. Here we take t to represent time rather than generations and we now have a discrete-time model with overlapping generations and a simple age structure. This general class of models has been used in the study of baleen whale populations (Allen and Kirkwood 1988) to set catch limits to prevent whale populations from being further depleted. Clark (1985) analyzed the stability properties of the delay-difference model and provided a bio-economic analysis. Botsford (1992) provides an analysis of complex dynamics in this model.

Adding age structure and overlapping generations is a stabilizing feature. As we vary the adult survival rate, we have a stable population level at high adult survival rates (Figure 3.13). In this case, we have set the parameter α to 3.5, well within the chaotic realm for the semelparous model with non-overlapping generations. An iteroparous life history in which some adults survive to contribute to the reproductive output of the population on more than one occasion confers an overall level of stability to the population when the survival rate is high. However, as survival rates decline, more complex underlying dynamics are revealed (Figure 3.13). In this case, factors that result in declining adult survival rates would be destabilizing.

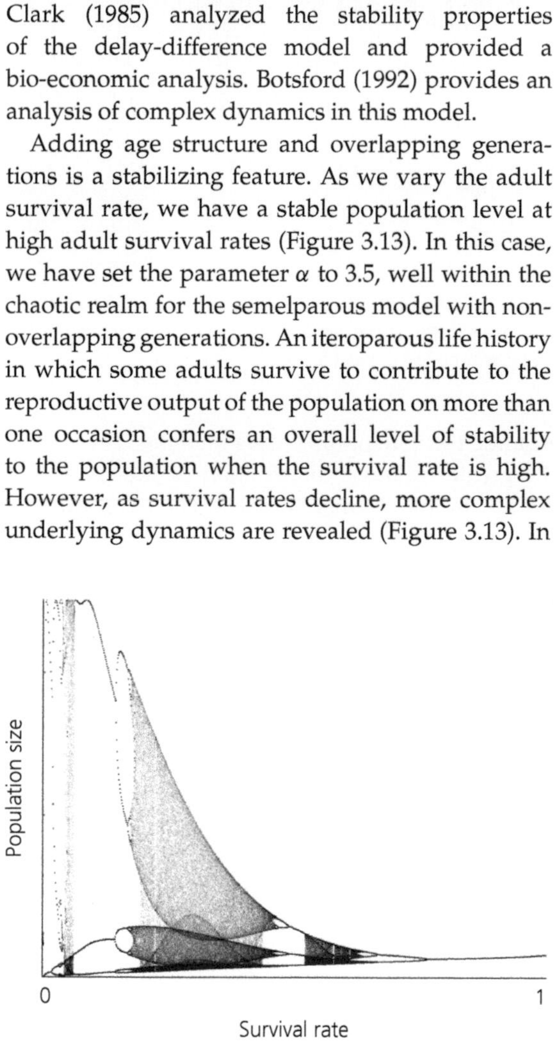

Figure 3.13 Bifurcation diagram for a delay–difference model incorporating a simple age structure. Here the survival rate is varied from 0 to 1 and the α parameter of the Ricker-logistic model is set to 3.5.

3.4 Matrix models

The matrix models described in Chapter 2 did not consider any form of density dependence. We next consider models that do incorporate some form of density dependence and exhibit a much broader range of dynamical behavior in age-structured populations. We also have greater latitude in explicitly expressing different forms of density-dependent processes relative to the simpler non-structured models described earlier. For example, we can include density dependence in survival rates, maturation schedules, and/or fertility. In practice, most attention has been given to survival rates, typically in the first year of life or first stage class, in these models.

3.4.1 Age-structured models

Leslie (1948) recognized the need to address the issue of unconstrained population growth and considered extensions to his basic matrix model to accommodate density dependence. The non-linear matrix models we will now consider can be written:

$$\mathbf{N}_{t+1} = \mathbf{L}_n \cdot \mathbf{N}_t \tag{3.38}$$

where $\mathbf{N}_{t+1}$ is again the vector of numbers at age for time $t+1$ and $\mathbf{L}_n$ is the matrix of age-specific fertilities and survival rates, one or more of which is now density dependent. In principle, any of the elements of the matrix $\mathbf{L}_n$ can be expressed as functions of density. The measure(s) of density used in these functions can include a number of forms ranging from the abundance of a single age (or stage) class to a weighted sum of the abundance over all age classes in the population (Caswell 2001):

$$N_t = \sum_a w_a n_{a,t} \tag{3.39}$$

where w_a is a weighting coefficient applied to each age class. In most applications, equal weights have been applied but if certain age classes exert a greater effect on survival, maturation, or fecundity, this can readily be accommodated (albeit at the cost of greater analytical complexity for evaluating stability).

Interestingly, many of the best-known early applications of non-linear matrix models were developed for exploited aquatic populations (Usher 1972; Jensen 1976; Levin and Goodyear 1980; DeAngelis et al. 1980; Cohen et al. 1983). In this section we describe several different density-dependent survival functions and ways of incorporating age-specific effects into these models. Levin and Goodyear (1980) developed a birth-pulse matrix model in which the census is taken just before reproduction. They applied a Ricker-type survival rate during the first year of life. Below, we illustrate this case for a simple three age-class model. Framed in terms of our earlier notation in Chapter 2, the transition matrix takes the form:

$$\mathbf{L}_n = \begin{pmatrix} s_0 e^{-cN} f_1 & s_0 e^{-cN} f_2 & s_0 e^{-cN} f_3 \\ s_1 & 0 & 0 \\ 0 & s_2 & 0 \end{pmatrix} \tag{3.40}$$

where s_0 is the density-independent survival rate, c is a coefficient, N is population size and the f_i is fecundity. In Figure 3.14a we show that at low levels of fecundity, the population trajectories converge to stable levels. However, at higher levels of fecundity more complex dynamics emerge (Figure 3.14b). Caswell (2001) provides a comprehensive treatment of non-linear dynamics in age-structured populations including the emergence of chaos.

DeAngelis et al. (1980) developed a birth-pulse model with the census occurring immediately after breeding. Intraspecific density-dependent effects were assumed to occur only among individuals in the first year of life such that the base level of survival s_0 declines with increasing initial cohort size N_0:

$$\frac{s_0}{1 + cN_0} \tag{3.41}$$

where c is a density-dependent coefficient. This model was applied to life-history data of five fish species and used to predict the return time to equilibrium following perturbations. The longer-lived species, striped bass (*Morone saxatilis*), had longer return times than the shorter-lived species

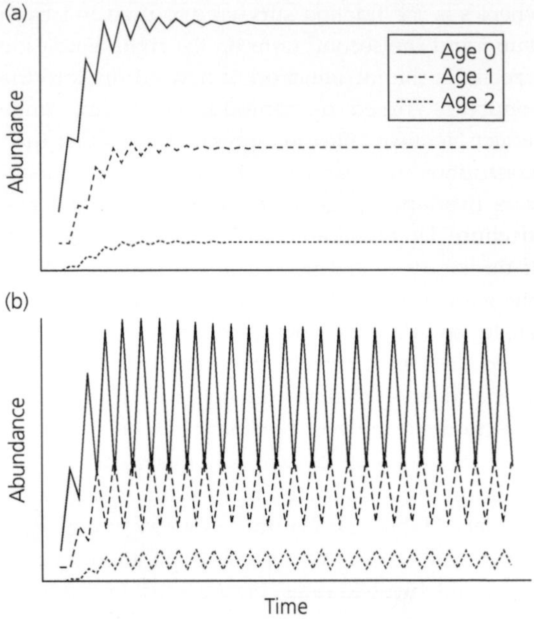

Figure 3.14 Population trajectories by age-class as a function of time for an age-structured matrix model with compensatory survival of the progeny for two levels of fecundity.

Atlantic menhaden (*Brevoortia tyrannus*) because of the greater relative contribution of the older age classes in the former species. The reproductive contribution of the older ages introduces a time lag that delays the compensation acting during the first year of life to return the population to equilibrium.

Usher (1972) elaborated on an alternative density-dependent survival formulation first suggested by Pennycuick et al. (1968). Pennycuick et al. proposed a sigmoidal function with a declining survival rate with increasing total population size. Here, we express this model in terms of a basal density-independent survival rate (s_0) and a density-dependent component:

$$\frac{s_0}{1 + e^{-b+cN_t}} \tag{3.42}$$

where b and c are constants N_t is the total population size at time t. With increasing magnitude of the density-dependent coefficient, the population can exhibit oscillations (Usher 1972).

We can apply the same type of graphical stability analysis described in Chapter 2 for a semelparous species with four age classes. Paulik diagrams for situations in which compensatory dynamics in the first year of life are important are shown in Figure 3.15. In Figure 3.15a, we depict a population which is growing and eventually stabilizes. In contrast, in Figure 3.15b, a decrease in the survival rate in the fourth quadrant results in an overall decline in the population, indicated by the continual drop in successive generations. Of course, density dependence can occur at one or more life-history stages and not just during the first year of life. It is also possible that one or more life stages might exhibit depensatory dynamics, further complicating the story.

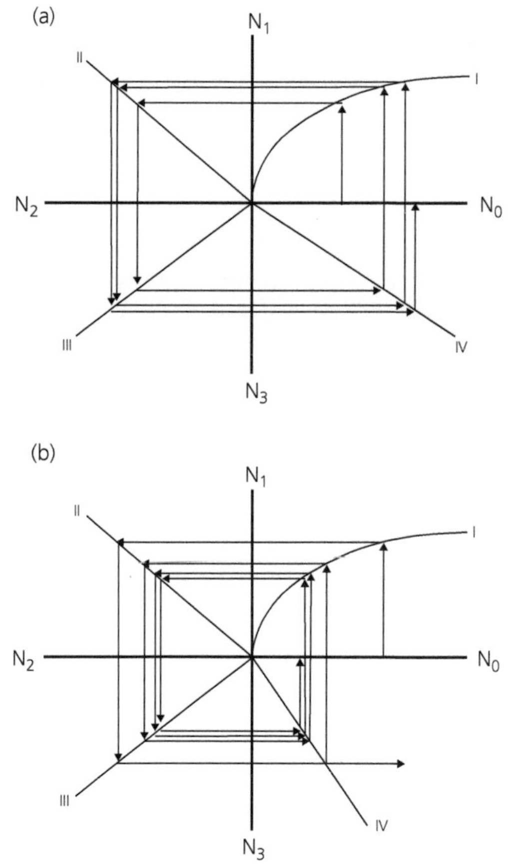

Figure 3.15 Paulik diagram for a population with four age classes with a compensatory process operating during the first year of life (a). Same population with a decreased rate of survival or reproduction in the quadrant IV resulting in a population collapse (b).

3.4.2 Stage-structured models

Stage-structured non-linear matrix models have generally been accorded less attention in the literature. However, Neubert and Caswell (2000) provide an insightful analysis of a two-stage model (juveniles and adults) that can embody semelparous and iteroparous life histories and density-dependent effects on fertility, growth, and survival. The projection matrix can be written:

$$\mathbf{L}_n = \begin{pmatrix} s_0\,(1-m) & \mathcal{F} \\ s_0 m & s_1 \end{pmatrix}$$

where s_0 is the probability of survival of juveniles, m is the fraction of surviving juveniles that mature to become adults within the time interval, and $\mathcal{F}$

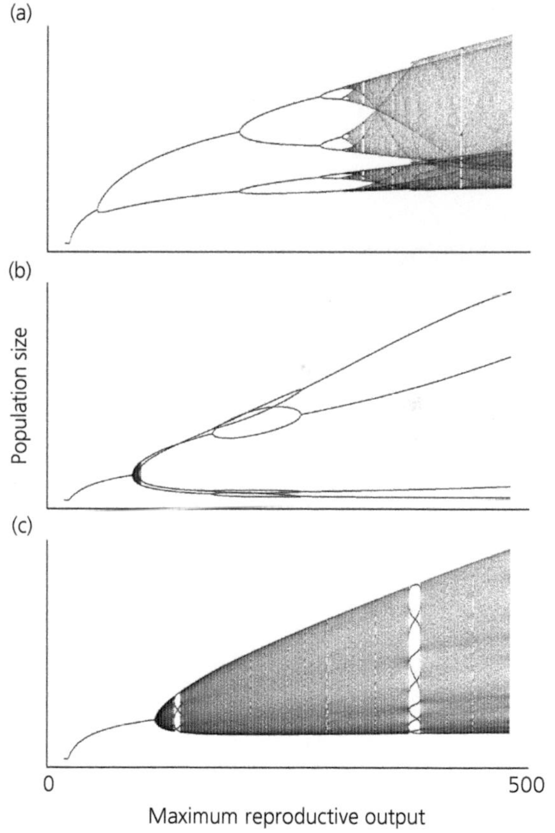

Figure 3.16 Bifurcation diagrams for the Neubert–Caswell nonlinear stage-structured matrix model for (a) density-dependent fertility, (b) density-dependent individual growth, and (c) density-dependent juvenile survival, as a function of maximum reproductive output.

is the number of juveniles at time $t+1$ produced per adult at time t. For the semelparous case, $s_1 = 0$ and $s_1 > 0$ for iteroparity. Maturation can occur during a single time period or can be extended over several time periods. These maturation strategies are referred to as precocious and delayed respectively (Neubert and Caswell 2000). Density dependence was sequentially incorporated in the reproduction, growth, and survival terms using an exponential form $[\exp(-bN)]$ where b is a coefficient and N is the total population number at each time step (in the examples shown, b is set to 1 in each case). As we have seen, this form results in an over-compensatory response. It will by now come as no surprise that complex dynamics can readily emerge in this system as the fertility ($\mathcal{F}$) term increases. Bifurcation diagrams are shown in Figure 3.16 for three cases: (a) density-dependent fertility, (b) maturation rate, and (c) juvenile survival as a function of the maximum reproductive output. Here we see that the form of density dependence strongly affects the way in which complex dynamics is manifest in highly non-linear models.

3.5 Summary

The observation that no population can grow indefinitely and that most populations persist on ecological timescales implies that mechanisms of population regulation exist. Feedback mechanisms include competition for limited resources, cannibalism, and predation rates that vary with density. Density dependence occurs when *per capita* birth or death rates depend on population density. Density dependence is *compensatory* when the population growth rate decreases with population density and *depensatory* when it increases. The logistic model incorporates density dependence as a simple linear function. A population exhibiting logistic growth will reach a stable population size. Non-linear density-dependent terms can give rise to multiple equilibria. With discrete-time models or time delays in density-dependent regulation, the approach to equilibrium may not be smooth—complex dynamical behavior is possible. Density-dependent feedback processes can compensate, up to a point, for natural and anthropogenic disturbances; beyond this point a population will collapse.

Additional reading

Topics covered in this chapter are reviewed in Wilson and Bossert (1971; see their Chapters 1 and 3); Roughgarden (1998; Chapter 4) and Gotelli (2008; Chapter 2), Vandermeer and Goldberg (2003; Chapters 1, 3, and 4)), and Case (2000; Chapter 5). Each provides very instructive accounts of these issues. More advanced treatments of these topics can be found Kot (2001; Chapters 1, 4, 5, and 24) and Caswell (2001; Chapter 16) and Botsford et al. (2019).

CHAPTER 4

Interspecific Interactions I: Predation and Parasitism

4.1 Introduction

Interactions among species shape the fabric of all ecological communities. Predation and parasitism are dominant forms of interspecific interaction in aquatic ecosystems. Other ecological interactions such as competition and mutualism are also extremely important. In this chapter, we focus on simple two-species models of predation and parasitism. We take up the topic of competition and mutualism in Chapter 5. Predation has been accorded the most attention by far in the literature on interspecific interactions in fishery ecosystems. A number of books devoted solely to this topic have been published (Ivlev 1961, Stroud and Clepper 1979, Zaret 1980, Kerfoot and Sih 1987, Gerking 1994). However, rich information resources do exist for other forms of interactions in aquatic ecosystems and attention is increasingly being given to quantifying parasitism, competition, and mutualism in these systems.

The processes of predation and parasitism hold important elements in common (Mittelbach 2005). Both predators and parasites derive nutrients and energy by attacking their prey and hosts respectively. They also can exert important regulatory control of the abundance of their hosts/prey, which suggests that models of parasitism and predation might have a number of features in common. Indeed, as we shall see, models borrowed directly from predation theory have been applied to parasites and vice versa. There are also, of course, marked differences in key elements of these two processes. Parasites are almost always much smaller than their hosts, often by one or more orders of magnitude[1]. In contrast, predators are typically larger (often substantially) than their prey. A number of ecosystem size spectrum models are in fact built around this principle. Individual predators kill and consume many individual prey items over their lifetimes. An individual parasite typically derives its energy from a single host (or perhaps multiple hosts if different parasitic life stages attack a sequence of hosts). The infection may or may not be lethal. These distinctions must be considered in model development.

In classical single-species models of harvested species, predation and disease are typically combined into a single "natural mortality" term. In most instances, natural mortality is treated as a constant over time and size or age. As we shall see, we can expect natural mortality to be neither time nor age/size invariant. Although explicit consideration of predation and disease in fisheries assessment and management is unquestionably challenging, it is also critically important. Predators and prey are often jointly exploited in fishery ecosystems (in many, but not all, cases by different fleet sectors). Differential harvesting of predators and prey can strongly alter the structure of aquatic ecosystems, affecting overall patterns of productivity. Perhaps most importantly, management strategies directed at predators and prey in isolation potentially work at cross-purposes in any attempt to specify

[1] In Section 4.3.2, however, we will examine an important exception to this "rule" in fishery ecosystems that has held important consequences for major freshwater fisheries.

Fishery Ecosystem Dynamics. Michael J. Fogarty and Jeremy S. Collie, Oxford University Press (2020).
DOI: 10.1093/oso/9780198768937.003.0004

optimum yields for each individually (Fogarty 2014). Parasitism can result in mortality through widespread and often episodic outbreaks of disease that can vary in intensity over time and can differentially affect different host life stages. Sublethal effects of parasitism can also be extremely important, affecting not only the condition and marketability of exploited aquatic species but reducing reproductive output and fitness of parasitized species. Disease outbreaks in natural and aquaculture systems consequently can impose important economic costs (Lafferty et al. 2015). Interestingly, harvesting can under some circumstances reduce disease intensity by altering transmission rates (Dobson and May 1987). In a survey of trends in the incidence of disease in marine ecosystems, Lafferty et al. (2004) note that fishing may have resulted in an overall reduction in disease outbreaks, in contrast to trends in many unexploited taxa.

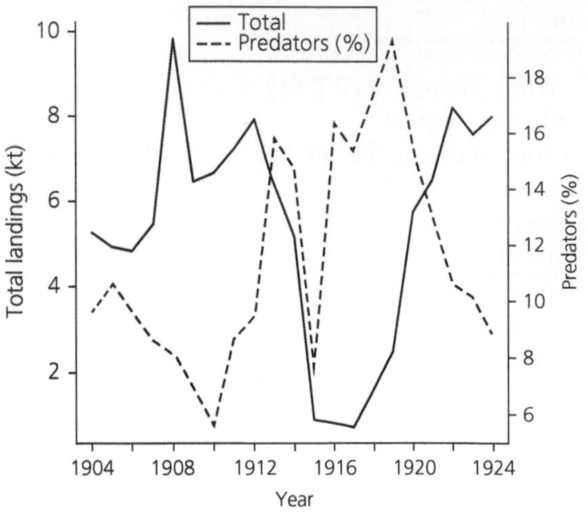

Figure 4.1 Total fishery landings (solid line) and the proportion of elasmobranch predators (dashed) at the port of Fiume in northern Italy during 1904–1924 (data from Scudo 1971). These observations provided insights for the subsequent development of simple predator–prey models by Vito Volterra (1926).

4.2 Predation

Heterotrophic organisms derive energy through predation on other organisms. Predation is a particularly dramatic form of interaction among species (or indeed within species where cannibalism is important). The dynamics of predator–prey systems are also of particular interest in our study of aquatic ecosystems because we will later be examining the role of humans as predators in these systems. One of the earliest analytical treatments of predation processes explicitly involved the impact of human harvesting activities on fish populations. Vito Volterra, the noted Italian mathematician, was approached by his future son-in-law Umberto D'Ancona, a marine ecologist, with an intriguing problem. D'Ancona was studying fluctuations of fish populations as reflected in catches in selected ports in northern Italy (D'Ancona 1954; Scudo 1971). D'Ancona noted that during the First World War when fishing effort in the Adriatic was sharply curtailed, the percentage of predator species increased markedly in the catch, suggesting a cyclical fluctuation (Figure 4.1). The intervention of the war and its effect on fishing provided an unplanned experiment revealing important clues to the interplay between harvesting and community structure in a fishery ecosystem. D'Ancona hoped that Volterra might be able to develop a theoretical treatment of predator–prey systems based on these insights. Volterra immediately set to work and published his famous predator–prey equations in a landmark paper in the journal Nature (Volterra 1926). He subsequently provided a more expansive treatment of this topic, reprinted in the influential journal of the International Council for Exploration of the Sea (Volterra 1928). Volterra retained his interest in theoretical ecology for the remainder of his career. In his seminal treatise, Alfred Lotka (1925) had independently derived similar predator–prey models, in this case motivated by plant–herbivore interactions. In the following, we begin by showing the development of very simple models that can replicate these types of observations.

Walters et al. (1986) provide an interesting and informative analysis of predator–prey dynamics with an example of interactions between Pacific cod, *Gadus macrocephalus*, and Pacific herring *Clupea pallasii*, populations in the Hecate Strait off British Columbia. Walters et al. (1986) examined this relationship based on time series of abundance for cod and herring. Diet composition data provided clear evidence of predation by cod on herring (Walters et al. 1986). These two species are embedded within a complex ecosystem with many factors affecting the abundance of both. Walters et al. (1986) found, however, that the observed patterns of cod and herring abundance were consistent with the hypothesis of a significant predator–prey interaction affecting recruitment of both species.

Here we focus on the relationship between abundance levels of adult herring and cod. These species appear to exhibit an out-of-phase cyclical pattern over the period of observation (Figure 4.2a). If we plot the relationship between cod and herring abundance, we find a distinctive counter-clockwise trajectory in the abundance levels of the two species when displayed in phase (or state) space (Figure 4.2b). A full analysis of these patterns would require a longer time series to determine if the apparent oscillations continue or exhibit dampening behavior. These observed patterns, however are diagnostic features of predator–prey interactions.

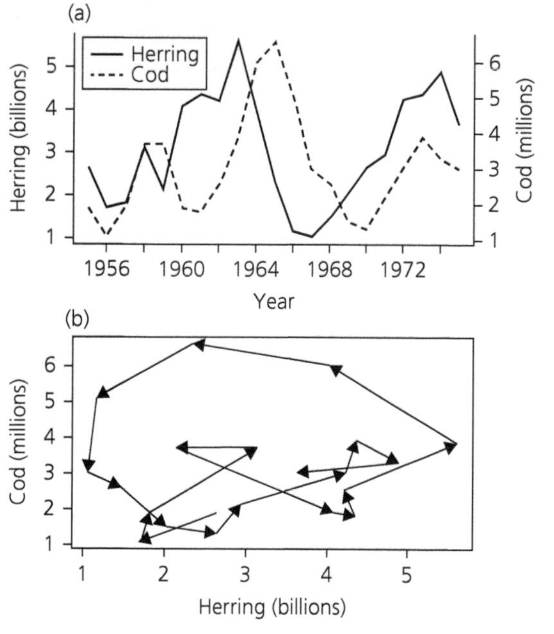

Figure 4.2 (a) Population trajectories of adult Pacific cod (*Gadus macrocephalus*) and herring (*Clupea pallasii*) and (b) phase-plane plot of trajectories of these species. (Data from Walters et al. 1986.)

4.2.1 Density-independent models in continuous time

We begin with a simple extension of our density-independent model (Equation 2.1) to the two-species case. The classical Lotka–Volterra predator–prey equations assume that all of the deaths of the prey are due to predation and that the birth rate of the predator is tied exclusively to the consumption of prey and its subsequent conversion to predator numbers. The model for the prey (N_1) is:

$$\frac{dN_1}{dt} = (b - a_{12}N_2)\,N_1 \tag{4.1}$$

where b is the *per capita* birth rate of the prey and a_{12} is the *per capita* mortality rate imposed by the predator on the prey[2]. For the predator (N_2) we have:

$$\frac{dN_2}{dt} = \left(c \cdot a_{12}N_1 - d'\right)N_2 \tag{4.2}$$

where c is the conversion efficiency of prey consumed into predator progeny and d' is the per capita death rate of the predator. In the absence of the predator, the prey will grow exponentially. In Section 4.2.2 we will consider the more realistic case where in the absence of the predator, the prey is regulated by density-dependent processes. We assume that the probability of encounter between predator and prey is a function of the product of the abundance of each. Here, we are invoking the law of mass action commonly employed in physical chemistry. Notice that the total number of prey consumed is proportional to their number (i.e. there are no limitations imposed on the predator by the time required to subdue and consume the prey and there are no satiation effects—we will relax this restriction in Section 4.2.3).

The population trajectories of predator and prey for this model exhibit out-of-phase periodic fluctuations (Figure 4.3a) as suggested by the observations of D'Ancona and Walters et al. This predator–prey

[2] For consistency, throughout this chapter and the next we will designate population numbers or densities as N_i where the subscript indicates the ith species, variously representing prey, predators, competitors, hosts, parasites, etc. This will permit easy extension to multispecies systems in Chapter 5 where i can take on numbers other than 1 or 2.

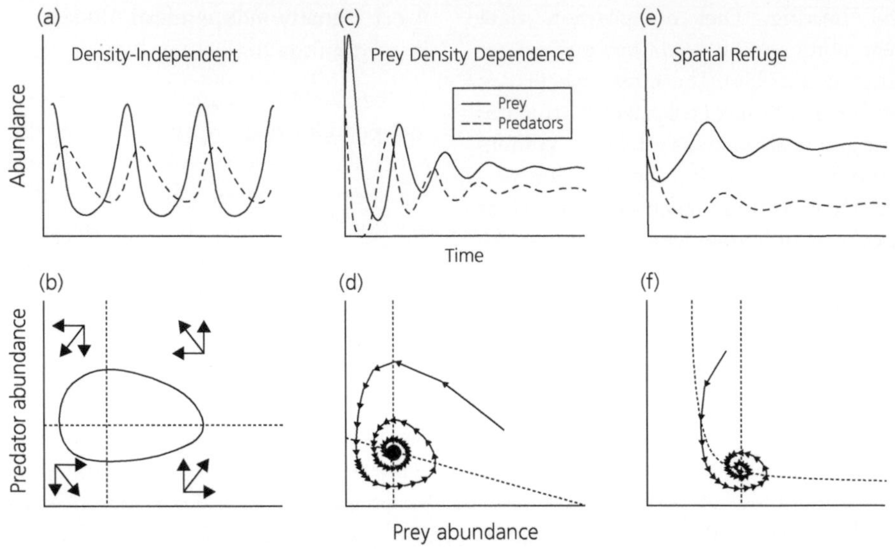

Figure 4.3 (a) Population trajectories in time for the density-independent Lotka–Volterra predator and prey model; (b) corresponding phase portrait for this system reflecting the oscillatory dynamics of the system; (c) population trajectories in time for Lotka–Volterra predator and prey model with prey density-dependence; (d) corresponding phase portrait for this system, showing approach to equilibrium; (e) density-independent Lotka–Volterra model with fixed number population refuge; and (f) corresponding phase portrait for this system, showing approach to equilibrium.

system exhibits a neutral equilibrium; any perturbations will result in a change in the trajectories of the predator and prey over time. It is also strongly affected by the initial conditions set by the starting numbers of predators and prey. Indeed, the amplitude and period of the cycles are entirely controlled by these initial conditions.

We find the points at which the net population growth rate is zero by setting the derivative of each equation to zero and solving. For the prey species these points are $N_1 = 0$ and

$$N_2 = \frac{b}{a_{12}} \tag{4.3}$$

and for the predator, we have $N_2 = 0$ and

$$N_1 = \frac{d'}{c \cdot a_{12}} \tag{4.4}$$

Because there is no density-dependence incorporated in the models for either predator or prey, these isoclines are independent of their abundance.

To see how the counter-clockwise trajectory emerges in phase space, we begin by plotting Equations 4.3 and 4.4 which provide intersecting isoclines. The isoclines for predator and prey divide the phase plane into four quadrants. We can then construct directional vectors for each quadrant to track the relative trajectories of predators and prey. Beginning in the upper right quadrant, both predator and prey are at high abundance. Here we predict that the predator population will increase further but the prey population declines because of the impact of the predator. The net vector of movement is toward the upper left quadrant (Figure 4.3b). In this quadrant, we have high predator abundance but low prey abundance; both the predator and the prey will decline and the net vector movement is toward the lower left quadrant. Here, we find low abundance of both predator and prey and we expect the prey to increase but the predator to decline, resulting in a net movement to the lower right quadrant (Figure 4.3b). In this quadrant, the prey is at high abundance while the predator level is low. The net vector movement is back toward the upper right quadrant (Figure 4.3b). We have therefore defined a counter-clockwise trajectory in phase space; the resultant vectors depict the general direction of movement. The actual trajectory followed by the population depends on the initial conditions and the specific parameter values as shown by the ovoid sketched in Figure 4.3b corresponding to the population patterns shown in the upper panel of this figure.

The overall population trajectories reflect a neutral equilibrium and any changes to either the initial conditions or the parameters of the model will result in different outcomes. Recall that the density-independent models described in Chapter 2 also did not exhibit stable equilibria. We next examine the implications of adding density-dependent processes to the model.

4.2.2 Density-dependent models in continuous time

We first consider simple extensions of the density-dependent models described in Chapter 3 to the multispecies case. We begin with the case of density-dependent regulation of the prey species and a density-independent structure for the predator. For consistency, we adopt the notation employed in Chapter 3 for density dependence in the prey birth rate. The model for the prey population is then:

$$\frac{dN_1}{dt} = (b_0 - b_1 N_1) N_1 - a_{12} N_1 N_2 \qquad (4.5)$$

where b_0 and b_1 are defined as for the single-species model in Chapter 3. For the predator we have as before:

$$\frac{dN_2}{dt} = (c \cdot a_{12} N_1 - d') N_2 \qquad (4.6)$$

The isocline for the prey population is now:

$$N_2 = \frac{1}{a_{12}} [b_0 - b_1 N_1] \qquad (4.7)$$

and the prey declines linearly with increases in the predator population. The isocline for the predator population is again:

$$N_1 = \frac{d'}{c \cdot a_{12}} \qquad (4.8)$$

Unlike the neutral equilibrium exhibited by the Lotka–Volterra model with no density dependence, we now have a stable equilibrium at the intersection of the isoclines for the predator and prey. Numerical solutions for the population trajectories in time for the predator and prey for this example are shown in Figure 4.3c, demonstrating an oscillatory approach to an equilibrium population level. The phase-plane portrait for this system reveals a spiral pattern in the approach to equilibrium (Figure 4.3d).

4.2.3 Refugia

We have noted that experimental studies repeatedly point to the importance of refuge areas for prey if both predators and prey are to persist. We can make simple modifications to the Lotka–Volterra predator–prey equations to explore this issue. In the following, we explore a scenario involving the protection of a fixed number of the prey population. Most typically, we think of spatial refugia but other forms of refuge (behavioral and others) can be envisioned. A Lotka–Volterra model with a constant number of protected prey can be written:

$$\frac{dN_1}{dt} = rN_1 - a_{12}\left(N_1 - N_1^P\right)N_2 \qquad (4.9)$$

where N_1^P is the number of protected prey and all other terms are defined as before. For the predator (N_2) we have:

$$\frac{dN_2}{dt} = c \cdot a_{12}\left(N_1 - N_1^P\right)N_2 - d'N_2 \qquad (4.10)$$

where d' is the death rate of the predator. The fixed number scenario could apply to situations in which the prey population uses a finite number of shelter sites associated with specific habitat characteristics (e.g crevices used by coral reef fishes and invertebrates). With this structure we now see a stabilization in the system and the characteristic oscillations of the Lotka–Volterra model give way to stable equilibria (Figure 4.3e). The phase portrait for this model is shown in Figure 4.3f.

4.2.3.1 The foraging arena

Walters and Kitchell (2001) introduced the concept of a foraging arena in which prey move between refugia and open sites in which they are vulnerable to predation. There is an on-going transition between vulnerable and invulnerable states for the prey depending on foraging choices made by individuals. For a simple system with one prey and one predator species, the rate of change of individuals in the prey vulnerable state, N_1^V, is given by:

$$\frac{dN_1^V}{dt} = v\left(N_1 - N_1^V\right) - v'N_1^V - a_{12}N_1^V N_2 \qquad (4.11)$$

where $v(N_1 - N_1^V)$ represents the rate of movement of individuals into the vulnerable state and $v'N_1^V$ is the transition of individuals from the vulnerable to the protected state. Note that the last term in Equation 4.11, the loss of vulnerable individuals to predation, invokes a mass action encounter mechanism and that handling time considerations are not included. It is assumed that the timescale of these transitions involving behavioral dynamics are fast relative to population level processes and that $dV/dt \sim 0$. We then have for the equilibrium:

$$N_1^V = \frac{vN_1}{v + v' + a_{12}N_2} \tag{4.12}$$

and the amount of prey consumed per predator is $\alpha_{12}N_1^V$. Notice that we now have a predator term in the denominator. We will return to the implications of this structure in Section 4.2.5. The foraging arena concept plays a critical role in the structure of the well-known Ecosim modeling package widely applied in fisheries applications throughout the globe and described in Section 13.3.2.

4.2.4 The functional feeding response

We have so far considered only the case in which the predator consumes prey in proportion to prey abundance. The amount of prey consumed in our simple model increases linearly with increasing prey abundance with a slope equal to the product of the per capita predation rate and the predator abundance level (resulting in a linear functional response). The translation of the number of prey consumed to predator abundance (the numerical response) in our simple model is also linear. Next, we consider the case in which the functional and numerical response is a non-linear function $f(N_1)$ of the number of prey consumed per predator.

The Russian fish ecologist, V.S. Ivlev (1961) demonstrated experimentally that consumption does not necessarily increase in direct proportion to prey abundance over a broad range of prey numbers but instead approaches a saturation level. He ascribed this pattern to satiation of the predator as its food requirements and energetic demands are met. Ivlev introduced a functional feeding response based on a mechanism of predator satiation resulting in an asymptotic form for the per-capita prey consumption by a predator. To describe this process, Ivlev developed a saturating model of the form:

$$f(N) = C_{max}\left(1 - e^{-\kappa N}\right) \tag{4.13}$$

where C_{max} is the maximum rate of prey capture per predator and κ controls the rate at which the predator is satiated.

It is clear that we need to consider a family of functional forms to represent predation processes. The simple linear form applied in the models earlier is predicated on the idea of a constant per-capita consumption rate of prey over the entire range of prey population sizes. For this linear functional feeding response, the number eaten increases with increasing prey population size (solid line in Figure 4.4a). The fraction of the prey population eaten remains constant, however (dashed line in Figure 4.4a).

Holling (1965) approached the problem from the explicit perspective of search and handling time, leading to additional non-linear functional feeding responses. For Holling's Type-II functional feeding response[3] (see Box 4.1) we have:

$$f(N_1) = \left[\frac{vN_1}{1 + vhN_1}\right] \tag{4.14}$$

where v is search efficiency of the predator and h is the handling time. The parameter v is sometimes identified as the attack rate by the predator. A representation of the per-capita consumption of prey as a function of prey population size for the Type-II functional feeding response is provided in Figure 4.4b (solid line) and the percentage of the prey population consumed per predator is indicated by the dashed line. At low prey abundance, the number of prey eaten declines while the percentage

[3] Holling specified a Type-I functional feeding response as a linear function from the origin conjoined with a horizontal line representing a saturation level. In the ecological modeling literature, the Lotka–Volterra linear functional feeding response is often referred to as a Type-I relationship even though it does not have the characteristic 'hockey-stick' form specified by Holling.

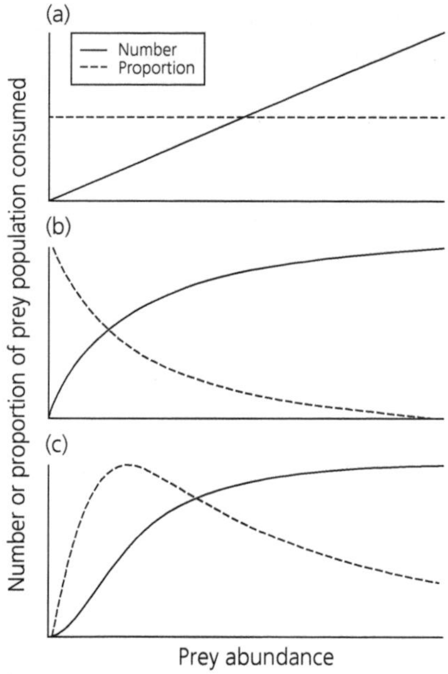

Figure 4.4 Per capita consumption of prey over a range of prey population sizes (solid line) and fraction of the prey population eaten (dashed line) for (a) Type-I functional feeding response, (b) Type-II functional feeding response, and (c) Type-III functional feeding response.

of prey eaten increases, which can destabilize the predator–prey dynamics. This saturating function has the same general shape as Ivlev's function (Equation 4.13).

Holling's Type-III functional feeding response can be written:

$$f(N_1) = \left[\frac{vN_1^2}{1 + vhN_1^2}\right] \qquad (4.15)$$

which again reaches an asymptotic level. However, unlike the Type-II functional response, there is an inflection at low prey abundance (Figure 4.4c solid line) and the percentage of the prey population consumed per predator is highest at an intermediate level of prey abundance (dashed line in Figure 4.4c).

We have assumed that predators do not interfere with each other. In Box 4.2 we relax this assumption. Now predator abundance appears in the denominator of the functional feeding response. We explore the full implications of this general issue in Section 4.2.5.

Box 4.1 The Mechanics of Predation

The total time (T) required to search for prey and to consume prey items is

$$T = T_s + T_h$$

where T_s is the the total time spent searching and T_h is the time spent handling prey. The handling time is:

$$T_h = hn_1$$

where h is the time required to handle and consume each prey item and n_1 is the number of prey encountered and captured by the predator:

$$n_1 = vT_sN_1$$

where v is the search efficiency and all other terms are defined as before. By re-arranging, we can determine the search time as:

$$T_s = \frac{n_1}{vN_1}$$

Substituting into the expression for the total time spent feeding, we have:

$$T = \frac{n_1}{vN_1} + hn_1$$

which can be written:

$$T = \frac{n_1}{vN_1} + \frac{vN_1hn_1}{vN_1}$$

(here, we've just multiplied the last term on the right-hand side by the identity vN_1 / vN_1) or:

$$T = n_1\left[\frac{1 + vN_1h}{vN_1}\right]$$

and we can calculate the feeding rate per predator as:

$$\frac{n_1}{T} = \left[\frac{vN_1}{1 + vN_1h}\right]$$

This result gives the so-called Type-II functional feeding response. As in the Ivlev model, individual consumption levels off at high prey abundance. The feeding rate reaches an asymptote due to the constraints imposed by searching for and handling prey. We can write this expression somewhat more compactly as:

$$\frac{n_1}{T} = \left[\frac{\omega N_1}{\delta + N_1}\right]$$

where $\omega = 1/h$ is the maximum feeding rate; $\delta = 1/vh$ the half-saturation constant (or the prey level at which the feeding rate is one half the maximum rate).

Box 4.2 Beddington Predator-dependent Functional Response

The total time (T) required to search for prey and to consume prey items in the presence of conspecific predators (Beddington (1975) is:

$$T = T_s + T_h + T_w$$

where T_w is "wasted" time due to interference among predators and all other terms are defined as in Box 4.1. T_s is the time spent searching and T_h is the time spent handling prey. The wasted time is given by :

$$T_w = ew\,(N_2 - 1)\,T_s$$

where e is the encounter rate between predators, and w is the time taken up in each encounter between predators. In Beddington (1975) formulation, the term for the number of predators is reduced by 1 to reflect the fact that predators do not encounter themselves. If for the moment we assume that handling time is negligible we have:

$$T = \frac{n_1}{vN_1} + ew\,(N_2 - 1)\,\frac{n_1}{vN_1}$$

where the first term on the right-hand side is again the time spent searching for prey. Following the approach in Box 4.1, we can write:

$$T = n_1\left[\frac{1 + ew\,(N_2 - 1)}{vN_1}\right]$$

and we can now calculate the feeding rate per predator as:

$$\frac{n_1}{T} = \left[\frac{vN_1}{1 + ew\,(N_2 - 1)}\right]$$

This result gives a predator-dependent analog of the prey-dependent Type-II functional response.

If we assume that the predator search efficiency is the same as the predator encounter rate (Turchin 2003) and that the number of predators is sufficiently large that we can ignore the fact that predators cannot encounter themselves; we can then simplify the expression above to give:

$$\frac{n_1}{T} = \left[\frac{vN_1}{1 + vwN_2}\right]$$

Finally, we can restore the effect of handling time to give a more general model:

$$\frac{n_1}{T} = \left[\frac{vN_1}{1 + v\left(hN_1 + wN_2\right)}\right]$$

Stevens (2009) notes that the Type-II and Type-III functional feeding responses expressed as number of prey consumed per predator (solid lines in Figure 4.4) can be difficult to distinguish in field data. The distinction lies in observations very near the origin which may not be well represented in empirical data. However, the proportion of prey consumed per predator as a function of prey abundance (dashed lines in Figure 4.4) is more easily distinguished throughout the range of potential observations, suggesting that attempts to discern the form of the functional feeding response could fruitfully focus on this representation.

We can now combine a logistic function for the prey population dynamics with different functional response terms. In the following, we will use the Type-II functional response to illustrate some key points.

If we insert the Holling Type-II functional feeding response into our model assuming density independence in both predator (N_2) and prey (N_1), we have:

$$\frac{dN_1}{dt} = bN_1 - \frac{\omega N_1}{\delta + N_1}N_2 \tag{4.16}$$

and

$$\frac{dN_2}{dt} = c\frac{\omega N_1}{\delta + N_1}N_2 - d'N_2 \tag{4.17}$$

Now net zero growth for the prey occurs at $N_1 = 0$ and

$$N_2 = \frac{b}{\omega}\,(\delta + N_1) \tag{4.18}$$

and for the predator at $N_2 = 0$ and

$$N_1 = \frac{d'\delta}{c\omega - d'} \tag{4.19}$$

We again have a vertical line for the predator isocline; the prey isocline now increases linearly as its abundance increases.

Next consider the case in which the prey exhibits density-dependent dynamics while the predator is not limited by its own abundance and exhibits a Holling Type-II functional response. We then have for the prey:

$$\frac{dN_1}{dt} = (b_0 - b_1N_1)\,N_1 - \frac{\omega N_1}{\delta + N_1}N_2 \tag{4.20}$$

while the predator remains as above. The prey isocline now is:

$$N_2 = \frac{\delta + N_1}{\omega} [b_0 - b_1 N_1] \tag{4.21}$$

which now gives a dome-shaped form.

4.2.4.1 Implications for stability

The shape of the functional feeding response has important implications for the stability of the predator–prey system, a conclusion anticipated by our evaluation of the Steele–Henderson model in Chapter 3. We again assume that the prey population exhibits logistic growth in the absence of a predator. A graphical evaluation of equilibrium points for the case in which we consider the effects of a predator on the prey under different functional responses reveals stark contrasts in possible outcomes. In Figure 4.5 we illustrate the effects of three different rates of consumption on the position and nature of the equilibrium points for each of three functional response curves. For a linear functional response, in this example, we see that for each of the levels of consumption examined, the lines representing the functional response intersect the prey production function to the right of the peak (Figure 4.5a). Each of these intersection points represents a stable equilibrium point. Notice that as we increase the consumption rate, we progressively reduce the prey population. Had we further increased the predator rate of consumption, we would no longer have an intersection point with the prey production curve when the consumption rate exceeds the intrinsic rate of increase of the prey. At this point, the prey population would be driven to extinction.

For the Type-II functional feeding response, we see that for the low and medium consumption rates, we have intersection points with the prey production curve (Figure 4.5b). These intersection points again represent globally stable equilibria. At the highest prey consumption rate depicted, however, notice that the functional response curve intersects the production function of the prey at two points. In this case, the intersection point to the right of the peak is stable. However, the lower intersection point is unstable. We would predict that if the prey population was driven below this point, a sudden population collapse would occur.

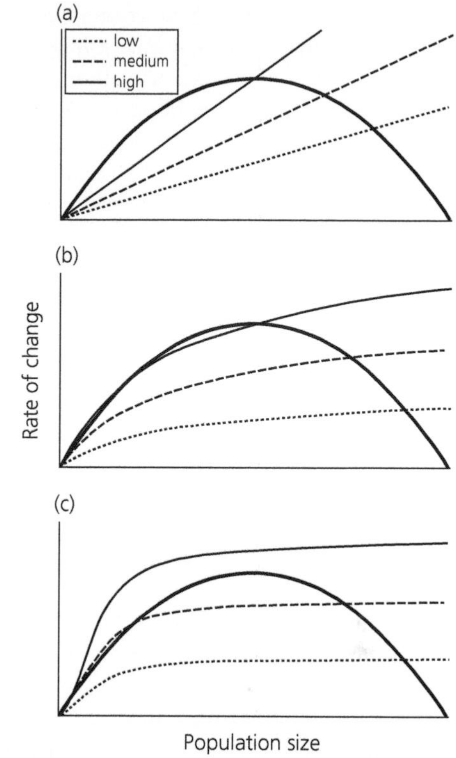

Figure 4.5 Logistic prey production function (dark solid line) and three levels of the rate of consumption (low, medium, and high) for (a) linear, (b) Type-II, and (c) Type-III and functional feeding responses.

For the Type-III functional response, we find that at the lowest consumption rate considered, a single stable intersection point exists (Figure 4.5c). However, at the medium consumption level, upper and lower stable equilibrium points exist with an intermediate unstable equilibrium (Figure 4.5c). In this case, if the prey populaton level is driven below this intermediate point, a collapse to the lower equilibrium point is predicted. Finally, for highest consumption levels depicted we have a single stable equilibrium point at a low prey population level. If consumption increased sufficiently beyond this rate, a population collapse ultimately would occur.

As we shall see in Chapter 11, virtually all traditional fisheries theory assumes that the functional response of humans as predators is linear. If this is not correct and a non-linear functional response more accurately reflects human harvesting strategies in at least some circumstances, then sudden changes in resource status could occur. We would be expecting well-behaved, stable behavior in the fishery and the rapid decline would be totally unanticipated. The history of fisheries suggests that this may not be uncommon.

4.2.4.2 Environmental effects

Environmental change, either natural or anthropogenic, can alter the predator functional response by changing activity levels and behavior in general. It can also directly affect the production functions of predators and prey. What are some consequences of these possibilities? In the following we will explore two examples, one dealing with the functional response and the second with shifting production regimes.

If a temperature increase results in an increase in consumption rates with increasing activity levels and metabolic demands, we can infer from Figure 4.5 that a population decline would result. This holds for each of the functional response terms. It is also possible that the shape of the functional response would change. Taylor and Collie (2003) demonstrated that temperature can, in fact, affect the shape of the functional response curves. In experiments in which post-settlement winter flounder, *Pseudopleuronectes americanus*, were preyed on by the shrimp *Crangon septemspinosa*, the functional response changed from Type III to Type II with an increase in water temperature from 10 °C to 16 °C (Figure 4.6). At the lower temperature, predator activity levels were reduced when prey density-levels were low. The functional response showed a clear inflection point characteristic of the Holling Type-III response (Figure 4.4c). With an increase to 16 °C, we see a Type-II response, with a rapid increase in consumption at lower prey densities. This finding suggests that we could experience a change from a system characterized by two stable equilibrium points to one with an upper stable and a lower unstable equilibrium point.

What if the prey production function changes? Under climate change, we can anticipate environmental shifts that change the overall productivity characteristics of aquatic systems. Persistent changes of this type are referred to as regime shifts. Returning to our example of a logistic production function for the prey and viewing the problem in state space, we can see that changes in either the predator or prey isoclines that affect where the point of intersection occurs, strongly alters the stability properties of the system. If the isoclines intersect to the right of the peak in the prey isocline, we have a stable equilibrium point (see Figure 4.7a). However, a quite different picture emerges when the predator and prey isoclines intersect to the left of the peak of the prey isocline. This can occur with an increase in the equilibrium prey size with increased productivity (assuming for the moment that the position of the predator isocline is unchanged). Here, unstable dynamics dominate such that the prey and predator populations may go extinct. An increase in nutrient supply and

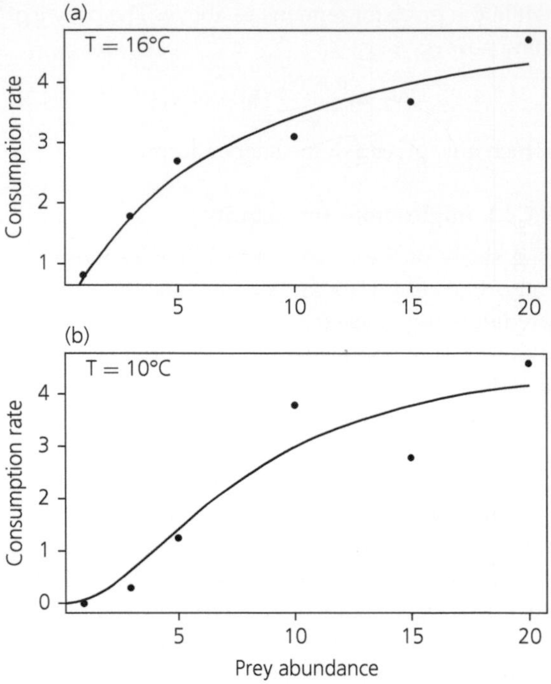

Figure 4.6 Consumption rate of shrimp *Crangon crangon* on post-settlement winter flounder larvae (no. consumed per day per square meter) as a function of prey abundance (no. per square meter) in experiments conducted at (a) 16 °C and (b) 10 °C. Data from Taylor and Collie (2003).

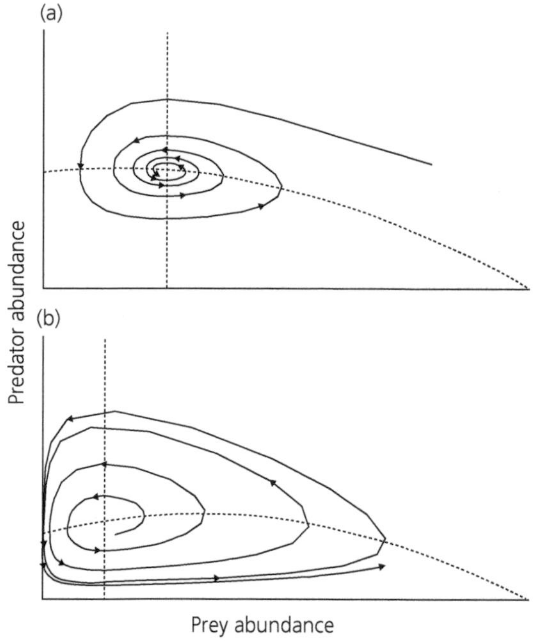

Figure 4.7 Predator–prey population trajectory in phase space with Holling Type-II functional feeding response and density-independence in the predator and density-dependence in the prey resulting in (a) stable equilibrium results when the isoclines intersect to the right of the maximum; (b) the prey population declines when the isoclines intersect to the left of the maximum. Vertical dotted line is the predator isocline. Domed-shaped dotted line is the prey isocline.

enhanced overall productivity could result in such a change. Rosenzweig (1971) referred to the resulting instability as the paradox of enrichment. It seems counter-intuitive that an increase in productivity can lead to a lack of stability and an apparent paradox. It is also conceivable that the position of the prey isocline and vertical predator isocline could both shift and still intersect the curvilinear prey isocline to the left of its maximum. We would then see that the trajectory in phase-space spirals outward (Figure 4.7b), ultimately, leading to the extinction of both predator and prey.

4.2.5 Predator dependence

In Section 4.2.4 we showed how we can extend the traditional Type II functional feeding response to accommodate interference among predators (see Box 4.2). It is not unreasonable to assume that predators can and do interact in ways that affect the outcome of the predation process. Predators can interfere with each other by effectively competing for food. They can also potentially cooperate in ways that facilitate the capture of prey. In the following, we will focus on predator interference. As we shall see, consideration of predator dependence holds important implications for the stability of the system and issues such as top-down vs bottom-up control of ecosystem dynamics. Skalski and Gilliam (2001) found that predator-dependent functional responses provided a better fit to 19 predator–prey systems than did the Holling Type-II prey-dependent model. Essington and Hansson (2004) reported that a predator-dependent functional response provided a better fit to cod and sprat interactions in the Baltic Sea than did prey-dependent functional response models, but not for cod–herring interactions. Below we review ways in which predator dependence has been taken into account in the modeling functional response.

Hassell and Varley (1969) introduced an early phenomenological model of mutual predator interference. The search efficiency (v) of the predator was taken to be an inverse function of its density:

$$v = QN_2^{-m} \tag{4.22}$$

where Q is defined as the "quest" constant and m is a shape parameter. The functional response is then:

$$f(N_1, N_2) = QN_1N_2^{-m} \tag{4.23}$$

(note that for $m = 0$ we have a simple linear functional response). This model is not valid for small predator population sizes and was intended only as an empirical descriptor. For the special case $m = 1$, the expression now involves the ratio of prey to predator abundance, giving a simple form of what is called ratio-dependence.

Beddington (1975) offered an alternative approach and viewed the problem as an extension of Holling's time budget analysis (see Box 4.2). Now, the total time dedicated to search includes "wasted" time due to encounters between predators competing for the same prey. For the case where handling time is negligible, Beddington's functional response can be written:

$$f(N_1, N_2) = \frac{vN_1}{1 + vwN_2} \tag{4.24}$$

(Turchin 2003 p. 85) where w is the time wasted during predator encounters and all other terms are defined as before. If handling time cannot be ignored we then have:

$$f(N_1, N_2) = \frac{vN_1}{1 + vhN_1 + vwN_2} \tag{4.25}$$

to cover both predator- and prey-dependent effects (Turchin 2003 p. 85). DeAngelis et al. (1975) independently proposed a similar functional form for the effect of predator interference. This function can be considered as a natural extension of the traditional prey-dependent functional response. With this formulation, we now see that the prey captured per predator declines monotonically with increasing predator numbers (Figure 4.8). For each predator level we see the characteristic asymptotic form for a Type-II functional response.

Ratio dependence, a special form of predator dependence, has been strongly advocated as an alternative to the traditonal prey-dependent functional response model (Arditi and Ginzburg 1989, 2012). In this approach, the ratio of prey to predator is substituted for the prey term in the traditional prey-dependent functional response. Arguments against this proposal have been most strongly articulated by Abrams (1994, 2015). The arguments for and against ratio-dependence are multifaceted; for a full exposition of the potential advantages and disadvantages of ratio-dependent models see Arditi and Ginzburg (2012) and Abrams (2015).

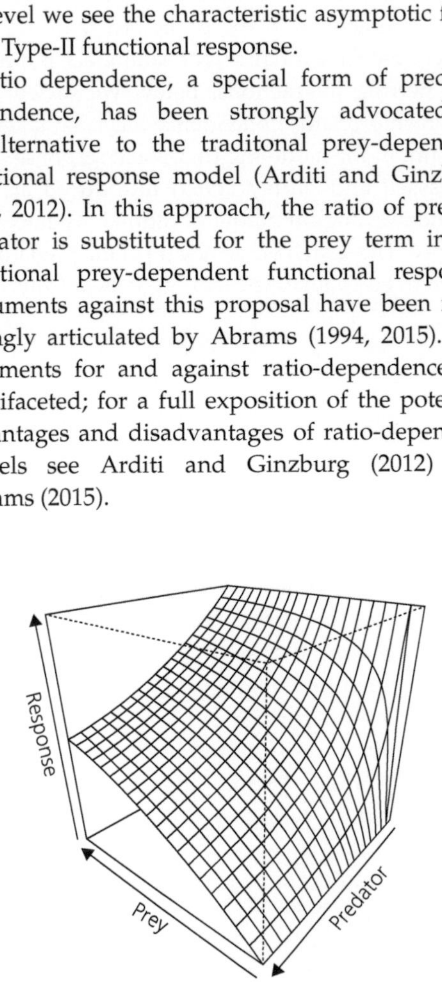

Figure 4.8 Predator-dependent functional feeding response in relation to predator and prey abundance.

4.2.6 Discrete-time models

We have seen in previous chapters that we can readily develop discrete-time analogs of population models framed as differential equations. Again, our motivation is to represent biological and ecological processes that do not necessarily operate continuously. Rather, they may exhibit seasonal or other patterns. We also have seen that discrete-time models can exhibit a remarkably diverse range of dynamical behaviors relative to their continuous-time counterparts.

4.2.6.1 Density-independent models

In the following, we describe a discrete-time predator–prey model specifying the encounter probabilities between predator and prey and build on this foundation to construct simple difference equation models. We start by assuming a random search pattern for the predator over a foraging area. As in our continuous-time models, we will assume that for a simple two-species system, the prey mortality is entirely attributable to the predator. If the encounter probability follows a Poisson process, then the chances of the prey escaping predation in the time interval is $exp(-a_{12}N_2)$ where a_{12} is again the per capita effect of the predator on the prey. Coupling this element with the birth rate of the prey we have:

$$N_{1,t+1} = bN_{1,t}e^{-a_{12}N_{2,t}} \tag{4.26}$$

where b is the discrete birth rate of the prey. The exponential term in Equation 4.26 is the prey survival rate (which is of course bounded between 0 and 1). Its complement, $1 - exp(-a_{12}N_2)$, is the fraction of the prey population consumed. If we multiply this term by the number of prey in the time interval, apply a conversion coefficient relating the number of prey consumed to the resulting number of predator births, and specify a predator mortality rate, we obtain for the predator population:

$$N_{2,t+1} = cN_{1,t}\left(1 - e^{-a_{12}N_{2,t}}\right) - d'N_{2,t} \tag{4.27}$$

where c again represents the conversion efficiency and d' is now the discrete death rate of the predator. This coupled model is a modification of the well-known Nicholson–Bailey host–parasitoid model. The zero growth isocline for the prey is:

$$N_1^* = \frac{b\,(1-d)\log_e b}{c\,(b-1)\,a_{12}} \quad (4.28)$$

and for the predator

$$N_2^* = \frac{\log_e b}{a_{12}} \quad (4.29)$$

Here we find that this model is inherently unstable. Rather than undertaking sustained oscillatory movements as in its continuous-time counterpart, the populations spiral out of control and ultimately collapse (Figure 4.9).

4.2.6.2 Prey density dependence

It is clear that some form of stabilizing mechanism is necessary if the species are to persist in these discrete-time models. Following Beddington et al. (1975), we can substitute an extension of the Ricker-logistic model for the prey to accommodate predation. The model for the prey can be written:

$$N_{1,t+1} = bN_{1,t}e^{-(\omega N_{1,t}+a_{12}N_{2,t})} \quad (4.30)$$

where ω is a compensatory term reflecting negative feedback for the prey population. The predator model remains as in Equation 4.27. With this modification we find that it is possible to obtain persistence of both predator and prey for cases of moderate birth rates of the prey (Figures 4.9c,d). This result parallels our example of the continuous-time case when density dependence is incorporated into the model for the prey (recall Figure 4.3c,d).

We have come to expect the possibility of a much richer array of dynamical outcomes in low-dimensional difference equation models than in their single-species counterparts. Beddington et al. (1975) in fact showed that extremely complex behavior can occur in this simple model. In Figure 4.10a, we show population trajectories for the predator and prey at higher levels of the prey birth rate than that depicted in Figure 4.9. If we now examine the phase-space representation of these data, we see the emergence of a complex attractor in state-space (Figure 4.10b). This complex geometric shape is called a strange attractor. We note that complex dynamics of this type does not generally occur in continuous-time models for two species. However, as we shall see in Chapter 6, for three or more species chaotic dynamics can appear in continuous-time multispecies models as well.

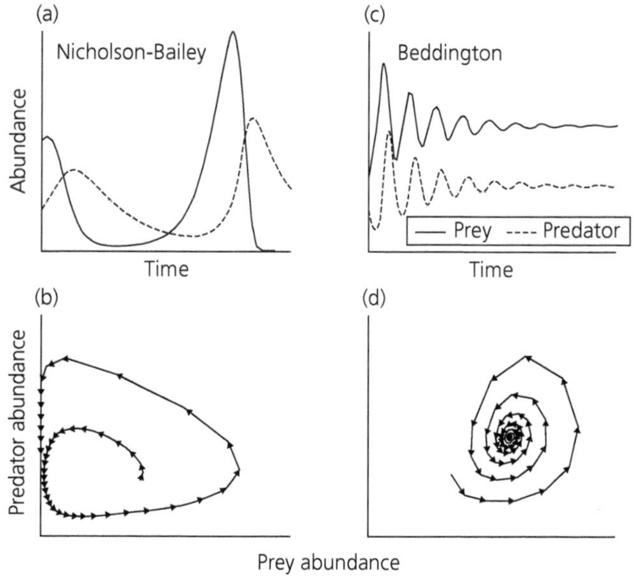

Figure 4.9 (a) Population trajectories for a Nicholson–Baily predator–prey model and (b) corresponding phase portrait for this system reflecting the unstable oscillatory dynamics of the system. (c) Population trajectories for the Beddington predator-prey model and (d) the phase portrait showing approach to a stable equilibrium.

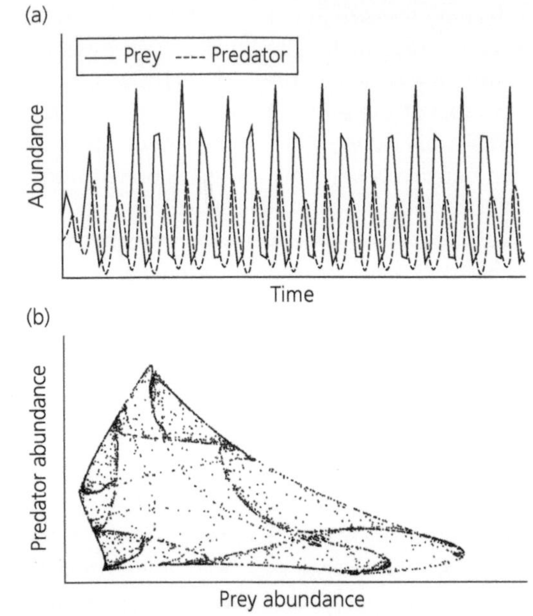

Figure 4.10 Phase-space representation of the Beddington discrete-time predator–prey model with prey density-dependence showing a resulting strange attractor.

We close by noting that as in our previous expositions of simple discrete-time models, the difference equation models described above strictly refer to semelparous species. As we have seen, we can construct higher-order difference equations to represent the iteroparous case. We will reserve further consideration of this issue for treatment of single- and multispecies harvesting models in Part III.

4.3 Parasitism and disease

Parasitism can be a major factor affecting the dynamics and regulation of animal populations (May and Anderson 1983). In the following, we will adopt May's categorization of parasitism into microparasites (e.g. bacteria, viruses, protozoans) that reproduce directly in the host, and macroparasites (e.g. helminths, crustaceans, etc.) that do not. This specification largely maps to other classifications such as microbial and "animal" parasites (Sindermann 1990). Infected hosts harbor some parasite burden level. When this burden is sufficiently high to cause significant impairment of the host, it is considered to be diseased. Lethal and non-lethal impacts of disease are possible, both with potentially important consequences for the dynamics of host populations. Although the impact of disease can easily be as dynamically important as predation and other forms of interspecific interactions (May and Anderson 1978), it has received comparatively less attention in the development of models of aquatic populations and communities than predation, although very sophisticated models of disease dynamics in harvested species have been developed (e.g. Hofmann et al. 2009).

Widespread documentation of massive mortality of aquatic populations due to disease is available (Sindermann 1990) and these episodic events can hold long-term population consequences. For example, perch (*Perca fluviatilis*) populations in Lake Windermere in the English Lake District were subject to a dramatic epizootic (epidemic) in 1976 (Figure 4.11a) resulting in mortality of approximately 98 per cent of the adult population (Langangen et al. 2011). The etiology of the disease outbreak remains uncertain but it apparently involved both primary and secondary infections (Craig 2015). The ramifications of this event for the demography of the perch population were dramatic. Ohlberger et al. (2011) documented an overcompensatory population response to this perturbation. Immature fish were less susceptible to the disease and, freed from cannibalistic control by adults, they increased in abundance following the epizootic event (Figure 4.11). Lower intraspecific competition among older fish resulted in high growth rates and reduced age at maturity. Collectively, these changes profoundly affected the demographic characteristics of the population. Ohlberger (2016) examined the selective pressure exerted by the epizootic and documented resulting genetic change in the perch population.

4.3.1 Models for microparasites

Anderson and May (1978) developed simple host–parasite models to explore the role of disease in regulation of host populations. The model for the host (N_1) can be written:

$$\frac{dN_1}{dt} = (b - d)\,N_1 - a_{12}N_2 \qquad (4.31)$$

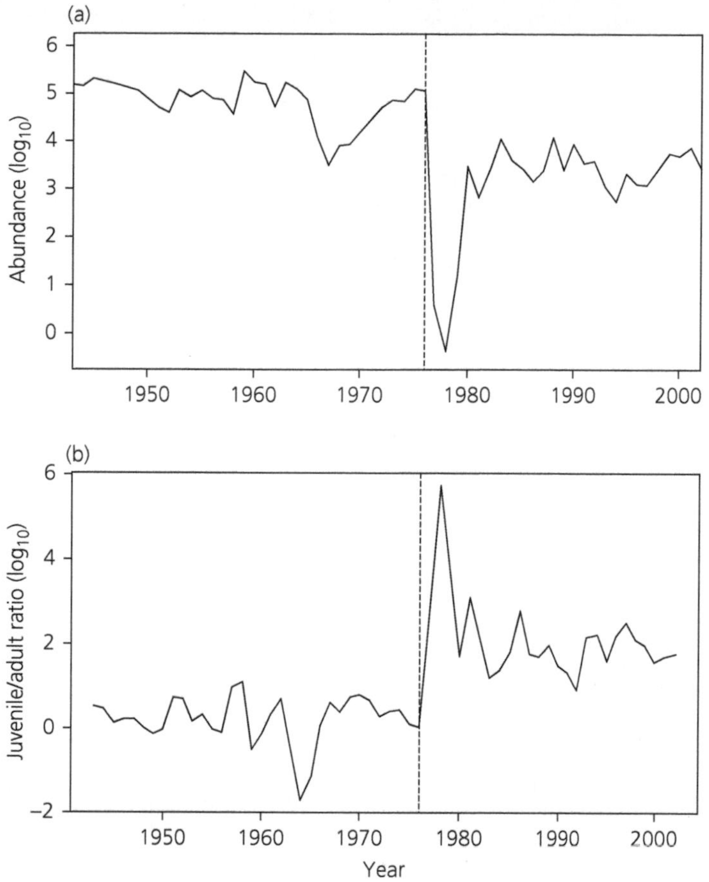

Figure 4.11 (a) Abundance of adult (age 3 and older) European perch (*Perca fluviatilis*) in the south basin of Lake Windermere, showing the effects of a massive epizootic in 1976 and (b) the ratio of juvenile to adult perch showing the effects of adult mortality and the subsequent increase in abundance of juvenile perch as a result of reduced cannibalism by adults. Data courtesy of Ian Winfield and Jan Ohlberger.

where b and d are the birth and death rates of the host population and a_{12} is the death rate inflicted by the parasite (N_2) on the host. Notice that this is similar to the prey model in the classic Lotka–Volterra predator–prey model but does not incorporate a mass action term for the effect of the parasite on the host. The model for the parasite is:

$$\frac{dN_2}{dt} = \left[\frac{\lambda N_1}{H_0 + N_1} - (d + d' + a_{12}) - a_{12}\frac{N_2}{N_1}\right] N_2 \tag{4.32}$$

where λ is the rate of production of the transmission stage per parasite, H_0 is a parameter controlling the efficiency of transmission of the parasite (smaller values of H_0 result in higher transmission rates), d' is the death rate of parasites within the host, and all other terms are defined as before. The predicted parasite burden per host at equilibrium for this model is:

$$\frac{N_2^*}{N_1^*} = \frac{b - d}{a_{12}} \tag{4.33}$$

and the equilibrium host population size is:

$$N_2^* = \frac{H_0 (b + d' + a_{12})}{\lambda - (b + d' + a_{12})} \tag{4.34}$$

For the parasite to regulate the host population under this model, its intrinsic growth rate must be positive and the following inequality must hold:

$$\lambda > (b + d' + a_{12}) \tag{4.35}$$

An illustration of population trajectories of the parasite and host under these conditions is shown in Figure 4.12. This simple model of microparasitism holds a number of features in common with the simplest versions of the Lotka–Volterra predator–prey model we examined earlier. In particular, the host population is not self-regulated and will increase exponentially if the conditions specified in Equation 4.35 are not met. We might then anticipate the possible emergence of periodic dynamics in this system, which is indeed what we find (Figure 4.12). Because of the conditions specified in the inequality above, the intrinsic rate of increase of the parasite is higher than that of the host and we see the expected reversal in the "consumer–resource" ratio relative to a classical predator–prey system. The population numbers of the parasite greatly exceed that of the host under these conditions (Figure 4.12).

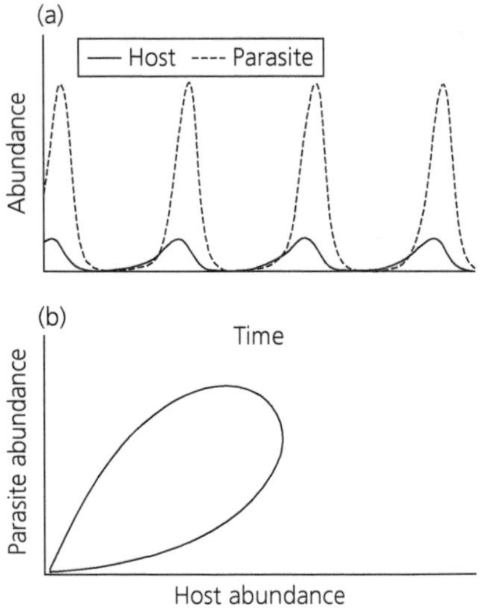

Figure 4.12 Population abundance of a simple host–parasite model showing (a) oscillatory dynamics and (b) phase portrait of host–parasite relationship.

4.3.2 Models for macroparasites

The models for microparasite–host dynamics include the important feature that if the host dies, its microparasitic biome also dies. For macroparasites, this is not necessarily true. A substantial catalog of macroparasites infecting fish and other aquatic organisms has been identified. The range of sublethal impacts on the host includes reduction in growth and condition, parasitic castration, and ovophagy, in which external eggs (principally of crustaceans) are eaten by the parasites. These impacts can each result in a reduction in fitness of the host organism. For economically important species, a number of macroparasites result in a reduction in product quality and marketability, which can impose significant economic losses.

A particularly dramatic form of macroparasitism has exerted devastating effects on fish populations in the Great Lakes. The sea lamprey, *Petromyzon marinus*, invaded the Great Lakes following the construction of a canal that allowed access from the Gulf of St. Lawrence. First sightings of sea lampreys were recorded in 1936 in Lake Michigan, 1937 in Lake Huron, and 1937 in Lake Superior. Juvenile lampreys are hematophagous ectoparasites that attach to the body of the hosts to remove blood meals. Sustained attachment ultimately weakens or kills the host. Lampreys prefer large-bodied fish as hosts and deeply affected lake trout *Salvelinus namaycush* populations in particular. Steep population declines occurred in all lakes within this system that harbored significant lamprey populations following the invasion. Although the joint effect of fishing and parasitism by lampreys has been implicated in the collapse of lake trout populations, the role of lampreys in the decline is unequivocal and resulted in massive efforts to eradicate this invader. The development and application of lampricides that affect the unusually long larval stage (up to 6 years) has proved effective in controlling lamprey populations but the effects on lake trout populations have endured. The constellation of pressures exerted by cultural eutrophication, overfishing, and a sequence of introductions of non-native species in the Great Lakes makes it difficult to disentangle the effects of any one factor on the dynamics of lake trout and other species.

In the following, we will focus on the lake trout–lamprey interaction as an example of a specialized host–macroparasite model. Jensen (1994) invoked the canonical form of the Lotka–Volterra predator–prey equations with density dependence to explore the implications of sea lamprey parasitism on lake trout. In the absence of fishing, the model is:

$$\frac{dN_{1,t}}{dt} = \left(\alpha_1 N_{1,t} - \beta_1 N_{1,t}^2\right) - a_{12} N_{1,t} N_{2,t} \quad (4.36)$$

and

$$\frac{dN_{2,t}}{dt} = \left(\alpha_2 N_{2,t-6} - \beta_2 N_{2,t}^2\right) + a_{21} N_{2,t-6} N_{1,t-6} \quad (4.37)$$

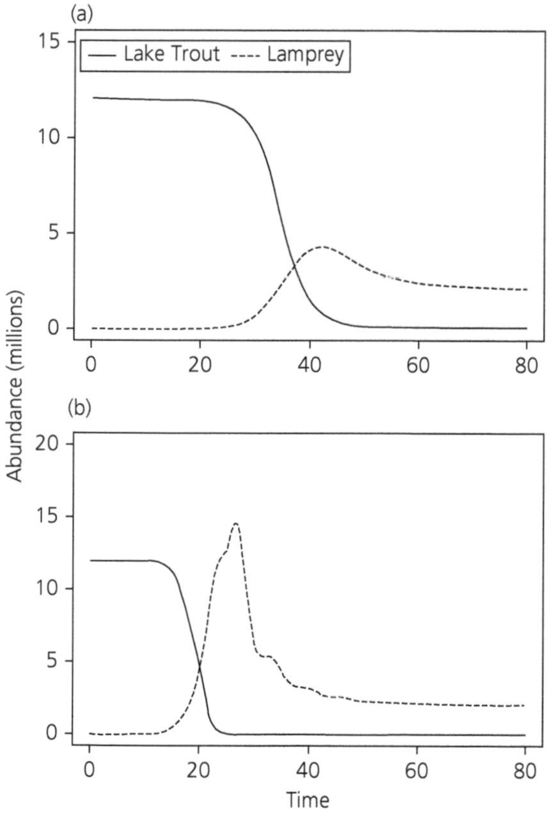

Figure 4.13 Simulation results based on the Jensen (1994) host–parasite model for two levels of the interaction term reflecting the effect of lamprey (*Petromyzon marinus*) on lake trout (*Salvelinus namaycush*) (a) $c_{12} = 5$ and (b) $c_{12} = 10$.

where a time delay of 6 years in the lamprey dynamics has been incorporated to reflect the extended larval period. Note that in this formulation, reproduction of the lamprey population (N_2) is not solely dependent on lake trout; it implicitly has alternative hosts for sustenance. Recall that in the simple Lotka–Volterra predator–prey models we examined earlier, the predator depends exclusively on the prey species and will go extinct if the prey is extirpated. Jensen developed parameter estimates for lake trout and sea lamprey to permit numerical solution of this system of equations. There is considerable uncertainty concerning the translation of energy derived from this form of parasitism into lamprey progeny. Jensen therefore examined a broad range of possible values of the parameter a_{21} in an attempt to explore the implications of this aspect of the host–parasite interaction. In the following, we show results for two values of this parameter. At the lower level of a_{21} examined in our example, the model shows a sharp decline in lake trout abundance as the lamprey population increases following a slow start, reflecting the long time delay related to the duration of the larval period (Figure 4.13a). The lake trout population is predicted to ultimately collapse and lamprey reaches an asymptotic level. At higher levels of the interaction term, however, the impact of lamprey parasitism on lake trout unfolds more quickly and the lamprey population initially overshoots its carrying capacity before settling down to a lower level (Figure 4.13b).

4.3.3 Epidemiological models

Models principally derived from human epidemiology have also been applied to aquatic populations. These models focus on the dynamics of the host population partitioned into three main elements: those susceptible to infection (S), infected individuals (I), and recovering individuals (R). These SIR models trace the course of epidemics in the host population. Unlike the models of parasitism described above, they do not explicitly model the dynamics of the parasite itself. The simplest SIR models do not consider the full dynamics of the host population but rather follow the course of a disease outbreak from its inception at some starting number of hosts.

We have noted that disease is a major threat to aquaculture operations. In this case, the course of a disease outbreak can typically be readily traced and remedial measures taken with antibiotic treatments and other options. In contrast, it is difficult to obtain detailed epidemiological data for aquatic populations in the wild. Periodic outbreaks of disease (principally phocine distemper virus, PDV), in seal populations have, however, been more amenable to analysis. Large-scale outbreaks of PDV occurred in 1988 and 2001, with resulting mass mortalities in harbor seal (*Phoca vitulina*) populations in northwestern Europe (Härkönen et al. 2006). On-going monitoring programs both prior to and following these outbreaks provided important opportunities for detailed analysis of the population trajectories over time (see Figure 4.14).

Grenfell et al. (1992) applied the following model (a variant of the standard SIR model with explicit consideration of death of infected individuals). Here we follow the course of an outbreak from its inception but not the full dynamics of the host population to represent the PDV outbreak among seals in East Anglia:

$$\begin{aligned}\frac{dN_S}{dt} &= -\phi N_S N_I \\ \frac{dN_I}{dt} &= \phi N_S N_I - \gamma N_I \\ \frac{dN_R}{dt} &= (\gamma) N_I\end{aligned} \tag{4.38}$$

where N_S is the density of susceptibles, N_I represents the density of infected individuals, and N_R is the density of recovered individuals; ϕ is the transmission coefficient, δ is the probability of dying due to infection and γ is rate at which individuals become resistant. This model traces the course of a single wave of infection affecting susceptibles in the population (notice that it considers deaths but not births). For the 1988 epizootic, the wave of infection lasted approximately 4 months (Grenfell et al. 1992). Although in our previous discussion and use of population models we have used population numbers and densities somewhat interchangeably, it is important to be clear that here we are strictly referring to population densities. Because disease transmission is a function of contact rates in relation to proximity, this issue is critical in the context of SIR models. For illustration, an example of the fraction of population susceptibles, infected, and recovered individuals from the inception of a non-lethal

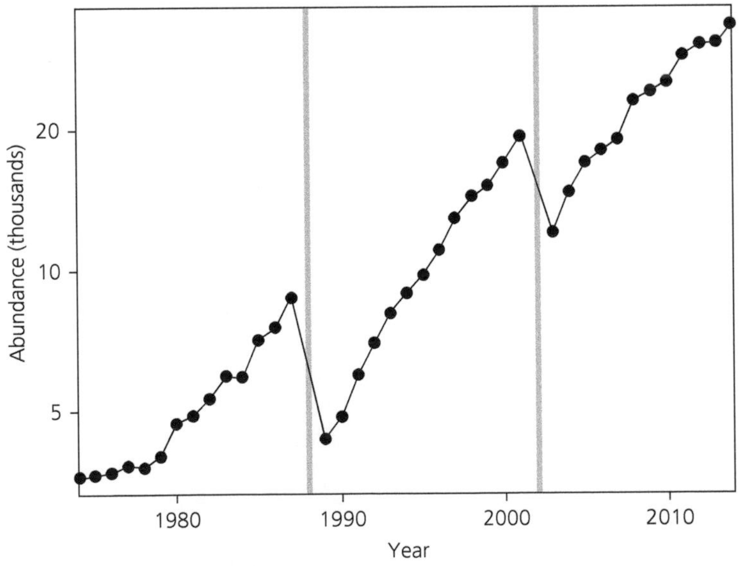

Figure 4.14 Effects on estimated harbor seal (*Phoca vitulina*) population numbers in northwestern Europe of the epizootics in 1988 and 2002. Vertical bars indicate outbreak years. (Adapted from Brasseur et al. 2018.)

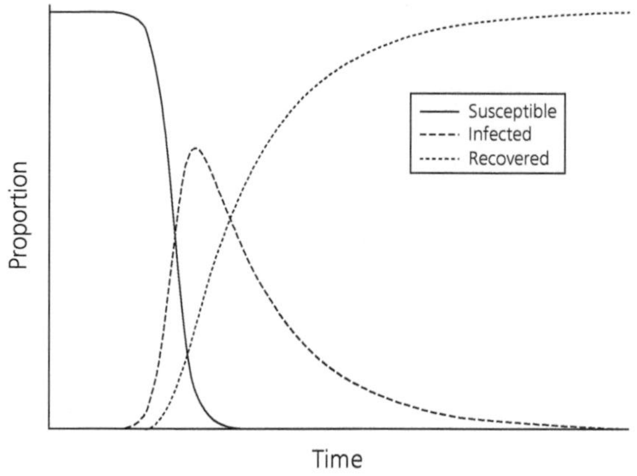

Figure 4.15 Fraction of host population susceptible, infected, and recovered throughout the course of an epidemic of a non-lethal disease following the simple SIR model. Recovered individuals are immune in this formulation.

epizootic event follows the characteristic pattern shown in Figure 4.15. The simple model structure shown in Equation 4.38 can be readily extended to a more dynamic representation of the host population by including renewal terms for births. For further details, see Vandermeer and Goldberg (2003) and Stevens (2009).

4.4 Summary

Predation and parasitism are dominant forms of interspecific interactions in aquatic ecosystems. Predation effects have been more commonly quantified in aquatic ecosystems than disease. Diet studies documenting predation are substantially more common that routine monitoring for disease in aquaculture systems. Incidence of disease in aquaculture systems is of course closely monitored and treated. Predator–prey dynamics are characterized by lagged cycles of prey and their predators, as exemplified by the Lotka–Volterra equations. Density-dependent regulation of the prey or predator population is required for stable coexistence of predator and prey populations. Predator–prey models are extended with the incorporation of non-linear functional responses, which can result in multiple equilibria. The behavior and dynamics of natural predators holds important insights in our consideration of human predation on aquatic resource species to be taken up in Part III of this book.

Disease outbreaks have wrought tremendous impacts on a very broad spectrum of aquatic species. For economically important species, these impacts include significant economic costs to fishing communities and aquaculture facilities. The significance of this issue in aquaculture operations where cultured species are held at high densities, in which high transmission rates prevail, has long been appreciated. It is no less important in capture fisheries, although monitoring the incidence of parasitism and disease in wild populations is considerably more complex and key processes more difficult to track.

Additional reading

Accessible accounts of the topics covered in this chapter are reviewed in Hastings [1996; Chapters 8 (Predation) and 10 (Disease)], Gotelli (2008; Chapter 6), Vandermeer and Goldberg [2003; Chapters 6 (Predation) and 7 (Disease)], Case [2000; Chapters 11 and 12 (Predation)]. Gerking (1994) provides a treatment of predation specifically related to fisheries issues.

CHAPTER 5

Interspecific Interactions II: Competition and Mutualism

5.1 Introduction

As we have seen in Chapter 4, direct evidence of interactions involving predation and parasitism is readily obtained in aquatic ecosystems. Examination of diet composition provides evidence of predatory–prey interactions and the occurrence of lesions and other external or internal markers can attest to the presence and impact of macro- and (some) microparasites. In contrast, detecting evidence of competition and mutualism can involve more subtle clues. For competition to occur, species must overlap in space and time and jointly use essential limited resources such as food and habitat. These conditions result in inverse population trajectories of the competing species (Link and Auster 2013). Controlled and replicated experiments can provide the most direct and compelling evidence for the importance of competition. Some of the classic demonstrations of interspecific competition have come from experimental studies of aquatic organisms (e.g. Gause 1934; Connell 1961; Werner and Hall 1977). However, experiments of this type can be difficult to carry out for larger aquatic species, particularly in marine environments. More progress has been possible in freshwater systems where experiments have proved tractable in laboratory settings and in replicated small ponds. In the case of mutualism, external signs of apparently beneficial interactions between individuals of different species can be directly observed but attempts at quantifying the effects of these interactions at the population level have been far less frequently undertaken.

If energy is limiting in an aquatic ecosystem, we can anticipate that it will be manifest in the effects of intra- and interspecific competition as reflected in changes in individual growth, patterns of resource utilization, and population levels. It may further be indicated in attributes such as greater stability at higher levels of ecological organization (communities or whole ecosystems) than for populations. Despite the challenges of detecting the effects of competitive interactions in aquatic ecosystems without formal experimental approaches, we can ask whether useful inferences might still be made by examining the effects of removal or addition of individuals on other species in the ecosystem. Fisheries clearly create contrasts in the relative abundance of different species in communities and ecosystems through the removal of individuals. Stocking and/or deliberate introduction of new species into a system to enhance recreational and commercial fishing opportunities has also been practiced. Can we take advantage of the changes induced by these addition/removal "quasi-experiments" to make inferences concerning species interactions? We can and do routinely record the magnitude of these interventions and resulting changes at the population and system levels, providing the foundation for inference. Jensen et al. (2012) convincingly argue that, with appropriate caution, important insights into ecological processes can be obtained through investigations of fishery interventions. While these addition/removals have not typically been undertaken to test specific hypotheses concerning interspecific interactions and are not replicated in any strict

Fishery Ecosystem Dynamics. Michael J. Fogarty and Jeremy S. Collie, Oxford University Press (2020).
DOI: 10.1093/oso/9780198768937.003.0005

sense, they are often repeated in different systems and can provide important insights into community dynamics through meta-analyses. We note that the full benefits of undertaking formal experiments at the whole system-level have been well articulated and strongly advocated in resource management (Holling 1978; Walters 1986). Adaptive management is centered on the idea of treating management interventions as formal, spatially replicated experiments (Walters 1986). The scientific value of this approach as a tool for advancing knowledge of the effects of alternative management choices is clear. However, perceived pragmatic constraints have hindered widespread adoption of the approach.

Here, we describe the classical models of competition and mutualism within a common framework based on the logistic model and some of its variants. As in Chapter 4, we again set the stage for consideration of multispecies assemblages in Chapter 6 by first examining models of two interacting species. We take advantage of information derived from formal experiments where possible and explore options for using quasi-experiments of the type described above as learning tools.

5.2 Competition

Interspecific competition is deemed to be one of the major forces structuring ecological communities (Giller 1984). The importance of intraspecific competition and compensation emerged as a central theme in Chapter 3. Here, we broaden the emphasis on competition to encompass interactions between pairs of species. We will broaden this focus further still to encompass interactions within multispecies assemblages in Chapter 6. We can partition competitive interactions into three major categories. Exploitative competition is a result of the joint effects of resource depletion by species sharing common resource needs and preferences. In turn, this results in an overall reduction of production and/or fitness of the species. Interference competition involves direct interactions, including aggression, among species vying for common resources. Preemptive competition reflects the role of timing in competitive interactions in which one species gains first access to resources (often space) in a way that precludes access by other species. Examples of each of these forms of competitive interactions are evident in many aquatic systems. Because of the wider opportunities for formal experimental tests involving larger organisms, particularly strong evidence of the role of competition in structuring freshwater ecosystems is available.

To motivate our exploration of competitive interactions, we turn to a striking example of competitive exclusion in a laboratory experiment pairing guppies (*Poecilia reticulata*) and a hybrid of two swordtail species (*Xiphophorus maculatus* and *X. hellerii*) (Silliman 1975). In these experiments, guppies outcompeted swordtails for limiting resources and the swordtails could not persist (Figure 5.1). Silliman (1975) was able to document higher productivity and greater competitive ability in the guppy population relative to swordtails. We will return to this issue in Section 5.2.6 and investigate potential mechanisms that can alter competitive outcomes. In the following sections, we review the classical models of

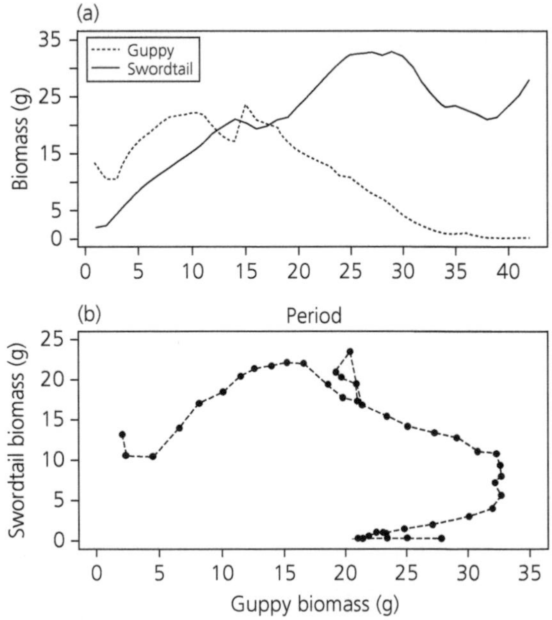

Figure 5.1 (a) Population trajectories for guppies and swordtails in the experiments of Silliman (1975) in control treatment showing competitive exclusion of swordtails and (b) phase-plane plot of trajectory of guppies and swordtails. [Silliman provided tabular data for biomass of guppies and swordtails averaged over three-week periods over the course of the experiment and this is used here].

competitive interactions, building on the structure of the single-species models described in Chapter 3. Is competitive exclusion the only possible outcome as in the guppy–swordtail example? If not, under what circumstances is coexistence possible? We first introduce the concept of the ecological niche and describe associated metrics that have been used to quantify patterns of resource utilization.

5.2.1 Competition and the niche

The intersection of patterns of resource use, interspecific interactions, and critical environmental variables defines the niche of a species or population. The niche essentially characterizes the place of a species in the economy of nature. Species or populations that share common resource needs and environmental preferences are potential competitors if resources are limiting. The intensity of competition is a function of resource abundance and the patterns of similarity in resource use among two or more species. Hutchinson (1957) differentiated between the "fundamental" niche and the "realized" niche. The former represents the broader biotic and abiotic dimensions of the environment occupied by the species or population in the absence of interference by other species. In contrast, the realized niche reflects the more restricted environmental space occupied by the species/population as influenced by interspecific interactions. To understand competition, we need to understand the commonality of resource needs, resource availability, and preferences among co-occurring species.

5.2.1.1 Niche metrics

A number of metrics have been proposed to quantify the range of resource use (niche breadth) of a species along one or more dimensions. Estimators of the overlap in utilization patterns between species have also developed to identify potentially competing species. Levins (1968) proposed one of the earliest measures of niche breadth (B):

$$B = -\sum_{j=1}^{n} P_j \log_e P_j \tag{5.1}$$

where P_j is the proportional utilization of resource j by a species/population. This measure is based on information theory. Levins also proposed an alternative measure:

$$B = \frac{1}{\sum_{j=1}^{n} P_j^2} \tag{5.2}$$

Together, these and related metrics of niche breadth describe patterns of resource use and can be important in quantifying the functional roles individual species/populations play in a given ecosystem (see Krebs 1999 for alternative measures and the strengths of each). An initial foray to test for interspecific competition involves estimating joint resource use by two species. Niche overlap can be expressed:

$$O = 1 - 0.5D \tag{5.3}$$

where

$$D = \sum_{j=1}^{n} |P_{a_j} - P_{b_j}| \tag{5.4}$$

and P_{a_j} and P_{b_j} is the proportional utilization of resource j by species a and b respectively. The difference in utilization patterns by two species is summed over n species to construct the final index. We can readily visualize these concepts for two species along a single resource axis (Figure 5.2). In

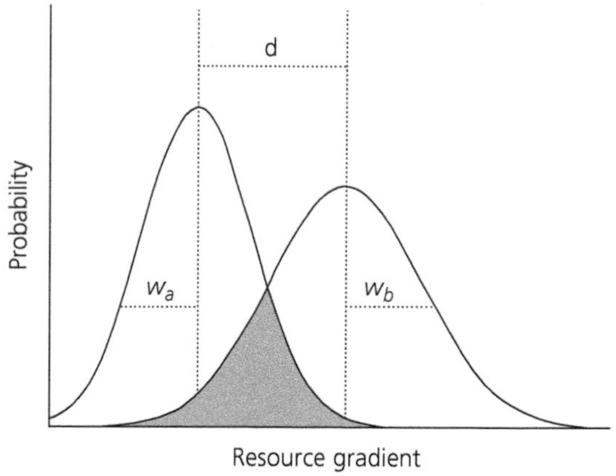

Figure 5.2 Utilization curves and overlap for a hypothetical two-species system along a single resource dimension. The niche widths for Species a and Species b are indicated as w_a and w_b and d represents the difference between the means of the two utilization curves. The overlap in resource utilization is represented by the shaded area.

this illustration, we show the simple case in which utilization patterns follow a normal distribution (with species b having a wider niche breadth than species a ($w_b > w_a$). The shaded portion of the figure represents the overlap in resource use for the two species. We note that application of these fundamental concepts is not constrained to normally distributed resources. Indeed, for factors such as dietary resources arrayed over a range of different prey taxa, we would necessarily employ a discrete probability distribution such as the multinomial to represent the process.

We can illustrate the computation of estimates of niche breadth and overlap using information on dietary overlap in co-occurring brown trout (*Salmo trutta*) and Arctic char (*Salvelinus alpinus*) in Swedish lakes. The diet of both species under allopatric and sympatric conditions is shown in Figure 5.3 (Nilsson 1963). When found in isolation, the proportional representation of six major prey items of these species is generally similar, differing principally in the relative contributions of large crustaceans/molluscs and insect larvae in the diets. In co-occurring populations of char and trout, the

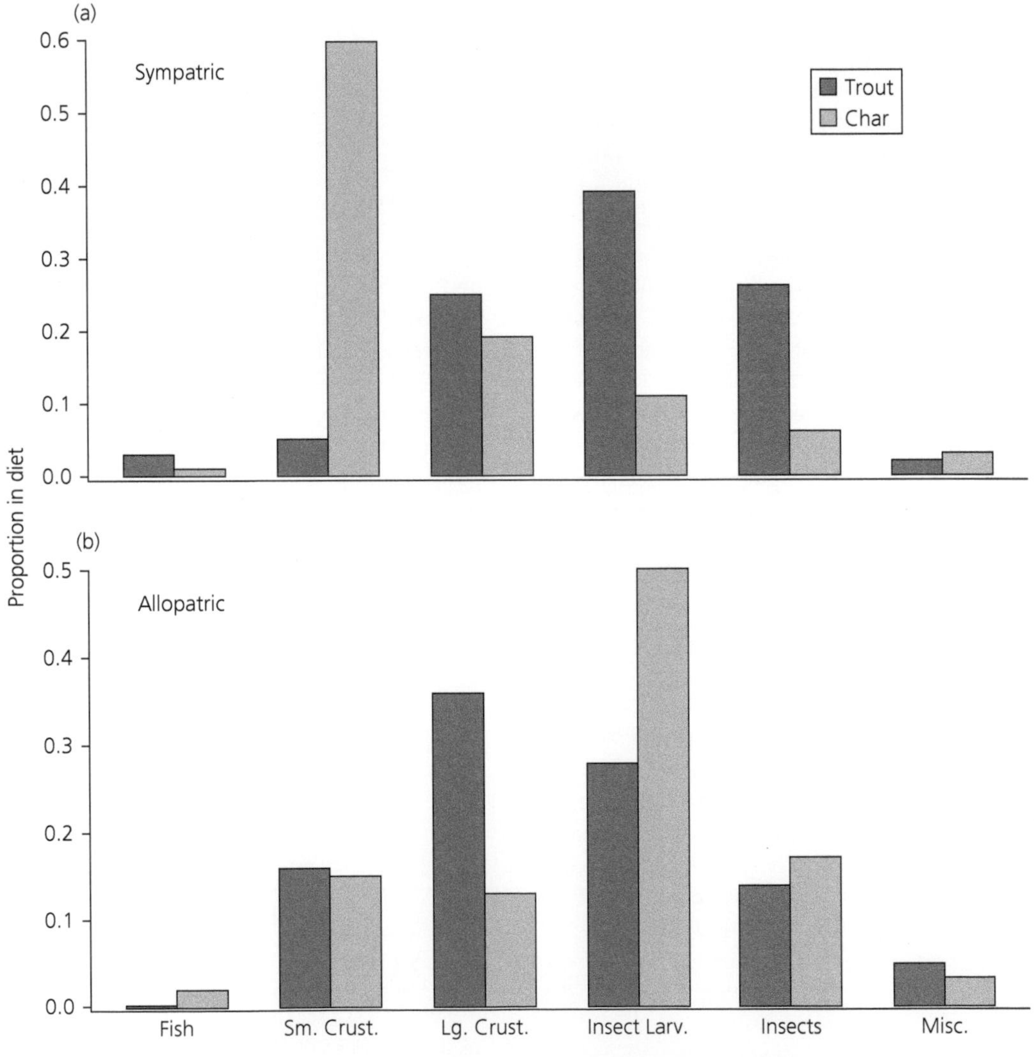

Figure 5.3 Diet composition of brown trout (*Salmo trutta*) and Arctic char (*Salvelinus alpinus*) in Swedish Lakes lakes under sympatric and allopatric conditions. The diet of both species under (a) allopatric and (b) sympatric conditions (Nilsson 1963).

contribution of insect larvae in the diet of char drops substantially while that of small crustaceans (principally gammarid amphipods) increases. The niche breadth decreases for both species under sympatry relative to the allopatric case, from 3.82 to 3.50 for trout and from 3.13 to 2.42 for char using Equation 5.1. Trout exhibits lower overall species selectivity for prey items than char and dietary shifts under sympatry are less pronounced than for char. Diet overlap is 0.44 for sympatric populations of trout and char but 0.75 for allopatric populations using Equation 5.4. The traditional measures of niche overlap do not explicitly account for resource availability and therefore cannot tell us whether competition is actually occurring. However shifting patterns of resource use under allopatric and sympatric conditions can provide important circumstantial evidence for competitive interactions if the ecosystems employed in the comparisons are otherwise sufficiently similar to permit valid inference.

5.2.2 Experimental evidence for competition

Here we describe approaches for detecting and quantifying competition with experimental approaches. We will consider both controlled and replicated experiments and what we have described above as quasi-experiments in which deliberate interventions involving reduction or addition of individuals and species have been made and quantities removed or added have been enumerated and recorded. This situation is common in many resource management settings. Although not formally replicated, it is sometimes possible to conduct meta-analyses to strengthen inference when quantified interventions have been undertaken in several systems simultaneously (or close in time). We emphasize, however, that these quasi-experimental manipulations must be interpreted with caution. Many other uncontrolled factors shape events in the field. Very few (if any) of the aquatic ecosystems we deal with remain uninfluenced by various forms of human intervention and none are immune to uncontrolled environmental disturbances.

5.2.2.1 Species removal experiments

Experiments involving the deliberate and replicated removal of one or more species are one of the most commonly employed designs in evaluating evidence for competition in ecosystems. Some of the classic studies of competitive interactions in aquatic ecosystems including freshwater (e.g. Werner and Hall 1977) and marine (e.g. Connell 1961) ecosystems have employed this approach. Hixon (1980) provided compelling evidence of competition between two congeneric surfperch species in a removal experiment in the Channel Islands off California. The striped surfperch (*Embiotoca lateralis*) occupied preferred shallow water habitat and excluded the black surfperch (*E. jacksoni*) from key food resources through aggressive encounters. In allopatric situations, both species will occupy the shallow water, food-rich habitat. When *E. lateralis* was experimentally removed from the shallow water region of rocky reefs where both species occur, *E. jacksoni* occupied the vacated space (Figure 5.4a).

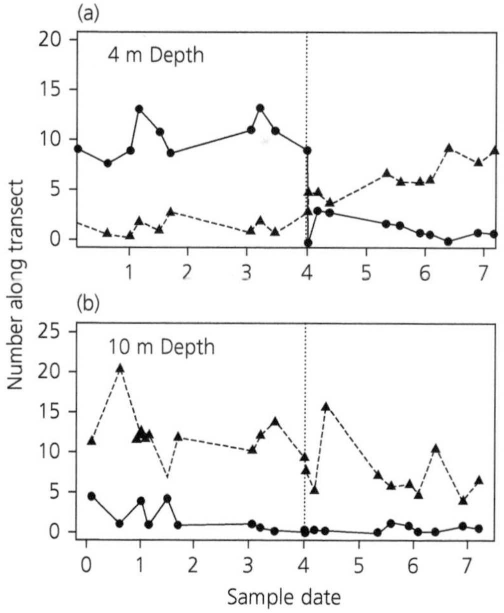

Figure 5.4 Effects of removal of (a) striped surfperch (*closed circles*) on habitat utilization of black surfperch (*triangles*) in shallow water and (b) removal of black surfperch in deeper water sites. Dotted vertical line indicates point of removal. Adapted from Hixon (1980).

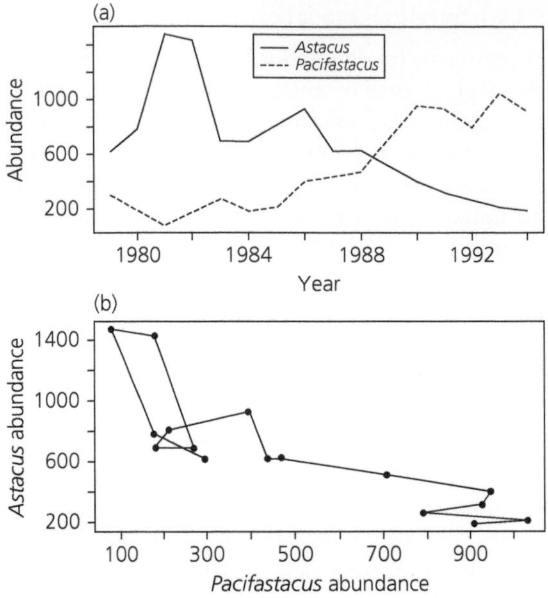

Figure 5.5 (a) Population trends of native crayfish species *Astacus astacus* following the introduction of the North American signal crayfish, *Pacifastacusleniusculus* in a small Finnish lake (Westman et al. 2002) and (b) phase-space representation of abundance of these species.

In contrast, when *E. jacksoni* was removed from its suboptimal habitat in deeper water, *E. lateralis* did not expand its distribution to encompass the deeper regions of the reef (Figure 5.4b). The effect of this experimental intervention was statistically significant (Hixon 1980). Experimental controls showed no significant changes for either species over the time course of the experiment. Hixon (1980) also employed habitat (algal cover) manipulations as part of the overall experimental design to isolate key aspects of the resource base.

5.2.2.2 Species addition "experiments"

As noted above, deliberate introduction of non-native species into aquatic ecosystems has been undertaken in attempts to increase fishery production. Lessons learned from the unintended consequences of past deliberate introductions and changing attitudes toward ecosystem manipulation in general have now strongly altered the prevalence of this practice in marine and freshwater ecosystems, although it still occurs and unintentional introductions continue. Among the unintended consequences of deliberate introductions have been apparent effects of competitive interactions between native and non-native species. For example, in response to a sequence of episodic large-scale mortality events due to disease in the crayfish *Astacus astacus* in Europe, the North American signal crayfish, *Pacifastacus leniusculus*, was introduced in the mid-twentieth century in thousands of lakes in over twenty countries. The larger size, higher reproductive output, and aggressive behavior of *P. leniusculus* led, however, to rapid replacement of *A. astacus* populations in a number of lakes in which introductions were made. The effect of introducing signal crayfish on the native crayfish species *A. astacus* in a small Finnish lake is shown in Figure 5.5 (Westman et al. 2002).

The apparent replacement of native species following introduction of exotics again provides circumstantial evidence of strong resource overlap and competition. Compelling evidence also emerges in cases in which patterns of resource use change in the presence of a competitor. There is widespread evidence for this phenomenon in aquatic systems (see Wootton 1998). Huckins et al. (2000) provide an important illustration of this effect with the introduction of redear sunfish (*Lepomis microlophus*) into a number of Michigan lakes to enhance fishing opportunities starting in the 1920s and continuing through the mid 1990s. Redear sunfish is not naturally sympatric with its congener, the pumpkinseed (*Lepomis gibbosus*). Both species are unusual among centrarchids in specializing on mollusks as prey. In a meta-analysis of the effects of stocking redear into 11 Michigan lakes for which pre- and post-introduction information on abundance was available, Huckins et al. were able to document an average decline of 54 per cent in abundance of pumpkinseed following stocking. In 13 reference lakes not subject to stocking of redear, pumpkinseed abundance actually increased an average of 60 per cent. Huckins et al. (2000) further showed a marked shift in diet composition for pumpkinseed for lakes in which redear were introduced and for which diet data were available. Both redear and pumpkinseed exhibit an increasing dependence on mollusks as prey with increasing size. In lakes stocked with redear, the representation

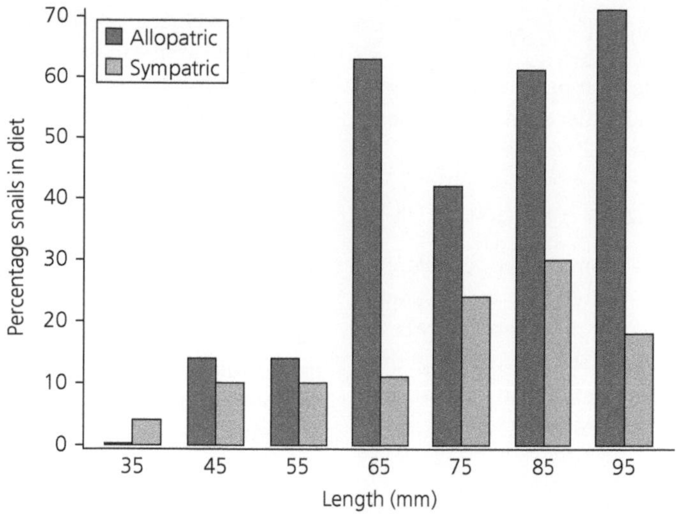

Figure 5.6 Shift in the representation of preferred gastropod prey in the diet of the pumpkinseed (*Lepomis gibbosus*) by size class under sympatry and allopatry with its introduced congener, the redear sunfish (*Lepomis microlophus*) in Michigan lakes (Huckins et. al. 2000).

in the diet of snails (the preferred prey of larger individuals), declined markedly in pumpkinseed diets (Figure 5.6). This effect was further tested in controlled laboratory experiments and the results support the inferences drawn from field studies (Huckins et al. 2000).

5.2.2.3 Estimating interaction strength

Resource overlap is neither a necessary nor sufficient condition for competition. The effects of interference competition may not be represented in estimates of resource overlap and as mentioned above, resources must be limiting for competition to occur. In this section, we will consider model-free estimates of interaction strength based on pre- and post-intervention population or density estimates. In Section 5.2.5 we will take up the topic of model-based estimates of competition coefficients. See McCallum (2008) for a thorough review of approaches to estimation of competition coefficients and interaction strength. These methods are based on species removal experiments. Here we wish to accommodate either removal or addition experiments Perhaps the simplest estimator of the effect of the removal or addition of a competitor is:

$$I_{ij} = \frac{|N_{i,0} - N_{i,1}|}{N_j} \tag{5.5}$$

where I_{ij} is the interaction index, $N_{i,0}$ is the abundance of the ith species at the start of the experiment or intervention (removal or addition of its competitor), $N_{i,1}$ is its abundance at the end of the experiment or intervention, and N_j is the abundance of the target species removed or added at the start of the experiment. In the case of a species addition experiment, this would be immediately after the introduction of the species. This measure can be computed for all non-target species in the experimental or natural community and a matrix of interaction strengths specified. Link and Auster (2013) proposed a more complex alternative index of interaction strength incorporating estimates of spatial and dietary overlap, the production-to-biomass ratios of potentially competing species, and information on relative body sizes.

5.2.3 Models of competition in continuous time

We can build on the framework described in Chapter 3 to construct models of competing species. The classical Lotka–Volterra competition equations for a two-species system can be developed as a simple extension of the logistic model (Equation 3.8) with the addition of terms for competitive interactions (Gause 1934):

$$\frac{dN_1}{dt} = (\alpha_1 - \beta_1 N_1) N_1 - a_{12} N_1 N_2 \tag{5.6}$$

where a_{12} represents the competitive effects of Species 2 on Species 1 and all other terms are defined as in Equation 3.8. For species 2 we have:

$$\frac{dN_2}{dt} = (\alpha_2 - \beta_2 N_2) N_2 - a_{21} N_2 N_1 \tag{5.7}$$

where a_{21} represents the competitive effects of Species 1 on Species 2. Analysis of this system of equations at equilibrium provides important insights concerning whether species obeying the Lotka–Volterra competition model can coexist. At equilibrium we have for Species 1:

$$\alpha_1 - \beta_1 N_1 - a_{12} N_2 = 0 \tag{5.8}$$

or

$$N_1 = \frac{1}{\beta_1} (\alpha_1 - a_{12} N_2) \tag{5.9}$$

Notice that when N_2 is zero, the single-species equilibrium point (α_1/β_1) will be attained. For Species 2 we have:

$$\alpha_2 - \beta_2 N_2 - a_{21} N_1 = 0 \tag{5.10}$$

or

$$N_2 = \frac{1}{B_2} (\alpha_2 - \alpha_{21} N_1) \tag{5.11}$$

and when $N_1 = 0$, Species 2 reaches its single-species equilibrium value (α_2/β_2). These lines represent points at which the population growth rate as a function of the other species is zero.

The Lotka–Volterra competition equations are phenomenological in that the resources for which the species compete are not explicitly represented. In Chapter 6 we will introduce consumer–resource models in which this issue is directly addressed in the context of food resources.

Examining different parameters values for the two species, we can find combinations in which one species or the other always prevails, some in which both species coexist, and others in which one species or the other survives depending on the parameter values and initial conditions. In the Silliman experiments, guppies won in the control treatment.

We can plot the isoclines for both species on the same graph to determine which species (if either) will outcompete the other. Figure 5.7a shows the case in which the isocline for Species 1 (solid line) lies entirely above that for Species 2 (dashed line). In this case, Species 2 will be outcompeted and Species 1 will reach its equilibrium point (α_1/β_1). Starting at any point in this phase space, note that for any point above the isocline of Species l, the abundance of Species 1 must shift to the left. Similarly, for Species 2, any point lying above its isocline must shift downward (the rules are of course reversed for the case where the point is below the isoclines for Species 1 and 2). The net change in position as the populations move toward the attractor is given by vector addition. We can determine that Species 1 will outcompete Species 2 when $\alpha_1/a_{12} < \alpha_2/\beta_2$ [i.e. $\alpha_1 < (\alpha_2/\beta_2)a_{12}$] and $\alpha_1/\beta_1 < \alpha_2/a_{21}$ [or $(\alpha_1/\beta_1)a_{21} < \alpha_2$]. In other words, Species 1 will win when its intrinsic rate of increase is greater than the effect of Species 2 on Species 1 multiplied by the single-species equilibrium level of Species 2 and when the intrinsic rate of increase of Species 2 is less than the effect of Species 1 on Species 2 multiplied by

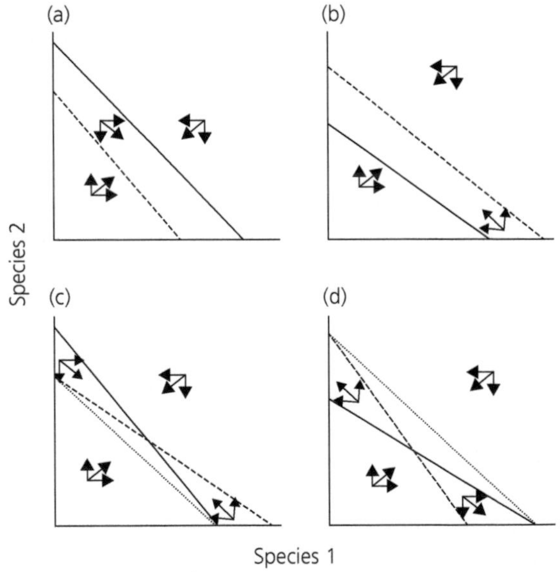

Figure 5.7 Phase-space representation of outcomes of competitive interactions for a two-species system (Species 1 solid line; Species 2 dashed line) for (a) competitive exclusion of Species 2 by Species 1 (b) competitive exclusion of Species 1 by Species 2 (c) stable coexistence with both species persisting (d) indeterminate outcome with one species persisting depending on initial conditions. Dotted line connects the single-species equilibrium values of the two species in panels (c) and (d).

the single-species equilibrium population size of Species 1.

For the case in which the isocline of Species 2 lies entirely above that for Species 1, Species 2 outcompetes Species 1 (Figure 5.7b). This will occur when $\alpha_2 > (\alpha_1/\beta_1)a_{21}$ and $\alpha_1 < (\alpha_2/\beta_2)a_{12}$. Figure 5.7c shows the case in which there is stable coexistence of the two species; this is possible when $\alpha_1 > (\alpha_2/\beta_2)a_{12}$ and $\alpha_2 > (\alpha_1/\beta_1)a_{21}$. Finally, Figure 5.7d depicts the case where the outcome of competition is indeterminate and depends on the starting conditions. The indeterminate case occurs when $\alpha_1 < (\alpha_2/\beta_2)a_{12}$ and $\alpha_2 < (\alpha_1/\beta_1)a_{21}$. In this situation, the point where the lines intersect represents an unstable equilibrium point. Whether the equilibrium point will be stable depends on whether the intersection point lies above the line connecting the single-species equilibrium points (dotted lines in Figure 5.7c,d). This result, however, holds only for the case of linear isoclines (see Section 5.2.3.1).

We can visualize the approach to equilibrium corresponding to each of the four cases above by starting at different initial points in the phase space and tracking the resulting trajectories. For the case corresponding to Figure 5.8a, Species 1 wins and the trajectories all converge to the point $N_1 = \alpha_1/\beta_1$, $N_2 = 0$ (Figure 5.8a). For the situation in which the isocline for Species 2 lies entirely above that of species 1, the trajectories converge to the point $N_1 = 0$, $N_2 = \alpha_2/\beta_2$, (Figure 5.8b). Coexistence is predicted for the case shown in Figure 5.8c; here, the trajectories converge to the point $N_1 = (\alpha_2 a_{12} - \alpha_1\beta_2)/(a_{12} a_{21} - \beta_1\beta_2)$, $N_2 = (\alpha_1 a_{12} - \alpha_2\beta_1)/(a_{12} a_{21} - \beta_1\beta_2)$ (Figure 5.8c). The indeterminate case in which one species will persist depends on the initial conditions (Figure 5.8d).

5.2.3.1 Non-linear isoclines

The linear isoclines for the Lotka–Volterra competition equations result directly from the linear per capita function used in the logistic model to describe the internal dynamics of each species. As we have seen in Chapter 3, non-linear per capita functions may be more appropriate for some species. If we now adopt the theta-logistic model as our foundation for the competition models (Gilpin and Ayala 1973) we have:

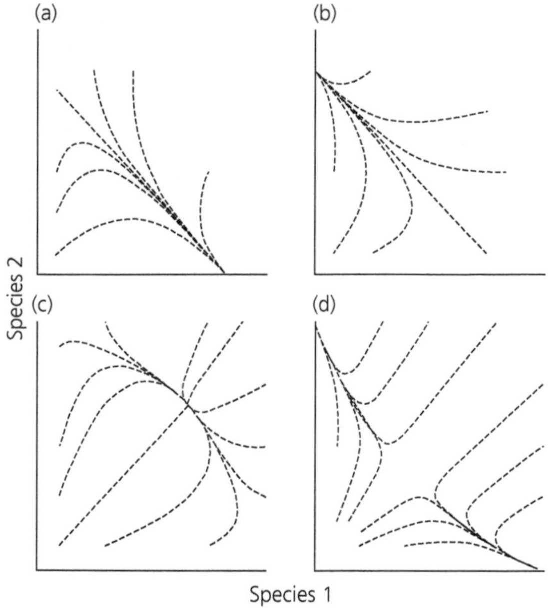

Figure 5.8 Population trajectories originating in different parts of phase-space for a two- species system corresponding to (a) competitive exclusion of Species 2 by Species 1 (b) competitive exclusion of Species 1 by Species 2, (c) stable coexistence with both species persisting, (d) indeterminate outcome with one species persisting depending on initial conditions.

$$\frac{dN_1}{dt} = \left(\alpha_1 - \beta_1 N_1^{\theta_1 - 1}\right) N_1 - a_{12}N_1N_2 \tag{5.12}$$

and

$$\frac{dN_2}{dt} = \left(\alpha_2 - \beta_2 N_2^{\theta_2 - 1}\right) N_2 - a_{21}N_2N_1 \tag{5.13}$$

where θ_i is the 'shape' parameter for species i; all other terms are defined as above. The logistic model is a special case with $\theta_i = 1$. We now have non-linear isoclines for Species 1:

$$N_1 = \left[\frac{1}{\beta_1}(\alpha_1 - a_{12}N_2)\right]^{\frac{1}{\theta_1 - 1}} \tag{5.14}$$

and Species 2:

$$N_2 = \left[\frac{1}{\beta_2}(\alpha_2 - a_{21}N_1)\right]^{\frac{1}{\theta_2 - 1}} \tag{5.15}$$

The isoclines are now curvilinear with θ controlling the degree of non-linearity. The isoclines bow outward for $\theta > 1$ and inward for $\theta < 1$.

5.2.4 Models of competition in discrete time

As we have noted in previous chapters, for species such as Pacific salmon with non-overlapping generations and seasonal patterns of reproduction, difference equation models are often the most appropriate choice. These factors are of course no less relevant in constructing multispecies models. To set the stage, consider the run size estimates for pink (*Oncorhynchus gorbuscha*) and chum (*O. keta*) salmon in the Skagit River feeding into Puget Sound shown in Figure 5.9. Pink salmon have a two-year life cycle and, in many river systems, exhibit a marked temporal pattern in which only one of two possible "lines" persist, resulting in pink salmon returning to these natal water bodies only in alternating years.

For the Skagit River population, pink salmon catches over the time period of observation occur in odd-numbered years. In contrast, chum salmon have a 4–5 year life cycle and all lines are represented. Ruggerone and Neilsen (2004) documented competitive interactions between pink salmon and sockeye (*O. nerka*), chinook (*O. tshawytscha*), coho (*O. kisutch*), and chum salmon throughout the North Pacific. Pink salmon were shown to be dominant over their salmonid competitors including chum salmon and this is reflected in the pattern of returns for the two species in the Skagit River (Figure 5.9). A clear inverse pattern of abundance is evident; in odd-numbered years, returns of chum salmon are low, particularly in very high pink salmon return years. The dominance of pink salmon has been attributed to successful competition for food resources (exploitation competition) rather than interference competition (Ruggerone and Neilsen 2004). Chum salmon alter their diet patterns when both species are present in the marine phase of their life cycle (in this case, in even years). In addition, alternating patterns in the age-at-maturation of chum salmon have been documented and hypothesized to be due to an evolutionary response to competition with pink salmon (see review in Ruggerone and Neilsen 2004). The observed pattern of returns for chum salmon reflects the collective impact of these ecological and evolutionary forces.

As we have seen, semelparous species such as Pacific salmon can be effectively modeled with

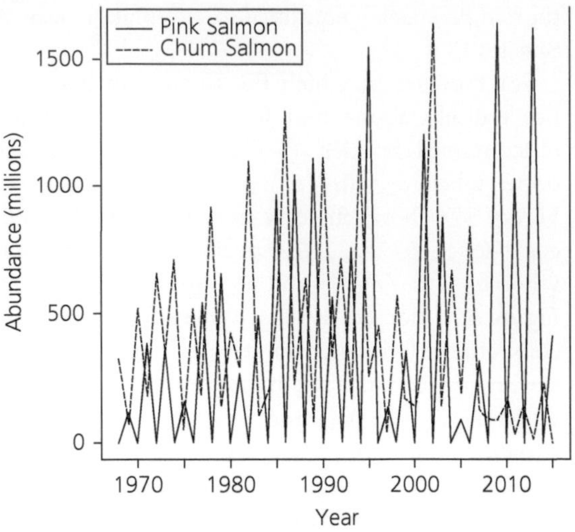

Figure 5.9 Run size estimates for pink (*Oncorhynchus gorbuscha*) and chum (*O. keta*) salmon in the Skagit River feeding into Puget Sound (courtesy Marissa Litz). Note that units on the *x*-axis are years and not generations.

discrete-time models. Leslie and Gower (1958) provided an early example of a discrete-time model for competing species (see Larkin 1963 for an extension of this model to exploited species). Hassell and Comins (1976) developed difference-equation models of two competing species and showed that, as in the single-species case, an extraordinary range of dynamical behaviors are possible. We start with a simple extension of the discrete logistic model to account for competitive interactions:

$$N_{1,t+1} = (1 + \alpha_1)\, N_{1,t} - \beta_1 N_{1,t}^2 - a_{12} N_{2,t} N_{1,t} \quad (5.16)$$

and

$$N_{2,t+1} = (1 + \alpha_2)\, N_{2,t} - \beta_2 N_{2,t}^2 - a_{21} N_{1,t} N_{2,t} \quad (5.17)$$

where the time step is one generation. At equilibrium $N_{1,t+1} = N_{1,t}$ and the zero growth isocline is now:

$$N_1^* = \frac{1}{\beta_1} (\alpha_1 - a_{12} N_2) \quad (5.18)$$

and similarly for Species 2 we have:

$$N_2^* = \frac{1}{\beta_2} (\alpha_2 - a_{21} N_1) \quad (5.19)$$

and we again have linear isoclines comparable to those found in continuous-time models.

As we have now come to expect with discrete-time models, very interesting dynamics can be exhibited by this competitive system, including periodic and aperiodic behavior. In Figure 5.10a we see in-phase periodicity of the species for the parameters chosen. However, a simple increase in the intrinsic rate of increase and compensatory mortality for Species 1 results in outcomes in which these functionally related species can appear to change synchronously or asynchronously over different stanzas of time (Figure 5.10b). This dynamical behavior gives rise to "mirage" correlations in which populations are correlated, anti-correlated, or entirely unrelated during different time periods (Sugihara et al. 2012) despite the fact that simple linear interactions between the species are an integral part of the model structure.

Many alternative model forms of course are possible. For example, a Ricker-logistic competition model can be written as:

$$N_{1,t+1} = N_{1,t}e^{\alpha_1 - \beta_1 N_{i,t} - a_{12}N_{2,t}} \tag{5.20}$$

and

$$N_{2,t+1} = N_{2,t}e^{\alpha_2 - \beta_2 N_{2,t} - a_{21}N_{1,t}}. \tag{5.21}$$

Applying the approach described above for the discrete logistic model, we again obtain zero isoclines directly comparable to the continuous-time case.

We will use the Ricker-based competition model to illustrate how changes in the magnitude of competition in a two-species system can alter the expression of complex dynamics. Inspection of Equations 5.20 and 5.21 shows that increasing the magnitude of the interaction term will reduce the growth rate of each population. To explore the implications of this change, bifurcation diagrams for Species 1 are shown in Figure 5.11 for increasing levels of competition from Species 2. As we increase the interaction strength, and holding all model coefficients constant except α_1 we see a clear change in the expression of dynamic complexity, with a diminution in chaotic dynamics for the choice of parameters used in this simulation.

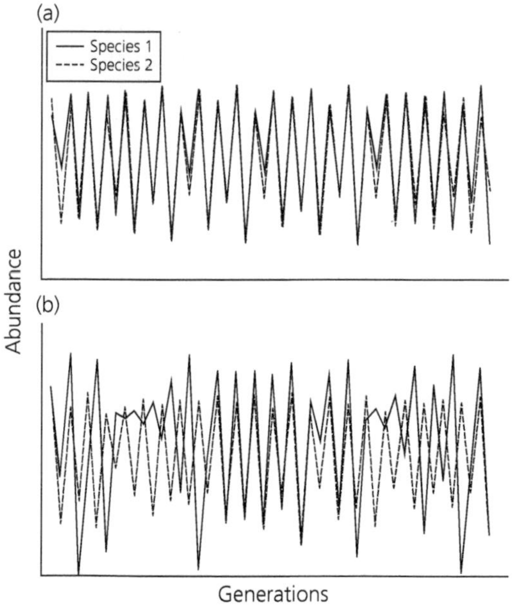

Figure 5.10 Population trajectories of the discrete logistic competition model of Sugihara et al. (2012) for two levels of the intrinsic rate of increase (a) 3.4 and (b) 3.8 for Species 1 and holding all other parameters constant.

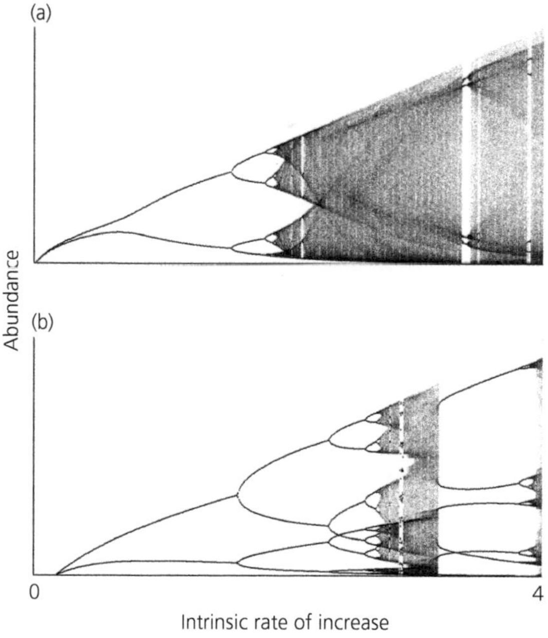

Figure 5.11 Bifurcation diagram for a two-species competitive system for Species 1 as a function of (a) lower effect of Species 2 on Species 1 and (b) higher effect of Species 2 on Species 1.

5.2.5 Model-based estimation of competition coefficients

Having now described models of competition in both continuous and discrete time, we can consider approaches to estimating competition coefficients (a_{ij}) using census data. As we have seen, experimental approaches have been fruitfully employed to test hypotheses concerning competitive interactions in aquatic systems. In instances where controlled and replicated experiments are not feasible, it may still be possible to extract important insights into potential competitive interactions when census data are available. We have seen hints of this in our consideration of the effects of species additions and removals in fishery ecosystems. In fact, this issue has received considerable attention in ecology as a whole. The strengths, and potential pitfalls, of relying on census data alone or in concert with ancillary supporting observations have been carefully examined (e.g. Laska and Wootton 1998; Wootton and Emmerson 2005; McCallum 2008). The general approach entails specification of a model defining the nature of the interactions and amenable to treatment as a time series. For example, adopting the Ricker-based competition model described in Equations 5.20 and 5.21 and linearizing, we have:

$$\log_e\left[\frac{N_{1,t+1}}{N_{1,t}}\right] = \alpha_1 - \beta_1 N_{1,t} - a_{12}N_{2,t} \qquad (5.22)$$

and

$$\log_e\left[\frac{N_{2,t+1}}{N_{2,t}}\right] = \alpha_2 - \beta_2 N_{2,t} - a_{21}N_{1,t} \qquad (5.23)$$

Given a time series of population estimates, it is possible to estimate the coefficients of this system of equations with regression techniques. To illustrate the approach in a simplified setting, we can apply the technique to the guppy–swordtail data described at the start of this chapter. Silliman (1975) had in fact estimated model coefficients from his data using analog computer techniques. Silliman employed a different model formulation specifically based on the Gompertz function with added interaction terms. The multispecies Ricker-logistic model is qualitatively similar. The fit of this model to the guppy–swordtail experiment (see Figure 5.1) is shown in Figure 5.12. Notice that for

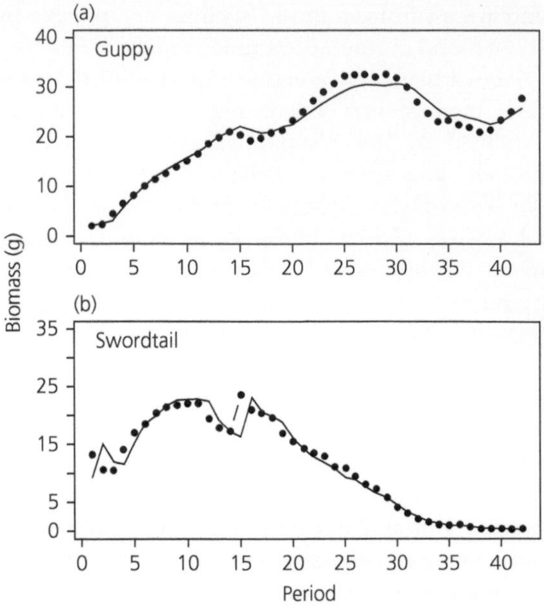

Figure 5.12 Observed (closed circles) and predicted biomass of (a) guppy and (b) swordtail in laboratory competition studies (Silliman 1975).

the guppy series, there were periods of consistent deviation from the model prediction (i.e. sustained patterns of positive or negative residuals). The final model fit therefore required estimation of additional parameters for autocorrelation in the residuals. The final parameter estimates for the coefficients indicated a significant effect of guppy abundance on swordtails. The interaction coefficient for the effect of swordtails on guppies was not statistically significant.

5.2.6 Altering competitive outcomes

In his classic competition experiments on aquatic micro-organisms, G.F. Gause considered the effect of dilution (or rarefaction) on competitive outcomes (Gause 1935 cited in Slobodkin 1961; Gause and Witt 1935). Gause modified the Lotka–Volterra equation to incorporate a rarefaction term in which a common loss or removal term was applied to both species. We now have

$$\frac{dN_1}{dt} = (\alpha_1 - \beta_1 N_1)\, N_1 - a_{12}N_2N_1 - mN_1 \qquad (5.24)$$

for Species 1 and

$$\frac{dN_2}{dt} = (\alpha_2 - \beta_2 N_2) N_1 - a_{21} N_1 N_2 - m N_2 \quad (5.25)$$

for Species 2; the common loss term is designated m and all other terms are defined as before. Now the isoclines for the two species are:

$$N_1^* = \frac{1}{\beta_1} (\alpha_1 - m - a_{12} N_2) \quad (5.26)$$

and

$$N_2^* = \frac{1}{\beta_2} (\alpha_2 - m - a_{21} N_2) \quad (5.27)$$

With the addition of the loss term we now represent the system in three dimensions as isoplanes (see Figure 5.13). At low or non-existent loss rates, Species 1 would outcompete Species 2 (note the non-intersecting isoclines along the right rear panel of Figure 5.13). However, with increasing rarefaction, the competitive outcome is changed to coexistence (see intersecting lines in left rear panel of Figure 5.13).

The motivation for Gause's consideration of the loss term was to offer additional insights into the importance of the intrinsic rate of increase in competitive outcomes. Notice that the addition of the loss term results in a reduction in the net rate of increase in the population. However, it can provide important general insights into the role of a non-selective predator as just one possible agent of removal from the populations. In this and the preceding chapter, we have deliberately focused on interactions between pairs of species in order to set the stage for broader considerations to follow. Gause's analysis shows just why it is so important to ultimately broaden our focus—increasing the dimensionality of the systems we consider opens avenues for alternative outcomes. In Chapter 6, we will expand our view to encompass multispecies systems in which several types of interactions simultaneously affect species within the community.

The guppy–swordtail experiments described in opening sections of this chapter included consideration of the effects of differential removal rates of each species (Silliman 1975). Although the experiment therefore departed from the non-selective removal process employed by Gause, Silliman showed that swordtails and guppies could coexist with differential removal rates (Figure 5.14). In the absence of exploitation on guppies, swordtails would otherwise be eliminated. Silliman (1975) was interested in not only engineering the persistence of both species, but by modifying the exploitation rates of both over time, generating insights into situations in which the joint yield of both could be optimized. He found in fact that it was possible to

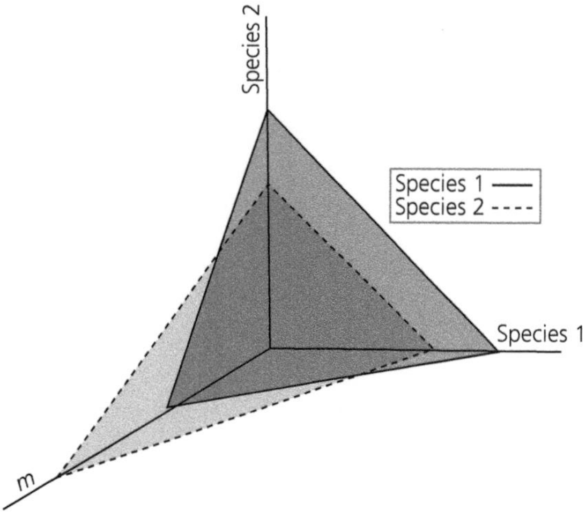

Figure 5.13 Isoplanes for a Lotka–Volterra competitive system with a common rarefaction (or loss) term applied to both species.

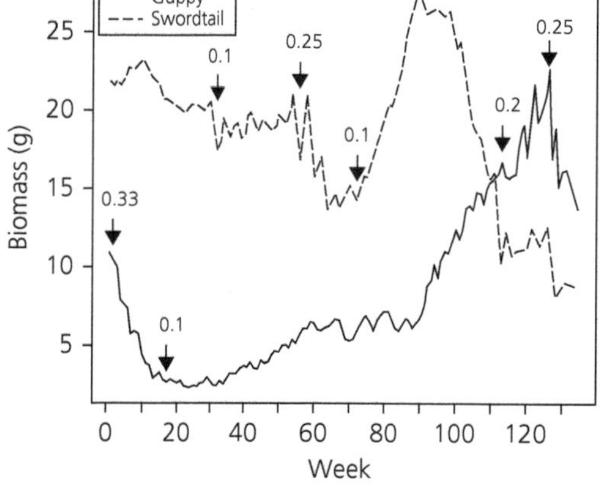

Figure 5.14 Biomass trajectories of guppy and swordtail in experimental treatments in which differential rates of removal (rarefaction) were applied to each species to maintain coexistence (after Silliman 1975). Target removal fractions and dates for the two species are indicated by arrows and associated numbers.

generate higher total yields involving both species than for either of the species alone.

5.2.7 The competitive production principle

Vandermeer (1989) advanced the concept of a competitive production principle in the context of agricultural yields involving "intercropping." Rather than adopting the traditional approach of monospecific agricultural practices, intercropping involves multispecies assemblages in which different species occupy different niches. These polyculture systems can potentially utilize a fuller spectrum of resources (e.g. nutrients) than any single species and provide higher total yields under certain circumstances. Vandermeer (1989) proposed a measure of the relative yield total (*RYT*) per unit area as the sum of the ratios of species under the two strategies. For a two-species system we have:

$$RYT = \frac{Y_1^p}{Y_1^m} + \frac{Y_2^p}{Y_2^m} \tag{5.28}$$

where Y_i is the yield and the superscripts p and m represent treatments involving polyculture and monoculture strategies respectively for the two species. One possible measure of whether the mixed-species strategy is superior to monocultures is if $RYT > 1.0$ (Vandermeer 1989). Setting RYT to 1.0 and rearranging, we have:

$$Y_1^p = Y_1^m - \left[\frac{Y_1^m}{Y_2^m}\right] Y_2^p \tag{5.29}$$

as an indicator of when the polyculture strategy is superior. If the yield for the polyculture system lies above this straight line, it outperforms monoculture as a strategy (Vandermeer 1989). These considerations of course hold direct relevance for aquaculture. Although monoculture strategies are common in aquaculture, mixed species strategies in ponds are also widely employed as part of integrated fish farming throughout the world.

These issues potentially hold direct relevance to natural fishery ecosystems. We have seen instances in which changes in resource utilization patterns of species in aquatic systems occur under sympatry relative to the allopatric case (see Section 5.2.1). This can result in a fuller utilization of available resources and potentially higher overall yield for the mixed-species system. We will return to this issue in Chapter 15. Viewed through a different lens, it also highlights the importance of maintenance of species diversity as a critical consideration in mixed-species capture fisheries.

5.3 Mutualism

Perhaps the most widely known examples of mutualism in aquatic ecosystems involve the reciprocal benefits enjoyed by cleaner fish and their "clients." The cleaner fish remove ectoparasites and improve the overall health of the client fish. In turn, the cleaner fish receives nutritional benefit. Here, our focus will be on protective mutualism in which one species derives protection from predators. This form of mutualism is facultative rather than obligatory—each species can persist in the absence of the other, albeit with a loss in productivity and/or fitness. Mixed species schooling (shoaling) is an underappreciated form of mutualism. It is recognized that single-species schools can afford protection from predators through an "escape in numbers" strategy. This mechanism holds for animal groups in general, including multispecies shoals (Goodale et al. 2017). Joint foraging in multispecies shoals is also recognized as important.

Although symbiotic relationships are widespread in aquatic ecosystems (e.g. Karplus 2014), the quantitative impact of the relationship is substantially less well known. Lynam and Brierley (2007) provide one of the few examples of an attempt to quantify the benefits of positive species associations in a fishery context. They related a measure of recruitment success (residuals from a stock-recruitment relationship) of the whiting (*Merlangius merlangus*) with an index of abundance of two jellyfish species of the genus *Cyanea* (*C. lamarckii* and *C. capillata*) in the North Sea. The jellyfish index was the natural logarithm of the maximum abundance of the two species in a standardized pelagic trawl haul. Higher survival of whiting recruits after adjustment for spawning stock size was positively related to higher jellyfish abundance (Figure 5.15). This relationship also holds when each species of *Cyanea* is treated individually. It is inferred that early-stage

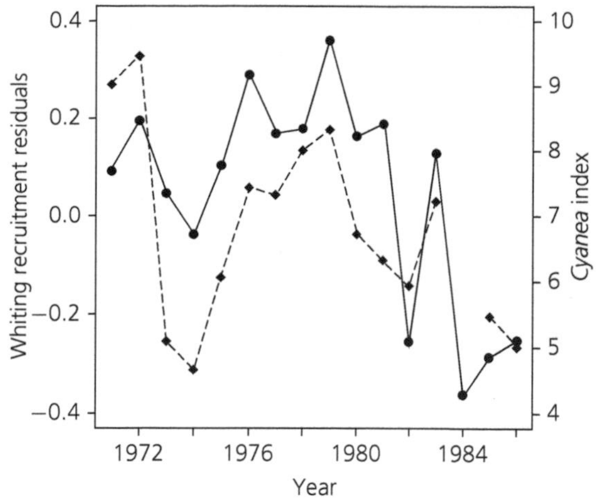

Figure 5.15 Relationship between whiting recruitment (adjusted for the effects of adult biomass) and the abundance of jellyfish (*Cyanea* spp.). Adpapted from Lynam and Brierley (2007).

whiting shelter among the tentacles of *Cyanea* to avoid predators. It is unknown if the presence of whiting confers any benefit to the jellyfish. The relationship may therefore be commensal rather than a case of mutualism.

Experimental tests involving experimental removal of cleaner fish have also been conducted, most notably in coral reef systems. A particularly compelling long-term experiment involving controlled removal of the cleaner wrasse, *Labroides dimidiatus*, from experimental sites at Lizard Island on the Great Barrier Reef was conducted over an 8.5 year period (Waldie et al. 2011). Two client Pomacentrid species in the treatment sites had significantly lower abundance and average size relative to control sites. Lower abundance was attributed to impacts on successful recruitment to these sites. General community-level effects, including reduced species richness and prevalence of "visitor" species were also observed. The adverse effect on client species with removal of cleaner fish indicates clear population benefits associated with mutualism. The cleaner species derives nutritional benefits from grooming and demonstrable effects on the demography of the client species were shown in this experiment.

5.3.1 Continuous-time models

We begin with an intuitive modification of the Lotka–Volterra system of equations to reflect positive relationships between two species:

$$\frac{dN_1}{dt} = (\alpha_1 - \beta_1 N_1) N_1 + a_{12} N_2 N_1 \tag{5.30}$$

and

$$\frac{dN_2}{dt} = (\alpha_2 - \beta_2 N_2) N_2 + a_{21} N_1 N_2 \tag{5.31}$$

where all terms are defined as in Equation 5.6 and 5.7 with the signs of the interaction terms now positive rather than negative. Following the same procedure we used to determine the equilibrium points for the competitive case we have for the isoclines:

$$N_1 = \frac{1}{\beta_1}(\alpha_1 + a_{12} N_2) \tag{5.32}$$

for Species 1; for Species 2 we have:

$$N_2 = \frac{1}{\beta_2}(\alpha_2 + a_{21} N_1) \tag{5.33}$$

We now find that our intuitive approach to specifying positive interactions in the Lotka–Volterra equations to represent mutualism holds an unsuspected challenge. For relatively low and/or asymmetric interaction coefficients ($a_{12}a_{21} < 1$), the isoclines intersect and we have the possibility of a stable equilibrium at the intersection point of the isoclines. However, when the product of the interaction terms is greater than one we have a highly unstable system in which the isoclines do not intersect and both species exhibit unbounded (and unrealistic) growth in what May (1975) memorably described as "an orgy of mutual benefaction." If we introduce the equivalent of non-linear functional response terms resulting in saturating benefits into the model of mutualism, we can avoid this problem.

5.3.2 Discrete-time models

We can of course readily adapt the discrete-time competition model described in Section 5.2.4 to account for mutualism:

$$N_{1,t+1} = (1 + \alpha_1) N_{1,t} - \beta_1 N_{1,t}^2 + a_{12} N_{2,t} N_{1,t} \quad (5.34)$$

and

$$N_{2,t+1} = (1 + \alpha_2) N_{2,t} - \beta_2 N_{2,t}^2 + a_{21} N_{1,t} N_{2,t} \quad (5.35)$$

At equilibrium $N_{1,t+1} = N_{1,t}$ and the zero growth isocline is now:

$$N_1^* = \frac{1}{\beta_1} [\alpha_1 + a_{12} N_2] \quad (5.36)$$

and similarly for Species 2 we have:

$$N_2^* = \frac{1}{\beta_2} [\alpha_2 + a_{21} N_1] \quad (5.37)$$

These are of course identical to the isoclines for the continuous-time case and the general lessons from that exercise hold.

We again see that for sufficiently high values of the intrinsic rate of increase (α_i) complex dynamics emerge. In Figure 5.16a we show an example of (a) synchrony in periodicity for the case of $a_{12}a_{21} < 1$ and a relatively low value for a_{12}. Maintaining the constraint $a_{12}a_{21} < 1$ but increasing a_{12} results in asynchrony in the fluctuations of the two species (Figure 5.16b).

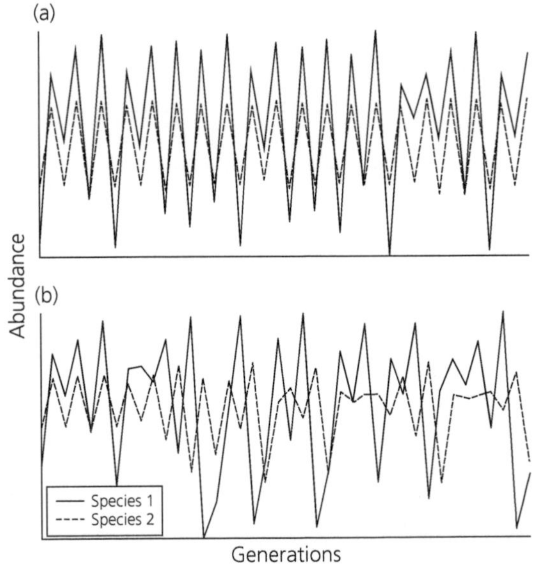

Figure 5.16 Example population trajectories for a discrete-time two-species system for (a) lower mutualism coefficient and (b) higher mutualism coefficient for the effect of Species 2 on Species 1.

5.4 Summary

Competition and mutualism are important forms of biotic interaction in aquatic communities. Quantification of the population and community-level effects of these interactions has historically been less common in fisheries analyses than predation. In part, this reflects the difficulties in conducting controlled experiments for larger-bodied organisms in aquatic environments. Documenting competition entails not only identifying patterns of shared resource use but evidence that these resources are limiting. Inferences concerning competitive interactions in non-experimental settings may be possible if histories of population change for putative competitors are available and quantifiable interventions involving the addition of a species (through deliberate or inadvertent introductions) or a differential reduction in abundance of the species through harvesting is undertaken. Care must be taken to account for other changes in the environment in these uncontrolled quasi-experiments.

Deterministic models of competitive interactions in continuous time for a two-species system predict four possible outcomes including stable equilibria in which only one species persists; an unstable equilibrium in which only one species persists depending on the initial conditions; and a stable equilibrium in which the species coexist. Competitive outcomes can be altered by external forces altering the abundance of the two species. As we have come to expect, much more complex dynamical behavior is possible for models framed in discrete time. In this case, we find that it is possible for "mirage" correlations to emerge in which the two populations can fluctuate in synchrony or appear anti-correlated or totally uncorrelated over different stanzas of time even though they are structurally related through interaction terms.

Mutualistic interactions are widely recognized in aquatic ecosystems but far less commonly quantified to date. Mixed-species schools of fish are known and this strategy may result in protection from predators and/or enhanced foraging success. Other specialized forms of mutualism and commensalism have been widely documented (e.g. cleaner–client relationships, situations in which fish are afforded protection through association with anthozoans etc.). Simple modification of the classical Lotka–Volterra competition models in which the signs of the interaction terms are changed from negative to positive to reflect mutualism reveal conditions under which unrealistic outcomes involving unrestrained growth of both species is predicted. In these cases, additional constraints involving saturating effects of mutualism are required.

Additional reading

Topics covered in this chapter are reviewed in Giller (1984), Roughgarden (1998; Chapter 14) and Gotelli (2008; Chapter 5), Vandermeer and Goldberg (2003; Chapter 8), and Case (2000; Chapter 14). Each provides very instructive accounts of these issues. More advanced treatments of these topics can be found in Kot (2001).

CHAPTER 6

Community Dynamics

6.1 Introduction

In Chapters 4 and 5, we examined the dynamics of pairs of interacting species. We now extend this framework to multispecies communities. For our purposes, a community will be defined as a group of co-occurring species in space and time. Broadening our focus to encompass multiple interacting species forces consideration of a number of issues beyond those we took up when examining species pairs. We now open pathways to the possibility of both direct and indirect interspecific effects, the interplay of different types of species interactions, and questions concerning the structure and function of species assemblages. A fundamental question in ecology is how communities with large numbers of interacting species persist. Does the addition of species make the community less resilient to external perturbations? Or do some species provide a "portfolio effect" by performing functionally similar roles in the ecosystem?

To motivate this topic, we start by examining empirical patterns in fish communities with a small number of interacting species. The fish assemblage of the Baltic Sea is relatively simple, with lower diversity than many other boreo-temperate marine ecosystems. It therefore offers a convenient stepping-stone from the two-species cases in the previous chapters to more complex communities. It will be instructive to consider the implications of multiple interaction pathways and the possibility of both direct and indirect effects on community dynamics. Consider the population trajectories of three dominant fish species in the Baltic: cod, herring, and sprat in three-dimensional space from 1974 to 2011 (Figure 6.1). The observed pattern suggests different regimes in the Baltic Sea. An early regime in the 1980s featured high cod and herring abundance and low sprat abundance. Since 1990, a more recent regime is characterized by high sprat abundance and lower cod and herring abundance. The dynamic range in herring population size has been considerably more restricted than that of cod and sprat over time.

Cod preys on both herring and sprat and the latter two species are potential competitors. In addition, herring, and particularly sprat, have been identified as predators of the early life stages of cod in the Baltic. This predator–prey reversal suggests that we will ultimately want to use models that include some level of age or stage structure to fully represent this assemblage of species.

In this chapter, we confront an issue that foreshadows a critical question occupying center stage in the remainder of this book—can we effectively manage the complexity and data requirements for EBFM? As we increase the number of species under consideration, the information requirements can become prohibitive. To address this issue, various measures of aggregation are often employed. In this chapter, we will discuss groupings based on taxonomic affinity or the roles of species in the ecosystem in terms of energy flow, habitat utilization etc. Aggregation by size rather than defined groups of species is also employed. This approach complements the species-based delineations because species of similar size often play similar roles in the transfer of energy in the system. Consideration of size-based methods leads to size-spectra, which can be applied, not just at the community level, but to the entire ecosystem, a topic that we pick up in Chapter 10. We also describe qualitative modeling techniques that may be applicable in situations where quantitative

Fishery Ecosystem Dynamics. Michael J. Fogarty and Jeremy S. Collie, Oxford University Press (2020).
DOI: 10.1093/oso/9780198768937.003.0006

approaches are difficult or impossible to implement. These methods are designed to allow inferences about system stability; we focus on the direction (and not the absolute magnitude) of change in the system as external perturbations affect it. In this approach we sacrifice any attempt at precision in favor of obtaining general insights.

6.2 Some attributes of communities

To set the stage for the remainder of this chapter, we first describe some attributes of interest of communities, including characterizations of species diversity, the importance of keystone predators in community dynamics, and defining functional groups within communities that play key roles in system structure and function. We also describe the potential role of compensatory processes operating at the community level and their relevance to the resilience of the ecosystem. Finally, we examine the central question of stability and complexity in community dynamics.

6.2.1 Species diversity

Determining the number of species present is a critical first step in characterizing a community. This key number is referred to as species richness. Fish are the most speciose of vertebrate fauna, with over 30 000 known species; projections suggest that up to 40 000 species may ultimately be identified. In contrast, approximately 5500 species of mammals are currently known and few new additions to this list are anticipated. The total known number of aquatic invertebrate species exceeds that of fish by a very substantial margin but the fraction of these species taken for human consumption is comparatively low.

Understanding how a community responds to natural and anthropogenic perturbations and, in particular, if any risk of losing species may ensue, is essential. As we have noted, human intervention in aquatic ecosystems is pervasive and this risk is very real. In his classic essay, *Round Pond*, the noted conservationist Aldo Leopold observed that "*To keep every cog and wheel is the first precaution of intelligent tinkering*" (Leopold 1949). That species richness must, at a minimum, be preserved is an essential consideration if exploitation or other forms of deliberate human intervention are to be undertaken. This fundamental concept is enshrined in fisheries and wildlife management and conservation law.

Understanding the factors affecting species richness in a specified region has long been a focal point of ecological research. Suggested determinants of species richness include how long the community has been in existence, the frequency and magnitude of disturbance regimes, and the overall area occupied. The size of the area occupied has repeatedly been shown to be an important factor in the expected number of species found in a community. This relationship typically follows a power function of the general form $S = \phi A^{\varphi}$ where S is species richness, A is area and ϕ and φ are coefficients. To illustrate this pattern, we show examples of double logarithmic plots for fish species in lakes and inland seas (Barbour and Brown 1974; Figure 6.2a) and coral reefs (Belmaker et al. 2007; Figure 6.2b). Barbour and Brown examined 70 lakes and enclosed seas in six of the seven continents and spanning nearly seventy degrees of latitude. Belmaker et al., in contrast, concentrated on coral reefs in a closely circumscribed region (the Gulf of Aqaba in the Red Sea). The overall slope of the species–area relationship for all the systems analyzed by Barbour and Brown was ~0.15. At a higher level of spatial resolution, higher slopes were found. For African lakes the estimated slope was ~0.35. These included the species-rich Great African Rift Valley lakes.

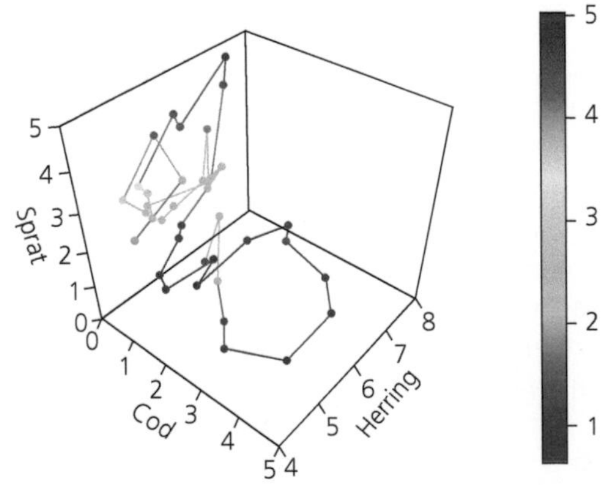

Figure 6.1 Three-dimensional phase portrait of abundance of Baltic Sea sprat (*Sprattus sprattus*; 10^8), cod (*Gadus morhua*; 10^6), and herring (*Clupea harengus*; 10^7) for the period 1974–2011. Data courtesy of Stefan Neuenfeldt. Color bar gives sprat abundance levels from 0 to 5×10^8 individuals.

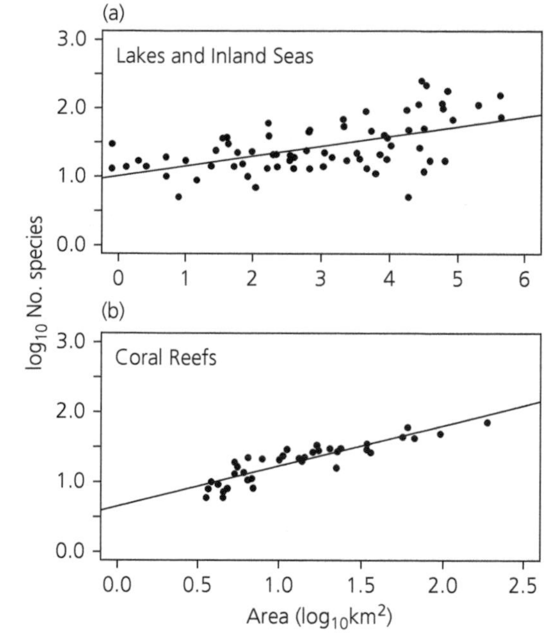

Figure 6.2 Fish species–area relationship for (a) 70 lake and inland seas (Barbour and Brown (1974) and coral reefs in the Red Sea (Belmaker et al. 2007).

Belmaker et al. reported a higher species–area slope of ~ 0.55, consistent with other coral reef systems. It is clear that in practical applications, higher spatial resolution of species–area relationships is important. Knowing the species–area relationship is intrinsically important. It has also been used to address pragmatic questions concerning how large aquatic reserves should be to afford sufficient protection at the population, community, and ecosystem levels (see Chapter 15).

Important latitudinal differences in species richness emerge in global examinations of species diversity. A map of known fish species richness in large marine ecosystems of the world is provided in Figure 6.3. Higher species richness in low latitudes and in upwelling areas is clear. The striking pattern of very high species richness in southwest Asia has been attributed to the very long history and relative stability of coral reef communities in the region.

We can go beyond the concept of richness to address the relative abundance of species in the community. A number of measures of species diversity that account for not only the number of

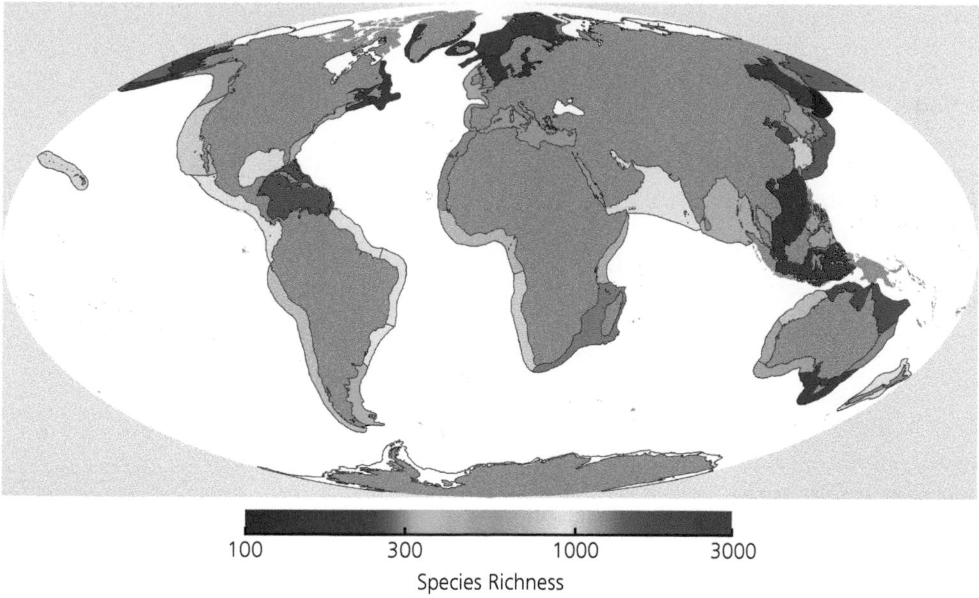

Figure 6.3 Species richness in Large Marine Ecosystems around the world (data source: Sea Around Us Project University of British Columbia). Map courtesy of Kim Hyde.

Box 6.1 Species Diversity

A number of measures of the species diversity in a community have been proposed. Below we provide formulae for three commonly used measures. Magurran (2004) provides comprehensive treatment of this topic and describes many more metrics of species diversity. One of the most commonly employed measures is the Shannon Index. This metric has foundations in information theory (diversity is equated with information in this context). The Shannon index is:

$$H' = -\sum_{j=1}^{n} (p_i)\left(\log_e p_i\right)$$

where p_i is the proportion of the ith species in a community of n species. This index takes into account both species richness and species evenness. In a comparison of two communities with equal species richness, the one with greater evenness will score more highly in this index. The evenness component of the Shannon index can be partitioned out by taking H'/log_eS where S is species richness.

The Simpson index is given by:

$$D = \sum_{j=1}^{n} \left(p_i^2\right)$$

where p_i is defined as above. This measure is based on the probability that any two individuals randomly drawn from an infinite pool of species will be conspecifics.The metric D decreases as diversity increases. The Simpson index therefore is commonly presented as either the complement or the inverse of D. In practice, because we are not sampling from an infinite pool of species, a finite correction factor is applied in calculating D.

The Berger–Parker index is an intuitively appealing measure of species dominance in which we take the ratio of the most abundant species and the sum of all individuals of all species (N):

$$d = \frac{N_{\max i}}{N}$$

This measure is commonly represented as $1/d$ to reflect diversity rather than dominance.

species, but relative community composition, have been proposed (see Box 6.1). For a comprehensive account see Magurran (2004). These metrics can serve a particularly valuable role in monitoring community change in response to natural and man-made disturbances. As an example, temporal variation in species richness and three common measures of species diversity are shown in Figure 6.4 for Lake Mendota, Wisconsin. This lake is part of an on-going Long Term Ecological Research program (http://lter.limnology.wisc.edu) for north temperate lake ecosystems. Hansen and Carey (2015) documented changes in both phytoplankton and fish assemblages in this lake during 1995–2011. The number of fish species recorded ranged from 20 to 28 during this time period. Although sampling was carefully standardized, some of the interannual variation is likely attributable, however, to false negatives (Hansen and Carey 2015). For illustration we computed the Shannon, Simpson, and Berger–Parker indices using a selected subset of species that were present for at least 14 of the 17 years of sampling reported by Hansen and Carey (Figure 6.4). These indices show similar patterns over time.

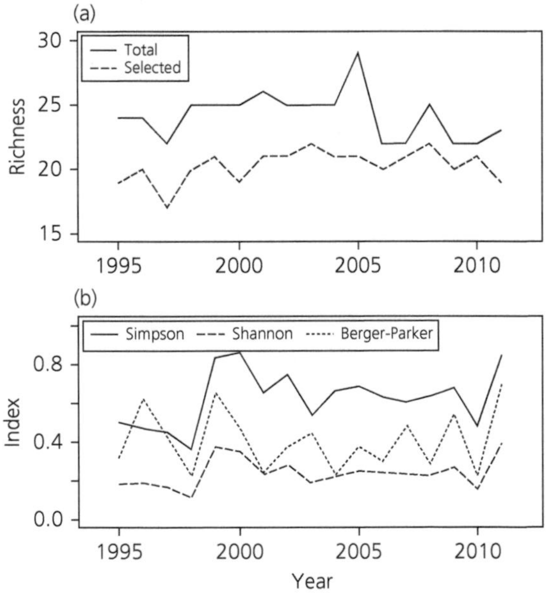

Figure 6.4 Measures of (a) species richness and (b) species diversity in Lake Mendota Wisconsin (data for species richness courtesy of North Temperate Lakes Long Term Ecological Research Program). Dashed line represents the subset of species included in the calculation of diversity indices in panel (b).

6.2.2 Keystone species and trophic cascades

A *keystone* species is one that has a disproportionately large effect in an ecosystem relative to its abundance. Keystone species serve a critical role in maintaining the structure of ecological communities. Perturbations affecting keystone species can have cascading effects throughout an ecosystem. Top predators often play a dominant role in aquatic communities and factors affecting these keystone species, including harvesting, can have large-scale repercussions for entire communities and ecosystems. The concept was first put forward by Robert Paine (Paine 1969) in his seminal work on tidepool communities and its importance has been widely recognized in other aquatic communities.

To illustrate this issue, we turn to the instructive example provided by Mittelbach et al. (1995, 2006) considering the consequences of extirpation of a keystone species in a lake ecosystem. Mittelbach et al. traced the ecosystem impact of the local extinction of a top predator, black bass (*Micropterus salmoides*), in Wintergreen Lake, Michigan following two consecutive severe winters in 1977 and 1978. These winterkill events resulted in the extirpation in the lake of both bass and an important planktivore, the bluegill (*Lepomis macrochirus*). Bass were reintroduced into the lake with the stocking of 700 juveniles in 1986 and standardized monitoring of the lake ecosystem was initiated. Bluegill remained absent from the lake until 1997 when 74 adults were deliberately released into the lake. The release of predation pressure on the remaining planktivores in the lake following the bass die-off resulted in high abundance levels of remaining members of this group (Mittelbach et al. 1995, 2006). The rapid resurgence of bass following its reintroduction precipitated a sharp decline in the planktivore assemblage (Figure 6.5). This change, in turn, resulted in a reorganization of the zooplankton community from one dominated by a guild of small-bodied cladoceran species under high planktivory to one dominated by large cladoceran species, particularly *Daphnia pulicaria*. During years dominated by the large-bodied *Daphnia* complex, phytoplankton abundance was reduced (as reflected in low chlorophyll levels) and water clarity was enhanced. This pattern affecting alternating trophic levels [high bass → low planktivores → high *Daphnia* → low phytoplankton] is called a *trophic cascade*. For a broad overview of trophic cascades in lake ecosystems see Carpenter and Kitchell (1996).

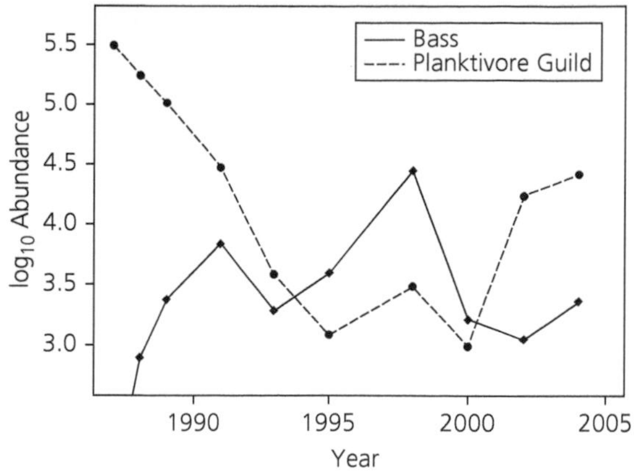

Figure 6.5 Estimated abundance of black bass and planktivorous fish in Wintergreen Lake Michigan (adapted from Mittelbach et al. 2005).

Parallel considerations have been evoked in marine ecosystems a result of fishing of top predators. Frank et al. (2005) documented changes in the Scotian Shelf large marine ecosystem off eastern Canada. Overfishing of a demersal fish community dominated by Atlantic cod (*Gadus morhua*) was linked to a dramatic increase in planktivorous fish and benthic macroinvertebrates, both of which are preyed on by the groundfish assemblage. Large-bodied calanoid copepods (notably *Calanus finmarchicus*) declined with the increase in the planktivorous fish guild. An increase in an index of phytoplankton abundance in turn was connected to the decline in large calanoid abundance. Again, a distinct alternating pattern emerges across trophic levels. Daskalov et al. (2007) provide another striking example of a trophic cascade, initiated in this case by overfishing of top predators coupled with the invasion of an alien ctenophore species in the Black Sea.

6.2.3 Guilds and functional groups

Many communities comprise large numbers of species, particularly in tropical ecosystems. It can be challenging to identify all the species, let alone elucidate and quantify their interactions. One prominent approach to dealing with complexity in ecological studies has been to focus on groups of species holding certain characteristics in common. Groups defined by taxonomic affinities or by their patterns of resource use are often used in this way. The former are sometimes referred to as *taxocenes*. Grouping species in *guilds* according to how they make a living (carnivores, herbivores, omnivores) and/or by their habitat (pelagic or benthic) is also commonly practiced. Species in the same guild often share certain characteristics, including size, trophic level, and morphological adaptations to feed on particular prey taxa. Thus they perform the same or similar roles in the community and are referred to as functional groups.

In these trophic webs, we expect potential competition for food resources *within* a guild and predator–prey interactions *among* certain guilds. Species within a guild are *fungible* to the extent that a decline in one species can be compensated by an increase in other species that play the same functional role. Compensation within functional groups on the northwest Atlantic shelf has been inferred from their stability over time (Auster and Link 2009).

The functional group approach is less useful when species can belong to more than one guild. Some species undergo ontogenetic shifts in their diets (Auster and Link 2009); others are omnivores. Either we must recognize the increased connectivity of the trophic web, or define a greater number of more exclusive guilds, with both choices increasing the resulting model complexity. For some purposes, grouping species by functional groups may not increase our understanding of community dynamics more than if the species were grouped at random (Rice et al. 2013). Many functional groups are dominated by a few species, such that analyses by functional group effectively describe the dynamics of the dominant species. In summary, the functional group approach is useful, but does not obviate the need to consider also the dynamics of individual species where possible. In Chapter 12 we will revisit the functional group concept in a management context.

6.2.4 Community compensation

In earlier chapters, we saw that compensatory processes are critical to the resilience of species to perturbations. Are there parallel considerations that are relevant to groups of species?

It has long been recognized that total overall landings from a fishery ecosystem may be considerably more stable than its parts in both freshwater (e.g. Regier and Hartman 1973) and marine (e.g. Sutcliffe et al. 1977) environments. Can the observation of relative stability in this important ecosystem service be attributed to the interplay of compensatory processes operating in these social-ecological systems? In unexploited communities there is evidence of compensatory processes operating at the community level (e.g. Tanner et al. 2009; Brown et al. 2016). We will refer to this as community compensation. The term density compensation is also used in this context (e.g. Tonn 1985; McGrady-Steed and Morin 2000). The underlying mechanism is resource limitation coupled with interspecific interactions, which then controls the carrying capacity of the overall ecosystem.

Do similar considerations hold for ecosystems that are strongly perturbed by various forms of human intervention? Walter and Hoagman (1971, 1975) provide abundance indices for nine key species in Green Bay, an embayment of northwestern Lake Michigan during 1929–69, a period during which a number of critical changes occurred. These encompassed the decline of lake trout (*Salvalinus namaycush*), the rise and ultimate control of sea lamprey (*Petromyzon marinus*), and culminated in an explosion of introduced plantivores [smelt (*Osmerus mordax*) and alewife (*Alosa pseudoharengus*)]. Despite wide fluctuations of individual species (Figures 6.6 a–c), the total abundance index remained remarkably stable during this four-decade period (Figure 6.6d). Walter and Hoagman (1971) reported evidence of significant biological interactions among a number of these species, suggesting that some form of compensatory process was operating. Subsequent events have continued the pattern of strong human intervention and there are now indications of reduced overall resilience. Important remedial actions have been undertaken.

6.2.5 Stability and complexity

In a highly influential paper, May (1972) examined the stability properties of model communities in

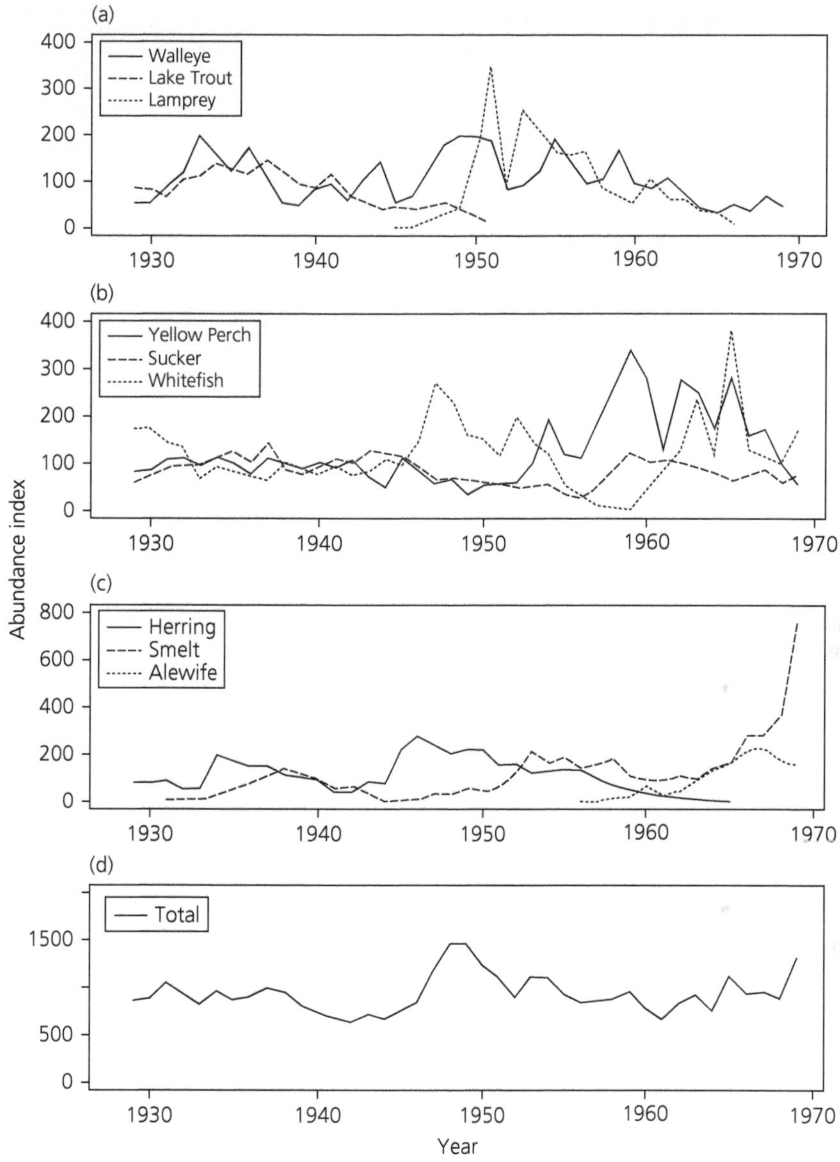

Figure 6.6 Abundance indices for (a–c) 9 species and (d) total abundance over a four-decade period in Green Bay in Lake Michigan (data provided in Walter and Hoagman 1971, 1975).

relation to their complexity. Earlier considerations of this question had focused on the potential benefits of higher species richness in a community when some members provide functional redundancy in key roles within the community, thereby providing a type of "insurance" effect. In May's simulations, complexity increased with the number of species and the number of interactions, or connectance, among them. He simulated randomly constructed communities and found that increasing the number of species, the degree of connectance, and the strength of the interactions, all decreased the probability that the system would be stable. It is important to note that in these simulations, some complex communities did persist. Subsequent research on this dynamic stability hypothesis considered food-web topology. Pimm and Lawton (1977) found that longer food chains have longer return times to equilibrium, implying that they would be less likely to persist in a fluctuating environment. Omnivory, leading to highly connected systems, was also generally found to be destabilizing. Gellner and McCann (2012) revisited the role of omnivory in system stability and noted that strong omnivory is destabilizing, leading to loss of species; low to moderate levels of interaction strength, in contrast, can permit species coexistence in complex food webs. Destabilizing forms of omnivory can operate either consistently or intermittently.

Can these simulation results be reconciled with the earlier views on the importance of species diversity in communities? In real ecological communities, we see the results of a filtered process in which species assemblages with low persistence have been eliminated. If initial community structures do involve random assemblages, a winnowing process presumably occurs and the filtered ensemble contains only the remaining resilient communities. We also see the outcomes of co-evolutionary processes in action. If the theory from model communities applies, we should expect communities with small numbers of species (such as the Baltic Sea and other boreal communities) to have high connectance with strong interactions. In contrast, communities with many species (e.g. the North Sea and other temperate fish communities) are expected to have low connectance or many weak interactions in order to be stable. Alternatively, a community with many species can have low overall connectance but some strong interactions among groups of species (May 1973), as in spatially structured tropical communities. These expected patterns are observed in many natural communities (McCann et al. 1998). A practical implication of this general result is that a given species may have strong interactions with a limited number of other species, which may limit the number of interaction terms that need to be estimated. On the other hand, if diffuse, indirect interactions are important, our ability to quantitatively predict the consequences of perturbations to food webs may be quite limited (Yodzis 2000).

A newer twist to this problem has emerged with the recognition that multispecies models comprising three or more species in continuous time can exhibit extraordinarily complex non-equilibrium behavior. As we have previously seen, even single-species models in discrete time can exhibit chaotic dynamics. For continuous-time models, chaotic dynamics generally do not emerge in models with less than three species. There is evidence that, for food chain models at least, the potential for complex dynamics increases with increasing numbers of species (see McCann 2012 for a comprehensive treatment of complex dynamics in food webs). In Section 6.4 we will take up the issue of dynamic complexity further.

6.3 Models of community dynamics

It is clear that we require a model structure that can accommodate different forms of interspecific interactions encompassing potentially very diverse assemblages of species. As in previous chapters, both continuous-time and discrete-time models have been developed and employed. Historically, a great deal of the treatment of multispecies models in theoretical ecology has been framed in continuous time (e.g. May 1973; Roughgarden 1975) with a strong focus on equilibrium analysis (most of these models cannot be solved analytically and numerical analysis is generally needed to examine dynamical patterns).

6.3.1 Continuous-time models

To start, we will retain the fundamental structure of the predation and competition models of Chapter 4. We begin with a Generalized Lotka–Volterra (GLV) system of equations:

$$\frac{dN_i}{dt} = \left[\alpha_i - \beta_i N_i \pm \sum_{j \neq i} a_{ij} N_j\right] N_i \qquad (6.1)$$

where all terms are defined as in Chapters 4 and 5 but are now extended to encompass an n-species community. This model can be expressed more compactly by recognizing that we can write the compensatory coefficient β_i (the effect of species i on itself) as a_{ii} giving:

$$\frac{dN_i}{dt} = \left[\alpha_i \pm \sum_{j=1}^{n} a_{ij} N_j\right] N_i \qquad (6.2)$$

Notice that here we are summing the interaction terms over all species pairs; this assumes that the effects of different interacting species on a focal species are independent. The sign of the coefficients, a_{ij}, indicates the general nature of the interaction (for caveats see Abrams 1987). For intra- and interspecific competition, the signs are negative. For a predator–prey pair, the prey exerts a positive effect on the predator through increased feeding, growth, and reproduction. Conversely, the predator exerts a negative effect on the prey by increasing its mortality rate. For predator–prey dynamics, we consider here only a linear functional feeding response, although later in the chapter we will expand our focus to encompass non-linear functional feeding responses for multispecies communities. Parasites exert a negative effect on their hosts while gaining benefits from the interaction. Mutualism is denoted by positive interaction signs for both species. Finally, we note that in Chapter 4 we concentrated on situations in which a predator derives energy for reproduction from a single prey. Here we have a case in which not only do multiple prey species contribute to predator reproduction but population growth of a predator is still possible in the absence of specified prey (if $\alpha_i > 0$). This implies that there can be an unspecified source of "other" prey that can provide energetic inputs supporting reproduction of a predator. At equilibrium, the model can be written in matrix form as:

$$\mathbf{\Lambda} + \mathbf{A} \cdot \mathbf{N} = 0 \tag{6.3}$$

where $\mathbf{\Lambda}$ is a vector of the intrinsic rates of increase, and $\mathbf{A}$ is a matrix of interaction terms. We are particularly interested in the matrix of interaction coefficients:

$$\mathbf{A} = \begin{pmatrix} a_{11} & a_{12} & \cdot & \cdot & \cdot & a_{1n} \\ a_{21} & a_{22} & \cdot & \cdot & \cdot & a_{2n} \\ \cdot & \cdot & \cdot & \cdot & \cdot & \cdot \\ \cdot & \cdot & \cdot & \cdot & \cdot & \cdot \\ \cdot & \cdot & \cdot & \cdot & \cdot & \cdot \\ a_{n1} & a_{n2} & \cdot & \cdot & \cdot & a_{nn} \end{pmatrix} \tag{6.4}$$

where the elements on the main diagonal are the intraspecific interaction terms, and all other terms represent the interspecific interactions. This is often referred to as the community matrix.

The vector of equilibrium population sizes is given by:

$$\mathbf{N}^* = -\mathbf{A}^{-1}\mathbf{\Lambda} \tag{6.5}$$

The equilibrium $\mathbf{N}^*$ is stable if all eigenvalues of $\mathbf{A}$ have negative real parts (Box 6.2). In this community context, stability means that all n species will persist over ecological timescales; otherwise one or more species will go extinct and the community will self-simplify.

Because we are now considering multiple types of interactions operating within communities we must consider their possible implications for community dynamics. Does the interplay of different interaction types overturn some of our previous conclusions concerning the outcomes of species interactions when examined separately as in Chapters 4 and 5? There are of course innumerable possibilities. Here we will focus on just two prominent examples that have been examined in some detail; both focus on the interplay of competition and predation in different forms. We will restrict our consideration to three-species systems for this purpose.

6.3.1.1 Intraguild predation

Species that derive energy over multiple trophic levels are very common. It is the exception rather than the rule that predators are restricted to one prey species or even prey type within a single lower trophic level. We have noted that omnivory can be destabilizing. Here we will focus on a special form of omnivory, *intraguild predation* (IGP), in which competitors also prey on each other (Polis et al. 1989; Holt and Polis 1997). This is common in different types of aquatic communities (e.g. Irigoien and de Roos 2011). In contrast to the case of a tri-trophic food chain in which an intermediate consumer species preys on a basal resource species and is in turn consumed by a top predator, the most basic form of IGP entails predation by the top predator on both the basal resource and intermediate consumer species. Here the top predator benefits by taking some prey directly from the basal resource without the attendant loss of energy involved in passing through the intermediate consumer. The predator of course is also then a competitor to the intermediate consumer. A particularly interesting and important case involves the implication of ontogenetic changes in diet for predation at different trophic levels. For example, most fish grow many orders of magnitude in size over their lifespan. The early planktonic life stages may feed on phyto- and/or zooplankton and

Box 6.2 Stability Analysis

To evaluate the stability of the generalized Lotka–Volterra Model:

$$\frac{dN_i}{dt} = \left[\alpha_i + \sum_{j=1}^{n} a_{ij}N_j\right] N_i$$

We first construct a matrix of partial derivatives of this function with respect to each of the individual species in the community (the Jacobian Matrix). For the case in which $i \neq j$, we have for the derivatives:

$$\frac{\partial dN_i/dt}{\partial N_j} = a_{ij}N_i$$

For the case in which $i = j$, the derivatives are:

$$\frac{\partial dN_i/dt}{\partial N_i} = \alpha_i + 2a_{ii}N_i + \sum_{j \neq i} a_{ij}N_j$$

For a three-species community, the Jacobian matrix, **J**, is constructed from the partial derivatives:

$$\mathbf{J} = \begin{pmatrix} \frac{\partial dN_1/dt}{\partial N_1} & \frac{\partial dN_1/dt}{\partial N_2} & \frac{\partial dN_1/dt}{\partial N_3} \\ \frac{\partial dN_2/dt}{\partial N_1} & \frac{\partial dN_2/dt}{\partial N_2} & \frac{\partial dN_2/dt}{\partial N_3} \\ \frac{\partial dN_3/dt}{\partial N_1} & \frac{\partial dN_3/dt}{dN_2} & \frac{\partial dN_3/dt}{dN_3} \end{pmatrix}$$

evaluated at the equilibrium abundances given by:

$$\alpha_i = -\sum_{j=1}^{n} a_{ij}N_j$$

Substituting for α_i above, the partial derivative for $i = j$ becomes:

$$\frac{\partial dN_i/dt}{\partial N_i} = -\sum_{j} a_{ij}N_j + 2a_{ii}N_i + \sum_{j \neq i} a_{ij}N_j = a_{ii}N_i$$

because the first summation contains one more term ($i = j$) than the second. The Jacobian matrix for our three-species community is:

$$\mathbf{J} = \begin{bmatrix} N_1^*a_{11} & N_1^*a_{12} & N_1^*a_{13} \\ N_2^*a_{21} & N_2^*a_{22} & N_2^*a_{23} \\ N_3^*a_{31} & N_3^*a_{32} & N_3^*a_{33} \end{bmatrix}$$

The equilibrium point defined by the vector $\mathbf{N}^*$ is locally stable if and only if all the eigenvalues of **J** have negative real parts.

then transition to larger prey species which in turn prey on planktonic organisms. Here, adult fish can prey on the competitors of their juveniles. Walters and Kitchell (2001) have labeled this pattern a *cultivation/depensation* effect in which adults improve the survival of their progeny by consuming their competitors. We have encountered foundational elements of this concept in Section 4.2.5.1 and it will figure prominently in Chapter 13 when we take up the commonly used Ecopath with Ecosim model.

The basic IGP model is a special form of the GLV system of equations:

$$\begin{aligned} \frac{dN_1}{dt} &= [\alpha_1 - \beta_1 N_1 - a_{12}N_2 - a_{13}N_3]\, N_1 \\ \frac{dN_2}{dt} &= [-\alpha_2 + a_{21}N_1 - a_{23}N_3]\, N_2 \\ \frac{dN_3}{dt} &= [-\alpha_3 + a_{31}N_1 + a_{32}N_2]\, N_3 \end{aligned} \tag{6.6}$$

where N_1 is the basal resource species, N_2 is the intermediate consumer, and N_3 is the top-level predator (see Tanabe and Namba 2005). Here, intraspecific compensation is confined to the basal resource species. Notice that for the intermediate consumer and the top predator, the intrinsic rate of increase is negative. Recall from Chapter 3 that $\alpha = b_0 - d_0$; in this model we are assuming that births come exclusively from energy derived from modeled prey and the "left-over" component of α is taken to represent mortality. This model can generate stable dynamics in which all three species persist, alternative stable states in which one species is eliminated, and extremely complex dynamics including chaos is possible for parts of the parameter space. We will defer consideration of complex dynamics until Section 6.4. Using parameters provided by Tanabe and Namba (2005), we see that altering the value of a_{31} (the energy derived by the predator from consuming the basal prey resource, here taken to range from 0 to 1) and holding all other parameters constant, can result in the outcomes described above. At $a_{31} = 0.7$ and given the other parameter choices, we have stable coexistence of all three species (Figure 6.7a) following the transient behavior related to the

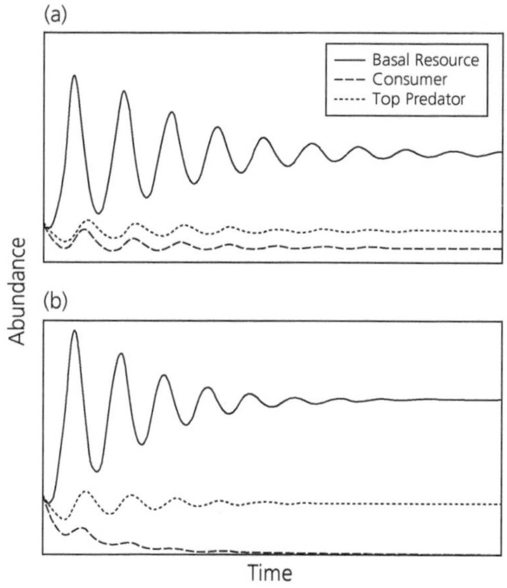

Figure 6.7 (a) Stable coexistence in a three-species intraguild predation system and (b) extinction of the intermediate consumer with an increase in the efficiency of prey utilization by the top predator. Model structure and parameter estimates from Tanabe and Namba (2005).

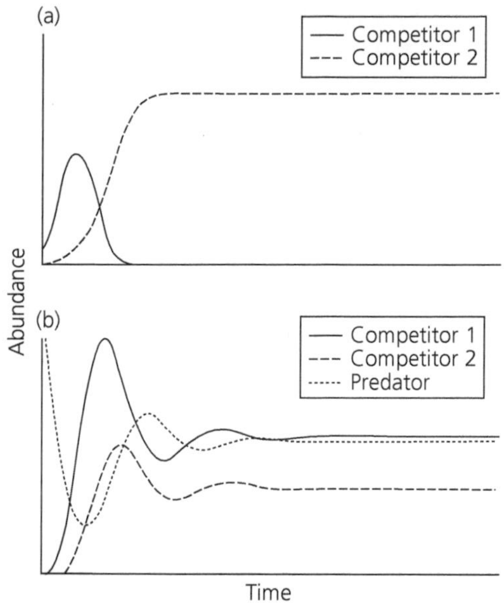

Figure 6.8 (a) Competitive exclusion in a two-species system using the model structure and parameters of Parrish and Saila (1970) and (b) effect of adding a predator resulting in stable equilibria for all three species.

initial conditions. However, increasing a_{31} to 0.9 is sufficient to eliminate the intermediate consumer and the new stable state now contains just the basal resource and the top predator (Figure 6.7b).

6.3.1.2 Competition-predation

We next consider the case in which predation can overturn instances of competitive exclusion. In this case, the predator consumes two prey items and therefore differs from the IGP case in involving only two trophic levels. We introduced this topic in Section 5.2.6. It is an issue of particular relevance because in Part III of this book we will be considering humans as predators and are interested in understanding the direct and indirect pathways through which we can alter exploited aquatic communities.

In an early treatment, Parrish and Saila (1970) examined the interface among predation, competition, and species diversity. Again, a special case of the GLV system of equations was used:

$$\begin{aligned}\frac{dN_1}{dt} &= [\alpha_1 - a_{11}N_1 - a_{12}N_2 - a_{13}N_3]\,N_1,\\ \frac{dN_2}{dt} &= [\alpha_2 - a_{22}N_2 - a_{21}N_1 - a_{23}N_3]\,N_2 \qquad (6.7)\\ \frac{dN_3}{dt} &= [-\alpha_3 - a_{33}N_3 + a_{31}N_1 + a_{32}N_2]\,N_3\end{aligned}$$

where N_1 and N_2 are competitors and N_3 is the predator. In this model, the intrinsic rate of increase for the top predator is negative and it depends on the energy derived from species N_1 and N_2 for population growth. Parrish and Saila made numerical studies of the case where in the absence of predation, one competitor species would exclude the other. For the parameter choices made by Parrish and Saila, it was shown that the persistence of the inferior competitor could be considerably extended with the addition of predation on both competitors. Cramer and May (1972) subsequently identified parameter sets in which stable equilibria of all three species could be attained. In Figure 6.8a we show the case of the two competitors in the absence of predation; Species 1 goes extinct for the set of parameters used in this simulation (Cramer and May 1972). Addition of a predator in Figure 6.8b results in stable equilibria for all three species.

6.3.1.3 Non-linear predation

In the preceding sections we considered only linear functional feeding responses. This allowed us to represent competition and predation in a structurally similar way in which the sign of the interaction coefficient, a_{ij}, conveyed information on the nature of the interaction (and potentially others such as mutualism). However, as we saw in Chapter 4, alternative forms of the functional feeding response must be considered. Most of the ecological literature employing non-linear functional response terms in multispecies models has focused on three-species systems. For example, Hastings and Powell (1991) examined the possibility of complex dynamics in a food chain comprising a basal resource species (N_1), an intermediate consumer (N_2) and a top predator (N_3) of the form:

$$\frac{dN_1}{dt} = \left[\alpha_1 - a_{11}N_1 - \frac{\omega_1 N_2}{\delta_1 + N_1}\right] N_1,$$
$$\frac{dN_2}{dt} = \left[c_2 \frac{\omega_1 N_1}{\delta_1 + N_1} - \frac{\omega_2 N_3}{\delta_2 + N_2} - d_2\right] N_2 \qquad (6.8)$$
$$\frac{dN_3}{dt} = \left[c_3 \frac{\omega_2 N_2}{\delta_2 + N_2} - d_3\right] N_3$$

where α_1 is the intrinsic rate of increase of Species 1 and a_{11} is an intraspecific interaction term; the ω_i and δ_i are coefficients of the functional feeding response of species i; the c_i are prey conversion efficiencies for species i; the d_i are mortality terms due to causes other than predation. Recall that we derive the functional response terms from fundamental consideration of search and handling time in the predation process. The form used in the Hastings–Powell model above can be directly related to these considerations.

Hastings and Powell (1991) were particularly interested in the conditions under which chaos emerges in this model. In this section, we provide an example in which stable equilibria for each of the three species is attained for certain choices of the parameters. We retain the same parameters used by Hastings and Powell but reduce the magnitude of the δ_1 coefficient. After an initial period of damped fluctuations, the species trajectories converge on fixed abundance levels (Figure 6.9).

For the case in which a predator exploits multiple prey types, we require a generalization of the functional feeding response term that reflects the fact that the saturation level is determined by multiple prey species which, in effect, attract the attention of the predator. We can recast the basic non-linear functional feeding response term to account for this factor and adopt the notation used in this chapter. For the multispecies analog of a Type-II functional feeding response we can write:

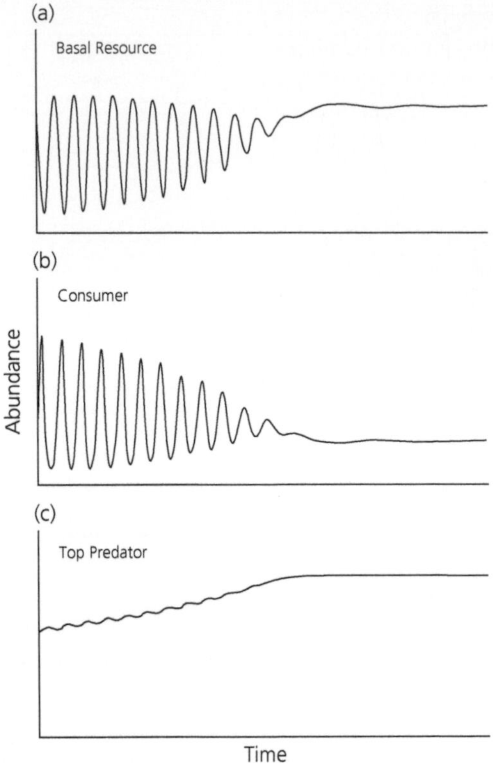

Figure 6.9 Stable coexistence in the three-species food chain model of Hastings and Powell (1991).

$$\left[\frac{a_{ij}N_j}{1 + \sum_j d_{ij}N_i}\right] \qquad (6.9)$$

where a_{ji} is the effect of prey species i on predator species $_j$ and the d_{ji} represent predator preference for different prey species. For a broader generalization of the multispecies disc equation, see Koen-Alonso and Yodzis (2005).

6.3.2 Discrete-time models

Perhaps the mostly commonly used first-order difference equation analog to the continuous-time GLV system of equations employs the standard Euler approximation:

$$N_{i,t+1} = N_{i,t} + \alpha_i N_{i,t} \pm \sum_{j=1}^{n} a_{ij} N_{j,t} N_{i,t} \qquad (6.10)$$

where, as before, we have retained the same symbols as for the corresponding continuous-time model with the understanding that they now represent finite rates of change.

An alternative representation based on first expressing the differential equation in terms of the relative derivative is:

$$N_{i,t+1} = N_{i,t} e^{\left(\alpha_i \pm \sum_{j=1}^{n} a_{ij} N_{j,t}\right)} \tag{6.11}$$

and again this version holds the advantage that it cannot result in negative population sizes (see Chapter 3).

To illustrate the use of the GLV model we return to the cod–herring–sprat complex in the Baltic Sea. These species have well-defined seasonal spawning cycles appropriate for the use of a model of community dynamics in discrete time. For this application we have used Equation 6.9. Previous applications of multispecies production models to this species complex have been provided by Sullivan (1991). Horbowy (2005) and Bauer et al. (2019) applied a production model of community dynamics in continuous time with explicit consideration of individual growth to these species. The original model formulation was expressed in terms of biomass rather than numbers because of the explicit fishery context in each of these cases.

These three species are of course exploited, and in our illustrative analysis using population numbers we have accounted for the removals by simply subtracting the catch in number for each species from the equation for that species. In Chapter 12 we will treat the issue of accounting for exploitation in much greater detail. The Baltic Sea model can be written:

$$
\begin{aligned}
N_{1,t+1} &= N_{1,t} + \left[\alpha_1 - a_{11}N_{1,t} - a_{12}N_{2,t} - a_{13}N_{3,t}\right] N_{1,t} \\
N_{2,t+1} &= N_{2,t} + \left[\alpha_2 - a_{22}N_{2,t} - a_{21}N_{1,t} - a_{23}N_{3,t}\right] N_{2,t} \\
N_{3,t+1} &= N_{3,t} + \left[\alpha_3 - a_{33}N_{3,t} + a_{31}N_{1,t} + a_{32}N_{2,t}\right] N_{3,t}
\end{aligned} \tag{6.12}
$$

where sprat, herring, and cod are Species 1, 2, and 3, respectively. The overall model fits are highly statistically significant (see Figure 6.10). Cod is found to exert statistically significant effects on herring and sprat. In contrast, no statistically significant effects of either herring or sprat on cod were evident in this analysis, although in both instances the signs of the coefficients were negative rather than positive. These pelagic fish are known to prey on the eggs and larvae of cod. Interestingly, the effect of sprat on herring was significant and positive. This suggests the possibility of an indirect effect in which circumstances favoring predation on sprat results in reduced predation on herring. Sullivan (1991) reported a similar result.

We note that, in general, simulation studies have indicated that when environmental stochasticity and/or observation error is present, it can be quite difficult to determine the interaction strength of biological interactions using empirical abundance or biomass data (e.g. Oken and Essington 2015). The lesson of course is not that the potential for real biological interactions can be ignored. Wherever possible, auxiliary knowledge of interspecific interactions should be employed to help inform the structure of the model to be tested. In some instances, this information may be sufficient to set prior distributions on the interaction terms in a Bayesian analysis.

In the Baltic example, we employed linear interaction terms that could represent different types of interactions within a single, simple framework. In instances where non-linear functional predation responses involve multiple prey taken by a predator, we need an alternative structure. In this case, the Baltic Sea model could take the form:

$$
\begin{aligned}
N_{1,t+1} &= \left[(1+\alpha_1) - a_{11}N_{1,t} - a_{12}N_{2,t} - \frac{a_{13}N_{3,t}}{1+d_{13}N_{1,t}+d_{23}N_{2,t}}\right] N_{1,t} \\
N_{2,t+1} &= \left[(1+\alpha_2) - a_{22}N_{2,t} - a_{21}N_{1,t} - \frac{a_{23}N_{3,t}}{1+d_{13}N_{1,t}+d_{23}N_{2,t}}\right] N_{2,t} \\
N_{3,t+1} &= \left[(1+\alpha_3) - a_{33}N_{3,t} + \frac{a_{31}N_{1,t}+a_{32}N_{2,t}}{1+d_{13}N_{1,t}+d_{23}N_{2,t}}\right] N_{3,t}
\end{aligned} \tag{6.13}
$$

where we have inserted a functional response based on Equation 6.9 into the production model. This form allows for competition between herring and sprat and incorporates joint predation on both pelagic species by cod.

6.3.2.1 Multispecies delay–difference models

As in the single-species case, we can introduce a form of age-structure through a simple modification of the multispecies production model described above. This change is important because there is a clear size dependence in predation processes. Most aquatic predators are gape-limited and prey on species substantially smaller than themselves. Size-dependent predation can hold very important

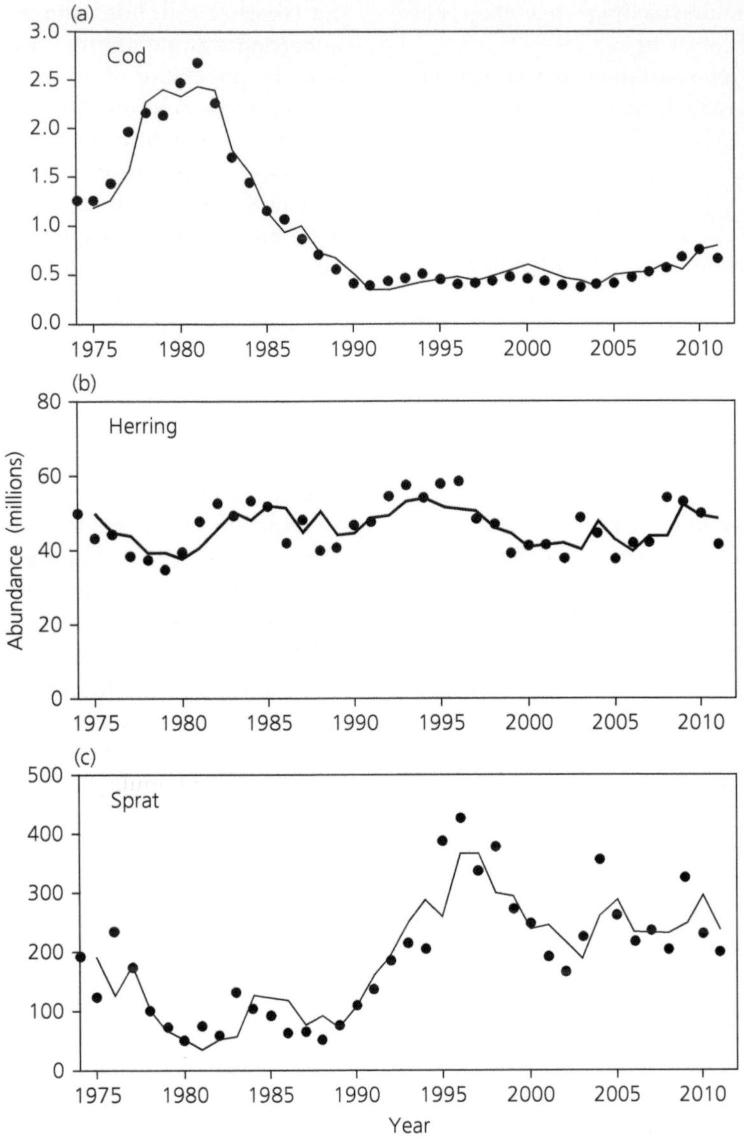

Figure 6.10 Observed and predicted population trajectories for (a) cod (*Gadus morhua*), (b) 1974 data point for herring (*Clupea harengus*), and (c) sprat (*Sprattus sprattus*) in the Baltic Sea using a generalized Lotka–Volterra model in discrete time.

implications for community dynamics through intraguild predation and other mechanisms. The general form of the model is:

$$N_{i,t+1} = s_{i,t}N_{i,t} + R_{i,t+1} \tag{6.14}$$

where $s_{i,t}$ is the annual survival rate of post-recruit individuals of species i, $R_{i,t+1}$ is the number of recruits at age r of species i at time $t_{i,t+1}$. Both the survival and recruitment terms incorporate multispecies interactions as described below (e.g. Collie and DeLong 1999). Basson and Fogarty (1997) explored the dynamical behavior of multispecies delay–difference models of this form incorporating only predation.

The post-recruit survival term now reflects two components—mortality due to interspecific interactions and to all other sources of natural mortality. The survival rate can be represented as:

$$s_{i,t} = e^{\left(-M1_i \pm \sum_{j=1}^{m} a_{ij} N_{j,t}\right)} \quad (6.15)$$

where $M1_i$ is the natural mortality rate of post-recruits from all natural sources other than inter-specific interactions and a_{ij} is the again the effect of species j on species i. The sign of the coefficient is negative for competition and the effect of a predator on a prey species; it is positive for the effect of a prey species on a predator.

Here, we apply a multispecies recruitment function as an extension of the Ricker model (e.g. Hilborn and Walters1992):

$$R_{i,t} = a_i N_{i,t-r} e^{\left(-b_i \pm \sum_{j=1}^{n} a_{ij} N_{i,t-r}\right)} \quad (6.16)$$

where a_i and b_i are intraspecific stock-recruitment parameters, and a_{ij} is the interspecific interaction-term.

6.4 Complex dynamics

For simple one-dimensional continuous-time models, the introduction of time delays or seasonality can result in complex dynamical behaviors, but chaotic behavior can generally only emerge in systems of three or more coupled differential equations characterized by sensitive dependence on initial conditions (e.g. Hastings et al. 1993). Lorenz's (1963) demonstration of chaos in simple models of atmospheric dynamics comprising three state variables was the first to show this dynamical behavior in continuous time models.

The dynamical properties of both the IGP model of Tanabe and Namba (2005; Equation 6.5) and food-chain model of Hastings and Powell (1991; Equation 6.7) described in Section 6.3 have been extensively examined and shown to exhibit chaotic dynamics for parts of the parameter space. McCann and Yodzis (1994) provided a re-parameterization of the Hastings–Powell model, which confirmed the possibility of complex dynamics with biologically realistic parameters based on allometric energetic principles. In Figure 6.9, we showed population trajectories that fall within the stable range of community dynamics.

By slightly increasing the parameter a_{31} in the Tanabe–Namba IGP model, we obtain bounded fluctuations in population levels of all three species (Figure 6.11 a–c). If we plot the trajectories of the species in three-dimensional space, we find that the highly irregular time series we observe resolve

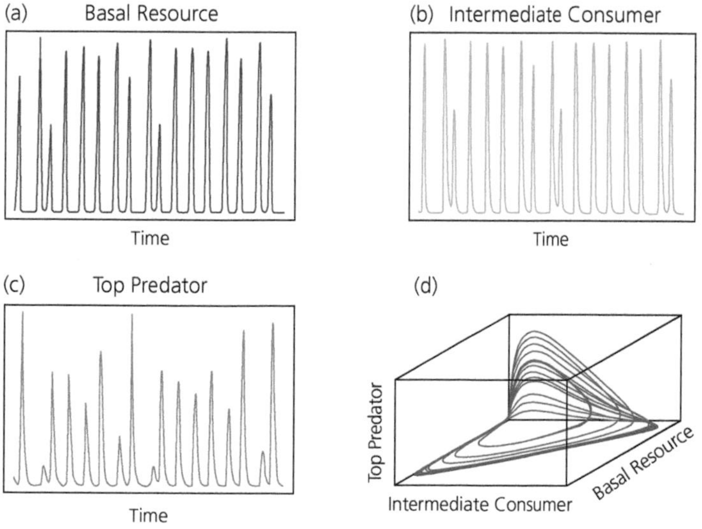

Figure 6.11 Complex dynamics in the intraguild predation model of Tanabe and Namba (2005) showing sbundance of all three species (a–c), and (d) the state-space representation of this three-species system. Adapted from Fogarty et al. (2016)

themselves into a well-defined geometrical object known as a "strange attractor" showing a hidden order underlying the population fluctuations (see Figure 6.11d).

In Chapter 14 we will show how we can construct "shadow" attractors based on empirical time series information to construct non-parametric models of complex systems and to develop short-term projections.

6.5 Size-spectrum models

In aquatic food webs, the trophic level of an organism is strongly related to its body size. Many vital processes scale with body size, including metabolism, predator–prey relationships, maturity, and fishing selectivity. Grouping individuals by body size (independent of species) therefore provides a complementary approach to trophic or functional groups. Calculations based on metabolic scaling relationships have been used to derive the expected distribution of biomass by size class (Kerr 1974). Size spectra, expressed as numbers (N) in geometric weight (w) intervals typically have slopes ≈ -1 (Kerr and Dickie 2001, Yurista et al. 2014), which implies that biomass (wN) is constant across size intervals (Sheldon et al. 1972).

We assume a power relationship between population numbers and individual body weight: $N = k_1 w^{-b}$, where k_1 is a constant and the exponent $b \approx 2$, which corresponds to a spectrum with slope $= -1$. Substituting $w = aL^3$ gives the analogous expression for body length (L): $N = k_2 L^{-3b}$, where $k_2 = k_1 a$. The total number in the length interval from L to mL can be derived by integration:

$$N_{tot}(L, mL) = \int_{L}^{mL} k_2 L^{-3b} dL = \left[\frac{k_2 L^{1-3b}}{1-3b} \right]_{L}^{mL} = \frac{k_2 \left(m^{1-3b} - 1\right) L^{1-3b}}{1-3b} \quad (6.17)$$

where m is a constant multiplier that defines geometric length classes. Taking logarithms, Equation 6.18 can be linearized:

$$\log\left(N_{tot}(L, mL)\right) = \log\left(\frac{k_2\left(m^{1-3b} - 1\right)}{1-3b}\right) + (1-3b)\log L \quad (6.18)$$

to obtain a linear size spectrum with slope $(1-3b)$. If $b = 2$, the expected slope is -5. With field data, we often don't know the length of each individual; instead, individuals are grouped in length classes of equal width (e.g. 5 cm). Box 6.3 shows how to construct a size spectrum from such length data.

We test these predictions with length-frequency data from standardized bottom-trawl surveys (Figure 6.12). In this example the data were grouped by 5-cm intervals, so it was necessary to correct the numbers as in Box 6.3. The slope of the North Sea size spectrum, -4.95, is very close to the predicted value of -5. In contrast, the Georges Bank spectrum is less steep, with a slope of -4. These size spectra illustrate important differences between the Georges Bank and North Sea fish communities, as noted by previous authors. The North Sea fish community is characterized by more numerous smaller individuals. In contrast, the Georges Bank has a broader spectrum of sizes, with more larger individuals and high abundance of intermediate sizes. The steeper slope for the North Sea has been attributed to higher fishing pressure and higher predation mortality on small-sized fish (Murawski and Idoine 1992).

When interpreting the slopes, one must bear in mind that observed spectra are often non-linear because all sampling gear is size specific,

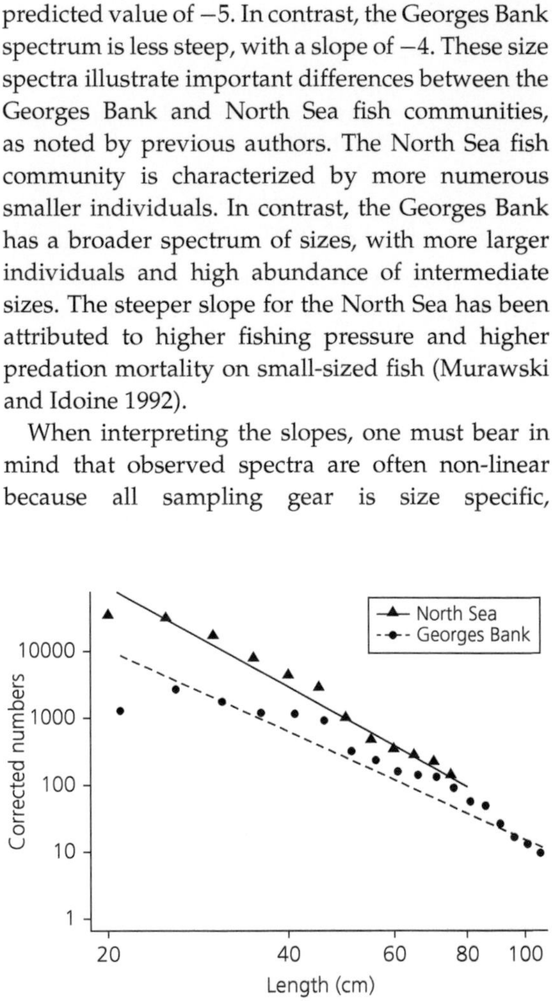

Figure 6.12 Size spectra of two temperate fish communities. The Georges Bank data (circles) are from the NEFSC autumn bottom-trawl survey, averaged over 1963–1972. The North Sea data (triangles) are from the International Bottom Trawl Survey, 1980–2004. The data were grouped by 5-cm length classes and corrected as in Box 6.3. Solid lines are the fitted regressions.

Box 6.3 Constructing a Size Spectrum from Length–Frequency Data

If length data are collected in intervals of equal width, the size spectrum (Equation 6.18) needs correcting to account for this difference in size intervals. Integrating numbers as Eq.6.18 but with length intervals (L_1,L_2) of equal width gives:

$$N_{tot}\,(L_1, L_2) = \int_{L_1}^{L_2} k_2 L^{-3b} dL = \left[\frac{k_2 L^{1-3b}}{1-3b}\right]_{L_1}^{L_2}$$

$$= \frac{k_2\left({L_2}^{1-3b} - {L_1}^{1-3b}\right)}{1-3b} = \frac{\left(\left(\frac{L_2}{L_1}\right)^{1-3b} - 1\right)}{1-3b} k_2 {L_1}^{1-3b}$$

The first factor on the right-hand side can be used to correct the numbers as:

$$\frac{N_{tot}\,(L_1, L_2)\,(1-3b)}{\left(\left(\frac{L_2}{L_1}\right)^{1-3b} - 1\right)} = k_2 {L_1}^{1-3b} \qquad (6.17)$$

Log transformation of these corrected numbers again gives a linear size spectrum with slope $1-3b$. The parameters b and k_2 can be estimated iteratively by linear regression or by non-linear regression.

communities are not at equilibrium, and because the spectrum within a trophic group may not be linear. In both ecosystems, the smallest and largest size classes appear under-represented, even though we limited the analysis to fish > 20 and < 130 cm. The Georges Bank spectrum, in particular, appears to be curved convex upwards. Size spectra within trophic groups have been reported to be dome-shaped (Kerr and Dickie 2001). In the fish size spectra, there is a suggestion of two domes with an inflection point around 50 cm (Figure 6.12). Most of the fish larger than 50 cm are piscivorous, such that these size spectra encompass at least two trophic levels.

6.6 Qualitative modeling approaches

A complete specification of the interplay among components of an ecosystem can be extremely difficult because of the large number of interactions to be quantified (scaling as the square of the number of system components) and uncertainties in the functional forms of some of the linkages. It is possible, however, to infer some important properties of the system with qualitative modeling approaches based on certain results in graph theory. These properties include the indirect effects of species on each other and the overall stability of the community. We find this approach particularly appealing because it starts with specifying a conceptual model of the system—an important initial step in any circumstance. This conceptual model can be expressed as a graph connecting interacting components of the system. The more complicated the system, the more valuable we feel this initial step will prove. In effect, it allows us to depict our current level of understanding of the system in a very intuitive way and to identify areas of uncertainty or the specification of alternative hypotheses of system structure encapsulated in alternative graphical representations.

One of the earliest applications of graph theory to aquatic ecosystems was by Parrish and Saila (1970) who extended the theory to encompass exploited systems. Here we describe the principal elements of graph theory as applied to multispecies communities but will return to its application in the qualitative modeling of marine ecosystems in Chapter 12. We begin with simple qualitative considerations of the interactions (positive, neutral, or negative) among pairs of species.

Consider the Baltic Sea example of one predator with two prey species examined earlier (Figure 6.13). In this case the two prey species don't interact directly but they affect the growth rate of each other through their common predator. The sign of the interaction between two species can be evaluated by multiplying the signs of the interactions joining them through the food web. A positive product indicates a positive indirect interaction—a negative product, a negative one. In Figure 6.13 the indirect effect of the prey species on each other is (−)(+) = (−), making them apparent competitors. Hence a reduction in the abundance of herring would be expected to enhance the population growth rate of sprat, and vice versa.

We now formalize these concepts to allow an analysis of qualitative stability of the system. First we require some basic definitions (Yodzis 1989). The nodes of a graph are designated vertices. The links (or lines) connecting vertices are called edges. We define a path as one or more connecting links starting at one variable and ending in a second vertex such that it does not traverse any vertex twice. The length of a path is defined by the number of vertices along the path. A loop is any path that returns to its starting vertex without crossing any intermediate vertices more than once. Two loops are considered to be conjunct if they share at least one element in common. Disjunct loops have no elements in common.

It is possible to represent the community matrix described earlier in this chapter as a directional graph (or a signed-digraph). The general approach has been called "loop analysis" (Levins 1974, 1975; Puccia and Levins 1985). Methods of qualitative stability analysis examine both the direction (sign) of the interactions and their magnitude. These stability analyses assume equilibrium conditions and the stability of the system at equilibrium depends on the nature of the feedbacks at each

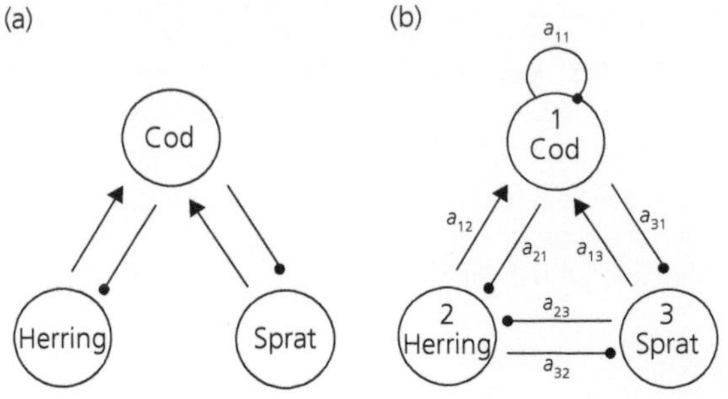

Figure 6.13 Loop diagram for a simple food web with one predator and two prey species patterned after the Baltic Sea cod–herring–sprat community. The pointed arrows indicate positive effects of prey on their predators. Circles indicate the negative effects of predators on their prey or competitive interactions. In (b) each link is labeled with a coefficient, which corresponds to an element in the community matrix. The self loop (a_{11}) indicates density-dependence.

Box 6.4 Determining Feedback in Qualitative Models

Here we illustrate the steps involved in determining feedback and stability. We will use a simple three-species food web comprising a predator with two prey species. The Baltic Sea model will serve as a template. This case includes direct competition between herring and sprat and compensatory control of cod by cannibalism. The feedback at the first (individual species) level is the sum over all loops of length 1:

$$F_1 = (-1)^2 L\,(1,1) = -a_{11}$$

which of course is stable (all other self-loops are zero). The feedback at the second level ($k = 2$) is:

$$F_2 = (-1)^2 L\,(1,2) + (-1)^3 L\,(2,2)$$

$L(1,2)$ is the sum over all pairwise interactions between species. $L(2,2)$ is the sum of all possible products of two links that form two disjunct loops. Since there is only one self-loop in this example $L(2,2) = 0$. Therefore:

$$F_2 = (-1)^2[(a_{12})(-a_{21}) + (a_{13})(-a_{31}) + (-a_{23})(-a_{32})]$$

or

$$F_2 = -(a_{12}a_{21}) - (a_{13}a_{31}) + (a_{23}a_{32})$$

which indicates that the second level will be stable if the feedback generated by competition between the two prey species is less than the effects of the predator on the prey. The feedback at the third level ($k = 3$) is:

$$F_3 = (-1)^2 L\,(1,3) + (-1)^3 L\,(2,3) + (-1)^4 L\,(3,3)$$

$L(1,3)$ is the sum of all loops of length three. $L(2,3)$ is the sum of all possible products of two disjunct loops, in which one of the loops is of length 2 and the other is a self loop. Finally $L(3,3) = 0$ because there are not three disjunct loops in this example. Therefore:

$$F_3 = (-1)^2[(-a_{23})(-a_{31})(a_{12}) + (-a_{21})(a_{13})(-a_{32})] + (-1)^3[(-a_{11})(-a_{23})(-a_{32})]$$

or

$$F_3 = (a_{23}a_{31}a_{12}) + (a_{21}a_{13}a_{32}) + (a_{11}a_{23}a_{32})$$

Since feedback at level 3 is positive, the overall system is unstable. In general, predator–prey interactions tend to stabilize communities, whereas competition tends to destabilize them (Li and Moyle 1981).

level (see Puccia and Levins 1985 for an overview). Recall that the negative feedback inherent in compensatory processes was critical to stability at the population level. Loop analysis extends these considerations to the community level. For the system to be stable, the overall feedback must be negative at each level in the system. In general, the feedback at each level can be defined as:

$$F_k = \sum_{m=1}^{k} (-1)^{m+1} L\,(m, k) \qquad (6.19)$$

where $L(m,k)$ is the sum over all possible products of k links that form m disjunct loops. For a stable system, we require that the feedback at all levels must be negative and that the feedback at lower levels be greater than that at higher levels (Puccia and Levins 1985). In Box 6.4 we work through the steps in computing the feedbacks and stability based on the Baltic Sea example.

Loop analysis has been used to evaluate the consequences of species introductions in oligotrophic lakes (Li and Moyle 1981). Such lakes have relatively simple communities that are supported by low, stable nutrient bases. Introduction of the peacock bass (*Cichla ocellaris*) into Lake Gatun in Panama destabilized the community, resulting in the local extirpation of 11 species of native fish. Loop analysis showed that these communities are destabilized by increased competition for a limited nutrient base (Li and Moyle 1981).

6.7 Summary

Ecological theory indicates that increasing the number of species, the number of interactions, and the strength of these interactions all tend to make communities less stable. Conversely, stability is enhanced by strong intra-specific density dependence, low connectivity, or weak trophic links. These theoretical predictions are borne out in many fish communities. Organizing species into groups according to size, function, or diet composition can reduce the dimensionality of fish community models. Analyses of fish communities from around the world lend support to the prediction of strong compensation within functional groups, with weaker predator–prey links among groups. Size spectra describe the distribution of individuals across size classes irrespective of their species. Allometric scaling relationships can be used to predict the slope and intercepts of size spectra. The slope of the size spectrum has been proposed as a community metric, but its application is complicated by the fact that many observed spectra are curvilinear. Qualitative approaches can be used to assess the indirect effects of species on each other and the overall stability of the community. However, as the connectance of the food web increases, qualitative analyses become indeterminate, unless the relative strengths of these interactions are known. As a result, the effect of perturbing one element of a food web on the other components is often unpredictable.

Additional reading

Topics covered in this chapter are reviewed in Pielou (1969; Chapter 18), Puccia and Levins (1985), Yodzis (1989; Chapter 7), Hilborn and Walters (1992; Chapter 14), Case (2000; Chapter 15), Magurran (2004), Walters and Martell (2004; Chapter 11), Stevens (2009; Chapters 7 and 10), and McCann (2012; Chapter 5). Morin (1999) and Mittelbach (2012) offer comprehensive treatments of the field of community ecology.

CHAPTER 7

Spatial Processes

7.1 Introduction

Up to this point, the population processes, and the models we use to describe them, do not directly account for the spatial structure of the population(s). It is well known that aquatic populations exhibit heterogeneous spatial structures as a result of specific habitat requirements, schooling and other behavior patterns, and ecological factors such as the presence or absence of predators. Geographic subareas may act as sources of new recruits or sinks where mortality dominates (Kritzer and Sale 2006). Are the simpler models considered previously necessarily inappropriate? We shall see that the answer to this question depends greatly on the rates of exchange between subgroups. High rates of exchange result in populations that can generally be treated effectively as homogeneous units. In contrast, populations that are characterized by low to moderate rates of exchange may require explicit consideration of spatial structure. The increasing use of spatial management strategies (long-term closure areas, Aquatic Protected Areas, etc.) motivated the extension of theory for spatially structured populations and the development of spatially explicit models (Collie et al. 2014). In this chapter we consider spatial patterns and processes, and the development of spatially explicit population models. We will take up the topic of spatial management strategies in Section 15.3.2.

It will be useful to introduce some definitions before proceeding. As we noted earlier, a *population* is a self-reproducing group of conspecific individuals that inhabit the same geographical area and are reproductively isolated from other populations. A *subpopulation* is a semi-independent, self-reproducing group of individuals with limited but measurable exchange of individuals with other components of a population. A *metapopulation* comprises a number of subpopulations distributed over space, linked through dispersal processes. Mixing of subpopulations can occur by advection of the egg and larval stages and by active migration of juveniles and adults. A *spawning component* is a segment of a population that occupies a distinct spawning area at a particular time of year. *Philopatry* is the tendency of an animal to remain in or return to the area of its birth. In the fisheries literature, the term *stock* is often taken to be synonymous with subpopulation. However, it is also sometimes defined more broadly as a resource exploited by a particular fishery in space and time. In this context, it may therefore represent a pragmatic simplification of more formal population-related concepts.

7.1.1 Patterns of distribution and abundance

Populations with patchy distributions occur at a gradient of densities from high to low. How will the spatial distribution change as total population size varies? Will the population distribution contract toward its center of gravity, or will high-density areas be hollowed out like a donut? Answering these questions is important, especially for harvested populations. A population that contracts as its abundance is reduced by harvesting may still occur at high densities in its core habitat. As we shall see in Chapter 11, this can present difficulties in correctly interpreting information on catch rates in exploited populations.

It is also important not to confound abundance–area relationships with distribution shifts due to climate change. It can be difficult to disentangle

Fishery Ecosystem Dynamics. Michael J. Fogarty and Jeremy S. Collie, Oxford University Press (2020). © Michael J. Fogarty & Jeremy S. Collie 2020.
DOI: 10.1093/oso/9780198768937.003.0007

the two. If the effect of climate change is primarily on a species' productivity and total abundance, the distribution of a cold-water species would contract under warming conditions and vice versa for a warm-water species.

7.2 Spatial distribution of single populations

Spatial distribution patterns of aquatic organisms can vary substantially over time in response to the interplay of a potentially large number of factors including resource availability, climate change, and fishing pressure. To set the stage, in Figure 7.1 we show the abundance of adult Atlantic cod in the North Sea for two selected years demonstrating very different distribution patterns as represented in research vessel surveys. In 1983, cod were widely distributed throughout the North Sea (Figure 7.1a). However, by 2017, their distribution in the southernmost part of the range was sharply diminished (Figure 7.1b) although cod were still found throughout the survey area. External drivers such as exploitation and climate change are known to have affected cod distribution and abundance in the North Sea (Perry et al. 2005). However, a broad spectrum of anthropogenic and ecological factors, both internal and external to the population, are likely to contribute to observed distribution patterns. Here, we will first explore some empirical measures of distributional change in aquatic populations. We then turn to models incorporating spatial structure.

7.2.1 Measures of distribution and dispersion

A number of measures of distribution and dispersion have been proposed to summarize the spatial characteristics of populations. These provide convenient metrics with which to measure changes over time. Rindorf and Lewy (2012) examined the sampling properties of several commonly employed indices based on abundance and distribution patterns in aquatic ecosystems including (a) the area occupied by the population, (b) Lloyd's index of patchiness (Lloyd 1967), (c) Gini index based

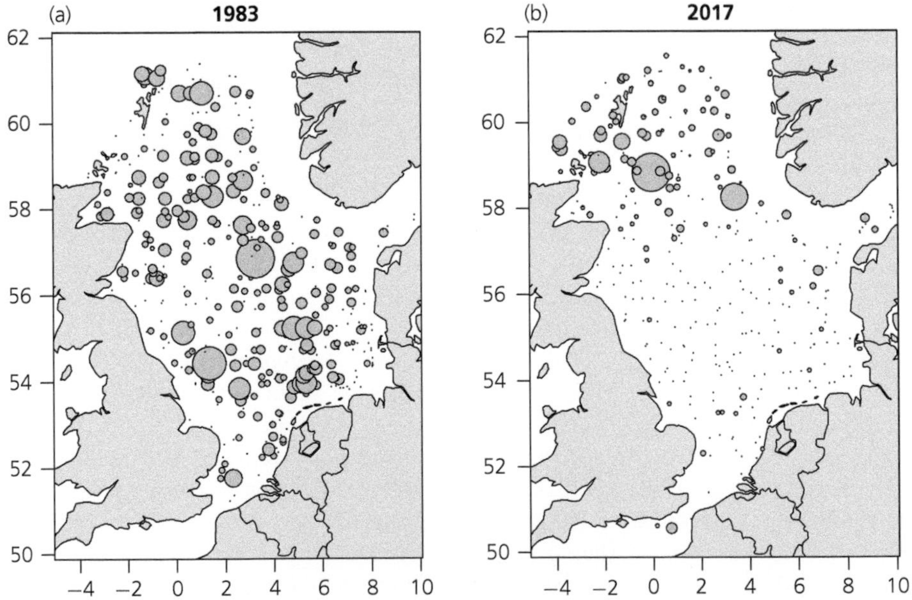

Figure 7.1 Spatial distribution of adult Atlantic cod (*Gadus morhua*) in the North Sea (a) 1983 and (b) 2017 based on research vessel surveys conducted under the auspices of the International Council for Exploration of the Sea (ICES). Bubble size is scaled to the highest abundance level in each year and is proportional to abundance. Points indicate samples with no cod in the sample. Data courtesy of Anna Rindorf.

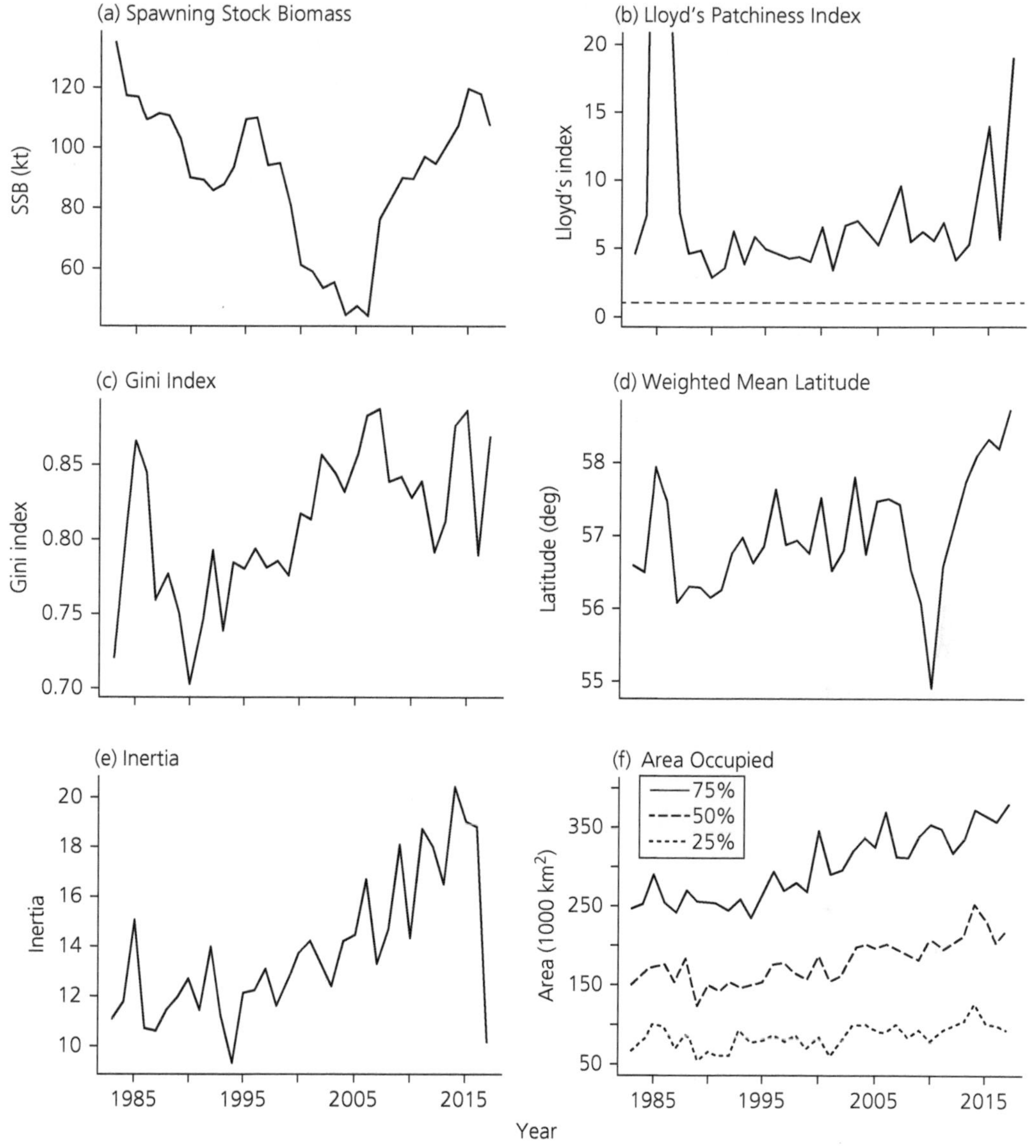

Figure 7.2 Spatial metrics of Atlantic cod adult abundance in the North Sea showing (a) spawning stock biomass from the ICES assessment, (b) Lloyd's index of patchiness, (c) Gini Index based on Lorenz curve, (e) mean latitude of cod weighted by abundance, (d) inertia (distance to center of gravity), and (f) area occupied by percentiles of the population, estimated with a kernel-density smoother. Data courtesy Anna Rindorf.

on the Lorenz curve, (d) center of gravity of the population, and (e) the distance of any sample from the center of gravity. In the following, we illustrate these measures for the North Sea cod example shown in Figure 7.1. The estimated biomass of the cod population decreased by half from 1983 to 2005 and has increased since then (Figure 7.2a). These changes in abundance must be considered when interpreting changes in cod distribution patterns.

We note that evaluating abundance–area relationships is complicated when total abundance is estimated by sampling rather than by census, which is the general case for aquatic populations.

The measures of abundance and area occupied are often calculated from the same samples, such that the area occupied can be biased by sampling error. The indices described in the rest of this section can be used to infer how spatial distribution changes with population size, but care should be taken because some of the indices can be biased, especially at low mean abundance ($\bar{n}$<10). Zero observations can be either sampling zeros where the probability of detection is low, or structural zeros where the species does not occur. The area occupied will be biased low if sampling and structural zeros are confounded. Only the proportion of structurally empty areas, Lloyd's index, and indices of the distance to the center of gravity are unbiased at all levels of abundance (Rindorf and Lewy 2012).

7.2.1.1 Area occupied

The estimated area occupied by a population provides a simple and intuitive measure of its spatial distribution. It is typically represented as the proportion of samples containing some minimum number of individuals of the species of interest. There are many possible ways of estimating this proportion. One approach is to construct an empirical cumulative probability distribution of the number caught at each station and determine the 95th percentile. If the number of samples taken in each year varies over time, one can correct for sample size by taking the ratio of the estimated number of stations at the 95th percentile to the total number of stations sampled in a given year. Rindorf and Lewy (2012) describe alternative measures including ones in which the proportion of zero catches are treated explicitly. In our application to North Sea cod, we integrated the area occupied from the kernel density estimates (see Section 7.2.1.6 below).

7.2.1.2 Lloyd's index

A number of important descriptors of distribution attributes have been developed based on the relationship between the sample mean and variance within a defined region. Although these metrics do not explicitly consider spatial coordinates, they do lead to insights into the probability of observing a specified number of individuals at any location by making a connection with the first two moments of the Poisson and Negative Binomial distribution. In these analyses, the focus is principally on characterizing randomly vs patchily (aggregated) distributed populations.

Lloyd's index of patchiness is a prominent example of this general approach:

$$L_P = 1 + \frac{s^2}{\bar{n}^2} - \frac{1}{\bar{n}} \tag{7.1}$$

where $\bar{n}$ is the mean number of individuals per sample and s^2 is the sample variance. For a randomly distributed population, described by the Poisson distribution, $s^2 = \bar{n}$ such that Lloyd's index is one. A patchy distribution, described by the negative binomial distribution, has $L_P = 1 + 1/k$, which increases with decreasing size parameter k (see Box 7.1). Estimates of L_p for North Sea cod are shown in Figure 7.2b. The estimates are all above 1, indicating patchy distributions. Prominent spikes in the L_p index are evident, particularly during the early part of the series. These spikes are attributable to large catches of cod in a small number of stations. Patchiness has increased since 1990, peaking again in the most recent year.

7.2.1.3 Lorenz curves

Lorenz curves plot the cumulative distribution of the samples in ascending order, $n_1, n_2, \ldots n_I$. Originally designed to evaluate income inequality in human populations (Lorenz 1905), this descriptor has also been applied in a number of fields including ecology. The diagonal line represents the hypothetical case in which all samples are equal (line of perfect equality). Departures from this reference line represent the case in which samples are not evenly distributed (shaded area in Figure 7.3). In the context of population data, this represents clustering or clumping with some samples having greater abundance levels. A variety of indices have been based on quantiles of the Lorenz curve. The Gini index, commonly used in socio-economics, is defined as twice the area between the Lorenz curve and the diagonal. This can be expressed:

$$G = \frac{\sum_{i=1}^{I-1} i\,(I-i)\left(n_{i+1} - n_i\right)}{(I-1)\sum_{i=1}^{I} n_i} \tag{7.2}$$

Box 7.1 Discrete Probability Distributions for Spatial Statistics and Models

We will take advantage of several discrete probability distributions in our treatment of spatial statistics and models. Two of these, the Poisson and Negative Binomial distributions are central to spatial metrics based on the ratio of the variance to the mean for a population. The Binomial distribution underlies our treatment of the random walk process. Discrete distributions are appropriate for this work because we will be dealing with count data (number of individuals) rather than continuous variables. For a good entrée to this topic see Bolker (2008). Note that the term probability distribution is used in a very specific way to designate the probability of an event and it differs from the colloquial use of the term distribution to designate an arrangement of objects of interest.

Poisson

The Poisson distribution describes the probability of a specified number of events (x) occurring in a fixed interval of time (or space). These events occur at a known constant rate and are independent of the time since the last event. Under these conditions, the pattern of individuals in space is random. The probability density function is:

$$P_r(X = x) = \frac{e^{-\lambda}(\lambda)^x}{x!}$$

where λ is the average number of events per time period; the variance is identical to the mean. This distribution is also sometimes written in terms of the rate constant; for our purposes, the key remains (the mean and variance) are the same. Our first-cut metric for a randomly distributed random variable therefore has a mean-to-variance ratio of 1.

Negative Binomial

The Negative Binomial is used to describe clumped arrangements of organisms. In most ecological applications it is derived from a Poisson distribution in which the rate coefficient is not constant but rather follows a gamma distribution. The probability density function for the Negative Binomial distribution is:

$$P_r(X = x) = \frac{\Gamma(k+x)}{\Gamma(k)x!}\left(\frac{k}{k+\mu}\right)^k\left(\frac{u}{k+\mu}\right)^x$$

and its mean is given by:

$$E(X) = \mu$$

and the variance is:

$$Var(X) = \frac{\mu + \mu^2}{k}$$

Binomial

The Binomal distribution describes a process in which a discrete random variable can take on one of two states. In invoking this process for the random walk model in one spatial dimension, these represent movement to the right or to the left of the current position. The probability density function is:

$$P_r(X = x) = \binom{N}{x} p^x(1-p)^{N-x}$$

with mean:

$$E(X) = pN$$

and the variance is given by:

$$Var(X) = p(1-p)N$$

For a large number of individuals, the binomial can be approximated by a normal distribution.

where I is the number of samples and n_i, the number of individuals in a sample. The Gini index ranges between 0 (all samples are equal) and 1 (all individuals are in one sample). Application to the North Sea cod example shows consistently high values, again indicating a concentration of individuals at a small number of stations (Figure 7.2c). After an initial peak, the Gini index has increased since 1990, similarly to Lloyd's index.

7.2.1.4 Center of gravity

The center of gravity is calculated as the abundance-weighted mean location (e.g. Murawski 1993). For example, the mean latitude is:

$$\overline{lat} = \frac{\sum_i lat_i \log(n_i)}{\sum_i \log(n_i)} \tag{7.3}$$

here n_i is abundance at station i. Weighting by the logarithm of abundance, for non-zero samples, provides an unbiased estimate of the mean location (Rindorf and Lewy 2012). The corresponding mean longitude can be calculated to provide the mean location in Cartesian coordinates. The mean location is often referred to as the center of gravity (C) of the spatial distribution. For a stock distributed along a coastline that does not run in a strictly north–south or east–west orientation, it is useful to

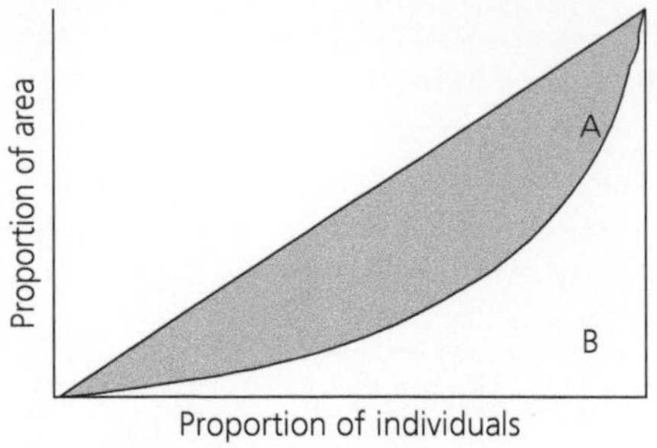

Figure 7.3 A hypothetical Lorenz curve for cumulative number of individuals sampled against cumulative area sampled (as proportions of total). The shaded area (A) is the departure from a uniform distribution. A geometric interpretation of the Gini index (G) is the ratio of the area A to A+B where B is the area under the Gini curve.

rotate the coordinate system to provide along-shore and onshore–offshore axes (Nye et al. 2009; Bell et al. 2014b). In Figure 7.2d we show the position of the mean latitude for adult cod in the North Sea from 1983 to 2017. We see an overall increase in the mean latitude over time, apart from a dip in 2010 when high survey catches occurred in the English Channel. Perry et al. (2005) had earlier documented this northward shift in relation to increasing water temperatures in the North Sea.

7.2.1.5 Geographical spread

Measures of the geographical spread of the population can be based on the average distance, dist(i,C), from an individual to the center of gravity (Murawski and Finn 1988). The average squared distance to C is:

$$d^2 = \frac{\sum_{i=1}^{I} n_i[dist\ (i, C)]^2}{\sum_{i=1}^{I} n_i} \tag{7.4}$$

where sums are taken over the number of stations, I. Expressed as a variance, this distance provides a measure of dispersion. We see an increase in d^2 in North Sea cod (Figure 7.2e).

Contour ellipses can be calculated based on the bivariate normal distribution. The area of the contour ellipse, A_p, containing p percent of the population can be estimated as:

$$A_p = C_p\sqrt{s_x^2 s_y^2 (1-\rho)} \tag{7.5}$$

where $C_p = \chi^2(2)p\pi$ (chi-squared statistic with two degrees of freedom and probability p), s_x and s_y are the standard deviations in the x and y directions, and ρ is the correlation coefficient (Rindorf and Lewy 2012). Confidence ellipses for adult cod in 1983 and 2017 are shown in Figure 7.4. The 2017 center of gravity (intersection of the ellipse axes) was displaced to the northwest relative to the reference position in 1983.

7.2.1.6 Kernel density estimators

Confidence ellipses based on the bivariate normal distribution assume a symmetrical distribution along any axis through the population, which may not be the case. If the spatial pattern of the population is discontinuous, with two or more areas of concentration, more general smoothing methods, such as kriging, can be used. The Kernel Density Smoother, a non-parametric method, allows for irregular spatial distributions by calculating the smallest area containing a given percentile of the population (Simonoff 1998); the smallest percentile chosen identifies the core area and, at the limit, the center of gravity. The kernel density estimates for adult cod in the North Sea in 1983 and 2017 are shown in Figure 7.5. The spatial distributions are asymmetric and in 2017 become discontinuous. The area occupied was calculated by integrating the area within the contours corresponding to percentiles of the population. In the North Sea cod example, there is no evidence of density-dependent habitat selection, as would be predicted by the basin model (see Section 7.4.1.1). Instead, the area occupied increased as the center of gravity shifted north (Figure 7.2f). This pattern occurs because the trailing edge of the distribution in the southern North Sea shifts more slowly than the leading edge in the north, effectively stretching the spatial distribution.

7.2.2 Climate and distribution

Any of the above methods can be applied to temporal series of abundance data. As coastal waters warm, many populations are shifting in space (e.g. Perry et al. 2005, Nye et al. 2009). The rate of movement of the center of gravity has been

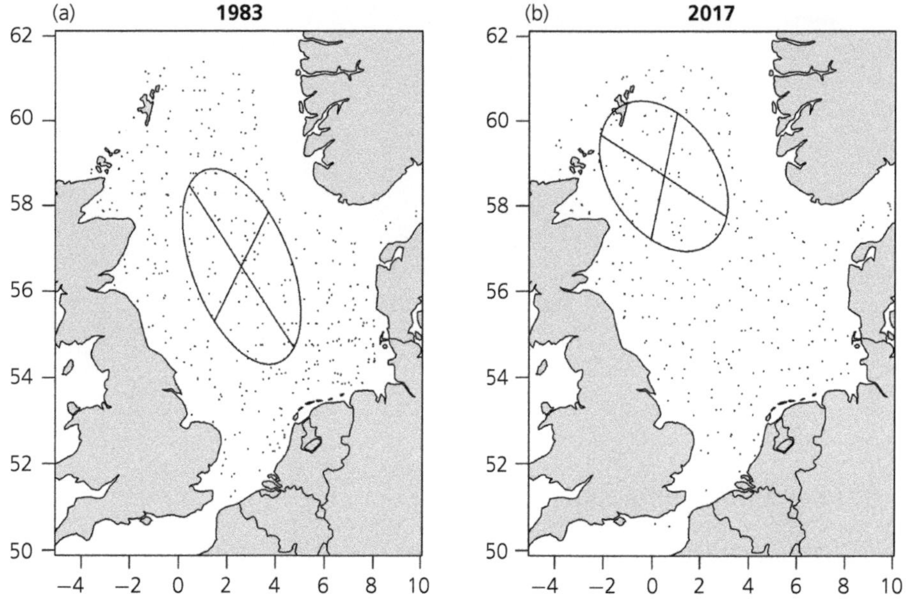

Figure 7.4 Confidence ellipses representing the center of gravity of adult cod (*Gadus morhua*) in standardized research vessel surveys in (a) 1983 and (b) 2017 in the North Sea.

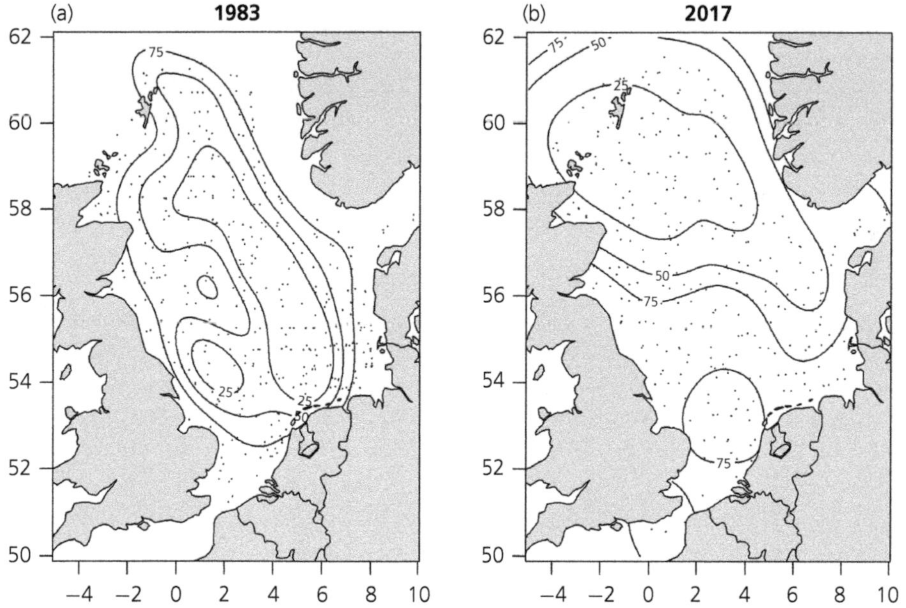

Figure 7.5 Kernal density estimates of total abundance of adult cod in standardized research vessel surveys in (a) 1983 and (b) 2017 for areas with non-zero observations in the North Sea.

related to climate velocity (the rate and direction of change of temperature isotherms) for assemblages of North American marine fish populations (Pinsky et al. 2013). Overall, climate velocity was found to account for approximately 40% of the variance in changes in the center of gravity for marine taxa in this study (Pinsky et al. 2013; Table 1). Intuition might suggest that distributional shifts would, in general, be poleward under global warming, and this is in fact observed in a number of cases. However, factors such as hydrographic flow patterns, topographic constraints, and other factors can result in a much richer set of outcomes at regional scales of interest (Pinsky et al. 2013). The unaccounted-for component of variance reveals that additional drivers play important roles in distribution shifts. These may include changes in resource availability, habitat factors, density-dependent processes (Bell et al. 2014b), and spatial patterns of fishing effort (Frank et al. 2018; Adams et al. 2018).

7.3 Models of movement and dispersal

The spatial metrics explored above are static snapshots of realized distribution patterns. They provide important insights into what has happened in the past. The underlying causes of any observed changes in distribution cannot, however, be directly inferred from these metrics alone. Nor can they (necessarily) be used to forecast future change. To complement these analyses, we require a more dynamic setting in which movements of the organisms are explicitly considered in addition to other relevant factors. Shifts in the distribution of aquatic populations over time can occur as a consequence of differential mortality in response to anthropogenic and natural drivers or directed movement, again in response to forcing factors.

We begin by considering a very simple algorithm to explore the consequences of random choices in movement in which an organism moves one unit of specified length in each time step. The length of a step is given by the product of velocity (V) and the length of a time period (τ). To introduce basic concepts, we focus on one spatial dimension in which the organism moves at random in either a positive or negative direction with equal probability. Because at each step only one of two discrete outcomes is possible, we have a process that can be described using a binomial distribution. We further assume that direction of movement at any time is independent of the direction at any previous time step (i.e., there is no memory in the process). In Figure 7.6 we show the outcome of a "random walk" model of this type. We track the position of

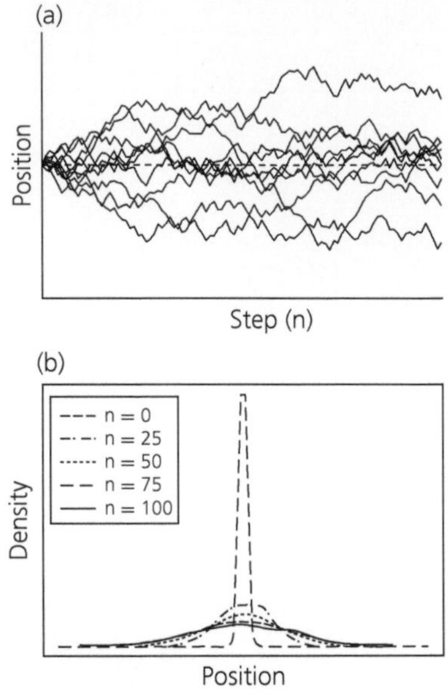

Figure 7.6 (a) Outcome of a random walk process for 25 individuals released from a point source in one spatial dimension after 100 time steps. The expected mean displacement is zero (horizontal dashed line). (b) Empirical probability distributions of 1000 individuals released from a point source after five time steps ranging from 0 to 100.

an initial number of $N = 25$ individuals released at a point source and observe their increasing spread with successive steps in the process (Figure 7.6a; see also Box 7.2). If we examine the density of this initial population over the spatial domain at selected time intervals, we see that the spread of the population from the initial point source progressively widens and approaches a bell-shaped distribution (Figure 7.6b). In fact, the underlying binomial distribution can be approximated by a normal probability density when N is large. We see immediately that the variance of the distribution will increase with time as shown in the numerical example in Figure 7.6b.

We recognize of course that most organisms do not move entirely at random as in the illustration above. Instead, we typically expect a directional component to the dispersal process in addition to the possibility of random movement. Further, so far we have introduced a fixed starting number of individuals but not tracked the dynamics of a population model incorporating births and deaths. To make this transition, we turn to population models incorporating terms for advection and diffusion

Box 7.2 The Random Walk in One Dimension

Here we describe the basic elements of a random walk process in one dimension. We are interested in the movements of an individual organism where it moves a fixed distance at each step (or time interval) in one of two directions with equal probability. The location along this single dimension at the nth step depends only on its previous position and the distance (d) moved in a positive or negative direction:

$$x_{i,n} = x_{i,n-1} + d$$

and the average position of individuals is:

$$\bar{x}_n = \frac{1}{N}\sum_{i=1}^{N} x_{i,n} = \frac{1}{N}\sum_{i=1}^{N} x_{i,n-1} + d$$

$$= \frac{1}{N}\sum_{i=1}^{N} x_{i,n-1} + \frac{1}{N}\sum_{i=1}^{N} d$$

Because the organism moves in one of the two directions with equal probability at each step, the average displacement over all individuals is zero even although any individual will likely have moved from the point source. Therefore the expected position over all individuals does not change with successive time steps. We are next interested in quantifying the spread of the individuals at successive time steps. For this, we will determine the variance over all individuals. We typically compute a variance as the sum of the squared differences between a variable of interest and the average over all observations of that variable. In our case, the average is zero and variance is given by:

$$x_{i,n}^2 = (x_{i,n-1} + d)^2 = x_{i,n-1}^2 + 2dx_{i,n} + d^2$$

which can be written:

$$\bar{x}_{i,n}^2 = \frac{1}{N}\left[x_{i,n-1}^2 + 2dx_{i,n-1} + d^2\right]$$

or

$$\bar{x}_{i,n}^2 = \frac{1}{N}\sum_{i=1}^{N}\left[x_{i,n-1}^2 + 2dx_{i,n-1} + d^2\right]$$

$$= \frac{1}{N}\sum_{i=1}^{N} x_{i,n-1}^2 + \sum_{i=1}^{N}\frac{1}{N}2dx_{i,n-1} + \sum_{i=1}^{N}\frac{1}{N}d^2$$

The sum of all displacements is zero and so the middle term of this expression is zero and we are left with:

$$\bar{x}_{i,n}^2 = \frac{1}{N}\sum_{i=1}^{N} x_{i,n-1}^2 + \sum_{i=1}^{N}\frac{1}{N}d^2$$

and the variance is :

$$\bar{x}_{i,n}^2 = x_{i,n-1}^2 + d^2 = nd^2$$

The variance therefore increases with each step.

to represent movement. Models of this type have a rich pedigree in the ecological and fisheries literature. The basic form of the partial differential equation for a population model with a one-dimensional spatial component (x) is:

$$\frac{dN}{dt} = g(N)N + D\frac{\partial^2 N}{\partial x^2} - V\frac{\partial N}{\partial x} \qquad (7.6)$$

where $g(N)$ is the *per capita* rate of change of the population (N), D is a diffusion coefficient, and V is a velocity coefficient. The diffusion component reflects random dispersal and the advection component represents directional movement. An illustration of spread in a population from a single point in space and in which $g(N)N$ is given by the Ricker-logistic model is illustrated in Figure 7.7 for the case in which movement is non-directional (no advection). We see that the population density again declines from the point source as individuals disperse. Restricting our consideration to a single

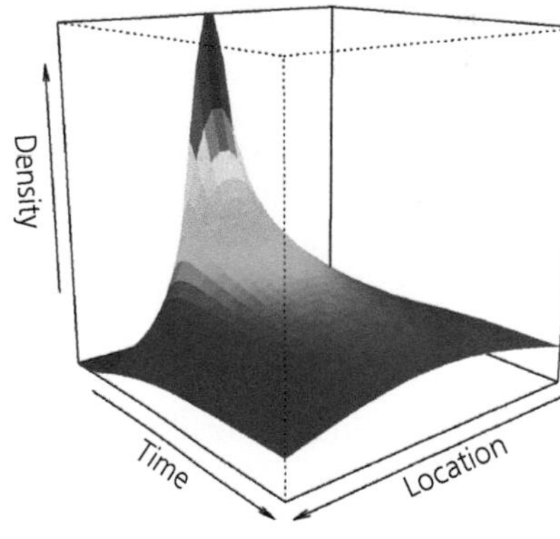

Figure 7.7 Density of a population obeying a Ricker-logistic model in one spatial dimension after 100 time steps.

spatial dimension provides a simple entré to this issue that may also serve as a useful approximation for movement along a coastline or river. More generally of course we require at least two dimensions to characterize movement, which can be readily accommodated. In the following section we will address the issue of directional change in relation to gradients of abundance.

7.4 Spatial population models

Models of movement and dispersal in both discrete and continuous spatial coordinate systems have been developed (see Kot 2001). Population models incorporating discrete spatial representations typically require the specification of probabilities of movement among spatial units within specified time periods. These probabilities can reflect both directional and random dispersal components. Models incorporating continuous spatial representations can accommodate fine-grained characterizations of habitat and other features. They too can incorporate directed and random components of movement. Both spatial modeling types have been paired with population models in discrete and continuous time. In the following, we highlight selected classes of spatial population models that have been employed in fisheries research.

7.4.1 Models in continuous time and space

7.4.1.1 The basin model

MacCall (1990) built on the concept of an ideal free distribution (IDF; Fretwell and Lucas 1970) to develop a "basin" model of dynamic patterns in the geography of aquatic organisms. The IDF construct is predicated on the idea that the number of organisms in a habitat patch is a function of the resources available in that habitat. Organisms are assumed to be able to determine the quality of each habitat and they choose to forage in the patch with the highest quality (the "ideal"). It is further assumed that animals are capable of moving freely from one patch to another without interference. The IDF concept invokes density-dependent habitat selection in which population size and local density are important factors influencing choice of habitat and hence the relative distribution of the population among habitats (MacCall 1990). One of the most commonly observed phenomena associated with density-dependent habitat selection is expansion and contraction of the range of a population, or differential utilization of marginal habitat with changes in population abundance.

Following MacCall (1990), we begin by adopting a modification of the logistic model as a starting point for further analysis. Specifically, we can modify our treatment of the logistic model in Chapter 3 to account for habitat-related differences in productivity:

$$\frac{dN_h}{dt} = (\alpha_h - \beta_h N_h) N_h \tag{7.7}$$

where h is a habitat index and all other terms are defined as in Chapter 3. We can define the realized per capita rate of increase as:

$$g^*(N) = \frac{1}{N}\frac{dN_h}{dt} = (\alpha_h - \beta_h N_h) \tag{7.8}$$

which is a function of suitability in habitat h. This is illustrated in Figure 7.8 for two habitats (A and B) differing in their habitat-dependent per capita rates of change; here we will assume that the compensatory coefficient β in both habitats is the same (MacCall 1990). The intersection points with the ordinate give the intrinsic rates of increase for the two habitats (α_A and α_B respectively); the lines intersect the abscissa at the equilibrium population sizes (α_A/β_A and α_B/β_B) for each habitat.

The original Fretwell–Lucas theory postulates that multiple discrete habitats that may be ordered by basic suitability. As the density of individuals increases in a habitat, realized suitability decreases

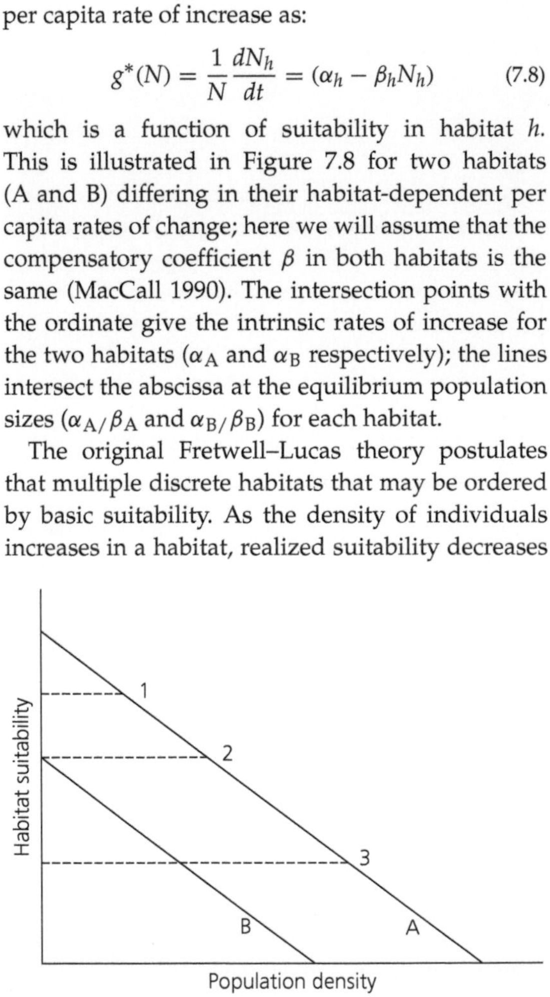

Figure 7.8 Two habitats, A and B differing in suitability and carrying capacity. The intersections between the solid and broken lines indicate population densities and corresponding suitabilities according to the ideal free distribution. Adapted from MacCall (1990).

due to density-dependent effects. In Figure 7.8 we show three horizontal lines representing increasing levels of population density and the corresponding decline in habitat suitability as measured by the realized per capita rate of change. Initially, individuals occupy habitats with the highest available suitability (see dashed line 1 in Figure 7.8), but as the realized suitability of these habitats declines due to increasing population density, other previously less-suitable unoccupied habitats become equally attractive and are colonized (line 2). The ideal free distribution is characterized by an equal realized suitability in all occupied habitats (line 3). Basic habitat suitability can be interpreted as "fitness" or "reproductive value", which should be manifested as marginal changes in the intrinsic rate of increase of the population in each habitat (MacCall 1990). The basin model provides a metaphor for suitability along a continuous range of habitat types (MacCall 1990). If habitat suitability (realized per capita growth rate) is depicted graphically as increasing in a downward direction, habitats can be described as a continuous geographic suitability topography having the appearance of a basin (Figure 7.9). According to the ideal free distribution, the population will fill this basin as if it were a liquid under the influence of gravity. The total carrying capacity over all habitats is shown by the dotted line in Figure 7.9, corresponding to the point where the per capita rate of change is zero. At the deepest part of the basin, the per-capita rate of population change is highest, representing the "ideal" habitat. In our example, Habitat B, at the margins of the basin, is less favorable than Habitat A (the interior region bounded by the two demarcation points for Habitat B at the periphery). Although for simplicity we have depicted the basin as fixed with a smooth surface, in practice it can have an irregular topography and can change over time (MacCall 1990). While the basin concept is intended to serve as an analogy, it has several properties that can be connected to the distribution of real populations:

- the depth of the basin at any point is a measure of basic habitat suitability;
- the steepness of the basin sides reflects the distributions of suitability across habitats;
- the surface will be approximately level due to the ideal free distribution;

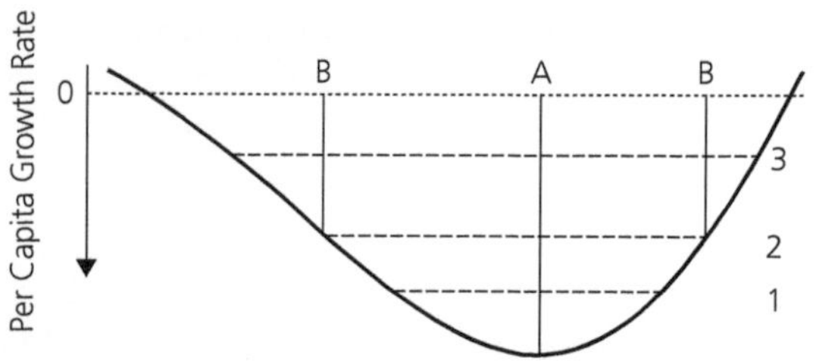

Figure 7.9 Transect through a continuous geographic fitness topography. Realized suitability is equal in all occupied habitats due to the ideal free distribution. A total population size of carrying capacity is reached when the per capita growth rate is zero (dotted line). The letters A and B and numbers 1–3 correspond to cases shown in Figure 7.8 . Adapted from MacCall (1990).

- the shoreline defines the range of the organism;
- the depth of the liquid is proportional to local density and is also a measure of realized habitat suitability;
- the total volume of liquid is functionally related to total population size;
- the addition of a single individual would have the same effect, wherever it is added to the population (a drop will raise the water level wherever it is added).

The basin model predicts that populations will tend to contract toward the most favorable habitat as their abundance is reduced in response to environmental change or by fishing. Bioeconomic theory predicts that fishers will preferentially exploit the most productive areas first. Accordingly we can expect a connection between the insights garnered through the basin model, and the behavior of fishers. Catch per unit effort (CPUE) at the center of the population (the depth of the liquid) will not decline as fast as the total abundance (the volume of liquid in the basin). Thus the model accounts for the tendency for CPUE to remain high even as the stock is fished down. Also, catchability will increase as population size decreases. In Chapter 11 we explore this phenomenon in greater detail.

We next consider movement patterns in the context of the basin model to consider more explicitly how the basin would be filled. MacCall (1990) adopted a modification of Equation 7.6 in which a habitat-specific logistic model is coupled with advection and diffusion terms. In this case, the advective term reflects directional movement toward the most favorable habitat available. In MacCall's formulation, the velocity term is replaced by a viscosity coefficient (essentially the inverse of velocity). The completed model is then:

$$\frac{\partial N}{dt} = (\alpha_h - \beta_h N_h) N_h + D\frac{\partial^2 N}{\partial x^2} - v^{-1}\frac{\partial g^*(N)}{\partial x}\frac{\partial N}{\partial x} \tag{7.9}$$

where v represents the flow per unit gradient in realized habitat suitability and all other terms are defined as before. This structure introduces an interesting twist. In order for an equilibrium to exist, the advection and diffusion terms must balance. This requires that the higher productivity in the optimal habitats results in flow from these habitats to lower productivity habitats. The high productivity habitats then serve as source areas for lower productivity habitats (sinks). The equilibrium surface is then a straight line as in the representation in Figure 7.9 (dotted line) but will be lower in the high productivity regions (optimal habitat) and higher in the marginal areas (recall that in Figure 7.9 the *per capita* growth rate is depicted as increasing as the bottom of the basin is approached).

MacCall considered a number of alternative population models to augment the logistic model used in framing the initial structure of the basin model described above. These include the theta-logistic introduced in Chapter 3 and a Ricker stock-recruitment model. Recall that we encountered the Ricker survival function in Section 3.4.1; we will see this model again in Chapter 9. The habitat-specific Ricker models can be written:

$$R_h = \mathcal{E}_h \exp\left(a_h - b_h \mathcal{E}_h\right) \tag{7.10}$$

where R represents the recruits resulting from eggs $\mathcal{E}$; a is the rate of increase at low density (≡ basic habitat suitability), and b is a compensatory coefficient. The per capita recruitment rate expressed as a logarithm is:

$$\log\left(R_h/\mathcal{E}_h\right) = a_h - b_h \mathcal{E}_h \tag{7.11}$$

which can be considered a measure of realized habitat suitability. According to the Ricker model, the degradation of habitat suitability is linear with density. If b_h is assumed constant among habitats, the result is a family of parallel per capita recruitment lines analogous to the representation of the logistic case in Figure 7.8. The geographic pattern of density should "map" the geographic pattern of suitability. The basic suitability of habitats (a_h) may be determined by combinations of density-independent factors, and the most favorable combinations of these factors are expected to occur near the center of the species range.

MacCall (1990) applied the basin model to the northern anchovy (*Engraulis mordax*) population off the west coasts of the United States and Mexico. Plankton surveys conducted under the California Cooperative Offshore Fisheries Investigations (CalCOFI) program provide an excellent opportunity to examine and apply the theory of density-dependent habitat selection because of the extensive data on anchovy larvae (and other species) and because of the unusually large dynamic range of abundance and distribution exhibited by northern anchovy larvae. For example, the estimated larval biomass ranged from 20000 mt in 1952 to approximately 700000 mt in 1965 (see Figures 1.12 and 1.13 in MacCall 1990). At low abundance, anchovy larvae were restricted to the southern California Bight and pockets along the Baja Peninsula. At high abundance, larval density increased in all these locations and the distribution spread along the coast and to deeper water as would be anticipated under the basin model. In 2011 the anchovy population collapsed from natural causes (there was no fishery), and the distribution again contracted to the nearshore waters (MacCall et al. 2016).

In this application, MacCall used the Ricker-type spawner-recruit model described above rather than the logistic population model. The rationale is to use the abundances of early larvae as an index of spawning abundance at each station, and to use spawner abundance to "map" the suitability basin. MacCall fit the habitat-specific Ricker model to time series of larval abundance from the CalCOFI survey. Suitability had to be assumed time invariant to permit parameter estimation. The predicted habitat suitability contours (Figure 7.10) compare well with the observed spatial distributions at low and high abundance. Estimated habitat suitability does not correspond in an obvious way to hydrographic parameters, except that the central part of the habitat may be a retention area. Perhaps adult anchovy are spawning in areas

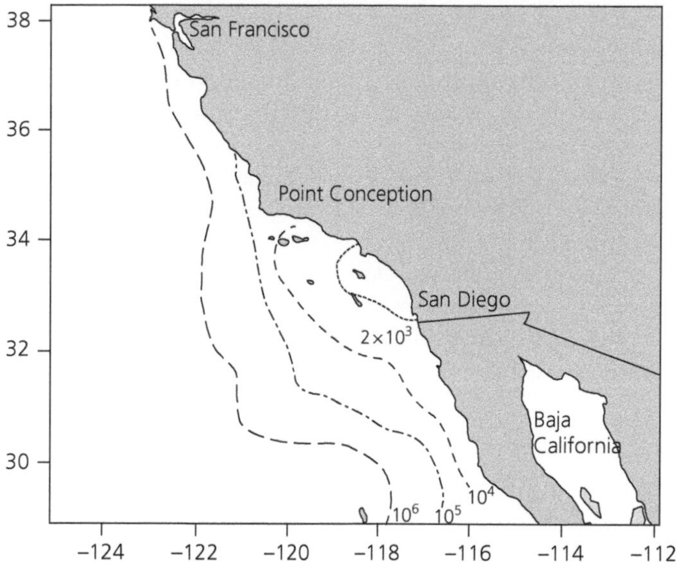

Figure 7.10 Estimated spawning habitat suitability contours for the central population of northern anchovy, showing the nominal edge of the larval distribution over a range of spawning biomass from 2000 to 1000 000 (mt). Adapted from MacCall (1990).

where their larvae will initially be retained in coastal waters.

There is a risk of circularity when using the realized distribution of animals to infer habitat suitability. If we assume that spawners distribute themselves so as to obtain equal spawning success, the realized distribution would be a measure of habitat suitability. However, it is exceedingly difficult to independently measure habitat suitability or localized spawning success, to rigorously test the hypothesis. To address this issue, Blanchard et al. (2005) computed temperature-dependent growth rates of North Sea cod as an index of fitness and considered temperature as an index of habitat suitability. Blanchard et al. reported good correspondence between the basic predictions of the basin model and the observed distribution of North Sea cod in relation to bottom temperatures when applying this independent measure of fitness.

7.4.2 Models in continuous time and discrete space

Perhaps the most widely known type of continuous time–discrete space model for aquatic populations is based on the concept of metapopulations (see Kritzer and Sale 2006 for marine examples). A metapopulation is a collection of subpopulations existing in a larger geographic area (Hanski and Simberloff 1997). Exchange between the subpopulations prevents the establishment of separate autonomous populations. Subpopulations may or may not be genetically distinct, depending on the degree of exchange and specificity of the genetic marker examined.

7.4.2.1 Metapopulation models

Levins (1969) provided an analytical framework for considering the dynamics of metapopulations. The simplest metapopulation models consider that each subpopulation is equal in size and longevity, existing at either local carrying capacity or empty (extinct) at any specified point in time. The metapopulation exists in dynamic balance between extinction and recolonization. The subpopulations are assumed to be linked with equal probability of exchange regardless of position. This assumption means that the actual spatial arrangement of the subpopulations is not important. Levins modeled the fraction of the patches occupied (p) as:

$$\frac{dp}{dt} = c\left(1-p\right)p - \varepsilon p \qquad (7.12)$$

where c is the rate at which individuals colonize from an occupied patch and ε is the rate of extinction in a patch. Here, the overall colonization level depends on the number of occupied (p) and unoccupied patches ($1-p$); extinction is a linear function of the number of occupied patches. Notice that the colonization function is quadratic and therefore similar to the logistic model (but framed in terms of patches rather than number of individuals in the population). The time-dependent solution of the Levins model is therefore sigmoidal. The equilibrium is given by:

$$p^* = 1 - \frac{\varepsilon}{c} \tag{7.13}$$

and for the metapopulation to persist we require that $c > \varepsilon$ (Figure 7.11). The proportion of patches occupied is zero for all $c < \varepsilon$ and increases as c exceeds ε.

Several variants on the basic Levins model have been proposed. Gotelli considered the case in which recolonization does not depend on occupied patches within the model domain but receives a subsidy from an external source. For example, if we were to model a nearshore population defined by an outer depth limit, we might envision that a connection could exist with a deeper water population outside of the defined model domain. If this deeper population served as the source of replenishment for the nearshore component we have:

$$\frac{dp}{dt} = c^+(1-p) - \varepsilon p \tag{7.14}$$

where c^+ is the rate at which individuals colonize from offshore and ε is the rate of extinction in a patch. Gotelli referred to this as a "propagule rain" system—the patches within the metapopulation domain are showered by propagules from the external source in a way that is independent of the number of occupied patches. This is an apt metaphor for our inshore–offshore example where eggs and larvae are dispersed from offshore and settle within the inshore patches. The model is no longer logistic-type and rises to an asymptote without an inflection point. The equilibrium is now given by:

$$p^* = \frac{c^+}{c^+ + e} \tag{7.15}$$

The Gotelli propagule rain model provided an alternative structure for the colonization component of the Levins model. Noting that extinction may not be a simple linear function of the number of occupied patches, Hanski (1982) proposed a "core-satellite" model in which the extinction component is also quadratic:

$$\frac{dp}{dt} = c(1-p)p - \varepsilon p(1-p) \tag{7.16}$$

where the coefficients c and ε are again defined as in the Levins model. Here, the rationale is that as p increases, propagules are increasingly likely to reach all patches and extinction events will decline to zero. Increasing patch occupancy therefore provides a "rescue" effect and the number of unoccupied patches declines. Core patches can provide the propagules to replenish satellite patches. As we might anticipate given this structure, the

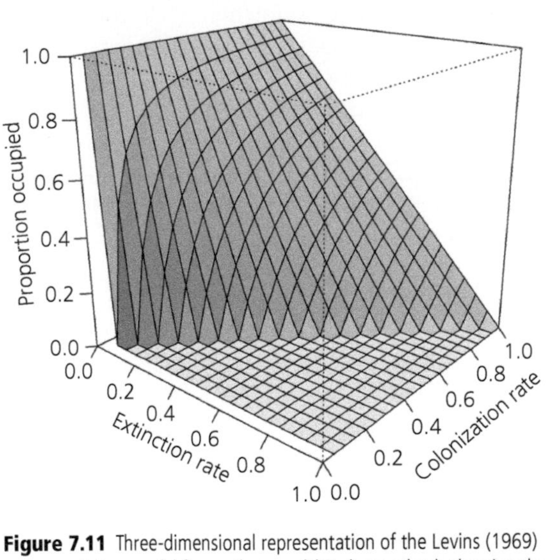

Figure 7.11 Three-dimensional representation of the Levins (1969) metapopulation model for northern cod (*Gadus morhua*), showing the proportion of occupied areas (p) for combinations of migration and extinction rates. Modified from Smedbol and Wroblewski (2002). The bisector of the extinction-migration plane (bottom panel) is defined by equal colonization and extinction rates. Patch occupancy is zero when the colonization rate is less than the extinction rate.

equilibrium is now given by $p^* = 1$ where all patches are occupied.

The original Levins model has undoubted heuristic value. With the alternative structures proposed by Gotelli and by Hanksi, these variants can accommodate different assumptions concerning colonization and extinction which can, in turn, be tested against observation.

Smedbol and Wroblewski (2002) applied the Levins metapopulation model to the population of northern cod (*Gadus morhua*) with spawning areas representing the patches in the modeling framework described above. Historically, spawning of northern cod has occurred in a series of offshore banks off Newfoundland and Labrador (Figure 7.12) and in inshore embayments. Smedbol and Wroblewski identified putative subpopulations of northern cod on the basis of timing and location of spawning, life-history characteristics, and genetic analysis. Differences in microsatellite DNA allele frequencies have been detected between spawning components (Ruzzante et al. 2000). Since microsatellite DNA fragments are assumed to be non-coding, these differences in allele frequencies do not result from natural selection. Therefore some local process of retention or natal philotropy at some life stage is necessary to maintain the subpopulation structure. Cod are known to exhibit fidelity to their spawning areas.

Based on the criteria identified above, Smedbol and Wroblewski proposed five principal subpopulations: (1) Saglek, Nain, and Makklovik–Harrison Banks, (2) Hamilton, Belle Isle–Funk Island Banks and Bonavista Corridor, (3) Northern Grand Bank, (4) Labrador Bays, and (5) Northeast Newfoundland Bays (Figure 7.12; see Smedbol and Wroblewski (2002) for detailed description). Based on historical records, Smedbol and Wroblewski determined that most of the bank and embayment subpopulations were occupied prior to onset of heavy exploitation and set the colonization parameter at $c = 0.9$. Given a lifespan of 25–30 years for northern cod, the extinction parameter was set at 0.03–0.04 per year (the inverse of the life-span range) with the premise that the subpopulation would go extinct if no individuals successfully spawned during the life span. In Figure 7.13 we show the time-dependent solution of the Levins model with these coefficients (solid line); the equilibrium patch-occupancy is 0.9 for these parameters. For comparison, we also show the results for the Gotelli propagule rain model with c^+ set to 0.3 and an extinction rate of 0.03 as in the original Smedbol and Wroblewski analysis. The propagule rain model does not follow the sigmoidal path for the Levins model but rather quickly increases to its equilibrium value of 0.909 (dashed line in Figure 7.13). In contrast, the Hanski core-satellite model retains the basic logistic structure and therefore is sigmoidal but with an equilibrium value of 1.0 as described above (dotted line in Figure 7.13).

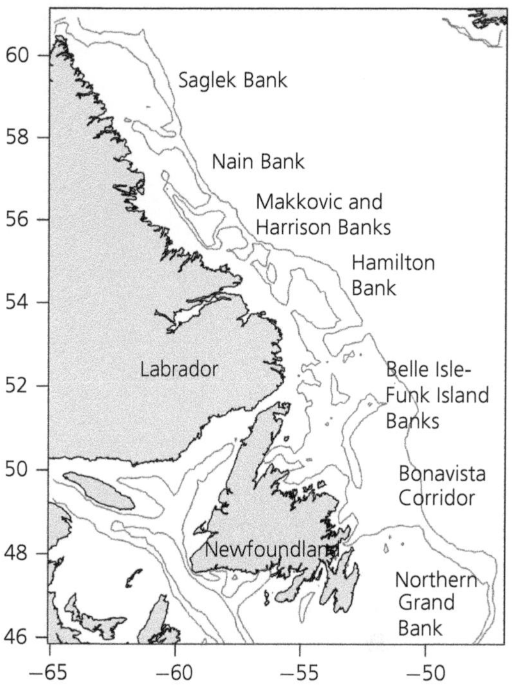

Figure 7.12 Map of Labrador and Newfoundland showing major offshore features defining putative offshore subpopulations of northern cod (*Gadus morhua*) (see text for description of groupings of banks and other features used in delineating subpopulations).

These metapopulation models can readily accommodate additional terms representing externally-

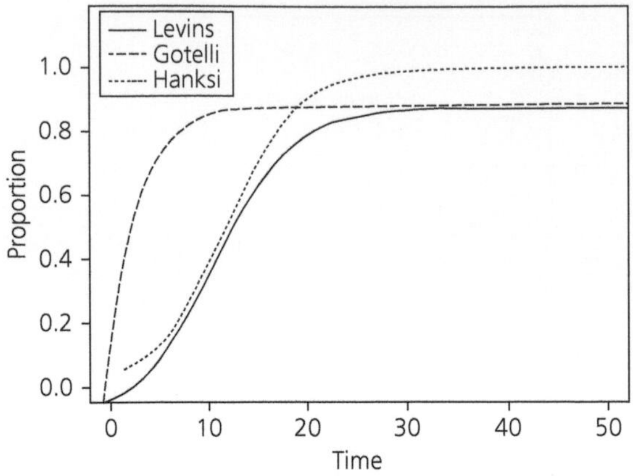

Figure 7.13 Time-dependent solutions of the (a) Levins model applied to the northern cod metapopulation for an unfished metapopulation with $c = 0.3$ and $\varepsilon = 0.03$ (solid line), (b) Gotelli propagule rain model with $c^+ = 0.3$ and $\varepsilon = 0.03$ (dashed line) and (c) Hanksi core-satellite model (dotted line).

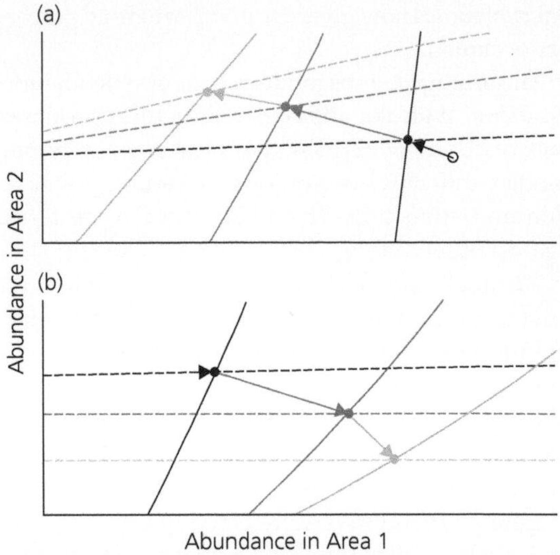

Figure 7.14 Isoclines for the spatial production model (Eq. 7.4). The solid lines are the isoclines for Area 1 and the dashed lines the isoclines for Area 2. A pair of isoclines is plotted for each of three cases. Points mark the intersection of each pair of isoclines; arrows trace the change in equilibrium abundances from one case to the next. (a) Equal migration rates. Open circle: $\delta_{12} = \delta_{21} = 0$; Black lines: $\delta_{12} = \delta_{21} = 0.1$, Dark gray lines: $\delta_{12} = \delta_{21} = 0.5$; Light gray lines: $\delta_{12} = \delta_{21} = 1$. (b) Unequal migration rates. Black lines: $\delta_{12} = 0.1$, $\delta_{21} = 0.2$; Dark gray lines: $\delta_{12} = 0.01$, $\delta_{21} = 0.5$; Light gray lines: $\delta_{12} = 0.001$, $\delta_{21} = 1$.

driven losses such as harvesting (Smedbol and Wroblewski 2002) or habitat destruction (Kareiva and Wennergren 1995). We will reserve further consideration of these topics until Chapter 13.

7.4.2.2 Spatially explicit production models

We can consider the implications of immigration and emigration in the traditional production-model framework developed in Chapter 3. Consider a population comprising two subareas. If the internal dynamics of each subpopulation are described by a logistic function, one possible representation of this system is:

$$\frac{dN_1}{dt} = (\alpha_1 - \beta_1 N_1) N_1 - \delta_{12} N_1 + \delta_{21} N_2$$

and (7.17)

$$\frac{dN_2}{dt} = (\alpha_2 - \beta_2 N_2) N_2 - \delta_{21} N_2 + \delta_{12} N_1$$

where δ_{12} is the rate of emigration from Area 1 to 2 and δ_{21} is the rate of immigration from Area 2 to 1 (Quinn and Deriso 1999, Eq. 10.4). This coupled pair of equations is analogous to the predator–prey model from Chapter 4 (Equation 4.6) and they can be analyzed with the same methods. The isocline for Area 1 is:

$$N_2 = \frac{1}{\delta_{21}} (\delta_{12} - \alpha_1 + \beta_1 N_1) N_1 \qquad (7.18)$$

and for Area 2:

$$N_1 = \frac{1}{\delta_{12}} (\delta_{21} - \alpha_2 + \beta_2 N_2) N_2 \qquad (7.19)$$

The intersection of the two isoclines defines the joint equilibrium. In the examples shown in Figure 7.14, the subpopulation in Area 1 is larger and less productive ($\alpha_1 = 1$, $\beta_1 = 0.05$) than Area 2 ($\alpha_2 = 2$, $\beta_2 = 0.2$). In Figure 7.14a the emigration rates are equal between the two areas ($\delta_{12} = \delta_{21}$). With a low migration rate, the equilibrium abundances are very close to their levels with no migration, such that the two areas are effectively separate populations. As the migration rate increases, the equilibrium abundances in the two areas converge and essentially become a single, pooled population.

The case of unequal emigration rates is shown in Figure 7.14b where the smaller, more productive Area 2 has a higher emigration rate than Area 1. Again with low migration rates, the equilibrium abundances are very close to the levels with no migration (20,10). As the emigration rates become more unequal, Area 2 becomes more of a source area, and Area 1 becomes more of a sink (gray isoclines in Figure 7.14b). In the extreme case (not shown), the subpopulation in Area 1 would not be self-reproducing (i.e. $\alpha_1 < \delta_{12}$) and would be entirely supported by immigration from Area 2.

7.4.3 Models in discrete time and space

7.4.3.1 Delay–difference model

Dispersal processes of different life stages of aquatic organisms can show marked contrasts. For many aquatic species, pelagic egg and larval stages can be transported in currents and circulation patterns as passive particles, or, perhaps more frequently, by an interplay of active behaviors and hydrographic transport mechanisms. For species with a sessile stage, this may be the sole means of dispersal. For other species, active swimming of juveniles and adults is important, encompassing both local movements and directed migratory behavior. It may be useful therefore to consider simple models that partition the life cycle into two or more age classes or stages reflecting these different modes of dispersal (see Wing et al. 1998; Fogarty 1998). The basic model for a population comprising two life stages (say pre-adults and adults) and two subpopulations in discrete time and space can be written:

$$\begin{aligned} N_{1,t+1} &= (1-\delta_1)\, s_1 N_{1,t} + \delta_2 s_2 N_{2,t} \\ &\quad + (1-\gamma_1)\, g_1\left(N_{t+1-r}\right) + \gamma_2 g_2\left(N_{t+1-r}\right) \\ N_{2,t+1} &= (1-\delta_2)\, s_2 N_{2,1} + \delta_1 s_1 N_{1,t} \\ &\quad + (1-\gamma_2)\, g_3\left(N_{t+1-r}\right) + g_4 \gamma_2\left(N_{t+1-r}\right) \end{aligned} \tag{7.20}$$

where $N_{i,t}$ is the number of adults in the ith subpopulation at time t, δ_i is the proportion of migratory adults, γ_i is the proportion of pre-adults (including eggs, larvae, and juveniles) and s_i is adult survival. The first two terms on the right-hand side of both expressions in Equation 7.20 represent the movement and survival of adults, whereas the third and fourth terms $g_i(N_{t-r})$ are functions relating to adult population size and recruitment from different sources. Note that for generality, N_{t+1-r} is deliberately not indexed by subpopulation but instead can represent a weighted average of the two groups to account for mixing during the spawning period (Fogarty 1998); r is the reproductive delay. The stock-recruitment relationships consider transport of larvae between subpopulations and potentially include abundance terms for both groups, depending on the sequence of critical life-history events and the timing of density-dependent processes. This basic structure provides a metapopulation model framed in terms of population number rather than number of occupied patches, with considerable flexibility concerning dispersal processes. It can accommodate concepts such as propagule rain and core-satellite dynamics. In an interesting application of a delay–difference metapopulation model, Wing et al. (1998) considered the case in which two subpopulations contribute to a combined larval pool which is then subject to redistribution to the subpopulations in relation to hydrodynamic transport processes. This provides a variant on the theme of propagule rain.

7.4.3.2 Full age-structured models

In many populations, migration rates depend on age or size. Swimming ability and speed tends to scale with body length, such that larger individuals undertake longer migrations. Quinn and Deriso (1999) provide equations for an age-structured population with migration.

$$N_{a+1,t+1,i} = s_{a,i}\left[N_{a,t,i}\left(1-\theta_{a,i\to j}\right) + N_{a,t,j}\theta_{a,j\to i}\right] \tag{7.21}$$

where $N_{a,t,i}$ is the number at age a, time t, in region i, $\theta_{a,i\to j}$, is the *proportion* migrating from region i to region j, and $s_{a,i}$ is the annual survival rate. Note that migration is assumed to occur at the start of the year. The life cycle can be completed with a second equation to define the numbers at age 1:

$$N_{1,t+1,i} = s_{0,i}\sum_a f_{a,i} N_{a,t,i} \tag{7.22}$$

as with the age-structured model in Chapter 2. Alternatively, a density-dependent stock-recruitment function could be used (see Chapter 9). Quinn and Deriso (1999, Section 10.1.3) show how these equations can be combined into multi-region matrix models. In the simple case with two areas, the vector of abundances, $\mathbf{N}_t$, is the concatenation of numbers at age in each area. The projection matrix, $\mathbf{M}$, has a block structure comprising four submatrices:

$$\begin{bmatrix} 0 & f_{2,1} & f_{3,1} & 0 & 0 & 0 \\ s_{1,1}(1-\theta_{1,1}) & 0 & 0 & s_{1,2}\theta_{1,2} & 0 & 0 \\ 0 & s_{2,1}(1-\theta_{2,1}) & s_{3,1}(1-\theta_{3,1}) & 0 & s_{2,2}\theta_{2,2} & s_{3,2}\theta_{3,2} \\ 0 & 0 & 0 & 0 & f_{2,2} & f_{3,2} \\ s_{1,1}\theta_{1,1} & 0 & 0 & s_{1,2}(1-\theta_{1,2}) & 0 & 0 \\ 0 & s_{2,1}\theta_{2,1} & s_{3,1}\theta_{3,1} & 0 & s_{2,2}(1-\theta_{2,2}) & s_{3,2}(1-\theta_{3,2}) \end{bmatrix} \tag{7.23}$$

The submatrices on the diagonal specify the age-specific fertilities and survival of animals that do not migrate, while the off-diagonal blocks specify the corresponding rates for individuals that migrate. Once the projection matrix has been specified, it can be used to project the population abundance by time and areas according to: $\mathbf{N}_{t+1} = \mathbf{M}\times\mathbf{N}_t$. The analytical methods introduced in Chapter 2 can be used to calculate the eigenvalues, eigenvectors, sensitivities, and elasticities.

The example shown in Table 7.1 is patterned after a species like Atlantic menhaden (*Brevootia tyrannus*). Area 1 is an inshore nursery area, while Area 2 is an offshore feeding area. Survival at age 1 and 2 is assumed to be higher inshore than offshore; conversely the survival of adults (age 3+) is assumed to be higher offshore where there are fewer predators. The emigration rates (θ_a) are specified such that age-1 fish remain inshore and age-2 fish migrate offshore, where

Table 7.1 Migratory matrix model comprising two areas and three age classes (Equation 7.23). Age 3 is a plus group consisting of ages 3 and older. The juvenile survival rate, s_0, was adjusted to obtain a stationary population ($\lambda_1 = 1$). Juvenile survival in Area 2 is one half that in Area 1. The first subtable defines the annual survival (s_a), fecundity (f_a), and emigration (θ_a) rates by age and area. The population projection matrix comprises four 3×3 submatrices. The third subtable projects the population for 5 years. Modified from Quinn and Deriso (1999) Table 10.3.

		Area 1		$s_0 = 0.211$	Area 2		$s_0 = 0.105$
			Age			Age	
		1	2	3+	1	2	3+
	s_a	0.5	0.7	0.5	0.3	0.5	0.8
	f_a	0	2	3	0	3	4
	θ_a	0.2	0.7	0.8	0.8	0.3	0.1
M	Age		Area 1			Area 2	
	1	0	0.423	0.635	0	0.476	0.635
Area 1	2	0.400	0	0	0.400	0	0
	3	0	0.210	0.100	0	0.210	0.050
	1	0	0	0	0	0.079	0.106
Area 2	2	0.060	0	0	0.060	0	0
	3	0	0.350	0.640	0	0.350	0.720
	Year	Age 1	Age 2	Age 3+	Age 1	Age 2	Age 3+
	0	100	50	20	100	50	20
	1	70	80	24	6	12	62
	2	94	31	25	8	5	92
	3	89	41	14	10	6	95
	4	89	40	16	10	6	94
	5	89	40	16	10	6	94
Stable age distribution		35.0%	15.6%	7.2%	4.1%	2.3%	37.7%
Reproductive value		7.9%	17.1%	23.5%	7.9%	18.2%	25.2%

they mostly remain as adults. The fecundity of adults is assumed to be slightly higher offshore than inshore, reflecting better feeding conditions and faster growth offshore. Of the eggs spawned offshore, 75% are assumed to drift to the inshore nursery area, which explains the non-zero elements in the top right of the matrix. The 25% of larvae retained offshore experience a survival rate that is one-half the inshore juvenile survival rate.

Starting with equal numbers at age inshore and offshore, the population vector quickly equilibrates, with juveniles concentrated inshore and adults offshore (Table 7.1). Offshore adults have the highest reproductive value by virtue of their high survival rate and higher fecundity. Age-2 fish inshore have the second highest reproductive value because they are likely to survive and migrate offshore. Note that actual juvenile survival rates are orders of magnitude lower while fecundities are an order of magnitude higher, but since these two terms are multiplied to obtain the fertilities (Equation 7.7), the negative and positive exponents would cancel out. The advection of larvae from offshore spawning areas to inshore nursery areas is thought to be an important determinant of Atlantic menhaden recruitment (Checkley et al. 1988). The fraction of larvae advected each year could be made a random variable to simulate interannual variability in the menhaden population.

7.5 Summary

Aquatic populations are patchily distributed. The full implications of this statement for the dynamics of these populations depend very strongly on movement and dispersal patterns. The characteristically heterogeneous distribution of exploited aquatic species is of course essential to harvesting strategies employed by fishers. It can also present important challenges to management when species distributions contract to core habitat areas and these concentrations can be readily located and exploited. The types of models described in this chapter provide an initial framework for considering the dynamics of spatially structured populations.

For many aquatic organisms the location of spawning and larval retention areas are critical determinants of population and subpopulation structure, particularly in marine environments. For freshwater environments, the size and system type (lakes, ponds, streams, rivers) of course play key roles in defining population structure in relation to movement and dispersal.

Spatial persistence of subpopulations seems to require that density-dependent regulation is stronger than dispersal. This finding is similar to that for species interactions, whereby stability requires density dependence to be stronger than interspecific interactions. There is an underlying equivalence between spatial models and the simpler non-spatial models introduced in Chapters 2–5. Because of their underlying quadratic form, conserving half the population is equivalent to protecting half the areas.

Realistic representation of spatial processes in models of aquatic populations is an evolving art. Quantifying movement and connectivity of aquatic species entails special challenges. Spatially explicit models should account for exchange among subpopulations in relation to their size, distance, and degree of separation. We know the general tendencies of larval movement, but not the dispersal patterns at the scales necessary to describe fully metapopulation connectivity. Improved understanding of larval retention mechanisms could be derived from recent advances in circulation models and particle tracking. Advances in genetic sequencing allow more samples to be analyzed and smaller differences to be detected. Spatially explicit simulation models have been developed but parameterizing the exchange rates is an on-going challenge. Mark-recapture studies have been an important tool in fisheries science for a broad range of uses including estimating population size, determining growth rates, and establishing movement and migration patterns for life stages amenable to tagging. The last provides a rich resource to draw on for quantifying connectivity among subpopulations. It is clear that each of the issues identified above require greater understanding of animal behavior. In many respects our understanding of behavioral mechanisms underlying topics ranging from habitat selection to migration patterns is still developing. Exploitation can exert direct and indirect effects on the behavior of exploited species in ways that increase the risk of population collapse. For example, the selective removal of older, experienced animals has been implicated in alteration of migratory pathways critical to replenishment of depleted areas (Petitgas et al. 2010; MacCall 2012).

Additional reading

Pielou (1969) provides an accessible introduction to the spatial statistics of populations. For an engaging treatment of random walk processes, see Denny and Gaines (2000, Chapters 5 and 6). Advanced treatment of spatial models in population ecology is given in Kot (2001). MacCall (1990) remains the standard for treatment of the basin model and associated topics. Gotelli (2008) and Stevens (2009) provide readable accounts of metapopulation theory in their primers of ecology. Hanski (1998) gives an in-depth treatment of metapopulation ecology. For examples of marine metapopulations, see Kritzer and Sale (2006). Quinn and Deriso (1999; Chapter 10) describe fishery-related application of spatial population models.

PART II

Ecological Production

CHAPTER 8

Production at the Individual Level

8.1 Introduction

The acquisition and transformation of energy by living organisms lies at the heart of all production processes in ecological systems. In this chapter, we will decompose the elements of production at the level of an individual organism and cast models of these processes in bioenergetic terms. Here, production is the rate of elaboration of new biomass by an individual organism. We are interested in constructing energy budgets for individuals subject to mass-balance constraints (a balance between energy inputs and losses). As we shall see, the ubiquity of allometric processes is a recurring theme in describing key elements of individual production. Allometry is the study of how biological processes and other characteristics scale with body size. Metabolic ecology, which deals with fundamental allometric processes in metabolism, is now a rapidly evolving subdiscipline of ecology (Brown et al. 2004; Brown and Sibly 2012). Building on a long history of investigation into the importance of body size on metabolism, we can dissect the growth of individual organisms into inter-related elements described by simple allometric functions. For many fisheries applications, traditional treatments of individual growth center on the development of descriptors of size or weight as a function of age. For our purposes, an emphasis on the bioenergetics of growth is more appropriate. However, we also derive the benefit of obtaining growth models that can serve many purposes. In taking this approach, we return to the origins of the theory of organismal growth developed by Ludwig von Bertalanffy (1938, 1957). As we shall see, individual production of aquatic poikilotherms depends strongly on water temperature and a number of non-linear relationships between temperature and individual elements of production have been proposed. Given the rapid rate of global warming and its effects on aquatic ecosystems, strong changes in production can be anticipated.

The visible manifestation of production processes at the level of an individual organism is its change in weight as a function of age. The growth of Atlantic herring (*Clupea harengus*) off the northeastern United States illustrates dramatic changes in weight-at-age over seasonal and multi-annual time scales (Figure 8.1). The observed average weight of an individual herring reflects the translation of food intake and losses due to metabolic and other energetic expenditures into somatic and reproductive growth. The striking seasonal pattern of gonad development followed by rapid release of the gametes in the autumn is clearly manifest in the observed weights-at-age in herring (Figure 8.1). In the following we will address the fundamental features of growth processes in aquatic species.

8.2 Energy budgets for individual organisms

In constructing an energy budget for an individual, we can take advantage of a longstanding and rich tradition of research in this field (e.g. Brody 1945; Winberg1960). The recognized importance of understanding the basic physiology of animal growth in an ecological context has been essential to the development of production models relevant to management of both natural and aquaculture systems.

The production of an individual organism can be expressed as:

Fishery Ecosystem Dynamics. Michael J. Fogarty and Jeremy S. Collie, Oxford University Press (2020).
DOI: 10.1093/oso/9780198768937.003.0008

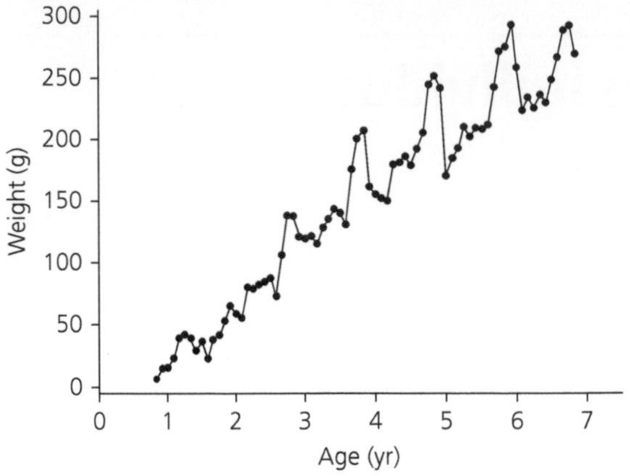

Figure 8.1 Mean weight-at-age for Atlantic herring derived from a monthly catch sampling program for the commercial fishery off the northeastern United States. (NEFSC unpublished).

$$P = P_S + P_R = I - R_D - E_F - E_N \quad (8.1)$$

where P is individual production, P_S is somatic growth, P_R is the elaboration of reproductive material, I is ingestion (consumption), R_D is the respiratory demand associated with metabolism, E_F is the egestion of solid waste (feces), and E_N is the excretion of nitrogenous waste products. Note that all terms must be expressed in the same units. Respiratory demand can be further partitioned into three components: standard metabolism (R_S) associated with routine maintenance; active metabolism (R_A) related to swimming and other activities; and specific dynamic action (R_{SDA}) reflecting the energetic costs of digesting prey. In general, each of these components is a function of body mass and environmental factors, most notably food supply and temperature. Production is a rate, often measured in units of kilocalories or kiloJoules (1 international kcal = 4.184 kJ) per day. Once an energy currency is adopted, it is important that all of the components of individual production be measured in the same units. It is not uncommon for biological production to be measured in units of biomass (or carbon) and then converted to energy. The rate of energy flow measured as kJ sec^{-1} (or kilowatts) is called power.

We can further define the Assimilation Efficiency (the proportion of ingested food available for growth) as:

$$A_e = \frac{I - E_F - E_N}{I} \quad (8.2)$$

and the Absorption Efficiency as:

$$A_b = \frac{I - E_F}{I} \quad (8.3)$$

both of which will help us convert measures of growth to metabolic rate processes. These efficiency measures can also be expressed as percentages.

In the following sections we will treat each of the elements of Equation 8.1 in turn. We begin with the development of models for individual growth from a bioenergetic perspective and then turn to consideration of reproductive processes. Because of the centrality of individual growth models in fishery science, we treat this topic in considerable detail (including consideration of weight- and length-based models). We then follow with treatments of ingestion, respiration, egestion, and excretion, and the development of full bioenergetics models incorporating each of these individual elements.

8.3 Growth

Ultimately the energy intake of an individual must meet its metabolic demands (Ursin 1979). Energy available for growth and reproduction after metabolic costs and other energy expenditures (e.g. swimming) have been met is "surplus" energy (Ware 1980). Many fish and aquatic invertebrates exhibit indeterminate growth—they can continue to grow after the age or size of sexual maturity is attained. The allocation of energy to reproduction and decreasing ingestion rates as a function of body size, however, often results in a deceleration of somatic growth. Indeterminate growth patterns in aquatic ectotherms are typically characterized by high plasticity contingent on environmental conditions. To illustrate different patterns of individual growth[1] in fish, we show the weight-at-age for

[1] In general, we do not have measures of the size-at-age of individuals over the lifespan. Rather, we typically infer growth patterns from cross-sectional data assembled from a number of individuals taken at the time of capture. Our estimates therefore include inter-individual variability in growth. However, longitudinal studies in which the distance between age marks (often "rings") in scales, otoliths, or other structures for individuals are measured can be done. It is then possible to

two congeneric species, yellowfin tuna (*Thunnus albacares*) in the Eastern Pacific and bluefin tuna (*Thunnus thynnus*) in the northwest Atlantic (Essington et al. 2001). Yellowfin tuna continues to grow rapidly throughout its lifespan of 5–7 yr (Figure 8.2a). In contrast, bluefin tuna are much longer-lived and show a reduction in the rate of growth at older ages (Figure 8.2b). It is clear we require models that can accommodate a spectrum of growth patterns. To meet this need, we begin by examining a general model for individual growth of the form:

$$\frac{dW}{dt} = f(W) \tag{8.4}$$

where $f(W)$ is a function describing the rate of change of body weight with age. One possible functional form for this model (von Bertalanffy 1938, 1957) is:

$$\frac{dW}{dt} = \eta W^m - \kappa W^n \tag{8.5}$$

where the first term on the right-hand side represents the buildup of body tissue (anabolism). The second term represents losses due to energetic costs of metabolic activity (sometimes designated as catabolism). Notice that each of these terms is in the form of a simple allometric function. If an independent estimate of the assimilation efficiency (AE) is available, we can derive an estimate of rate of ingestion by dividing the anabolic term by AE (e.g. Essington et al. 2001). We note that it is also useful to calculate the gross growth efficiency (the ratio of growth to consumption and often expressed as a percent).

In general, analytical solutions to Equation 8.5 are not possible except for certain choices of the exponents m and n. If $m = n = \varphi$, we have:

$$\frac{dW}{dt} = (\eta - \kappa)\, W^{\varphi} \tag{8.6}$$

and the solution is given by:

$$W_t = W_0 + [(1-\varphi)\,(\eta-\kappa)\,t]^{\frac{1}{1-\varphi}} \tag{8.7}$$

where W_t is the weight at time (age) t and W_0 is the initial weight. This model describes an exponential

"back-calculate" the growth history of an individual organism if the relationship between body size and the size of ageing structure is known. Because it is very labor intensive, this approach is often restricted to special studies rather than large-scale fishery analyses.

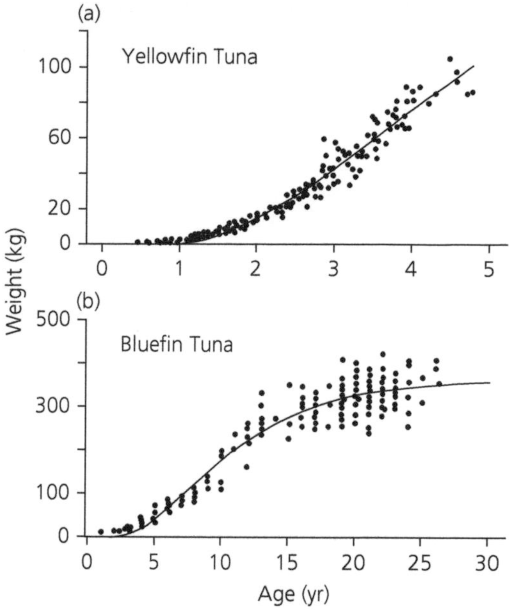

Figure 8.2 Weight-at age (kg) for (a) yellowfin tuna (*Thunnus albacares*) and (b) bluefin tuna (*Thunnus thynnus*) with fitted von Bertalanffy growth curves (adapted from Essington et al. 2001).

change in weight with increasing age. Notice that in fitting this model to an observed series of weights-at-age, we will not be able to separately identify values of η and κ. For the special case $\varphi \rightarrow 1$, the solution is:

$$W_t = W_0 e^{(\eta-\kappa)t} \tag{8.8}$$

This form does not have an asymptote and is appropriate for species exhibiting strongly indeterminate growth. Interestingly, although the observed weight-at-age for yellowfin tuna (Figure 8.2a) superficially appears to conform to a model of this type, it does in fact depart from actual exponential growth and requires a somewhat more complicated model.

We next turn to cases in which the allometric exponents differ. The case $n = 1$ in Equation 8.5 has been accorded special attention in the development of models of individual growth. The model then takes the form:

$$\frac{dW}{dt} = \left(\eta W^{m-1} - \kappa\right) W \tag{8.9}$$

(see von Bertalanffy 1938, 1957; Richards 1959; and Paloheimo and Dickie 1965). The case in which m can assume any positive value is sometimes referred

Box 8.1 Solution to the Generalized von Bertanffy Growth Function

We start with the basic form of the Generalized von Bertanffy Growth Model:

$$\frac{dW}{dt} = \left(\eta W^{m-1} - \kappa\right)W$$

This is a separable differential equation:

$$\frac{dW}{\left(\eta W^{m-1} - \kappa\right)W} = dt$$

Note that the general rule for the solution to the left-hand side of this equation is:

$$\int \frac{dx}{\left(a + bx^n\right)x} = -\frac{1}{an}\log_e\left(\frac{a + bx^n}{x^n}\right)$$

where we have $a = -\kappa$, $b = \eta$, and $n = m-1$. For the right-hand side we have

$$\int_{t_0}^{t_t} dt = t - t_0$$

and to simplify, we will let $t_0 = 0$. The model then becomes:

$$\frac{1}{\kappa\,(m-1)}\log_e\left(\frac{\eta W_t^{m-1} - \kappa}{W_t^{m-1}}\right) - \frac{1}{\kappa\,(m-1)}\log_e\left(\frac{\eta W_0^{m-1} - \kappa}{W_0^{m-1}}\right) = t$$

or

$$\log_e\left(\frac{\eta W_t^{m-1} - \kappa}{W_t^{m-1}} \cdot \frac{W_0^{m-1}}{\eta W_0^{m-1} - \kappa}\right) = \kappa\,(m-1)t$$

Taking antilogs and rearranging we have:

$$\eta - \frac{\kappa}{W_t^{m-1}} = \left(\eta - \frac{\kappa}{W_0^{m-1}}\right)e^{\kappa(m-1)t}$$

or

$$\frac{\kappa}{W_t^{m-1}} = \eta - \left(\eta - \frac{\kappa}{W_0^{m-1}}\right)e^{\kappa(m-1)t}$$

and after inverting and rearranging we obtain:

$$W_t = \left[\frac{\eta}{\kappa} - \left(\frac{\eta}{\kappa} - W_0^{1-m}\right)e^{-(1-m)\kappa t}\right]^{\frac{1}{1-m}}$$

In this model, $(\eta/\kappa)^{1/(m-1)}$ is the limiting or upper size. The above solution can be simplified by letting $W_\infty{}^{1-m} = \eta/\kappa$, $h = (\eta/\kappa - W_0{}^{1-m})W_\infty{}^{m-1}$, and $k = (1-m)\kappa$ to give:

$$W_t = W_\infty\left(1 - be^{-kt}\right)^{\frac{1}{1-m}}$$

to as the generalized von Bertalanffy function (for $m = 2/3$ we have the so-called specialized von Bertalanffy function). The solution is given by:

$$W_t = \left[\frac{\eta}{\kappa} - \left(\frac{\eta}{\kappa} - W_0^{1-m}\right)e^{-(1-m)\kappa t}\right]^{\frac{1}{1-m}} \tag{8.10}$$

(see Box 8.1). In this model, $(\eta/\kappa)^{1/(m-1)}$ is the limiting or maximum size (designated W_∞). The above solution can be simplified by letting $W_\infty{}^{1-m} = \eta/\kappa$, $h = (\eta/\kappa - W_0{}^{1-m})\,W_\infty{}^{m-1}$, and $k = (1-m)\kappa$ to give:

$$W_t = W_\infty\left(1 - he^{-kt}\right)^{\frac{1}{1-m}} \tag{8.11}$$

The shape of the curve for different levels of the parameter m is shown in Figure 8.3 (note that the sign of h will change depending on whether m is

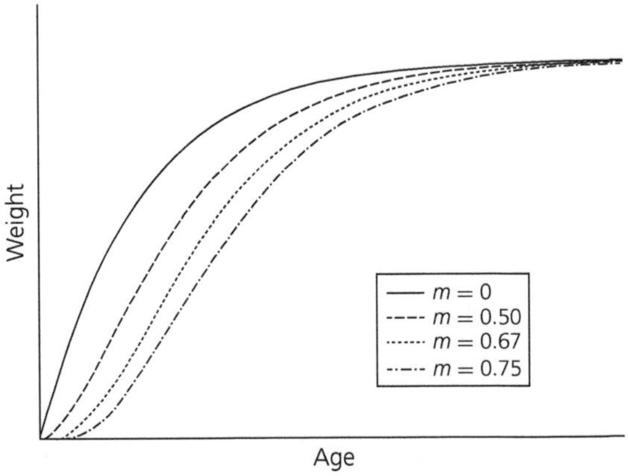

Figure 8.3 Body size (weight) as function of age for the von Bertalanffy growth function with the catabolic exponent set at $n = 1$ and the anabolic exponent m ranging from 0 to 0.75.

Box 8.2 Special Cases of the von Bertalanffy Growth Function

In the following, we describe special cases of the von Bertalanffy growth function corresponding to different choices of the exponent m. When $m = 2/3$, we have the specialized von Bertalanffy growth curve:

$$W_t = W_\infty\left(1 - be^{-kt}\right)^3$$

When $m = 0$, we have the monomolecular model:

$$W_t = W_\infty\left(1 - be^{-kt}\right)$$

which does not exhibit an inflection point. For $m = 2$, we have the logistic model:

$$W_t = \frac{W_\infty}{\left(1 + be^{-kt}\right)}$$

which exhibits the highest growth rate when W_t is one half the maximum size. Finally, when $m \rightarrow 1$, the model becomes the Gompertz function:

$$W_t = W_\infty e^{-be^{-kt}}$$

which is a double exponential curve with a maximum growth rate at W_∞/e.

greater than or less than 1). The weight-at- age of both yellowfin and bluefin tuna can be described by this model. In Box 8.2 we show model forms for different choices of m corresponding to several well-known growth functions.

8.3.1 Growth in length

In practice, it is far more common in fishery sampling programs to record body size in units of length than in weight. A principal motivation for large-scale age-determination programs in fisheries is to characterize the age composition of the catch and of the population. In most instances this is accomplished by at-sea or port sampling using a two-stage sampling program in which a large number of length measurements is made. Age determinations are then undertaken for a subsample of these measured individuals.

An illustration of observed length-at-age for several long-lived species representing a diverse array of taxa, body sizes, and life-history patterns is provided in Figure 8.4. The ocean quahog (*Arctica islandica*), is the longest-lived, non-colonial, aquatic organism known, with a sharply defined asymptotic size of less than 10 cm shell length (Figure 8.4a; Kilada et al. 2007). Sturgeon are among the most primitive fish species and can attain body lengths of several meters. The largest recorded individual was 7.2 m caught in 1827 in the Volga River. Semakula and Larkin (1968) provide length-at-age estimates for white sturgeon (*Acipenser transmontanus*) taken incidentally in the salmon gillnet fishery of the Fraser River in British Columbia (Figure 8.5b). The Hawaiian green sea turtle (*Chelonia mydas*) has an estimated lifespan of over 50 years and can attain a carapace length of over one meter (Zug et al. 2002; Figure 8.4c).

For our purposes we ultimately need to translate length-at-age to weight-at-age. One avenue to effect this transformation is to first establish the relationship between body weight and length and then to incorporate this representation into a model of size-at-age. Observations suggest that an allometric function of the form:

$$W = al^b \tag{8.12}$$

can be used effectively to model the relationship between length (l) and weight (W). For the special case in which direct proportional change in linear dimensions of body size (length, width, depth) is maintained, $b = 3$, we have a so-called isometric relationship. We note that the ratio of body weight to the cube of length is a commonly used measure of condition (often referred to as Fulton's K) although generalizations that do not require $b = 3$ are also used and are less restrictive. This special case has been assumed in a number of weight-transformed length models because of some of its simplifying properties in yield models under this constraint (e.g. Beverton and Holt 1957). However, the estimated length–weight exponent does depart from 3 for a number of species. For example, the mean estimate of b for over 70 species on the northeast U.S. continental shelf is 3.1; in general, we should not assume a priori that $b = 3$.

We can expect within-year differences in the length–weight relationship in relation to factors such as food availability, feeding activity, and reproductive status. We typically employ length–weight relationships that have been averaged

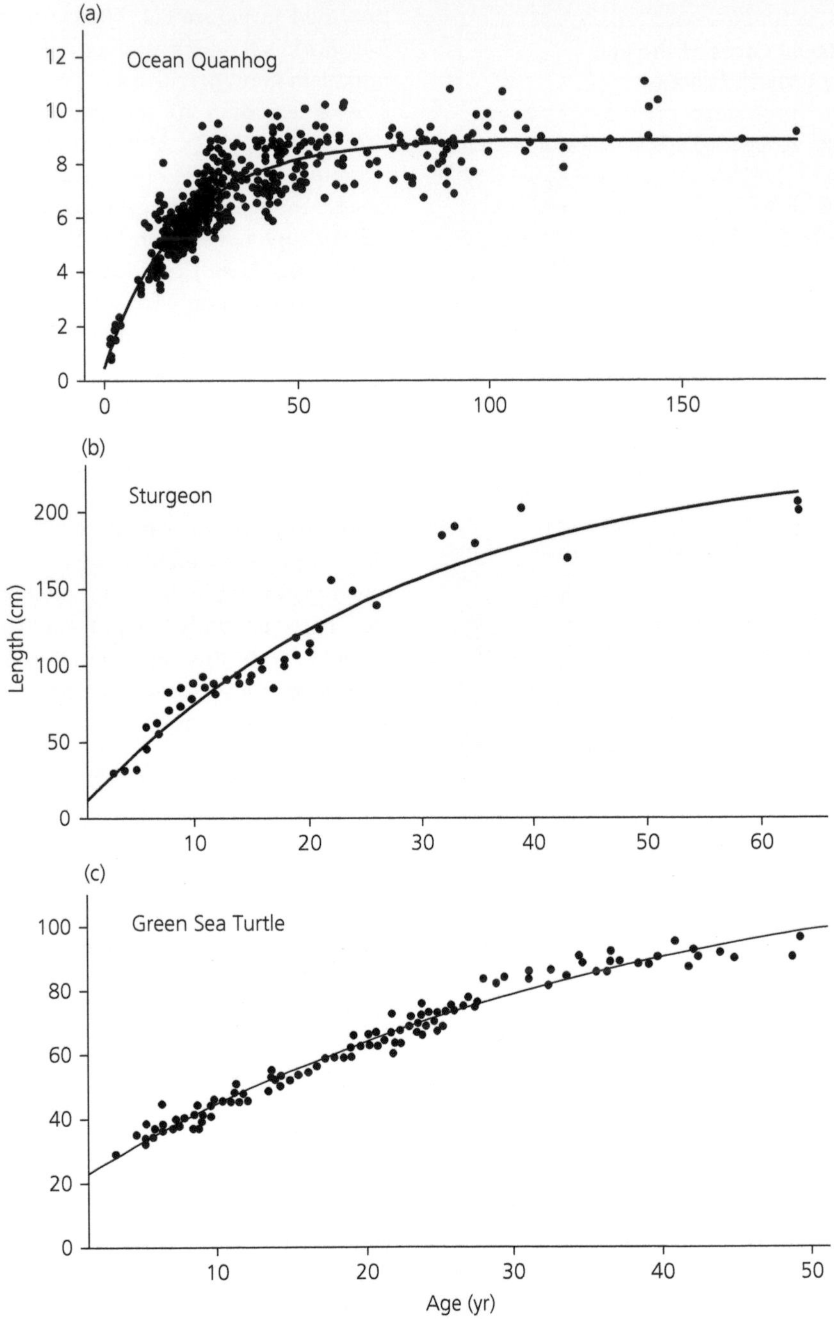

Figure 8.4 Growth in length for (a) Ocean quahogs (*Artica islandica*) on Sable Bank, (b) white sturgeon (*Acipenser transmontanus*) in the Fraser River, and (c) green sea turtles (*Chelonia mydas*) off Hawaii. Ocean quahog data courtesy of Raouf Kilada. Sturgeon data from Semakula and Larkin (1968). Green turtle data from Zug et al., (2002).

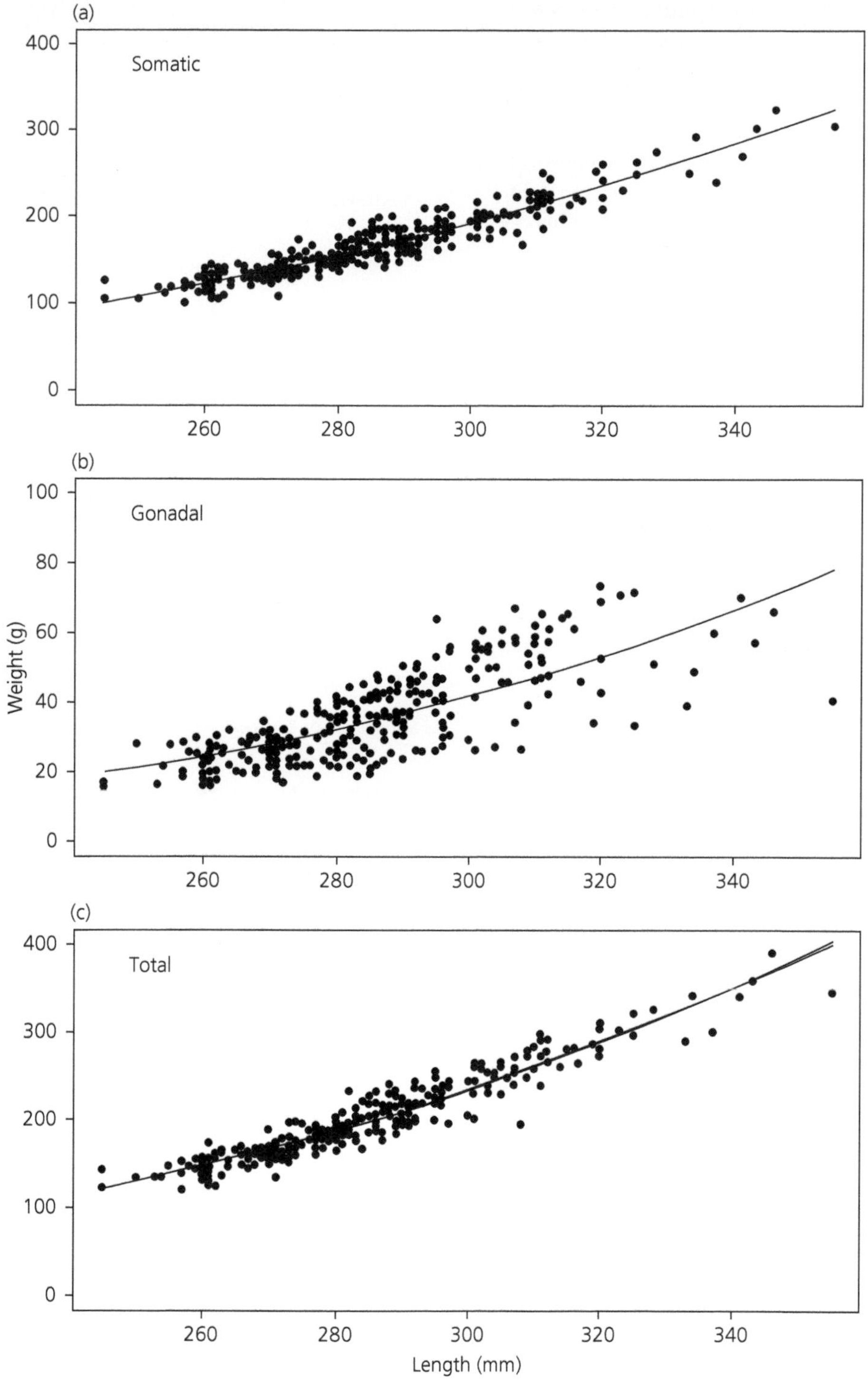

Figure 8.5 Relationship between length and (a) somatic, (b) gonadal, and (c) total weight for Atlantic herring (*Clupea harengus*) sampled during the spawning season off the Northeast Coast of the United States. (NEFSC unpublished).

over seasonal cycles. An example for which it has been possible to examine somatic and gonadal weights specifically during the spawning season is shown in Figure 8.5. For Atlantic herring off the northeastern United States, the estimated exponent for the somatic weight–length relationship is 3.07. Interestingly, the exponent for the gonadal weight–length relationship is much higher (4.19) for samples restricted to the spawning season.

Perhaps not surprisingly, the variability in gonad weight relative to length is substantially higher than the somatic weight–length relationship. The sampling program encompasses a time period in which gonadal development is rapid prior to spawning, accounting for the greater variability in the gonad weight–length relationship. The exponent of the total weight–length relationship is 3.19. Gonad weight during the spawning season accounts for 17.2% of total body weight and a very substantial energetic investment in reproduction is evident.

One of the simplest specifications of a growth function exhibiting decelerating growth in length can be written:

$$\frac{dl}{dt} = k\,(L_\infty - l) \tag{8.13}$$

where l is size (length), k is the growth coefficient, and L_∞ is the maximum (asymptotic) length. Notice that in this formulation, as l approaches L_∞, the rate of change in length approaches zero. The integrated form of this equation is used to represent the length-at-age of the species depicted in Figure 8.4. We can combine information from the derivative of the length–weight relationship (Equation 8.12) and Equation 8.13 to form the identity:

$$\frac{dW}{dt} = \frac{dW}{dl} \cdot \frac{dl}{dt} \tag{8.14}$$

(recall that the chain rule tells us that $dx/dt = (dx/dy)(dy/dt)$. We then have:

$$\frac{dW}{dt} = a \cdot b \cdot l^{b-1} k\,(L_\infty - l) \tag{8.15}$$

and noting from Equation 8.12, that $l = (W/a)^{1/b}$, we can write:

$$\frac{dW}{dt} = k \cdot L_\infty a \cdot b \cdot \left(\frac{W}{a}\right)^{\frac{b-1}{b}} - k \cdot b \cdot W \tag{8.16}$$

Making the substitutions $\omega = kL_\infty ba^{1/b}$ and $\phi = bk$ we obtain:

$$\frac{dW}{dt} = \left(\omega W^{\left(\frac{b-1}{b}\right)-1} - \varphi\right) W \tag{8.17}$$

and its solution is similar in form to Equation 8.10. For the special case in which $b = 3$ (the isometric form), we again obtain the specialized von Bertalanffy model.

Alternatively, we can integrate Equation 8.13 to give:

$$l_t = L_\infty \left(1 - he^{-kt}\right) \tag{8.18}$$

where now $h = exp(kt_0)$ and t_0 is the hypothetical age at which the length is zero. Unlike the weight-based model, this model does not exhibit an inflection point in growth at younger ages.

We can combine Equation 8.18 with the length–weight relationship (Equation 8.12) to give the von Bertalanffy model (Equation 811) in weight:

$$W_t = W_\infty \left(1 - he^{-kt}\right)^b \tag{8.19}$$

where $W_\infty = a\,L_\infty{}^b$.

8.3.2 Seasonal growth

Most applications of individual growth models as descriptors have used annual time steps. However, it also possible to construct growth models with finer temporal resolution. For example, a number of models of seasonal growth employing trigonometric functions to represent within-year growth have been developed (e.g. Pitcher and MacDonald 1973). One possible form (Hoenig and Hannamura 1990) is:

$$\frac{dl}{dt} = k\,(L_\infty - l)\left[1 - \vartheta \cos\left(\frac{2\pi}{\Omega\,(t - t_s)}\right)\right] \tag{8.20}$$

where ϑ determines the amplitude of the oscillation, Ω is the period, and t_s represents the phase. The solution is given by:

$$l_t = L_\infty \left(1 - e^{-\left[k(t-t_0) - \left(\frac{k\Omega}{2\pi}\right)\sin\left(\frac{2\pi}{\Omega(t-t_s)}\right) + \left(\frac{k\Omega}{2\pi}\right)\sin\left(\frac{2\pi}{\Omega(t_0-t_s)}\right)\right]}\right) \tag{8.21}$$

for the initial condition $l_0 = 0$. An illustration of the size-at-age trajectory is provided in Figure 8.6. For the case of $\Omega > 1.0$, this particular specification can result in a seasonal reduction in size. For cases where shrinkage in linear dimensions (e.g. total length) is not observed, appropriate constraints on

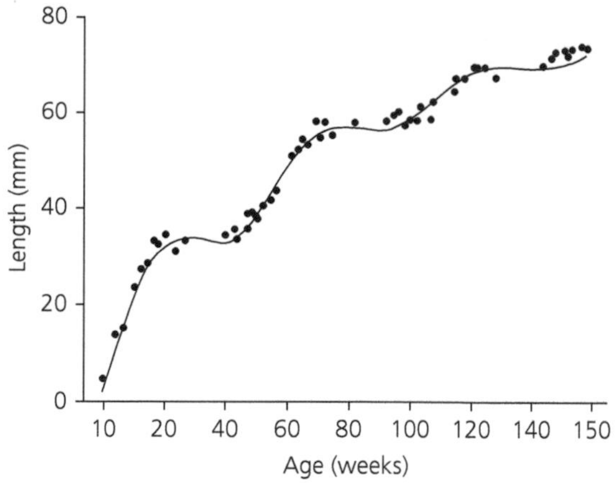

Figure 8.6 Seasonal growth of minnows (*Phoxinus phoxinus*) in a tributary of the River Thames (points). The fitted line is the seasonal growth model of Hoenig and Hannumara (1990). Data extracted from Pitcher and MacDonald (1973).

the parameter Ω must be applied for models of length-at-age. As evident in Figure 8.1 however, seasonal reductions in weight are entirely feasible and must be accommodated.

The seasonal growth in weight corresponding to Equation 8.21 is:

$$W_t = W_\infty \left(1 - e^{-k[t-t_0]-\left(\frac{k\Omega}{2\pi}\right)\sin\left(\frac{2\pi}{\Omega(t-t_s)}\right)} + \left(\tfrac{k\Omega}{2\pi}\right)\sin\left(\tfrac{2\pi}{\Omega}(t_0-t_s)\right)\right)^b \qquad (8.22)$$

Note that we have not invoked a specific mechanism driving the seasonal growth pattern although temperature and food availability are clearly critical factors. In Section 8.5 we will consider models in which an explicit connection with temperature is made.

8.3.3 Growth models in discrete time

We have so far considered only the case in which growth is modeled with differential equations. There are instances, however, where we wish to approximate the continuous time model(s) with a difference equation. In these cases, we are often interested in using annual (or in some cases, semi-annual) time steps for ease in computation. There are also important cases, however, in which growth in linear dimensions such as length is truly discontinuous. In the following we will examine examples of each of these cases.

Paloheimo and Dickie (1965) offer a specification of the bioenergetics growth model in discrete time. Their model can be expressed:

$$\frac{\Delta W}{\Delta t} = I - L \qquad (8.23)$$

where ΔW is the change in weight, I is energy intake, and L is energy loss over a specified time interval Δt. Paloheimo and Dickie (1965) suggest that a variant of the saturating consumption model of Ivlev (1965; see Chapter 4) and an allometric energy loss model can be combined to give:

$$\frac{\Delta W}{\Delta t} = c_{max}\left[1 - e^{-v'N'}\right] - \kappa W^n \qquad (8.24)$$

where c_{max} is the maximum rate of consumption, $v\prime$ is a measure of conversion efficiency and N' is prey concentration. Paloheimo and Dickie (1965) further suggest an alternative structure using an allometric function for ingestion, essentially resulting in a discrete-time version of Equation 8.5.

Other applications of discrete growth models have been developed to forge a relationship between delay–difference population models expressed in numbers (e.g. Allen 1971; Clark 1976; see Equation 3.33) and ones expressed in biomass (Deriso 1980; Schnute 1985). The weight at age $t+1$ can be expressed:

$$W_{t+1} = W_t + \rho\left(W_t - W_{t-1}\right) \qquad (8.25)$$

where $\rho = exp(-k)$ and it is clear that successive weight increments decline continuously with increasing age. In Box 8.3 we demonstrate the steps in deriving the corresponding expression for growth increments in length units. The derivation of Equation 8.25 follows the same sequence shown in Box 8.3 for length. If we are interested in tracking the weight starting with the age of recruitment ($t = r$) we can write the simple recurrence relationship:

$$W_t = W_r \frac{1 - \rho^{t-r+1}}{1 - \rho} \qquad (8.26)$$

an expression that we will encounter again in Chapter 11.

Box 8.3 Discrete Incremental Growth

Here we show how the von Bertanffy growth in length model (Equation 8.18) can be reconfigured to express incremental changes in length over discrete time periods (typically one year). We can write an expression for the length at age $t+1$ as:

$$l_{t+1} = L_\infty \left(1 - he^{-k(t+1)}\right)$$

and the difference between lengths at successive ages ($l_{t+1} - l_t$) is obtained by simply subtracting our earlier expression for l_t from that for l_{t+1} above to give:

$$l_{t+1} - l_t = L_\infty \left(1 - he^{-(kt+k)}\right) - L_\infty \left(1 - he^{-kt}\right)$$

which can be further simplified to:

$$l_{t+1} - l_t = L_\infty he^{-kt} \left(1 - e^{-k}\right)$$

We can rearrange the basic model for length at age as:

$$L_\infty - l_t = L_\infty h' e^{-kt}$$

and substituting this result into the previous expression, we have:

$$l_{t+1} - l_t = L_\infty \left(1 - e^{-k}\right) - l_t + l_t e^{-k}$$

We can simplify this further by now letting $\rho = \exp(-k)$ and expressing l_{t+1} as a function of l_t by next adding l_t to both sides of the equation to obtain:

$$l_{t+1} = L_\infty (1 - \rho) + \rho l_t$$

which is known as Ford's growth equation (Ford 1933). The Walford (1946) plot of l_{t+1} against l_t, has slope ρ and intercept $L_\infty(1-\rho$), from which L_∞ can be estimated, once is ρ known. Finally, subtracting

$$l_t = L_\infty (1 - \rho) + \rho l_{t-1}$$

from the expression above for l_{t+1} and simplifying gives the relationship between successive growth increments:

$$(l_{t+1} - l_t) = \rho \left(l_t - l_{t-1}\right)$$

which is also linear with slope ρ.

8.3.3.1 Discontinuous growth

A number of crustacean taxa (notably lobsters, crayfish, crabs, and shrimp) are among the most valuable and widely sought fishery resources. Crustaceans grow through molting processes in which the old exoskeleton is shed, the body expands through uptake of fluids over a relatively short period of time, and the new shell hardens, fixing the new size.

Crustacean growth expressed in terms of carapace length or carapace width can be expressed as a function of the probability of molting in a specified time interval (or alternatively, the intermolt duration) and the length increment per molt. Particularly in temperate and boreal regions, molting typically occurs in well-defined seasonal periods. Crustacean growth has been described with the classical continuous growth models described in previous sections[2]. However, specialized molt-process models capturing the fundamentally discontinuous nature of crustacean growth in linear dimensions have also been developed (for reviews, see Wahle and Fogarty 2006; Chang et al. 2012).

In general, both the intermolt duration and the molt increment is a function of pre-molt size, temperature, and other factors, including food supply, population density, and biological condition (including limb loss). Estimates of the molt increment from field data typically involve mark and recapture information for individuals at large at least one full molting period following tagging.

Approaches taken to determining the intermolt period necessarily differ for taxa exhibiting well defined seasonal molting periods (principally in temperate-boreal systems) and those exhibiting continuous, non-synchronous molting patterns in tropical and subtropical systems. Commonly used functional forms to describe intermolt duration include polynomial and exponential models (Wahle and Fogarty 2006; Chang et al. 2012). For species in seasonal environments, it is more common to estimate the probability of molting within a specified time period (typically one year). Characterization of the molt probability based on exponential, polynomial, and logistic functions have been widely used (Wahle and Fogarty 2006; Chang et al. 2012). An estimate of the intermolt duration can be obtained as the inverse of the molt

[2] We note that following the molt, tissue build-up proceeds and overall increases in weight are not sharply discontinuous. Accordingly, continuous-time models can be appropriately applied for weight-at-age models.

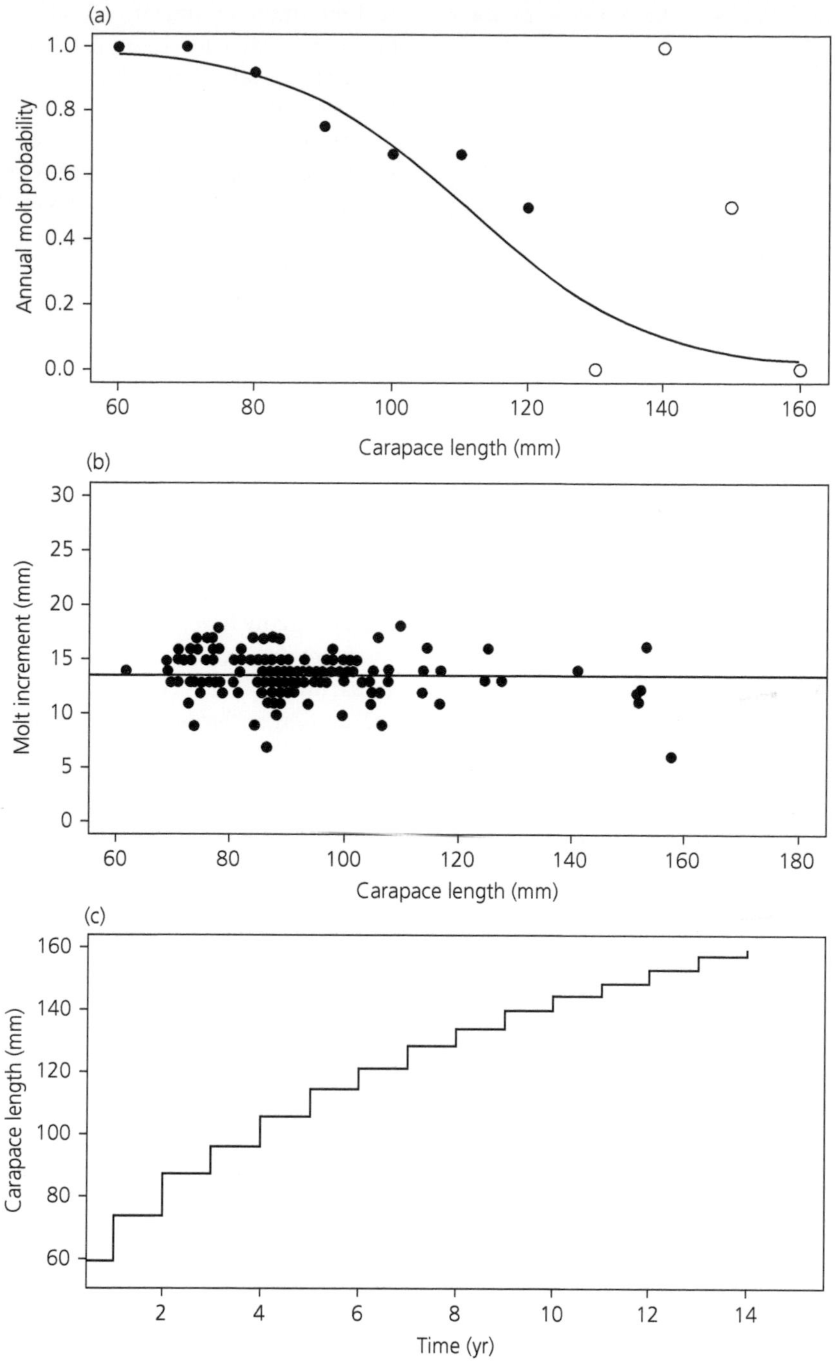

Figure 8.7 Molt process model for the American lobster (*Homarus americanus*). Information on (a) the length-specific probability of molting and (b) expected molt increment, is used to determine the (c) expected size at successive times with an initial size of 60 mm carapace length at coded time 0. Adapted from Fogarty and Idoine (1988).

probability if this can be determined with some precision.

We can specify a discrete growth model in which linear size can be represented in matrix form:

$$\begin{pmatrix} l_{1,t+\Delta t} \\ l_{2,t+\Delta t} \\ l_{3,t+\Delta t} \\ \cdot \\ l_{n,t+\Delta t} \end{pmatrix} = \begin{pmatrix} p_{1,1} & \cdot & \cdot & \cdot & \cdot \\ p_{2,1} & p_{2,2} & \cdot & \cdot & \cdot \\ \cdot & p_{2,3} & p_{3,3} & \cdot & \cdot \\ \cdot & \cdot & \cdot & \cdot & \cdot \\ \cdot & \cdot & \cdot & p_{n,n-1} & p_{n,n} \end{pmatrix} \cdot \begin{pmatrix} l_{1,t} \\ l_{2,t} \\ l_{3,t} \\ \cdot \\ l_{n,t} \end{pmatrix} \tag{8.27}$$

where the elements on the main diagonal or the transition matrix ($p_{i,i}$) represent the probability of a lobster remaining within the same class during the time interval. Note that $\sum_{i=1}^{n} p_{ij} = 1$ and $1 \geq p_{ij} \geq 0$. All the elements below the main diagonal represent the probability of increasing in size during the time interval. For simplicity in Equation 8.27 we have shown only the case in which an individual remains the same size or advances to the next size class in a specified period of time (the entries on the subdiagonal). In this representation, growth is constrained to the next largest size class. In this case, all elements except the main and subdiagonals are zero. We can allow for growth to non-adjacent size categories by incorporating non-zero elements beyond those on the main and subdiagonal. Shrinkage in size can be represented by non-zero elements above the main diagonal. Note that the elements of the main diagonal represent the complement of the molt-probability estimates. An example of this approach for female lobsters (*Homarus americanus*) on the outer continental shelf of the United States is shown in Figure 8.7. Mark–recapture information was used to determine the probabilities based on annual molt increments and molt probabilities. In Figure 8.7, we show the expected length-at-time; however, we can readily represent the full probability distributions for size-at-age (Fogarty and Idoine 1988).

8.4 Reproductive processes

In the preceding sections, we have not explicitly separated the issue of somatic and reproductive growth and have simply dealt with overall body size. We have inferred that the energetic costs of reproduction and other factors are reflected in a diminution in overall somatic growth. It can be inferred that for some species the cumulative costs of reproduction differ substantially between females and males, leading to strong sexual dimorphism in body size. We can illustrate some key points using growth estimates for the Steller sea lion (*Eumetopias jubatus*) in Alaska (Winship et al. 2001). Male sea lions attain body weights over twice that of females (Figure 8.8; Winship et al. 2001). Sexual selection has often been invoked in instances in which males compete for females and breeding territories. Large body size confers an important advantage in these contests which account for an important part of the reproductive energy costs for males. For females, the approach to an asymptotic size is sharply defined relative to that of males, presumably reflecting the very different selective pressures on males and females. In general, we can expect that growth patterns following maturation can take many forms within and between species. In the following, we describe models in which growth is partitioned into two stanzas, with the age-at-maturation as the demarcation point.

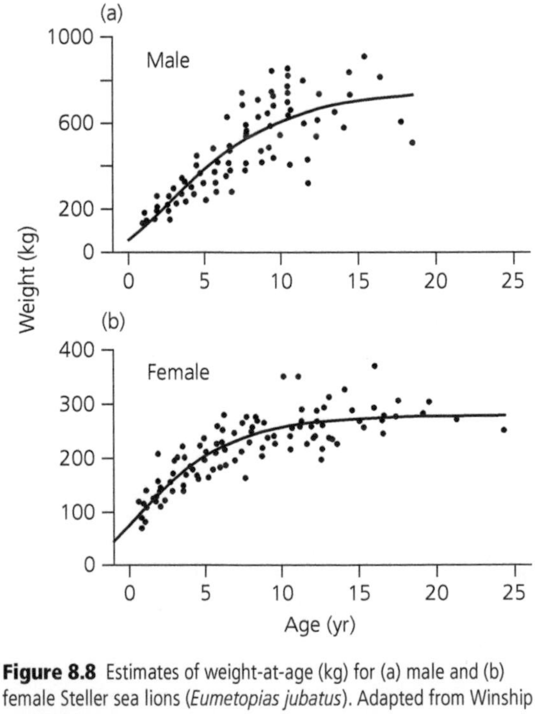

Figure 8.8 Estimates of weight-at-age (kg) for (a) male and (b) female Steller sea lions (*Eumetopias jubatus*). Adapted from Winship et al. (2001).

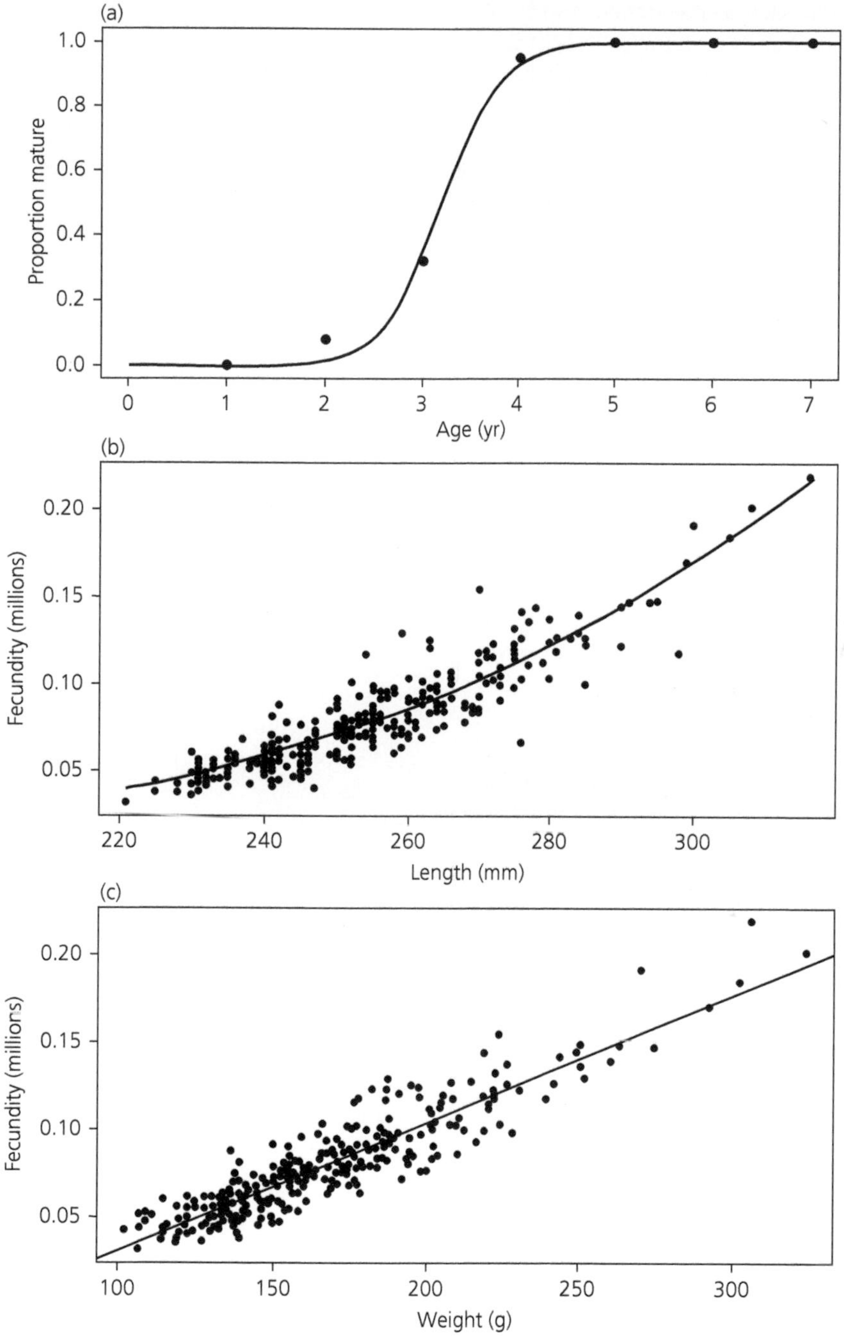

Figure 8.9 Estimated (a) age-at-maturity, (b) fecundity as a function of body length, and (c) fecundity as a function of body weight for Atlantic herring (*Clupea harengus*) off the northeast coast of the United States. (NEFSC unpublished).

We first need to define the demarcation point. For an individual organism, maturation is of course a binary process (immature or mature). There is, however, individual variation in the size- or age-at-maturation. Typically, an attempt to capture this variation is represented in the form of a maturity ogive (the cumulative probability of being mature as a function of age). We can readily model this relationship with a form of a logistic function:

$$p_m(a) = \frac{1}{1 + e^{-\alpha(a - a_{50\%})}} \tag{8.28}$$

where $p_m(a)$ is the proportion mature at age a, $a_{50\%}$ is the age at which half the individuals are mature, and α is a slope or shape coefficient. We can use logistic regression to estimate the parameters from the raw maturity data to properly account for the error structure (maturity is a binomially distributed random variable).

To provide an illustration, we return to Atlantic herring (Figure 8.9), which shows a transition to maturity over an approximate 2-yr time interval for the population as a whole. In the following, we will continue to frame our models in units of body weight. We can, however, relate fecundity to measures of body size (both length and weight) to provide a direct connection with reproductive processes. We can again express the relationship between fecundity and different measures of body size as an allometric function. We find that the exponent of the relationship between total body weight and fecundity is 1.44. It is typically assumed that the relationship between body weight and fecundity has an exponent of ~ 1. The allometric relationship obtained here indicates that, for this herring stock, larger females produce proportionately more eggs. In a meta-analysis involving 45 fish species, Barneche et al. (2018) reported that the relationship between body mass and fecundity was allometric with a slope of 1.24. These results hold implications for the importance of preserving larger females as reproductive stock.

8.4.1 Partitioning somatic and reproductive growth

Day and Taylor (1997) developed a two-stanza growth function for juveniles and adults that can be expressed as a modification of Equation 8.6:

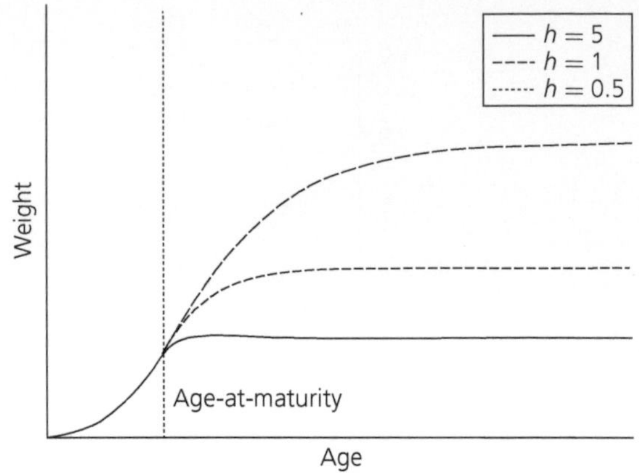

Figure 8.10 Day and Taylor growth model for three levels of the adult growth parameter h. This parameter controls the relative allocation to somatic and reproductive growth with higher h values representing proportionality greater allocation to reproduction.

$$\frac{dW}{dt} = p_s (\eta - \kappa) W^{\varphi} \tag{8.29}$$

where p_s is the proportion of resources devoted to somatic growth and all other terms are defined as before. Day and Taylor (1997) set $\varphi = 2/3$, corresponding to isometric growth. For juveniles, $p_s = 1$ and the solution is identical in form to Equation 8.7, resulting in a period of exponential growth in weight. For adults we have:

$$\mu = e^{-h'(t - a_{mat})} \tag{8.30}$$

where a_{mat} is the age at maturity and h' controls the relative allocation to somatic growth (notice that Equation 8.30 is necessarily bounded between zero and one). For higher levels of h', we approach determinate growth in which change in body size following maturation effectively halts. The solution to Equation 8.24 for the adult phase of the life history is then:

$$W_t = W_A + \left[(1 - \varphi)\frac{\eta - \kappa}{h'}\left(1 - e^{-h't}\right)\right]^{\frac{1}{1-\varphi}} \tag{8.31}$$

where W_t is the weight at time (age) t and W_A is the weight-at-maturity. We can then piece together the juvenile and adult models as shown in Figure 8.10. Lester et al. (2004) develop a related approach and explore its implications for optimal life-history strategies involving tradeoffs between growth and reproduction.

8.5 Temperature-dependent growth

Temperature affects virtually every aspect of the biology and ecology of aquatic organisms. For example, Elliot (1975) demonstrated strong thermal effects on growth of brown trout fed to satiation and reared under different temperature regimes in the laboratory. We note that increasing temperature also increases metabolic demand, with important implications if food availability is a constraint, as is likely under natural conditions. The specific growth rate [$(1/W)(dW/dt)$] of brown trout at four different initial body weights as a function of temperature is shown in Figure 8.11. The specific growth rate in this case is a non-linear function of temperature with a well-defined maximum for each of the size classes tested. Here we see that there is an optimum temperature beyond which the specific growth rate sharply declines. Having a clear understanding of the effect of temperature on growth rates assumes particular importance in light of escalating water temperatures with global warming. Growth rates affect not only individual production but vulnerability to predation, reproductive output, and many other factors. In the following, we will provide examples of how temperature effects have been incorporated into the classical models of individual growth. In Section 8.6, we will take up the theme of temperature effects in further detail with respect to the individual elements of production.

Several approaches to modifying the von Bertalanffy growth function to account for temperature effects have been developed. Mallet et al. (1999) expressed the von Bertalanffy growth coefficient k as a function of temperature:

$$k = k_{opt}\left[\frac{(T - T_{min})\,(T - T_{max})}{(T - T_{min})\,(T - T_{max}) - \left(T - T_{opt}\right)^2}\right] \tag{8.32}$$

where k_{opt} is the value of k at the optimum temperature (T_{opt}), and T_{min} and T_{max} are the minimum and maximum temperature levels at which growth occurs for the species/stock. The shape of the curve depends on where the optimum temperature lies between T_{min} and T_{max}. For optimum temperatures closer to T_{min}, the curve is backward-peaked while for optimum temperatures closer to T_{max} it is forward-peaked).

Walters and Essington (2010) address the issue of temperature-dependent growth in the context of the bioenergetically based representation of von Bertalanffy growth:

$$\frac{dW}{dt} = \eta W^m f_I(T) - \kappa W^n f_M(T) \tag{8.33}$$

where $f_I(T)$ is a function describing the effect of temperature on the anabolic component of growth and $f_M(T)$ describes the effect on the catabolic component. These multipliers are centered on the Q_{10} concept—a measure of the rate of change of a biological process of raising the ambient temperature by 10 °C. The temperature function for the anabolic component is:

$$f_I(T) = Q_I^{\frac{T-10}{T}}\left[\frac{e^{-g(T-T_m)}}{1 + e^{-g(T-T_m)}}\right] \tag{8.34}$$

Here, Q_I is the Q_{10} coefficient for anabolism, T is temperature, T_m is the temperature at which feeding drops to one half the predicted rate from Q_I, and g controls the rate at which ingestion declines as T_m is approached. Walters and Essington (2010) provided empirical estimates of Q_I for 5 species ranging from 2.81 to 9.71. Walters and Essington (2010) employ a power curve adhering to the standard Q_{10} representation as the multiplier for catabolic processes:

$$f_M(T) = Q_M^{\frac{T-10}{T}} \tag{8.35}$$

8.5.1 Physiological time units

We can approach the issue of integrating temperature information into growth models in a fundamentally different way. Brander (1995) developed a simple but effective approach by combining age and temperature observations to construct a hybrid timescale relevant to physiological processes. To illustrate the approach, we show the relationship between age and weight for five northeast Atlantic cod stocks (Figure 8.12a). Clear differences in elevation of the growth function (in this case a simple exponential model for ages 2–6) are evident. The growth pattern essentially follows a latitudinal gradient related to water temperature. Plotting the mean weights against the product of age and temperature (Figure 8.12b), however, resolves the

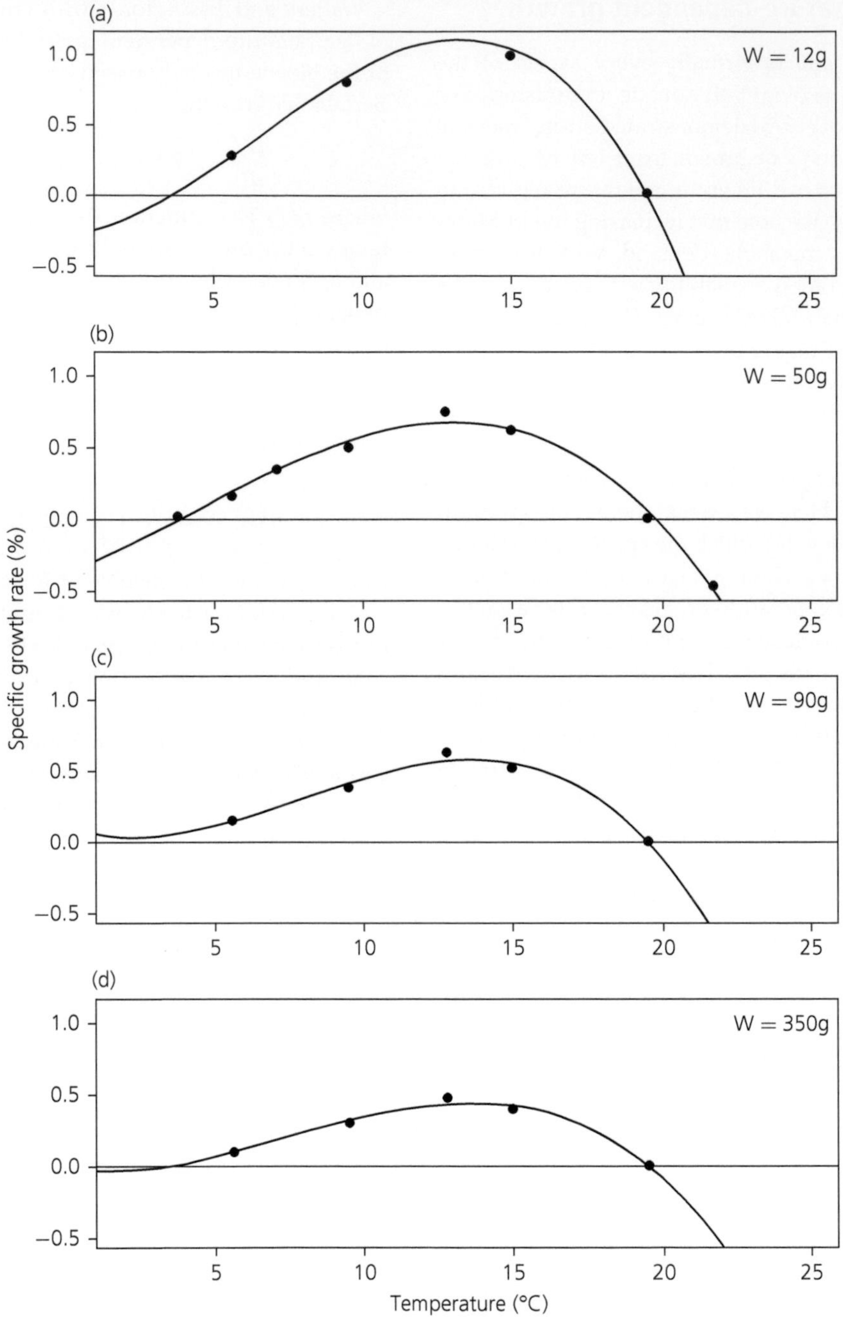

Figure 8.11 Specific growth rate of brown trout (*Salmo trutta*) as a function of temperature for approximate initial mean weights of (a) 12g, (b) 50g, (c) 90g, and (d) 350g. (Data extracted from Elliot 1975.)

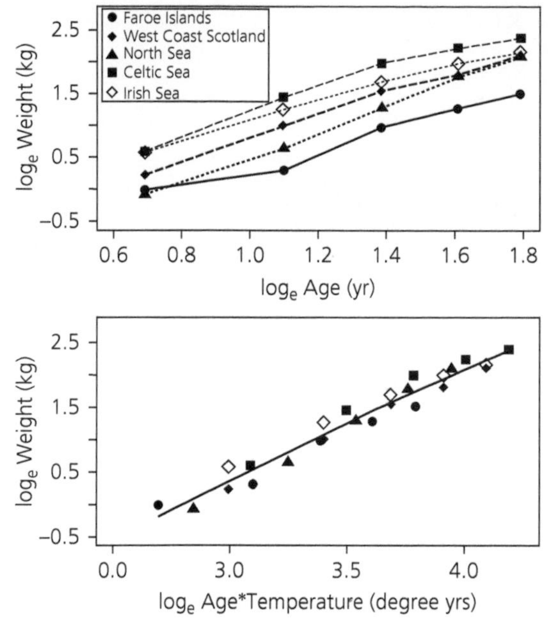

Figure 8.12 Relationship between (a) age and body weight (natural logarithms) for five northeast Atlantic cod (*Gadus morhua*) stocks and (b) the relationship between the product of age and mean ambient temperature and body weight for the same stocks. (Data from Brander 1995.)

differences among areas as reflected in the elevation of the curves to reveal a commonality in growth patterns following adjustment for temperature.

A related approach involves the specification of growing degree-days (Curry and Feldman 1987) commonly used in agricultural and other studies. In this metric, the growth can be expressed as a function of the cumulative temperature sum during a specified time interval. The number of growing degree-days is typically specified as:

$$GDD = \sum_{i=1}^{n} (T_i - T_{min}) \tag{8.36}$$

where T_i is the temperature on day i and T_{min} is a minimum threshold temperature below which growth stops. As with the Brander index, the degree-day measure can be directly substituted for more conventional measurements of time in growth models. For a review of this approach to a number of aquatic species, see Neuheimer and Taggart (2007).

8.6 Full bioenergetic models

In this section, we treat each of the components of individual production (ingestion, metabolic demand, egestion, and excretion) in greater detail. We have seen that the classical models of individual growth can be expressed as simple allometric functions collectively related to energy intake and losses. As noted in the Introduction to this chapter, full bioenergetic models invoke mass-balance constraints based on sound biological principles. This also permits us to estimate some unknowns given independent estimates of other components of the model. For example, we may wish to estimate growth as a function of ingestion and various sources of energy loss. Alternatively, given direct estimates of growth, we can solve for the consumption required to account for the observed growth. The accounting mechanisms employed follow a distinct sequence tracing the amount of food consumed; energy losses due to standard and active metabolism and specific dynamic action; and the elimination of waste products (Figure 8.13).

We can also estimate food ingestion rates as a function of independently derived estimates of growth, prey composition, and thermal history. Growth projections can also be made given information on key metabolic processes, temperature histories, etc. Here, we will frame the discussion within the context of a widely used bioenergetics model for aquatic organisms, the Wisconsin model, recently updated and made available as an R package *Fish Bioenergetics 4.0* (FB4) (Deslauriers et al. 2017). A compilation of over 100 bioenergetics models for 70 species developed using the Wisconsin framework have now been cataloged and made available (see *fishbioenergetics.org*). A user-friendly interface facilitates the use of the model to permit further analysis of the existing compilation or creation of new models.

8.6.1 Ingestion

The Wisconsin model defines an allometric function for ingestion in terms of a maximum ingestion rate adjusted by multipliers for the level of consumption ($0 > p_I > 1$) and a temperature function $f_I(T)$:

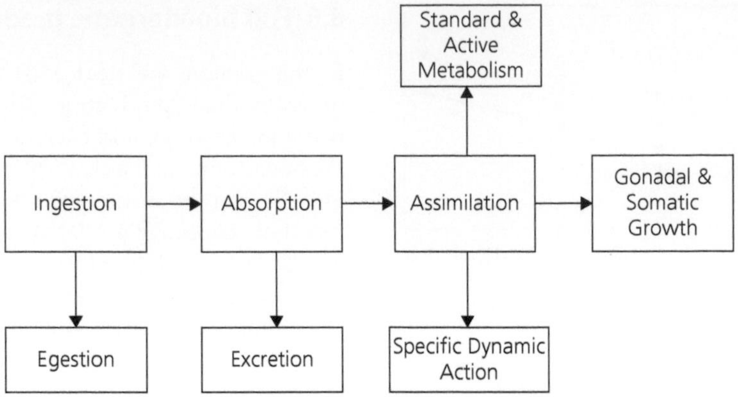

Figure 8.13 Flow diagram for the sequence of bioenergetics processes from ingestion of food to resulting somatic and gonadal production.

$$I = p_I a_I W^{b_I} f_I(T) \quad (8.37)$$

where a_I and b_I are allometric coefficients for ingestion. The multiplier p_I permits adjustment for the feeding level; when $p_I = 1$, consumption is at its maximum given the size of the organism and temperature conditions. Although p_I can be specified by the user, it can also be used as a "tuning" parameter to balance the overall model given a set of observed growth data.

Currently, four functional forms for the temperature term are available for use in FB4 including (1) exponential, (2) a dome-shaped function we will refer to as the Kitchell function, (3) a double-sigmoidal model, and (4) a cubic polynomial. The Kitchell function (Kitchell et al. 1974, 1977) is taken to be applicable to "warm-water" species. The sigmoidal function (Thornton and Lessem 1978) is deemed suitable for "cold-water" species, particularly at lower temperatures (Delauriers et al. 2017). This model comprises two conjoined logistic functions, the first describing the temperature effect from the lowest temperature at which ingestion occurs to some optimum temperature. The second represents the effect of temperature on ingestion from the optimum temperature to the highest temperature at which consumption occurs.

Specification of each of the temperature functions is provided in Box 8.4. In Figure 8.14 we illustrate the relative the shapes of these four functions. We scaled the models to match as closely as possible, given their different structural forms. The exponential model of course cannot capture a decay in food intake if an optimum temperature does exist. The remaining three models, which allow non-monotonic changes in the response to temperature are each capable of representing processes for which an optimum temperature exists but exhibit different levels of flexibility. Ultimately the choice of which model to use will rest on the availability of suitable empirical data to test the different options.

8.6.2 Respiration

Quantification of standard and active metabolism is most often expressed in terms of estimates of oxygen consumption. These estimates are typically acquired in respirometer studies in the laboratory. A significant body of literature indicates that the mass-specific respiration rate of aquatic organisms follows an allometric relationship. Respiration rates are directly measured based on oxygen consumption rates ($gO_2/g/d$) and standardized to one gram of body weight (BW) at 0 °C. Oxygen consumption can then be converted to energy units and used in balancing the overall bioenergetics model, which of course requires that all elements are specified in common units. A default value of 13.56 kJ/gO_2 is often applied for the oxygen to energy conversion although taxa-specific estimates are available that account for differences in body composition.

The Wisconsin model specifies the following function for total respiration:

$$R_T = a_R W^{b_R} f_R(T) \cdot A_R \quad (8.38)$$

where a_R and b_R are coefficients for the allometric respiration term, $f_R(T)$ is the temperature function

Box 8.4 Temperature Functions in Fish Bioenergetics 4.0

Fish Bioenergetics 4.0 offers a selection of several different functions representing the effect of temperature (T) on ingestion and respiration. All four options are available for ingestion. Respiratory processes currently only consider the exponential and Kitchell multipliers. The parameters of these functions will differ for ingestion and respiration. In the following, we provide subscripts (i) to account for these differences.

1. *Exponential.* The exponential multiplier is:

$$f(T) = e^{Q_i T}$$

where Q_i is the Q_{10} coefficient for the ith process (ingestion, egestion, respiration).

2. *Kitchell.* The Kitchell multiplier takes the form:

$$f(T) = \left[\frac{T_{max,i} - T}{T_{max,i} - T_{opt,i}}\right]^X e^{\left[X - \left(1 - \frac{T_{max,i} - T}{T_{max,i} - T_{opt,i}}\right)\right]}$$

where T_{max} is the temperature beyond which ingestion or respiration halts, T_{opt} is the optimum temperature and X is given by:

$$X = \left\{\frac{\left[\log_e Q_i \left(T_{max,i} - T_{opt,i}\right)\right]^2}{400}\right\}$$

$$\left\{1 + \left[1 + \left(\frac{40}{\log_e Q_i \left(T_{max,i} - T_{opt,i} + 2\right)}\right)^{0.5}\right]^2\right\}$$

3. *Double Logistic.* The 'lower' logistic function for temperatures equal to or lower than the optimum is:

$$K_{L,i} = \frac{c_{L,i} e^{[G_{L,i}(T - T_{L,i})]}}{1 + c_{L,i}\left(e^{[G_{L,i}(T - T_{L,i})]} - 1\right)}$$

where $T_{L,i}$ is the lower water temperature at which temperature dependence is a small fraction $c_{L,i}$ of the maximum rate and $G_{L,i}$ is given by:

$$G_{L,i} = \frac{1}{T_{opt,i} - T_{L,i}} \cdot \log_e \frac{0.98\left(1 - c_{L,i}\right)}{0.02 \cdot c_{L,i}}$$

where $T_{opt,i}$ here is the temperature at which ingestion is 98% of the maximum rate .

The 'upper' logistic function for temperatures equal to or above the optimum is:

$$K_{U,i} = \frac{c_{U,i} e^{[G_{U,i}(T_{U,i} - T)]}}{1 + c_{U,i}\left(e^{[G_{U,i}(T_{U,i} - T)]} - 1\right)}$$

where $T_{U,i}$ is the level at which temperature dependence is some reduced fraction $c_{U,i}$ of the maximum rate, and $G_{U,i}$ is given by:

$$G_{U,i} = \frac{1}{T_{L,i} - T_{M,i}} \cdot \log_e \frac{0.98\left(1 - c_{U,i}\right)}{0.02 \cdot c_{U,i}}$$

where $T_{M,i}$ is the temperature greater than or equal to $T_{opt,i}$ at which dependence is still 98% of the maximum.

4. *Cubic Polynomial.* The exponential cubic polynomial form is:

$$f(T) = e^{\left(a_1 T + a_{2,i} T^2 + a_{3,i} T^3\right)}$$

where a_1– a_3 are coefficients.

for respiration, and A_R is a multiplier reflecting activity levels (swimming) affecting respiration. In FB4, the temperature term can be represented by either the exponential or Kitchell functional forms. The activity term can be set to a constant [typically a fixed multiplier of specific metabolism (Winberg 1960)]. It can also be specified as a function of swimming speed and temperature:

$$A_R = e^{\upsilon V} \tag{8.39}$$

where υ is a coefficient and V is swimming speed (cm s^{-1}). If swimming speed depends on body weight above some cutoff temperature, we have:

$$V = a_{V+} W^{b_V} f_V(T) \tag{8.40}$$

where a_{V+} is an intercept term when temperature is above the cutoff value and b_V is the exponent over all temperature values. The case in which swimming velocity is a function of mass and temperature below a cutoff temperature is represented by:

$$V = a_{V-} W^{b_V} f_V(T) \tag{8.41}$$

where $a_{V-}$is an intercept term when temperature is below the cutoff value and b_V is defined as in Equation 8.40. Finally, swimming velocity can be set to a specified fraction of the metabolic rate (the Winberg coefficient) although this constant can be subject to considerable uncertainty.

To finalize our accounting of metabolic loss terms, we need to account for Specific Dynamic Action

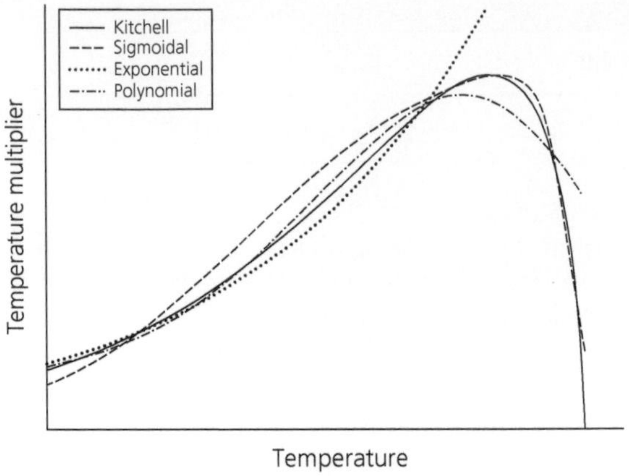

Figure 8.14 Shape of four temperature multipliers used in FB4 including (1) Kitchell, (2) Sigmoidal, (3) Exponential, and (4) Cubic polynomial functions. For this illustration, we used the Kitchell model as the base and fit the remaining functions to mimic this function as closely as possible.

(SDA). SDA is the amount of energy expenditure above the basal metabolic rate due to the cost of processing food for use and storage. In FB4, SDA is specified as:

$$SDA = s_R\,(I - E_F) \qquad (8.42)$$

where s_R is a specified fraction of the assimilated energy, with values typically ranging from 0.15 to 0.20. The mean value in the current compilation of bioenergetics models in FB4 is 0.16.

8.6.3 Egestion and excretion

Losses due to excretion and egestion are typically modeled as functions of ingested food and temperature.

Egestion is expressed:

$$E_F = p_F I \cdot f_F(T) e^{a_F \cdot p_I} \qquad (8.43)$$

where p_F is the proportion of ingested food evacuated, a_F is a coefficient for the dependence of egestion on feeding level, and p_I (proportion of maximum ingestion) is a calculated coefficient used to balance the energy budget. The equivalent expression for excretion is:

$$E_N = p_N\,(I - E_F) \cdot f_N(T) e^{a_N \cdot p_I} \qquad (8.44)$$

where p_N is a constant proportion of energy lost to excretion and a_N is a coefficient for the dependence of excretion on feeding level. A power function for temperature is employed for both excretion. and egestion. Adjustments for indigestible prey can also be made (Deslauriers et al. 2017).

8.6.4 Energy density of predators and prey

Again, to invoke mass-balance constraints consistent with the laws of thermodynamics, we require a common energy currency. Meta-analyses have revealed that the energy content of fish tissue can be expressed as an allometric function of body mass measured as percent dry weight:

$$E_D = a_E DW^{b_E} \qquad (8.45)$$

where E_D is energy density (J/g wet weight), a_E and b_E are allometric coefficients, and DW is the percent dry weight in the sample (e.g. Hartman and Brandt 1995; Johnson et al. 2017). We note that energy density often also varies seasonally in higher latitudes, linked to well-defined seasonal pulses of productivity. Brey et al. (2010) provide a recent and invaluable compilation of estimates of body mass, elemental composition, and energy content for over 3000 aquatic species encompassing bacteria, plants, and animals (see http://www.thomas-brey.de/science/virtualhandbook).

8.6.5 Yellow perch in Lake Erie

In their classic study, Kitchell et al. (1977) analyzed the bioenergetics of yellow perch in Lake Erie, tracking growth from the time of larval settlement to age 5. Here, we recap this analysis using original parameters of this study as catalogued in FB4. The model operates on daily time steps. The simulation starts following the estimated time of larval settlement as per Kitchell et al. (1977). Young-of-the-year fish were assumed to consume exclusively invertebrate prey. During the second year of life perch were assumed to eat one half invertebrate prey and one half fish prey; all older fish were taken to be exclusively piscivorous. Fish aged two and older were assumed to be sexually mature. Temperature was taken to follow the same seasonal cycle for each year of the simulation using the thermal series for

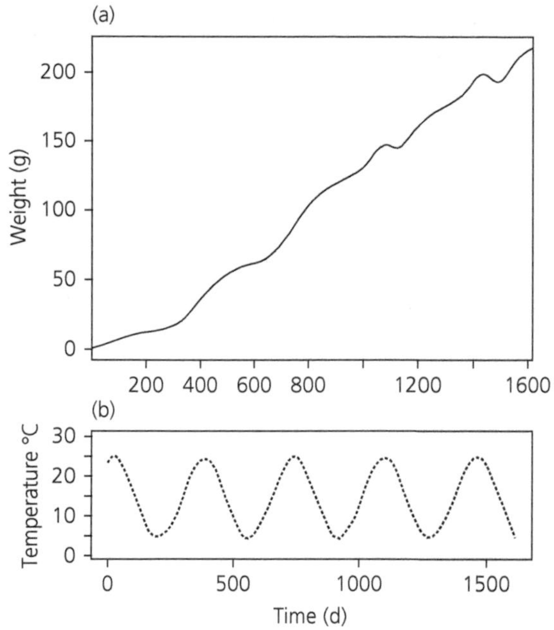

Figure 8.15 (a) Simulated growth of yellow perch (*Perca flavescens*) in Lake Michigan from the time of larval settlement (designated time 0) and (b) corresponding annual temperature cycle employed in simulations using the *Fish Bioenergetics 4.0* software.

the western basin of Lake Erie. The temperature multiplier for ingestion and respiration followed the Kitchell model. Simulated weight-at-age and the temperature series are shown in Figure 8.15. The trajectory of body mass clearly shows seasonal effects on growth, particularly for the younger ages, but no evidence of an approach to an asymptotic size over the age classes represented.

8.7 Metabolic ecology

The critical importance of body mass and temperature to metabolic processes of aquatic organisms has emerged throughout this chapter. Allometric functions coupled with temperature modifiers play near-universal roles in characterizing the elements of production in marine, estuarine, and freshwater species. The exploration of these issues has a deep tradition in the aquatic sciences. Metabolic ecology (Brown et al. 2004, Brown and Sibly 2012) broadens the scope of this endeavor to encompass a much wider array of organisms and seeks to place metabolic processes at the very core of ecological understanding. The issues covered in this chapter provide substantial support for this basic premise.

A more restricted (and controversial) issue is the question of whether the exponents of the ubiquitous allometric functions in metabolic ecology assume certain expected values across broad taxonomic categories after adjustment for temperature (Brown and Sibly 2012). Under the proposed Metabolic Theory of Ecology (MTE), the expected value of these exponents conforms to a "one-quarter scaling rule" with a value of 0.75. When expressed as mass-specific values, the expected value of the exponents is −0.25 (Brown and Sibly 2012).

Essington et al. (2001) examined information for 37 fish stocks derived from either bioenergetic or pollutant tracer analyses and found that the exponent m averaged 0.80 for both types of analyses, with estimates derived from tracer studies higher than those from bioenergetics analyses. The mean estimate for the bioenergetics analyses alone was lower at $m = 0.70$. The range from the two sources of information therefore brackets the expected value under allometric theory.

Essington et al. (2001) also estimated the parameters of the loss function and found that for both types of analysis, the exponent n averaged 1.0, higher than the value of 0.75 expected under metabolic theory. In this case, Essington et al. (2001) considered energy losses due to reproductive output and locomotion, resulting in higher estimates for n once these factors were taken into account.

As with compilations for other taxonomic groups (see the contributions in Sibly et al. 2012), we find some departures from the expected values for the exponents under MTE. Nonetheless, it is a valuable exercise to examine the extent to which broad-scale patterns may (or may not) emerge to help focus attention on unifying principles. It is worth noting that these compilations often represent estimates for older fish and information on the larval and early juvenile stages are less commonly available.

8.8 Summary

Models of individual growth commonly used in fisheries and ecological research can be built around simple allometric functions representing the build-up of body mass (anabolism) and metabolic loss terms incorporating the effects of respiration, egestion, and excretion. From a bioenergetics perspective, body weight is a natural choice for the response variable in these models because it can be readily recast in terms of energy. However, many growth models in the fisheries literature are expressed in terms of body length. We have accordingly shown how to translate between length-based and weight-based growth models. We have also illustrated extensions of the basic model framework to incorporate seasonal growth. We further address the special case of discontinuous growth in crustacea. This taxonomic group includes many of the most economically important species in global capture and culture fisheries.

Temperature affects virtually every dimension of the biology and ecology of aquatic organisms. We show how modifications of traditional models of individual growth can be augmented to account for temperature effects. We further describe alternative approaches based on physiologically relevant temperature-timescales that can be directly employed in growth models.

The development of "full" bioenergetics models considering each of the individual elements of production is a natural culmination of the issues described above. Perhaps the most widely known and applied bioenergetics model for fishery systems is the Wisconsin model, Fish Bioenergetics, now in its fourth incarnation. By invoking mass-balance constraints the bioenergetics approach offers important avenues for estimating elements of production that can be difficult to obtain otherwise.

Finally, we touch on the theme of Metabolic Ecology. An attempt is now underway to understand possible mechanisms leading to scaling laws reflected in common values of the exponents of the allometric relationships in metabolic processes. A number of meta-analyses show values of the exponent of about 0.75 for relationships between body mass as modified by temperature. Exponents of around -0.25 have been found for a number of mass-specific relationships in metabolism. An examination of information from aquatic species (principally fish) provides estimates near (but not necessarily exactly at) these expected values.

Additional reading

Pitcher and Hart (1982; Chapter 4) describe fish nutrition, growth, and production. Wootton(1998) provides an extremely informative treatment of production in fish populations. *Fish Bioenergetics* (Jobling 1994) remains a standard in the field and is well worth consulting. A comprehensive catalog of models of growth and fecundity along with important information on estimation procedures can be found in Quinn and Deriso (1999).

CHAPTER 9

Production at the Cohort and Population Levels

9.1 Introduction

As we turn from the issue of production at the level of the individual organism to population level processes, we confront the challenge of understanding one of the central issues in ecology—the nature of regulatory control and its implications for population stability and resilience. In preceding chapters we have seen that compensatory processes can provide important stabilizing mechanisms in population dynamics. This issue is particularly important in understanding how populations may respond to stressors, including exploitation. In this chapter we will go beyond our earlier consideration of compensatory processes to focus on specific regulatory mechanisms operating on recruitment, individual growth, and reproductive output.

In the following, we will adopt the classical framework proposed by Russell (1931) as a roadmap for characterizing production of aquatic populations. The change in biomass (B) of a population can be described as the sum of gains and losses:

$$\frac{dB}{dt} = Recruitment + Growth - Mortality + Immigration - Emigration \quad (9.1)$$

The concept of recruitment was introduced in Chapters 2 and 3 in the context of age-structured matrix models. Here, we will delve into underlying recruitment processes and how we might incorporate specific regulatory mechanisms into these models. Individual growth was described in Chapter 8; it remains now to scale it to the population level and to consider the potential importance of compensatory growth processes. Similarly, we have dealt with mortality (or its complement, survival) in Chapters 2–4. In this chapter, we will focus on partitioning mortality into its component parts: natural mortality (including disease, predation, etc.), and fishing mortality. The importance of immigration and emigration of subpopulations was discussed in Chapter 7. In this chapter we will deal with closed populations. We will first treat the issue of cohort production (changes in biomass due to growth and mortality once the initial abundance of the cohort is established). We then broaden our consideration to encompass production at the population level with inclusion of recruitment processes in addition to growth and mortality.

9.1.1 Compensation and regulation in aquatic populations

The critical importance of understanding population regulation and its implications for production of exploited aquatic populations has long been a focal point of fisheries research, with particular attention placed on events during the early life-history period. Without some form of population regulation, sustainable equilibrium yields are not possible. Ricker (1954) and Beverton and Holt (1957) made seminal contributions to understanding the role of compensatory processes in recruitment dynamics, setting the stage for a rich research tradition in fisheries science. Density-dependent growth, reproduction, and mortality for post-recruits can also can be extremely important as regulatory processes (Rothschild and Fogarty 1998, Fogarty and O'Brien 2016; Andersen et al. 2015).

Fishery Ecosystem Dynamics. Michael J. Fogarty and Jeremy S. Collie, Oxford University Press (2020).
DOI: 10.1093/oso/9780198768937.003.0009

The inherent difficulty in empirically detecting population regulation in the face of environmentally driven variability is well recognized by all ecologists. The problem is particularly acute in understanding the dynamics of many exploited fish and invertebrate species subject to very high levels of recruitment variability. This issue is exacerbated by inevitable observation error in population estimates, which alone can easily obscure any underlying compensatory processes (Walters and Ludwig 1981). Given the typically high levels of variability in recruitment, in order to understand the nature of compensatory processes in aquatic species we must focus on evidence for underlying regulatory mechanisms, relying on a synthesis of monitoring, modeling, and detailed site-specific analyses (Rose et al. 2001). We strongly support this broad view and attempt to show how it can be employed in evaluating production at the population level. Plots of recruitment estimates against egg production or adult biomass are almost universally difficult or impossible to interpret without deeper understanding of underlying mechanisms and sources of variability.

9.2 Cohort production

Production of a cohort is determined by the interplay of growth and mortality of individuals born within a specified time period and a given area. Throughout this chapter we will follow convention and express cohort production in terms of biomass, although conversion to units of energy can be readily made (see Chapter 8). The weight of a cohort at any given time is a function of its initial number, losses due to various mortality events, and the mean weight of individuals in the cohort. In the following, we expand our earlier treatment of growth to consider density-dependent processes and then turn to further examination of mortality. The consideration of growth and mortality rates in this section will be revisited when we take up the topic of production at the population level in Section 9.3.

9.2.1 Growth

We provided an extensive treatment of the growth of individuals in Chapter 8. In the context of cohort production, we must expand these estimates to the total weight attributable to growth for the cohort as a whole. This can be done by multiplying the mean cohort biomass and the individual growth rate over a specified time interval to give production during that time period (see Section 9.2.3). Although far less commonly addressed in assessments of cohort (and population) production, we will also consider the possible effect of population size on individual growth. We saw in Chapter 8 that prey availability can strongly affect growth. Intraspecific competition can be a critical determinant of the food available to individuals. Compelling evidence that this is an important consideration emerged in early stocking experiments in small freshwater ponds (e.g. Swingle (1950); see Weatherley (1972) and Wootton (1998) for excellent reviews of this and related topics). Swingle (1950) focused on the phenomenon of stunted growth at high stocking densities and noted that mixed-culture systems comprising predators and prey can be effective in addressing this problem. Predators crop their prey and reduce intraspecific competition of the prey species. If an appropriate balance between prey and predators can be maintained, overall production can be optimized in these ponds (Swingle 1950). We will return to the general topic of balanced harvesting in Chapter 15. To illustrate the potential effect of population size on growth, we examine the mean weight-at-age for haddock (*Melanogrammus aeglefinus*) on Georges Bank. This population underwent a dramatic change in population size which persisted for several decades following a major perturbation induced by excessive fishing pressure by distant water fleets in the 1960s. The mean weight-at-age 2 of haddock increased by approximately 50% following a sharp reduction in cohort size after 1965 (Figure 9.1). Although the role of other environmental factors cannot be discounted, it is clear that the effect of cohort abundance deserves careful scrutiny.

Beverton and Holt (1957) provided an early treatment of density-dependent individual growth in their seminal treatise. Walters and Post (1993) subsequently evaluated the effect of food supply on individual growth in length as mediated by population density (see also Lorenzen and Enberg 2002).

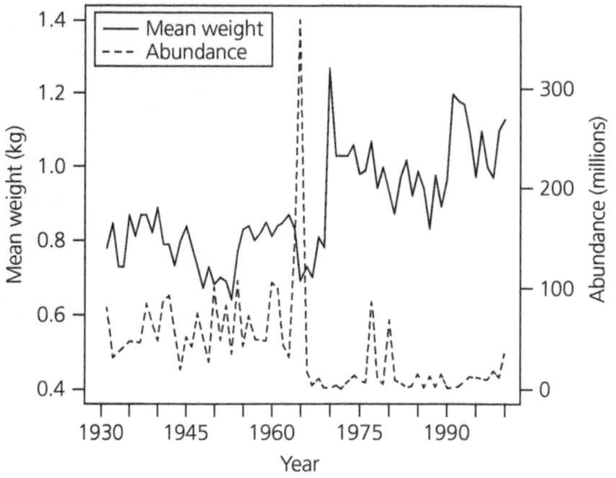

Figure 9.1 Trends in mean-weight-at-age 2 (kg) and estimated abundance-at-age 2 (millions) of haddock (*Melanogrammus aeglefinus*) on Georges Bank. Data from Clark et al. (1982) and NEFSC unpublished.

Perhaps the simplest modification to the growth models introduced in Chapter 8 to account for the effects of population size can be expressed:

$$\frac{dW}{dt} = \left(\eta W^{m-1} - \kappa\right)W - \delta N \qquad (9.2)$$

where δ is a coefficient, N is cohort abundance and all other terms are defined as in Chapter 8. Here we have chosen the specialized von Bertalanffy growth model to explore the implications of change at fixed levels of cohort abundance on individual weight-at-age. The solution is given by:

$$W_t = W_\infty\left(1 - he^{-(k+\delta N)t}\right)^{\frac{1}{1-m}} \qquad (9.3)$$

and examples of changes in weight-at-age for three levels of cohort abundance are shown in Figure 9.2. Here density affects the rate of growth. Alternative modifications could be made that result in an alteration of the asymptotic size (see Walters and Post 1993; Lorenzen and Enberg 2002). We note that if the rate of growth is slowed sufficiently, the asymptotic size will not be attained in the lifespan of an individual, thus exerting a secondary effect on asymptotic size.

The importance of density-dependent individual growth remains an underappreciated issue in understanding population regulation in fish (Lorenzen and Enberg 2002). Andersen et al. (2017) note that the relative importance of post-recruitment regulatory processes, including density-dependent growth and maturation, can be connected to factors related to spatial extent of the populations. Marine and freshwater ecosystems are likely to differ substantially in this regard. Most freshwater populations occupy clearly delineated and often relatively closed regions. In contrast, marine populations often occupy broader and more open areas. Density-dependent processes that occur later in life hold important implications for management; harvesting directed at older fish under these circumstances is suboptimal (Andersen et al. 2017).

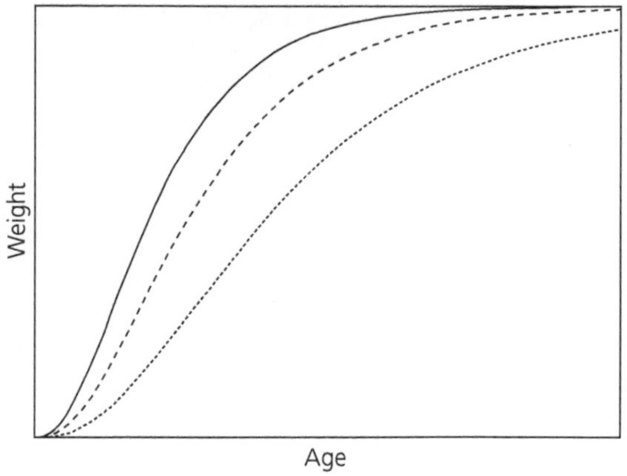

Figure 9.2 Weight-at-age for a density-dependent individual growth model at three levels of density.

Changes in growth rates can be related to a number of factors ranging from changes in food availability and/or overall system productivity to increased intraspecific competition for food. Appropriate management responses can differ markedly depending on whether the changes are related to density-dependent or density-independent factors. Accordingly, greater attention to compensatory growth is warranted.

9.2.2 Mortality

We have now encountered the role of mortality in population dynamics at a number of points, starting with our investigation of simple models in Chapters 2 and 3 and continuing on with consideration of factors such as predation and parasitism/disease (Chapter 4). Here, we will expand this treatment to encompass aspects of estimation and quantification

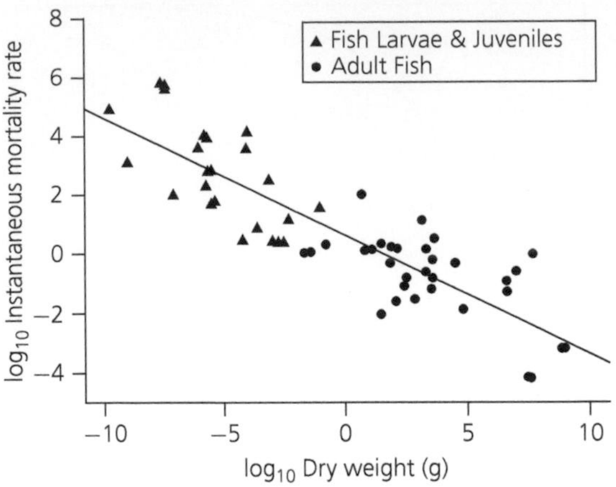

Figure 9.3 $\log_{10}$ instantaneous annual mortality rate as a function of $\log_{10}$ dry weight (g) for marine fish. (Adapted from Peterson and Wroblewski 1984.)

of mortality rates partitioned into different sources. Traditional single-species approaches typically partition mortality into two components, natural and fishing mortality. Multispecies models further partition natural mortality into predation mortality and "other" mortality. We wish to encourage further partitioning of specific components of natural mortality where possible, including the effects of disease and adverse environmental conditions. The most important lesson to absorb from the examples to follow is that natural mortality is anything but constant.

We begin by noting that overall mortality is broadly determined by body size. Instantaneous mortality rates per year arrayed along a continuum of body sizes are shown in Figure 9.3 for a number of marine fish populations (Peterson and Wroblewski 1984). Note the very high levels of mortality during the larval and juvenile stages of fish shown in this meta-analysis.

In the remainder of this chapter, we will be using a common framework to represent rates of mortality attributable to different sources. The finite rate of mortality due to a specific factor (predation, disease, fishing etc.) is given by the ratio of the number dying due to the mortality agent in a specified time period to the number at the start of the period. It can therefore be interpreted as a probability. The finite rates of mortality are not additive (an individual can succumb only to one direct cause). For example, for the case of two sources of mortality during a specified time period we have:

$$P(m_T) = P(m_1) + P(m_2) - P(m_1)P(m_2) \tag{9.4}$$

where $P(m_T)$ is the probability of death due to all sources of mortality during the time interval. If we now have three agents of mortality, the expression now becomes:

$$\begin{aligned} P(m_T) = {} & P(m_1) + P(m_2) + P(m_3) - P(m_1)P(m_2) \\ & - P(m_1)P(m_3) - P(m_2)P(m_2) \\ & + P(m_1)P(m_2)P(m_3) \end{aligned} \tag{9.5}$$

and so on. We can readily convert finite rates to instantaneous rates. If these sources of mortality operate independently, the instantaneous rates are additive. We have now specified time in sufficiently small intervals that the probability of an individual being subject to more than one source of mortality is negligible. The instantaneous rate of total mortality (Z) can be expressed as:

$$Z = -log_e\,[1 - P(m_T)] \tag{9.6}$$

If the sources of mortality are not independent, a subtle (and more difficult to resolve) issue arises. An individual weakened by disease or senescence may be more vulnerable to predation, extreme environmental conditions, etc. Unfortunately, it is rare to have sufficient information to tease these elements apart. This issue is explicitly recognized in the context of predation in donor control models in which predators crop vulnerable individuals that would otherwise die due to other causes such as disease or senescence (Pimm 1982).

9.2.2.1 Predation mortality

The finite rate of mortality attributable to predation is given by the ratio of the number of prey consumed during the time interval and the number at the start. In Chapter 12 we will encounter multispecies models in which predation mortality and population size can be estimated. In many smaller freshwater systems, direct enumeration may be possible using non-lethal sampling techniques. In larger freshwater and marine systems a number of different fishery-independent and fishery-dependent sampling systems are employed. These range from standardized surveys using net-based, hydroacoustic, and other sampling tools to large-scale mark–recapture studies. For now, it suffices to recognize that we can obtain estimates of abundance at regular time intervals. (see Hilborn and Walters 1992; Quinn and Deriso (1999).

To illustrate the main elements involved in estimating the predation mortality rate experienced by a prey species, we will focus on the highly simplified case of a single prey and single predator species in which all predator size or age classes have identical preferences for the prey. In Chapter 12, we will relax these restrictions to encompass multiple predators and prey with different suitability functions for different prey species and size or age classes. We will adhere to convention and designate the instantaneous predation mortality coefficient $M2$:

$$M2 = \frac{I\psi N_{pred}}{N_{prey}} \tag{9.7}$$

where I is *the per capita* ingestion rate of the predator, ψ is the suitability coefficient of the prey for the predator N_{pred} is the population size of the predator species, and N_{prey} is the amount of the prey species available to the predator. The per capita ingestion rate (or ration) can be determined by several methods including estimation based on coefficients of the von Bertalanffy growth model and measures of assimilation efficiency (Essington et al. 2001; Aydin 2004; Temming and Herrmann 2009; Walters and Essington 2010; Wiff et al. 2015; see Chapter 8), empirical estimators integrating life-history parameters and laboratory experiments (Pauly 1986; Palomares and Pauly 1998), expanded application of the classical bioenergetics approach (Kitchell et al. 1974; Deslauriers et al. 2017), and estimates employing field observations of diet composition and quantity and laboratory experiments of gastric evacuation rates in multispecies assessment models (e.g. Anderson and Ursin 1977; Sparre 1991; Magnusson 1995). We will reserve treatment of multispecies assessment models until Chapter 12. Using the anabolic component of the generalized von Bertalanffy growth model (see Equation 8.5) we have (Essington and Walters (2001)):

$$I = \frac{\eta W^m}{A_e} \tag{9.8}$$

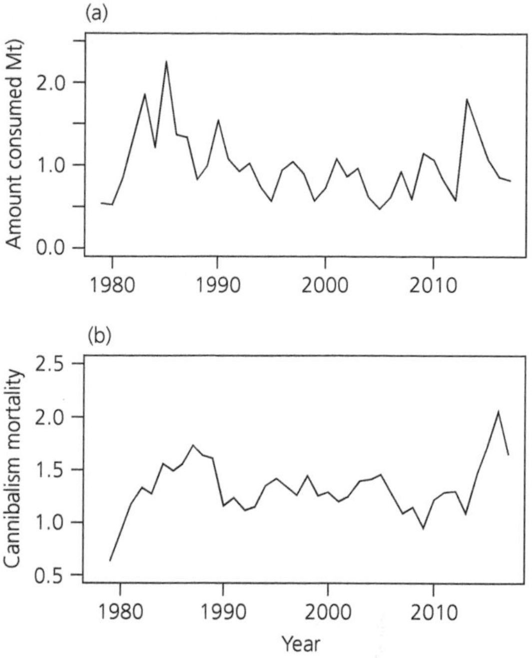

Figure 9.4 Estimates of (a) consumption of walleye pollock (*Gadus chalcogrammus*) in the Eastern Bering Sea by conspecific adults and (b) predation mortality of age 1 pollock attributable to cannibalism (J. Ianelli, personal communication).

In Figure 9.4a we show estimates of the biomass of walleye pollock consumed by conspecifics in the Eastern Bering Sea (in this case estimated using a full multispecies model: Holsman and Aydin 2015; J. Iannelli personal communication). The biomass of pollock lost to intraspecific predation is greatest by far for the youngest age classes (Holsman and Aydin 2015). This pattern of higher vulnerability of younger, smaller individuals is repeated for many species in instances of both intraspecific and interspecific predation. The estimated instantaneous mortality rate of age-1 pollock due to cannibalism by adults is shown in Figure 9.4b.

9.2.2.2 Parasitism/Disease

We have seen that mortality events due to disease can result in very dramatic changes in abundance over short intervals of time. For the 1988 outbreak of PDV in East Anglia described in Chapter 4, the finite rate of mortality due to disease was estimated to be 0.48 based on direct counts of dead seals adjusted for sighting probabilities (Grenfell et al. 1992). This translates into an instantaneous rate of 0.65.

Pacific herring (*Clupea pallasii*) in Prince William Sound provides an interesting case study. Pacific herring in the Sound was subject to an outbreak of viral hemorrhagic septicemia virus (VHSV) in the winter of 1992–93, resulting in a sharp decline in adult herring biomass (Marty et al. 2003; Figure 9.5). This epizootic had been preceded by another strong perturbation, the Exxon Valdez oil spill in 1989, although no causal association has been identified. A monitoring program was established in 1994 to determine the subsequent incidence of disease as reflected in gross pathology (visible lesions etc.), and microscopic examination of tissue (Marty et al. 2003). This effort revealed the presence of VHSV and also *Icthyophonus hoferi*, a pathogen implicated in several important epizootics that had decimated herring populations in the North Atlantic (Sindermann 1970). Marty et al. (2003) developed an estimator of the survival rate (the complement of the finite mortality rate) comprising a base-line survival rate s_0 modified by a multiplier incorporating an index of the prevalence of different disease vectors ($x_{i,t}$) along with an empirically determined coefficient of lethality for the ith vector (c_i):

$$s_t = s_0 \left(1 - \sum_{i=1}^{n} c_i x_{i,t}\right) \tag{9.9}$$

Note that here the survival rate declines linearly with increasing prevalence of one or more pathogens. The survival rate can be translated into an instantaneous mortality rate by taking $-\log_e(s_t)$. Estimates of the mortality rates attributable to disease of herring in Prince William Sound are provided in Figure 9.5b (Marty et al. 2003).

9.2.2.3 Other mortality

We have concentrated on two sources of natural mortality that have been intensively investigated in a number of aquatic ecosystems and that can be quantified. Not all potential sources of natural mortality events are monitored and there remains a need to account for "residual" natural mortality after accounting for known agents of mortality. In multispecies models, it is common to partition instantaneous total mortality (Z) into three components: other mortality ($M1$), predation mortality ($M2$), and fishing mortality (F). Here disease is considered under the broad umbrella of other mortality. We have seen that disease-related mortality can be highly episodic and, in some cases, can result in catastrophic changes in population size. Agents of mortality that can be considered as components of $M1$ also include lethal anoxic events and extremes of environmental conditions. Massive fish kills associated with anoxic events and harmful algal blooms are routinely observed and reported in local media. Typically, these reported events occur in freshwater or estuarine waters where the proximity to intensely populated areas virtually assures that they will be noted. Less commonly observed, but no less important, are events that occur in offshore waters. A particularly striking event occurred in 1882 at the edge of the continental shelf off the northeastern United States (Cushing 1982a). A massive die-off of tilefish (*Lopholatilus chameleonticeps*) was reported to have

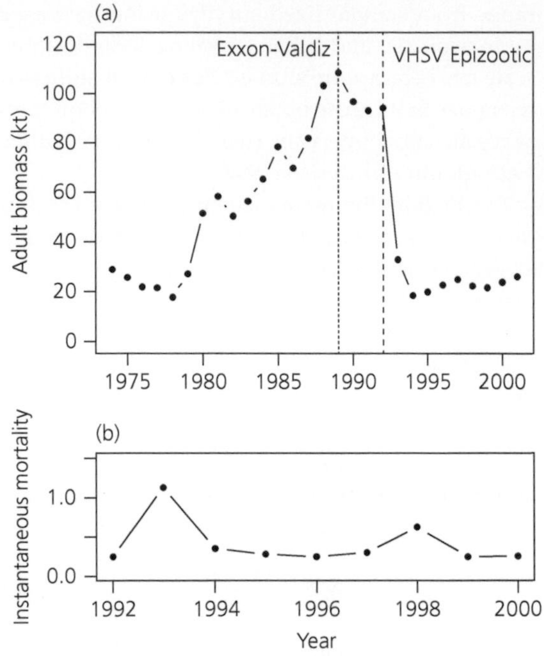

Figure 9.5 Trends in (a) adult biomass (kt) of Pacific herring (*Clupea pallasii*) in Prince William Sound. Dotted vertical line marks the timing of the Exxon Valdez oil spill in 1989. Dashed vertical line indicates the outbreak of viral hemorrhagic septicemia virus (VHSV) in 1992. (b) Estimates of disease-related mortality 1992–2000. (Marty et al. 2003).

involved up to one billion fish (Collins 1884 cited in Marsh et al. 1999). Despite its remote location from shore, an event of this magnitude was virtually assured of being observed from fishing vessels and other ship traffic. This event has been attributed to an intrusion of very cold Labrador Shelf water into the offshore reaches of the Middle Atlantic Bight (Marsh et al. 1999). Values assigned to the *M*1 component of natural mortality are almost invariably (and understandably) assigned a fixed value. It is clear however, that we must consider major components of *M*1 to potentially vary widely and in some cases result in catastrophic episodes of mortality.

9.2.3 Estimating cohort production

We can now build on the foundation above to estimate production of aquatic species at the cohort level. Here, we will deal with the elaboration of body tissue though growth of the cohort as a whole (total cohort production) and net cohort production in which mortality is considered. In the following, we will be principally dealing with total mortality rather than mortality partitioned into different sources. We begin with the instantaneous growth framework of Ricker (1946). The rate of change of cohort biomass can be written:

$$\frac{dB}{dt} = \frac{d\,(N \cdot W)}{dt} = N\frac{dW}{dt} + W\frac{dN}{dt} \tag{9.10}$$

where N is the number in the cohort and W is individual weight. The first term on the right-hand side of Equation 9.10 specifies the growth component and the second specifies the mortality. Biomass is of course given by the product of number and weight ($B = N \cdot W$). Recall that we have previously provided expressions for dW/dt (Chapter 8) and dN/dt (Chapter 2). For the simple case of constant growth (G) and mortality rates (Z), we can combine these two elements to obtain:

$$\frac{dB}{dt} = (G - Z)B \tag{9.11}$$

The solution is given by:

$$B_t = B_0 e^{(G-Z)t} \tag{9.12}$$

and the cohort biomass will increase exponentially over the lifespan when $G > Z$. The mean biomass over a unit time interval (say one year) for Equation 9.12 is:

$$\overline{B} = \frac{B_0}{G - Z}\left[e^{(G-Z)t} - 1\right] \tag{9.13}$$

The rate of change of cohort production (P) can be written:

$$\frac{dP}{dt} = \frac{dB}{dt} - W\frac{dN}{dt} = N\frac{dW_t}{dt} \tag{9.14}$$

We can focus on the last term in Equation 9.14, and integrate over a specified time interval:

$$P = \int_{t_1}^{t_2} N_t \frac{dW_t}{dt} dt \tag{9.15}$$

Notice that the integral contains N_t and we are accounting for change in population numbers through mortality over the time period of integration (again assuming immigration and emigration are negligible or balance). The growth component is of course represented in the derivative term in Equation 9.15. For the case of exponential growth and mortality over the time period, the production is simply:

$$P = G\overline{B} \tag{9.16}$$

where the average biomass term is given by Equation 9.13. This is often referred to as the instantaneous growth method.

In practice, we often estimate the growth and population size of a cohort at discrete intervals using empirical estimates of standing stock and individual weight. We can replace the integral in Equation 9.15 with the sum of the product of mean number and the weight increment during each interval:

$$P = \sum_t \overline{N}_t \Delta w \tag{9.17}$$

This is referred to as the increment summation method (Crisp 1984). A mathematically equivalent expression can be written:

$$P = \sum_t \Delta N\, \overline{w}_t \tag{9.18}$$

where $\Delta N = N_t - N_{t+1}$. This is referred to as the increment removal method. Note that because we are employing direct empirical estimates of changes in numbers and individual weights over each time

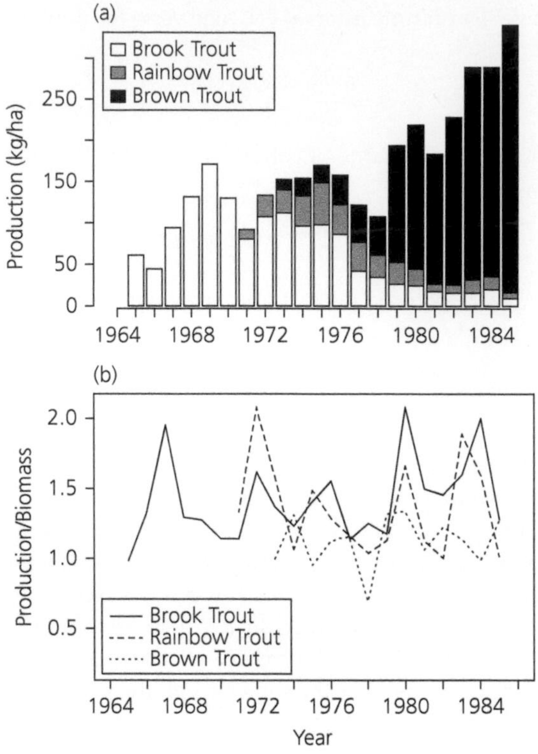

Figure 9.6 Trends in (a) annual production of three trout species in Valley Creek, Minnesota and (b) the production-to-biomass ratio for these species. Data from Waters (1999).

interval, the estimate of realized production implicitly captures the effects of any compensatory processes affecting growth and mortality.

Waters (1999) provided an interesting production analysis for Valley Creek, Minnesota, revealing a striking pattern of species replacement and changes in production over a two-decade period. Waters (1999) estimated the population densities of salmonids using mark–recapture experiments. Growth rates were based on age determination methods using annual markings on fish scales in subsamples of the populations over a two-year period. Production was estimated using the Ricker instantaneous growth–mortality formulation (Equation 9.16). At the start of the study in 1965, the stream was characterized by a low-diversity assemblage of fish species dominated by the brook trout, *Salvelinus fontinalis*. The only other fish species recorded were the sculpin *Cottus cognatus* and the brook lamprey *Lampetra appendix*. A series of flood events starting in 1966 exerted strong impacts on the brook trout population, principally through habitat disturbance. Flooding further facilitated the first entry of rainbow trout (*Oncorhynchus mykiss*) from private ponds into the stream in 1967. Brown trout (*Salmo trutta*) invaded in 1971, ultimately replacing brook and rainbow trout in the stream (Figure 9.6a). In contrast to the clear trends in production of each of the three species, their production-to-biomass ratios were relatively constant (Figure 9.6b). This result has been often observed in other water bodies (Waters 1977). In general, the P/B ratio of brown trout was somewhat lower than that of brook and rainbow trout. The estimated overall production of the three species, however, increased as brown trout attained dominance. It appears that the apparent replacement sequence was strongly related to initial habitat disturbance. The ultimate dominance of brown trout, however, is consistent with patterns following introduction or natural invasion by this species in North America (Budy and Gaeta 2017) and elsewhere. Brown trout tend to be larger-bodied and more aggressive than species they replace, suggesting a possible role of competitive dominance in these outcomes.

9.3 Population production

Populations are of course composed of a collection of cohorts. In our treatment of cohort production we did not consider dynamic processes affecting the entry of new recruits to the population. Rather, we computed cohort production given an observed or assumed number of recruits. Explicit consideration of recruitment processes is necessary to allow for a more dynamic view of production at the population level. We can take advantage of the treatment of individual growth and of mortality provided above and complete the story with consideration of recruitment processes to evaluate production at the population level.

For many aquatic species, recruitment is the dominant and most variable component of production. Recruitment is defined as the number of individuals of a cohort surviving to a specified life stage or age. Various forms of mortality winnow the initial number of viable eggs produced by a population to those ultimately surviving to the recruitment stage. Recruitment processes reflect the interplay of external forcing mechanisms such as predation and physical drivers in the environment that affect demographic rates, and stabilizing (compensatory) mechanisms exhibited by the population.

Recognition of high levels of variability in recruitment of many aquatic populations has engendered important debates over the question of whether there is a relationship between the reproductive output of a population and resulting recruitment. This question of course holds important implications for management. The production of viable eggs by a population provides the raw material for recruitment. For a closed population, the number of recruits can never exceed the number of viable eggs produced. In this sense, there is unquestionably a structural relationship between egg production and recruitment. Variability in growth and mortality rates during the pre-recruit stages, however, strongly shapes the number of individuals surviving to the recruitment stage. Given the reproductive strategy of many teleosts (high fecundity and little parental investment in survival of the young), high levels of variability in recruitment are inescapable when growth and mortality rates vary (Fogarty et al. 1991). From this perspective, the question is not whether there is some form of fundamental relationship between egg production and recruitment but rather how variable the relationship becomes given variability in demographic rates and processes and the time frame over which they operate prior to recruitment. Quite small levels of variability in mortality rates, for example, translate into substantial levels of recruitment variability. This does not imply that the fundamental structural relationship between egg production and recruitment can be ignored nor does it obviate the need for ensuring adequate egg production by the population.

We can illustrate some key points with an examination of a long-term research program on brown trout in Black Brows Beck in northwestern England (Elliot 1994). This study afforded a unique opportunity to examine recruitment processes of this species. High-resolution and precise sampling was possible in this relatively small stream in a way not feasible in many large fishery ecosystems. Elliot (1985, 1994) measured egg production and the resulting number of individuals over a sequence of life stages. We can clearly trace the evolution of increasing levels of variability with increasing time from the egg stage through successive life history stages (Figure 9.7). The relationship between the number of eggs produced and the number of age 0+ juveniles (parr) in April/June is characterized by low levels of variability and clear evidence of compensatory processes operating very early in the life history (Figure 9.7a). If we next examine the relationship between egg production and number of age 1+ fish of the same cohort in August/September of the following year, we see higher levels of variability in the number surviving to this stage (Figure 9.7b). Finally, extending the time period from the egg stage to the number of age 2+ survivors in May/June reveals higher variability still (Figure 9.7c). The cumulative effect of environmental influences on survival over increasingly longer time frames translates into increasing variability at different life stages, ultimately obscuring the critical role of egg production in the overall process. The key issue is understanding the nature of the compensatory processes at work and the effect of exogenous forcing on the population.

In the following, we will consider not only the deterministic "skeleton" of recruitment models representing alternative compensatory mechanisms but will also explore approaches to understanding and dealing with processes contributing to high levels of variability in recruitment.

9.3.1 Deterministic recruitment models

The classical models of recruitment of aquatic populations are built around the core concept of compensation in demographic rates and processes. In the following we will illustrate the development of models incorporating different forms of compensatory processes including density-dependent mortality within a cohort, cannibalism by adults on progeny, the role of density-dependent growth of pre-recruits and its implications for vulnerability to predation, and bioenergetic processes. We will also consider the role of compensatory processes affecting the overall egg production by the population.

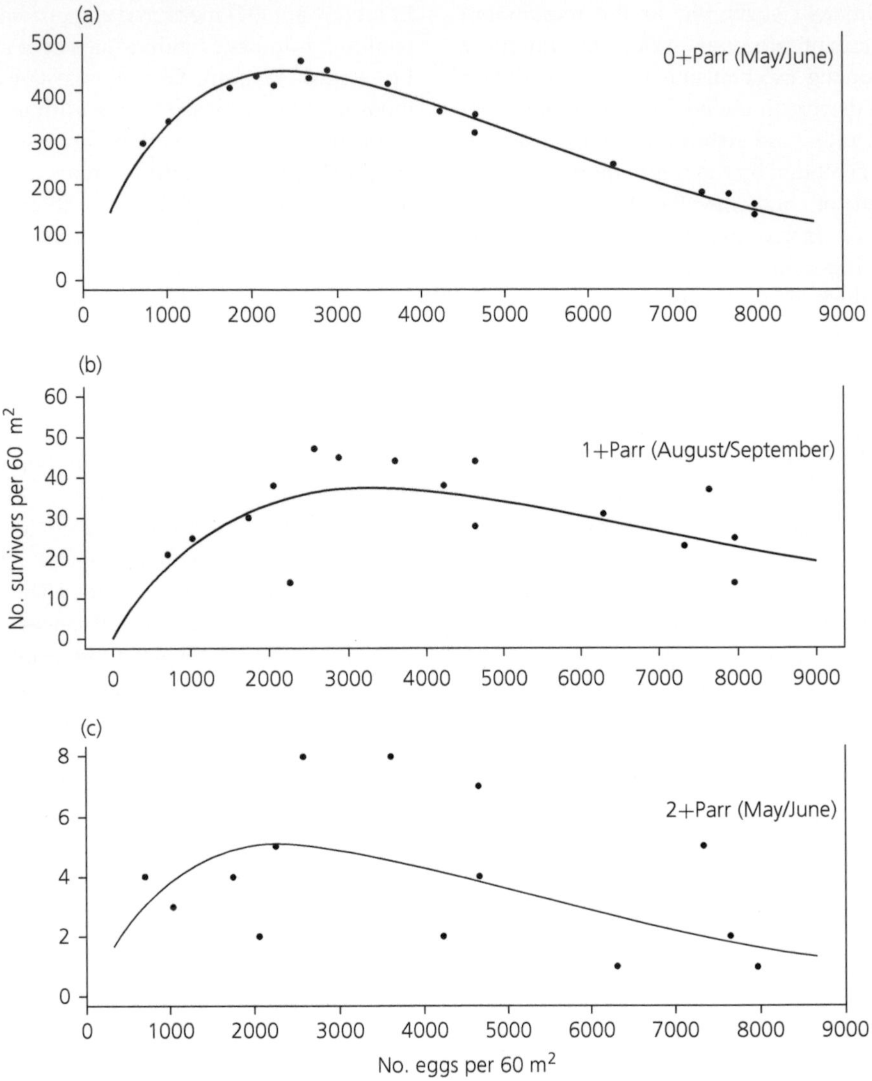

Figure 9.7 Number of surviving individuals of brown trout (*Salmo trutta*) cohorts at three successive life history stages as a function of initial egg production at Black Brows Beck from 1967-83 (data from Elliot 1985 Table 2). Stages: age 0+ parr in May/June of natal year; age 1+ parr in August/September in second year of life, and age 2+ parr in May/June of third year of life.

Although this list does not exhaust the range of possibilities for compensation and regulation during the pre-recruit stage(s), it does encompass the mechanisms underlying the best-known recruitment models. In the following, we will provide examples of the biological and ecological processes underlying different recruitment models to motivate the model development. We frame our exposition in terms of the relationship between the production of viable eggs and resulting recruitment.

In many respects, this perspective returns to the origins of fishery recruitment theory (Ricker 1954; Beverton and Holt 1957). The importance of specifically accounting for reproductive processes in the production of fish populations is a central theme of the contributions found in Jakobsen et al. (2016). It has, however, often typically been necessary to employ a proxy measure of true reproductive output using adult biomass. If this relationship is linear, spawning biomass can be taken as a proxy. Rothschild and Fogarty (1989) explored the implications when the assumption of a simple proportional relationship between viable egg production and adult biomass is not met.

9.3.1.1 Null model

Throughout this chapter we will take a density-independent model as our null hypothesis. For a closed population, the relationship between viable egg production and recruitment will be a straight line through the origin. It is important to emphasize that the appropriate null model is not a horizontal line indicating no relationship between egg production and recruitment. Taken to its logical conclusion, this would imply that recruits would be produced in the absence of egg production. Adopting such a null model would obviously entail high risk to the population (e.g. Fogarty et al. 1992, 1996).

We begin with the simple observation that for a closed population, the number of individuals in a cohort (N) can only decline over time:

$$\frac{dN}{dt} = -\mu N \tag{9.19}$$

The solution to this differential equation is given by:

$$N_t = R = \mathcal{E} e^{-\mu t} \tag{9.20}$$

where N_t is the number of recruits (hereafter R), $\mathcal{E}$ is the number of viable eggs, μ is the instantaneous rate of mortality during the pre-recruit period, and t the time interval between reproduction and recruitment. In this case, R is simply the product of the proportion surviving from the egg to the recruit stage and N_0, which gives a simple linear relationship between egg production and recruitment with slope $e^{-\mu t}$ (Figure 9.8). For a closed population, the relationship of course goes through the origin. For metapopulations with interchange among subpopulations, the relationship may not pass through the origin (e.g. for a sink population receiving a subsidy from a source population).

The null recruitment model implies that there are no constraints on the number of recruits produced, leading to unrealistic predictions of unrestrained population growth. We can readily extend the density-independent recruitment model to incorporate various types of compensatory processes affecting growth, reproduction, and survival during the pre-recruit phase. We view recruitment models not simply as heuristic guides to the shape of the egg production-recruitment relationship but as the elaboration of testable biological hypotheses concerning different compensatory mechanisms.

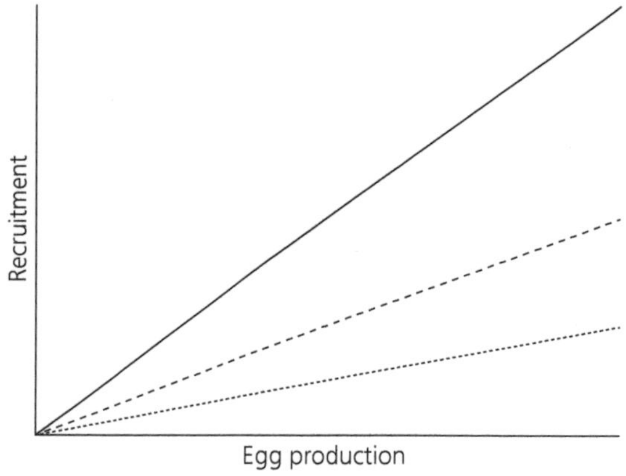

Figure 9.8 Density-independent model relating recruitment and egg production for three levels of the density-independent mortality rate.

9.3.1.2 Intra-cohort competition

In situations where members of a cohort compete for limiting resources (food, space, etc.) density-dependent mortality may be critically important. Investigations of plaice *Platessa platessa* from the larval settlement stage through the first year of life have proved very instructive. Iles and Beverton (1998) provide an important synthesis of estimated mortality rates of 0-group plaice for 15 locations in the northeast Atlantic. Mortality rates as a function of cohort density for the longest individual study, Gullmar Bay in Sweden, are shown in Figure 9.9. A linear increase in mortality with increasing cohort size is clearly evident.

Our simple null model can be extended to account for a linear increase in mortality with increasing cohort density by making the substitution $\mu = (\mu_0 + \mu_1 N)$ where μ_0 is the instantaneous rate of density-independent mortality and μ_1 is the coefficient of density-dependent mortality (Beverton and Holt 1957). The model for the rate of decay of the cohort can then be expressed:

$$\frac{dN}{dt} = -(\mu_0 + \mu_1 N) N \tag{9.21}$$

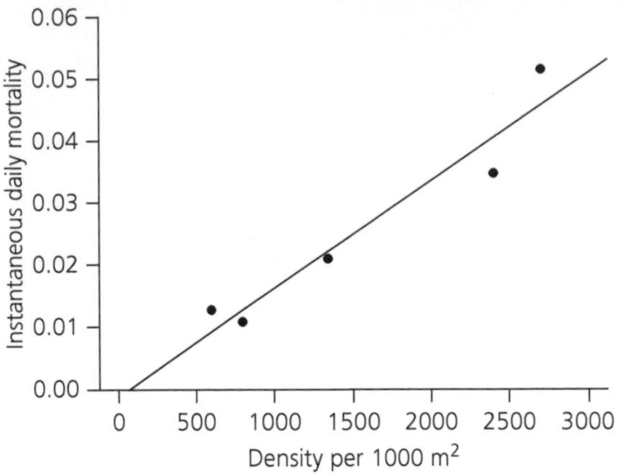

Figure 9.9 Density-dependent mortality of 0-group plaice (*Pleuronectes platessa*) in Gullmar Bay, Sweden (Iles and Beverton 1998).

Note that this model simply indicates that the per capita rate of change of cohort size (dN/Ndt) declines linearly with increasing cohort size (N). Recognizing that Equation 9.21 has the same quadratic form as the logistic equation (Equation 3.8), we can solve it with integration by parts as in Box 3.1 after changing the signs of the coefficients. The solution gives the well-known Berverton–Holt model:

$$R = \left[\frac{1}{\mathcal{E}}e^{\mu_0 t} + \frac{\mu_1}{\mu_0}\left(e^{\mu_0 t} - 1\right)\right]^{-1} \tag{9.22}$$

where R is the number of recruits (N_t). We can simplify this expression by letting $b = (\mu_1/\mu_0)[\exp(\mu_0 t) - 1]$:

$$R = \left[\frac{1}{\mathcal{E}}e^{\mu_0 t} + b\right]^{-1} \tag{9.23}$$

and further by letting $a = \exp(\mu_0 t)$ to give:

$$R = \left[\frac{a}{\mathcal{E}} + b\right]^{-1} \tag{9.24}$$

This model describes an asymptotic relationship between egg production and recruitment. Figure 9.10a illustrates the general form of the curve and how it changes as the density-independent mortality term increases. The effect of increasing this mortality term is to shift the curve downward and to reduce the slope at the origin. Figure 9.10b shows the effect of changing the density-dependent mortality rate.

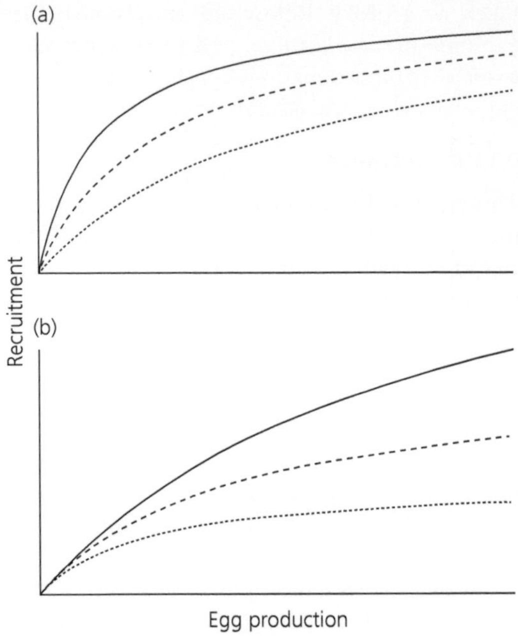

Figure 9.10 Beverton–Holt model relating recruitment and egg production for three levels of (a) the slope at the origin parameter and (b) the density-dependent mortality coefficient.

We further note that intra-cohort cannibalism could also result in a model of this general form. In this chapter, we will refer to this asymptotic form as a compensatory recruitment model and will distinguish it from "over-compensatory" models in which recruitment actually declines at higher levels of egg production, although some authors define these terms differently.

9.3.1.3 Cannibalism by adults

Cannibalism has been shown to be an important population regulatory mechanism in many fish populations (Dominey and Blumer 1984). Ontogenetic shifts in diet are observed in fish populations and adult fish are more likely to be cannibalistic. It has been hypothesized that cannibalism can in fact play an important role of energy transfer within a population if the costs of reproduction are surpassed by the energy gained by later consumption of the progeny (see Longhurst 2010 for a recent review). Some fish species, notably gadoids, are known to exhibit high levels of intraspecific predation. We turn again to the example of walleye

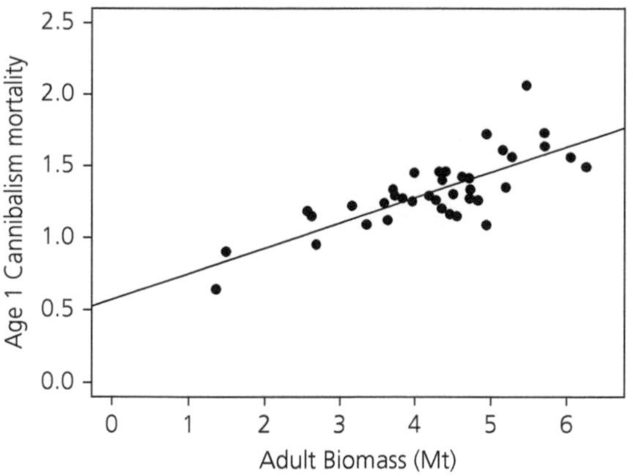

Figure 9.11 Intraspecific predation mortality of age 1 walleye pollock (*Gadus chalcogrammus*) as a function of adult biomass in the eastern Bering Sea (J. Ianelli personal communication).

pollock in the Bering Sea to illustrate some key points. The relationship between mortality on age-1 pollock generated by cannibalism and adult biomass is shown in Figure 9.11. We see a clear linear relationship between mortality due to cannibalism and the adult pollock population.

To represent intraspecific predation by adults on pre-recruits, we can let $\mu = (\mu_0 + \mu_2 S)$ and the model can now be specified:

$$\frac{dN}{dt} = -(\mu_0 + \mu_2 S)N \tag{9.25}$$

where μ_2 is the coefficient of "stock-dependent" mortality (Harris 1975), and S is a measure of the adult population (typically biomass). Note that S is treated as a constant in the integration. Here, the per capita rate of change declines linearly with adult population size. Some segments of the adult population such as larger, older individuals may contribute more to cannibalism, and the index of the adult population used can and should reflect this fact when possible (Link et al. 2012).

The solution to this equation is:

$$R = \mathcal{E}e^{-(\mu_0+\mu_2 S)t} \tag{9.26}$$

which can be further simplified to:

$$R = a\mathcal{E}e^{-bS} \tag{9.27}$$

where $a = \exp(-\mu_0 t)$ and $b = \mu_2 t$. This recruitment function results in a dome-shaped curve. As the density-independent mortality term increases, the curve is again shifted downward and the slope at the origin is reduced (Figure 9.12a). As the compensatory mortality term increases, the curve is shifted downward and it becomes more highly convex (Figure 9.12b).

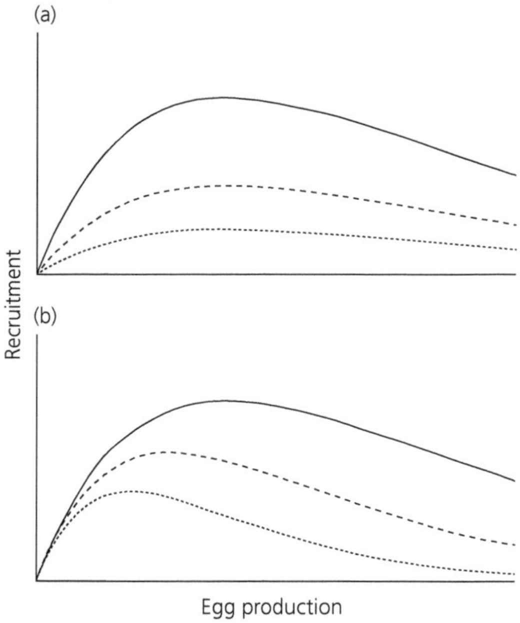

Figure 9.12 Ricker model relating recruitment and egg production for three levels of the (a) slope at the origin parameter and (b) the compensatory mortality term assuming $\mathcal{E} \propto S$.

9.3.1.4 Size-dependent processes

Compensatory recruitment models based on size-specific mortality rates have also been developed to reflect the interaction of compensatory growth and mortality rates. If smaller individuals are more vulnerable to predation, then density-dependent factors that affect the time required to grow through a window of vulnerability to predation will have a direct effect on recruitment (see Houde 2016 for an overview). In particular, size can have critical effects on vulnerability when the ratio of predator to prey size is relatively low (Miller et al. 1988). Accordingly, density-related effects on growth can have potentially important implications for survival rates even if mortality itself is independent of density.

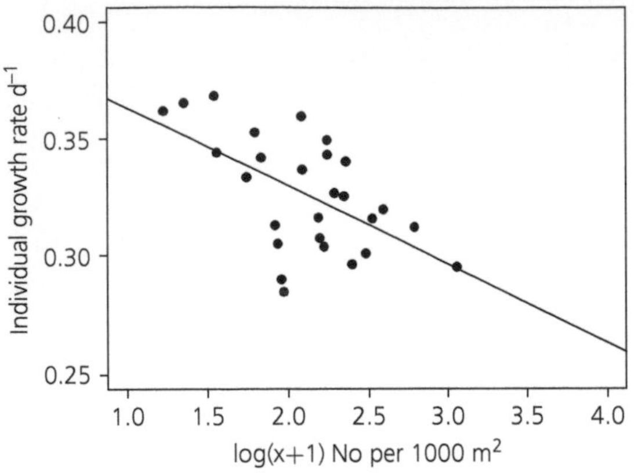

Figure 9.13 Individual growth rate of larval southern bluefin tuna (*Thunnus maccoyi*) as a function of cohort density. (Jenkins et al. 1991).

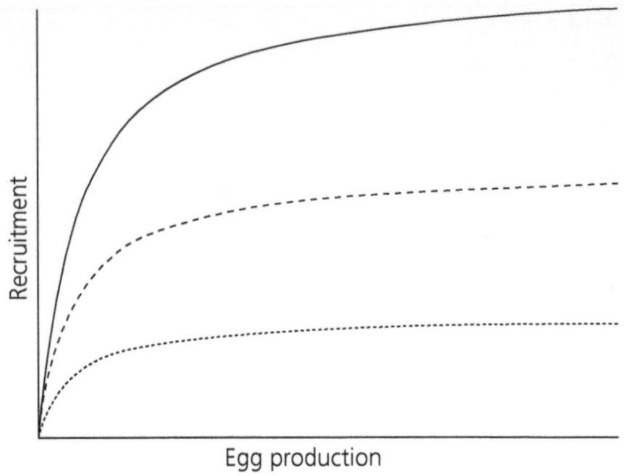

Figure 9.14 Shepherd–Cushing model relating recruitment and egg production for three levels of the parameter K indicating the level at which density-dependent growth dominates.

An example of density-dependent larval growth for southern bluefin tuna (*Thunnus maccoyii*) is provided in Figure 9.13 (Jenkins et al. 1991).

It is possible to directly model growth processes and their interaction with mortality during the pre-recruit stage. Consider a model for individual growth in weight:

$$\frac{dW}{dt} = g(W) \tag{9.28}$$

where $g(W)$ is a compensatory function for individual growth. If the mortality rate is size-dependent then we have:

$$\frac{dN}{dt} = -\mu(W)N \tag{9.29}$$

and dividing Equation 9.29 by Equation 9.28 we have:

$$\frac{dN}{dW} = -\frac{\mu(W)}{g(W)}N \tag{9.30}$$

to give the rate of change in cohort size as a function of change in weight. This model has been discussed by Werner and Gilliam (1984). Without further specification of the functions $\mu(W)$ and $g(W)$, it is not possible to determine the functional form of this size-based recruitment function. However, if the growth rate is taken to depend on the number in the cohort and the mortality rate to be density-independent, then the recruitment function will generally be compensatory. Large cohorts grow more slowly and remain within the vulnerable (high mortality) pre-recruit period longer.

Shepherd and Cushing (1980) assume that $g(W) = G^*/(1 + N/K)$ where G^* is the maximum growth rate, N is cohort number, and K is a constant related to the abundance of food. It is further assumed that the mortality rate μ is independent of density. When $N = K$, the growth rate is exactly one half of the maximum rate. Separating variables in Equation 9.30, we can then write the model as:

$$\frac{dW}{W} = -\frac{G^*}{\mu}\frac{dN}{\left[1+\frac{N}{K}\right]N} \tag{9.31}$$

and the solution is:

$$\log_e\left(\frac{W_1}{W_0}\right) = -\frac{G^*}{\mu}\log_e\left[\frac{(K+\mathcal{E})\,N_1}{(K+N_1)\,\mathcal{E}}\right] \tag{9.32}$$

where the number of viable eggs produced ($\mathcal{E}$) is the lower limit to integration on the right hand side of Equation 9.31. Exponentiating and letting $A = exp\{-(\mu/G^*)\log_e(W_1/W_0)\}$, the model becomes (after rearranging terms):

$$R = \frac{A\mathcal{E}}{1+(1-A)\,\mathcal{E}/K} \tag{9.33}$$

giving an asymptotic relationship between total egg production and recruitment (here, the number surviving to some specified weight class ($R = N(W_1)$; see Figure 9.14).

9.3.1.5 Compensatory reproductive output

We have so far concentrated on density-dependent or stock-dependent mortality operating during the early life-history stages as a compensatory mechanism in recruitment dynamics. However, important compensatory mechanisms affecting recruitment are not limited to this period (Rothschild & Fogarty 1998; Fogarty & O'Brien 2016; Andersen et al. 2015). For example, at high population levels, food resources may be limiting and less energy may be allocated to reproduction by the mature stock. Consider the case in which the mean fecundity declines exponentially with increasing spawning biomass:

$$f = c \cdot e^{-dS} \tag{9.34}$$

where c is the mean fecundity level at low spawning biomass levels (S), and d is the rate of decline in fecundity with increasing biomass. Craig and Kipling (1983) provide an example of a relationship of this form for pike in Lake Windermere (see Figure 9.15; note the logarithmic scale for the fecundity term). Suppose further that the survival rates during the pre-recruit phase are density independent. We can write the recruitment model as:

$$R = c'\mathcal{E}e^{-dS} \tag{9.35}$$

where $c' = c\,[exp(-\mu t)]$ and we note that the product of f (Equation 9.34) and S gives the egg production. This model is identical in form to a Ricker model but generated by a very different mechanism.

Ware (1980) combined elements of an over-compensatory model of egg production with a compensatory pre-recruit survivorship function to generate a flexible three-parameter recruitment model. Invoking bioenergetic principles, Ware postulated that the mean weight-at-maturity decreases exponentially with increasing adult population size and declining per capita food supply, resulting in a corresponding decline in egg production. Ware's relationship between egg production and spawning biomass can be written:

$$\mathcal{E} = c''Se^{-d''S} \tag{9.36}$$

where c'' and d'' are constants. Ware then combined this stock-dependent fecundity model with density-dependent pre-recruit mortality (Beverton–Holt type) to give:

$$R = \left[\frac{e^{\mu_0 t}}{c''Se^{-d''S}} + \frac{\mu_1}{\mu_0}\left(e^{\mu_0 t} - 1\right)\right]^{-1} \tag{9.37}$$

where we have simply substituted Equation 9.36 for N_0 in Equation 9.22. We can simplify by letting $\beta' = (\mu_1/\mu_0)[\exp(\mu_0 t) - 1]$, and $\alpha' = \exp(\mu_0 t)/c$. to give

$$R = \left[\frac{\alpha' e^{d''S}}{S} + \beta'\right]^{-1} \tag{9.38}$$

The shape of this recruitment function for several levels of the parameter d'' is shown in Figure 9.16. This model can represent a spectrum of shapes rang-

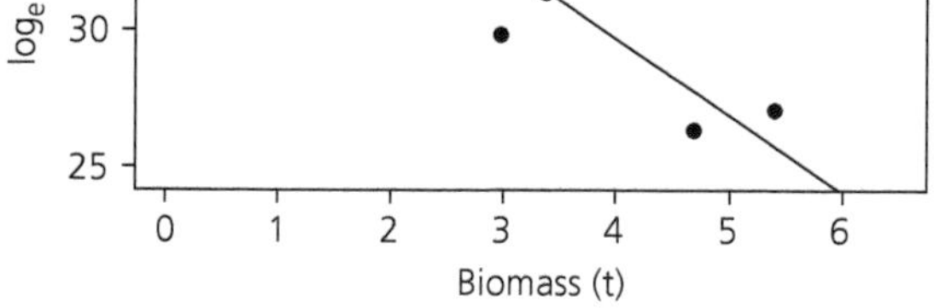

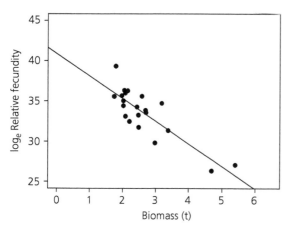

Figure 9.15 Relative fecundity of pike (*Esox lucius*) in Lake Windermere as a function of population biomass (Craig and Kipling 1983).

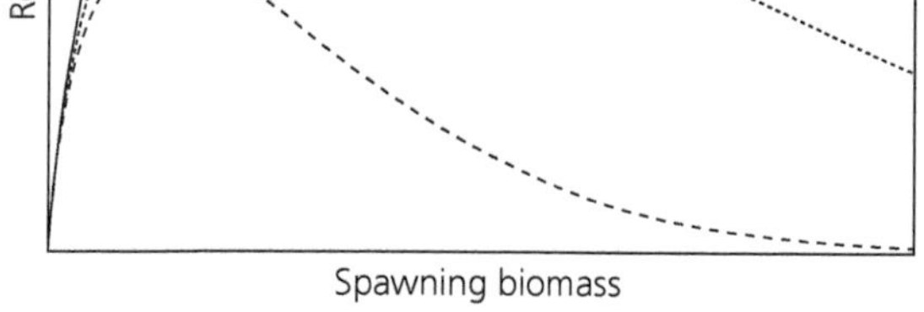

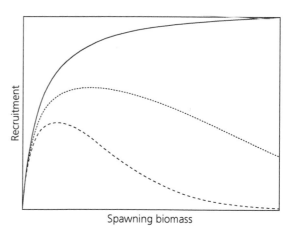

Figure 9.16 Bioenergetics-based model of recruitment as a function of Spawning biomass (Ware 1980).

ing from over-compensatory (Ricker) to asymptotic (Beverton-Holt) forms.

There are of course numerous alternative specifications of the model forms and combinations of mechanisms. The objective here has simply been to point out the potential mechanisms and to illustrate how they can be combined to model compensatory recruitment processes. Understanding the nature of the underlying processes can provide important insights into the form of the resulting recruitment functions.

9.3.2 Recruitment variability

Our treatment of stock recruitment relationships thus far has focused on alternative compensatory mechanisms that can confer some resilience to exploitation and other natural and anthropogenic stressors. However, it is clear that these bivariate deterministic relationships cannot capture the full implications of life in variable aquatic environments. The resulting variability in recruitment in many aquatic populations is not simply noise but rather reflection of the reproductive strategy of these species. It is a central issue in the population dynamics of many aquatic populations.

The models described in Section 9.3.1 treat recruitment solely as a function of egg production; other aspects of the biotic and abiotic environment are not explicitly considered. Yet, as noted earlier, recruitment is extremely variable, largely as a result of the effects of exogenous forcing factors. Key attributes of most teleost fishes (and many exploited aquatic invertebrates) include the production of a superabundance of eggs (Rothschild 1986), little or no parental investment in the progeny following hatching, and widespread dispersal of eggs and larvae in an advective environment. Most of these eggs are destined to perish. Rothschild suggests that the constellation of characteristics in this overall reproductive strategy can be viewed as a "sampling" process in spatially and temporally variable environments. If a sufficient number of the progeny encounter conditions favorable to survival, replacement of the adult population is possible despite very high overall mortality of the progeny. Longhurst (2010) provides further important insights on possible determinants of the "anomalously high" fecundity of teleosts, including the role of cannibalism on progeny in the transfer of energy in the population. See Houde (2016) for an excellent review of the hypotheses related to causes of recruitment variability in these species.

In the following, we will briefly review two principal approaches to addressing aspects of this problem. The first entails extending recruitment models to incorporate environmental covariates in an attempt to partition the overall variance in recruitment into definable sources. The second treats recruitment as a stochastic process in which key parameters (notably pre-recruit mortality) are viewed as random variables. We then obtain a probability distribution of recruitment levels for a given level of egg production rather than just point predictions. A full analytical treatment of this topic is beyond the scope of this work. However, key insights can be obtained through simulation studies. We finally note that our focus on high-fecundity species does not address a very different type of stochasticity relevant to species with relatively low fecundity and population sizes. Many marine mammals and other larger aquatic vertebrates fall into this category. Chance variation in the integer number of births and deaths in a population (demographic stochasticity) can hold very important consequences for populations reduced to low levels by human or natural disturbances (see Fogarty 1993b). This is a critical issue in conservation biology.

9.3.2.1 Recruitment models with environmental covariates

To the extent that specific environmental factors affecting recruitment can be identified and quantified, these can, with appropriate care, be incorporated into recruitment–stock formulations. This approach has received considerable attention as an extension to traditional recruitment models (see Hilborn & Walters 1992 for an overview and caveats). The approach should be framed with specific hypothesis concerning the effect of potential explanatory variables in mind and not as a large-scale data mining exercise. There are many potential environmental variables that could, in principle, be included in an analysis of this type and many may appear to be significant by chance alone. The problem is exacerbated by the fact that many of these variables can be autocorrelated and the effective error degrees of freedom in the analysis may be much lower than the nominal levels.

Consider a simple extension to account for additional physical or biological environmental variables in recruitment models:

$$R = f(\mathcal{E})\, e^{\sum_i \delta_i X_i} \qquad (9.39)$$

where δ_i is the coefficient for the ith environmental factor X_i. If we express the environmental terms as anomalies with mean zero, the recruitment multiplier term has an expected value of 1. To illustrate these points, we will revisit the analysis of Hare et al. (2010) investigating temperature effects on recruitment of Atlantic croaker (*Micropogonias undulatus*) on the eastern seaboard of the United States. Croaker is at the northern extent of its range in the Mid-Atlantic Bight (MAB). If temperature affects recruitment, we might expect that croaker recruitment and productivity in the MAB might be enhanced by warming temperatures.

Hare et al. (2010) used a Ricker-type recruitment function with a temperature covariate:

$$R = aSe^{-bS+\delta T} \qquad (9.40)$$

under the assumption of lognormal errors. For this illustration, we have reanalyzed the Hare et al. data, reserving the last five years of data for testing the forecast skill of the model. A comparison of the observed and predicted recruitment for the earlier part of the time series is shown in Figure 9.17. We also show the out-of-sample forecasts for the last 5 years of the recruitment series. Although these latter observations were not used in fitting the model, the forecasts based on the earlier part of the time series are in close accord with the observed levels (Figure 9.17).

9.3.2.2 Stochastic recruitment models

We can adopt a very different approach to dealing with recruitment variability by exploring the implications of exogenously driven variation in mortality and/or growth rates during the pre-recruit stage. We will refer to this as environmental stochasticity. For small populations, chance variation in the number of deaths in a given time interval (demographic stochasticity) without attribution to any specific environmental

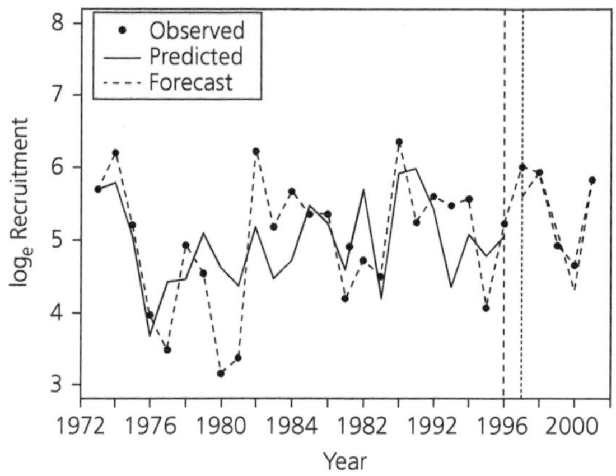

Figure 9.17 Recruitment ($\log_e 10^6$) of Atlantic croaker (*Micropogonius undulatus*) and predictions derived from a Ricker model with a temperature covariate term (1973–1996) and out-of-sample forecasts of recruitment (1997–2001).

driver can be an important consideration for endangered species. In the following, we will treat the simplest case and consider only the null (density-independent) recruitment model with environmental stochasticity.

To motivate the analysis to follow, we consider the classic example of Skeena River sockeye salmon in British Columbia (Shepard and Withler 1958), often used to illustrate the potential magnitude of recruitment variability in fish populations (Hilborn and Walters 1992). We see an increasingly wide scatter of recruitment values as the adult population size increases within the range of available observations (Figure 9.18). The null hypothesis of a density-independent relationship between stock and recruitment cannot be rejected for this series. A straight line through the origin (in this case represented by the median bisector of the data) is shown in Figure 9.18. Sissenwine and Shepherd (1987) proposed the median bisector of data in the stock-recruitment plane as a pragmatic solution to defining the null recruitment model in cases characterized by high variability and a resulting indeterminate form of the stock-recruitment relationship. We also show how the data can be partitioned in the stock–recruitment plane to provide empirical estimates of the probability of different ranges of recruitment for given ranges of adult population size (dotted lines in Figure 9.18). Hilborn and Walters (1992) demonstrate how this information can be used to develop a non-parametric (or model-free) representation of the stock–recruitment relationship for these data.

How might we account for the striking pattern of increasing variability with adult population size in the context of the recruitment models described above? We begin with the case in which the density-independent mortality term is assumed to be an uncorrelated, normally distributed random variable (a so-called white noise process). The assumption of normally distributed mortality rates can be justified under the Central Limit Theorem (see Hilborn and Walters 1992; Fogarty 1993b). If the mortality coefficient varies randomly during the pre-recruit phase, the overall mortality can be viewed as the accumulated sum of random variables over shorter time intervals. For the case of independent mortality rates and for a relatively large number of such intervals, the overall mortality rate will be normally distributed. The Central Limit Theorem also holds under much more general conditions for non-independent stationary processes (see Fogarty 1993b).

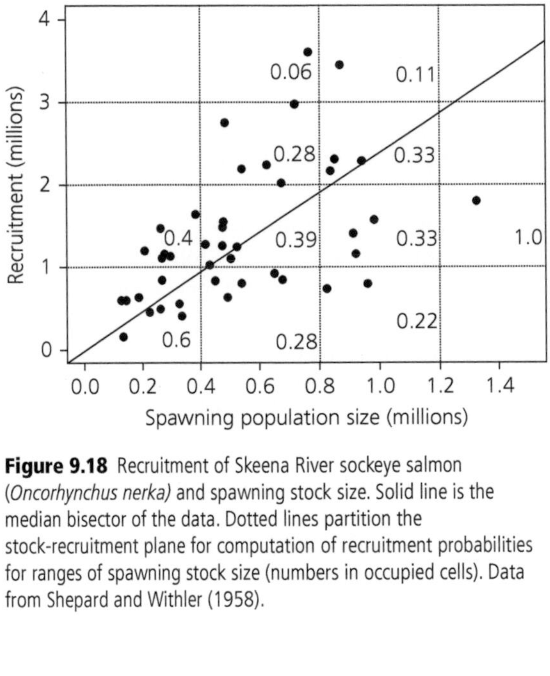

Figure 9.18 Recruitment of Skeena River sockeye salmon (*Oncorhynchus nerka)* and spawning stock size. Solid line is the median bisector of the data. Dotted lines partition the stock-recruitment plane for computation of recruitment probabilities for ranges of spawning stock size (numbers in occupied cells). Data from Shepard and Withler (1958).

Inspecting Equation 9.20 and now treating μ as a normally distributed random variable, we see that the survivorship term $\exp(-\mu)$ will be lognormally distributed (the exponential of a normal random variable is lognormal). We are interested in the probability distribution of recruitment for each level of egg production. We will therefore be multiplying the survivorship term by a constant (different values of $\mathcal{E}$); the resulting product remains lognormally distributed (see Fogarty 1993a,b for analytical results). We can evaluate the lognormal probability distribution for a range of egg production levels for specified levels of the mean pre-recruit mortality and its variance (see Figure 9.19). Each conditional distribution is of course positively skewed. With increasing egg production, the spread of the distribution increases as we saw with the Skeena River sockeye salmon example. We also find that the mean

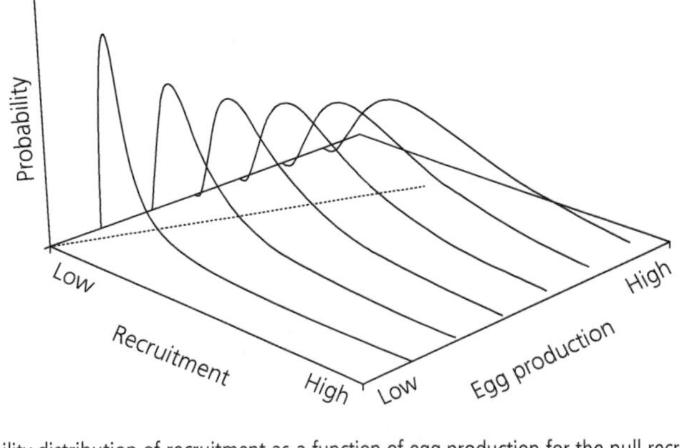

Figure 9.19 Conditional probability distribution of recruitment as a function of egg production for the null recruitment model.

Box 9.1 Stochastic Density-Independent Recruitment Model

Under the assumption that the density-independent mortality rate model is normally distributed, recruitment under the null density-independent model for a given level of egg production N_0 will follow a log-normal distribution:

$$P(R|\mathcal{E}) = \frac{R^{-1}}{\mathcal{E}} \exp\left[\frac{-(\log(R/\mathcal{E}) + \bar{\mu})^2}{2\sigma_\mu^2}\right]$$

where $\bar{\mu}$ is the mean density-independent mortality rate and σ_μ^2 is its variance (Fogarty et al. 1991; Fogarty 1993a,b). The shape of the distribution for several levels of egg production is illustrated in Figure 9.19.

If the assumption of normally distributed mortality rates holds, the mean recruitment for a given level of total egg production is:

$$R = \mathcal{E}e^{-\bar{\mu}+\sigma_\mu^2/2}$$

and we see that recruitment is higher in the stochastic case than in the deterministic case. The variance in recruitment for a given level of egg production is:

$$V(R) = \mathcal{E}^2 e^{-2\bar{\mu}+\sigma_\mu^2}\left(e^{-2\bar{\mu}+\sigma_\mu^2} - 1\right)$$

Accordingly, we can anticipate higher levels of risk to the population with increasing variability in recruitment. Under these conditions recruitment can decline by chance, possibly reaching critically low levels if mortality is highly variable. Notice that the expression for the variance of recruitment conditioned on egg production is a function of the square of the total egg production. We would therefore expect that populations with higher levels of egg production would exhibit higher absolute variability in recruitment for the null model.

The relative variability in recruitment as measured by the coefficient of variation (CV) for the null recruitment model is:

$$CV(R) = \left(e^{-2\bar{\mu}+\sigma_\mu^2} - 1\right)^{1/2}$$

recruitment is higher than the deterministic level by the factor $\exp(\sigma_\mu^2/2)$ (see Box 9.1). We further find that although the absolute variance in recruitment increases with increasing levels of egg production, relative variability (as measured by the coefficient of variation) is constant (Box 9.1). Fogarty (1993a) provides results for other recruitment functions including Ricker and Beverton–Holt.

Occasional strong year classes (represented by the tails of the distributions depicted in Figure 9.19) can be important in maintaining the population. This variation on the concept of a "storage" effect (Chesson 1984) may permit population persistence in a variable environment. It must be appreciated that harvesting can interfere with this mechanism by truncating the age distribution of females and reducing the number of lifetime reproductive opportunities (Fogarty 1993a; Longhurst 2002; Beamish et al. 2006). This perspective does not emerge in a consideration of purely deterministic processes. It highlights the need to specifically focus on the maintenance of robust age structures.

9.4 Summary

The dominant focus on production processes in fisheries science sets it apart from other areas of population ecology in which population numbers are the principal currency for analysis. The foundation established in this chapter sets the stage for Chapter 11 in which we will revisit the topic of production at the cohort and population levels in the context of harvesting strategies. We have extended the consideration of individual growth and mortality provided in earlier chapters and have attempted to broaden the context for understanding compensatory processes.

An understanding of how a fish population will respond to harvesting requires not only an accurate accounting of its effective reproductive output but an understanding of the relative importance of compensatory mechanisms operating at different points in the life cycle. Our emphasis on recruitment as a function of total egg production of the population rather than spawning stock biomass is intended to encourage further development of monitoring and research programs designed to broaden our understanding of reproductive processes (see contributions in Jakobsen et al. 2016) and estimating effective reproductive output.

We have highlighted a number of distinct compensatory mechanisms that underlie models of recruitment processes and shown ways in which researchers have focused on the evidence for different mechanisms that ultimately influence recruitment. Information of this type can inform choices for the appropriate deterministic "skeleton" of recruitment models. As we have seen, a number of different mechanisms can give rise to similar shaped recruitment curves. These can be partitioned into two major categories: compensatory mechanisms giving rise to asymptotic recruitment curves and over-compensatory mechanisms resulting in dome-shaped curves. These two principal categories hold very different implications for management strategies and the dynamics of the species. Fogarty and O'Brien (2016) provide further information on some alternative mechanisms and a treatment of multi-stage processes not covered here due to constraints on space.

Overlaying these considerations is the recognition that recruitment in aquatic species that produce large numbers of eggs with little or no parental investment in the progeny following spawning is highly variable. In this context, a dominant theme that emerges is that even relatively low levels of variability in growth and mortality can translate into quite high levels of recruitment variability. This variability in recruitment should be viewed as an integral part of the life-history strategies of many aquatic species and not simply "noise". Favorable environmental conditions for growth and survival of the early life stages resulting in strong recruitment can be an important factor in maintaining populations through a form of "storage effect" in which high recruitment in some years can allow persistence of the population through periods of lower recruitment (Fogarty 1993a). One potentially important mechanism may be an "escape in numbers" in which predators cannot crop the early life stages in proportion to their abundance (see Chapter 4 for discussion of saturating functional feeding responses). For populations characterized by alternate stable states of high and low abundance, strong year classes can also provide a way for populations to move from low to high states.

Additional reading

We highly recommend Wootton's (1998) treatment of production processes in fish populations as essential reading. Cushing 1995) provides an overview of production and population regulation in marine fish populations. Quinn and Deriso (1999) give a comprehensive treatment of recruitment models (their Chapter 3) and individual growth (their Chapter 4); see also Hilborn and Walters (1992; Chapters 7 and 13). Both of these provide good accounts of estimation issues in fitting recruitment and growth models.

CHAPTER 10

Production at the Ecosystem Level

10.1 Introduction

In this chapter, we continue the progression from populations to communities to all taxa in the ecosystem. Apart from the book-keeping problem of keeping track of many species, the ecosystem perspective introduces a new set of challenges. It is sometimes necessary to aggregate species into groups, especially at the lower trophic levels, which is necessary not only to reduce complexity, but because the data are often lacking to partition functional groups into their component species. A second challenge is that the spatial and temporal scales of processes affecting different trophic levels are vastly different. For example, plankton may have lifespans of days and the processes affecting their production may vary over 10s of km, whereas fish have lifespans of years and can range over 100s or 1000s of km (Figure 10.1). Averaging over larger scales can filter out important processes for plankton, yet fine-scaled description and modeling of fish populations on the scales relevant to plankton is generally not possible.

The "size-based" approach recognizes that most biological processes scale strongly with body size. It is therefore possible to describe important ecosystem processes by classifying the biomass of organisms by body size, ignoring the individual species. The distribution of biomass across size classes at a point in time is known as the size spectrum (see Section 6.5). Dynamic size-structured models describe the transitions from one size class to another. They provide an important complement to traditional species or taxa-based approaches. As we saw in Chapter 8, fundamental allometric principles are clearly evident in the production of individual organisms and the size spectrum approach builds on this foundation. In the context of ecosystem processes, predation exhibits strong allometry.

The "species-based" approach focuses on the dynamics of focal species that are of primary interest. The dynamics of the focal species are described in great detail, whereas the trophic levels above and below the focal species are aggregated and treated more as boundary conditions. In contrast, the "trophocentric" approach aggregates at the species level but considers all the trophic levels in some detail. The trophocentric approach emphasizes the supply of energy from lower trophic levels and the role of predation in regulating the production at each level. Because body size is strongly correlated with trophic level, the size-based and trophocentric approaches have much in common. However, the size-based approach assumes a monotonic progression of size classes, whereas the trophocentric approach recognizes the topology of the food web.

The species-based and size-based/trophocentric approaches can be considered as orthogonal perspectives of the ecosystem that are appropriate in different situations (Figure 10.2). The species-based approach is suited for understanding variability at the scale of the focal species (e.g. interannual variation), whereas the trophocentric approach is most useful for understanding the longer-term (decadal) changes in ecosystems. Whether these two perspectives can be viewed simultaneously with ecosystem models remains an open question. Modern computers can handle all the necessary calculations but there is always the risk of studying the model ecosystem in place of the real one. An on-going research challenge is how best to couple fine-scale models of the lower trophic levels (e.g. plankton and benthos) with more aggregated

Fishery Ecosystem Dynamics. Michael J. Fogarty and Jeremy S. Collie, Oxford University Press (2020). © Michael J. Fogarty & Jeremy S. Collie 2020.
DOI: 10.1093/oso/9780198768937.003.0010

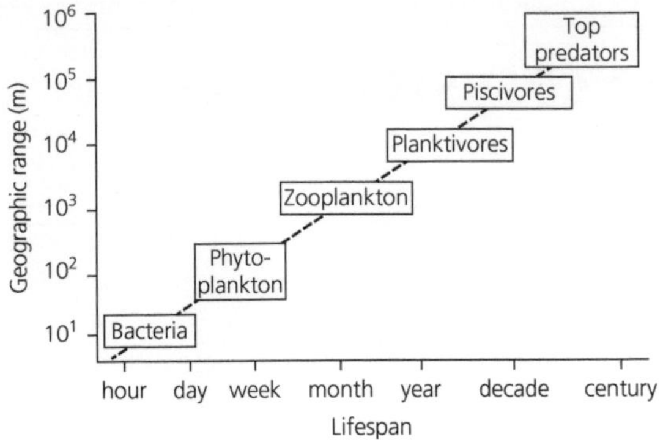

Figure 10.1 Time and space scales of aquatic organisms. Geographic range corresponds to the range of an individual during its lifespan, not the geographic range of the species. This is a modified version of the "Stommel diagram."

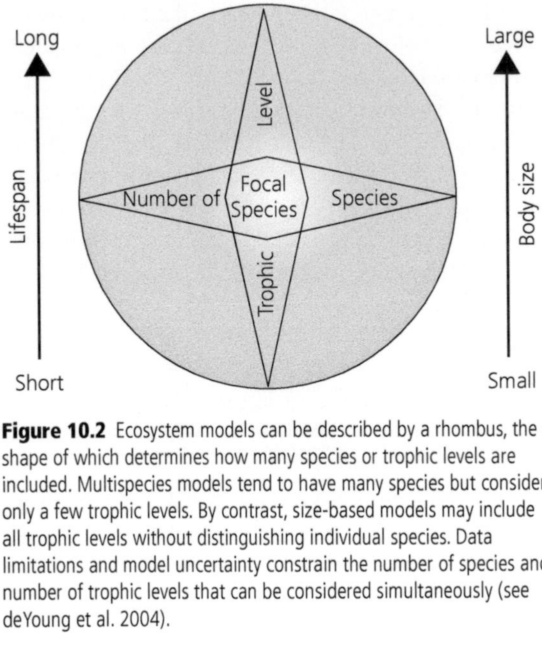

Figure 10.2 Ecosystem models can be described by a rhombus, the shape of which determines how many species or trophic levels are included. Multispecies models tend to have many species but consider only a few trophic levels. By contrast, size-based models may include all trophic levels without distinguishing individual species. Data limitations and model uncertainty constrain the number of species and number of trophic levels that can be considered simultaneously (see deYoung et al. 2004).

models of the upper trophic levels (e.g. fish and other top predators) into functional end-to-end models.

This chapter starts with a brief introduction to food webs and then describes linearized and mass-balance descriptions of ecosystems. Energy budgets for individual organisms were introduced in Section 8.2. Here we extend that concept to energy budgets of trophic compartments (species, functional groups, etc.) and the energy flow among compartments.

We illustrate the analytical machinery behind mass-balance models as a prelude to treatment of this topic in Chapter 13 in the context of harvesting models. We then take up the topic of models that directly incorporate biogeochemical considerations. Many models of lower trophic levels in limnology and oceanography adopt this approach and methods to extend these considerations to full ecosystem models are a very active area of research. We close with models that explore the uses of biomass spectra as alternative representations of ecosystems.

10.2 Food webs

Food webs comprise the basic structural elements of ecosystems. Species are linked through trophic interactions that ultimately control overall levels of ecosystem productivity. To characterize ecosystems, we can start by visualizing the food web and providing some simple descriptive statistics defining its structure. The number of species included and levels of taxonomic aggregation employed often reflect the specific research questions pursued. No documented web can be considered fully complete. Nonetheless, we can obtain important insights into the architecture of food webs using these studies. Compilations of studied food webs in terrestrial and aquatic ecosystems are now readily available online, providing important opportunities for comparative analysis of food web properties.

We will illustrate some basic points with a food web for the Benguela Current system off southwest Africa employed by Yodzis (1998). This web representation includes 29 species or species groups (S'), from bacteria and phytoplankton to marine mammals and seabirds. Yodzis was particularly interested in potential interactions between protected species and fisheries; the taxonomic resolution employed reflects this orientation with high specificity for upper trophic levels and much lower resolution for the lower trophic levels. By constructing simple diagrammatic representations of the trophic connections, we obtain an initial roadmap of energy flow within the ecosystem. In Figure 10.3, we show two alternative visualizations of food web structure for this ecosystem. A conventional hierarchical depiction arranged

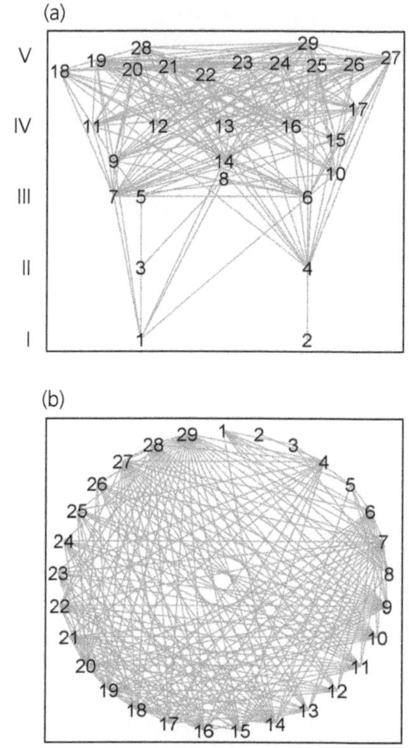

Figure 10.3 Benguela Current food web with 29 nodes: (1) phytoplankton, (2) benthic filter-feeders, (3) bacteria, (4) benthic carnivores, (5) microzooplankton, (6) mesozooplankton, (7) macrozooplankton, (8) gelatinous zooplankton, (9) anchovy, (10) pilchard, (11) round herring, (12) lightfish, (13) lanternfish, (14) goby, (15) other pelagics, (16) horse mackerel, (17) chub mackerel, (18) other groundfish, (19) hakes, (20) squid, (21) tunas, (22) snoek, (23) kob, (24) yellowtail, (25) geelbek, (26) whales and dolphins, (27) birds, (28) seals, (29) sharks. Roman numerals indicate trophic levels.

by trophic level is depicted in Figure 10.3a. In this representation, there are five principal trophic levels. Here, we have used a weighted average for the trophic position for each species or taxonomic group based on its diet composition. In aquatic ecosystems, most species, particularly at the higher trophic levels, draw on food resources from a number of different trophic levels. The breadth of dietary items of fish consumed over the lifespan is particularly striking, in accord with the dramatic change in size from the larval to adult stages. Larvae of many species feed on phytoplankton and small zooplankton. Following the transition to the juvenile stage the diet can encompass a much broader spectrum of prey including planktonic, benthic, and larger demersal and pelagic species. Piscivorous fish typically maintain a mixed-feeding strategy but with progressively higher concentration of fish in the diet with increasing size. Omnivory is a dominant feeding strategy for many fishes, particularly when viewed over the entire life cycle.

Although we deliberately chose the Benguela food web example for its manageable number of taxa, it still can be difficult to trace individual pathways through the food web. In some cases, a circular arrangement of the same web can offer advantages in discerning connections (Figure 10.3b). Tracing the links from any node at the circumference of the circle across to all other connected nodes can in some cases be easier than navigating the tangled skein of interactions in the hierarchical representation. For highly speciose webs, however, it is inevitably difficult to trace all of these linkages and their food web diagrams simply convey a sense of complexity.

There are 203 trophic links (L) in the Benguela web. We can begin to interpret this number by placing it in the context of the total possible number of links. The connectivity of a food web is defined as the ratio of the observed number of links to the total number possible. A commonly used measure of connectivity is $C = L/(S'(S'-1)/2)$. We can represent the food web as a matrix which in its simplest form is a symmetrical matrix composed of ones and zeros representing the presence or absence of an interaction between species or taxa (S'). The elements on the main diagonal represent self-links (representing cannibalism) which are not counted here; our interest centers on interspecific interactions. Because the matrix is symmetrical, we need only count the number of interactions above or below the main diagonal. The denominator of the measure of connectivity above captures these features. The connectivity of the Benguela food web is 0.5 using this metric. To place these observations in context, we note that Link (2002a) provided comparable average summary statistics from catalogs of terrestrial and aquatic food webs and for a list of 12 studies deemed to be comprehensive (representing all trophic levels). For the latter grouping, the number of species (nodes) in the webs ranged from 12 to 182; the number of links spanned 36–2366; and the connectivity estimates ranged from 0.43 to 0.66. The estimate of connectivity for the Benguela web is comparable for other reported estimates with similar number of species reported by Link (2002a). Estimates of connectivity can provide insights into the question of stability and complexity explored in Chapter 6 (see the review by Pimm 1982). In general, the probability of the system being stable declines sharply for increasing numbers of species for a given level of connectivity. More refined treatments of this question are possible if interaction strength estimates are available.

10.3 Energy flow and utilization

Plants use energy from sunlight to take up essential nutrients and convert carbon dioxide to organic matter, which forms the base of the food web. In aquatic ecosystems, single-celled phytoplankton form the foundation of the trophic pyramid. In many aquatic ecosystems multicellular plants also contribute to primary production in shallower freshwater and marine environments where sunlight reaches the bottom. In aquatic environments, light levels, nutrient availability, temperature, and grazing strongly influence primary production. In both freshwater and marine ecosystems, primary production is highest near the equator and declines with increasing latitude (see review by Ware 2000). Sunlight at wavelengths between 400 and 700nm is used in photosynthesis (photosynthetically active radiation, PAR). A satellite derived map of PAR measured over coastal and marine environments shows the highest levels of PAR at low latitudes (Figure 10.4a). In general the highest levels of primary production occur in nearshore waters and on the continental shelves (typically defined as waters less than 200m depth) and in upwelling areas (Figure 10.4b). Nutrient concentrations are highest in sunlit waters of the continental shelves, driven by hydrodynamic processes such as upwelling, and land-based runoff. Atmospheric deposition is also important throughout the Earth's surface.

An important distinction can be made between new and regenerated primary production. New production derives from utilization of nutrients made newly available for plant growth from outside the photic zone, largely through hydrodynamic processes from deeper water, although runoff from land can also be an important contributor in freshwater, inland, and nearshore waters. In contrast, regenerated production derives from the recycling of nutrients as a result of the decomposition of organic matter through bacterial action. This regenerated production passes through the microbial food web (Azam et al. 1983) and because it involves two or more additional trophic steps, much of the energy is dissipated before being able to contribute to growth of upper trophic level organisms. Small photoautotrophs [nanoplankton (< 2 microns) and picoplankton (2–20 microns)] are important elements of the pathways using regenerated nutrients in the microbial food web. Larger phytoplankton cells (20–200 microns) dominated by diatoms utilize newly available nutrients and are preyed on in the grazing/predator food web. Overall, regenerated production predominates in aquatic foodwebs. Steele et al. (2007) estimated that new production on Georges Bank accounted for approximately 30% of the total production. In an earlier analysis for the English Channel, Steele (2001) found that only about 15% of the total primary production was new production. Iverson (1990) argued that fish production relies principally on new production passed through the grazing/predator food web, with important implications for expected levels of yield in different ecosystems.

Primary production passes through the food web by consumption (C) and production (P) of higher trophic level species, which in turn can be grouped into trophic compartments. The fraction of the food consumed by each trophic compartment that is passed on to consumers at higher trophic levels (P/C) is called the trophic or growth efficiency. Trophic efficiencies are typically on the order of 10% but vary by trophic group and may be as high as 20 to 30% for the lower trophic levels. Moving up the trophic pyramid, body size increases with trophic level as the total production attenuates according to the trophic efficiency.

The energy content of organisms can be measured with bomb calorimetry and the energy flux expressed in kilocalories or Joules per unit time. Carbon is the essential building block of organic matter and can be used as a proxy for the more elusive concept of energy flow (Steele

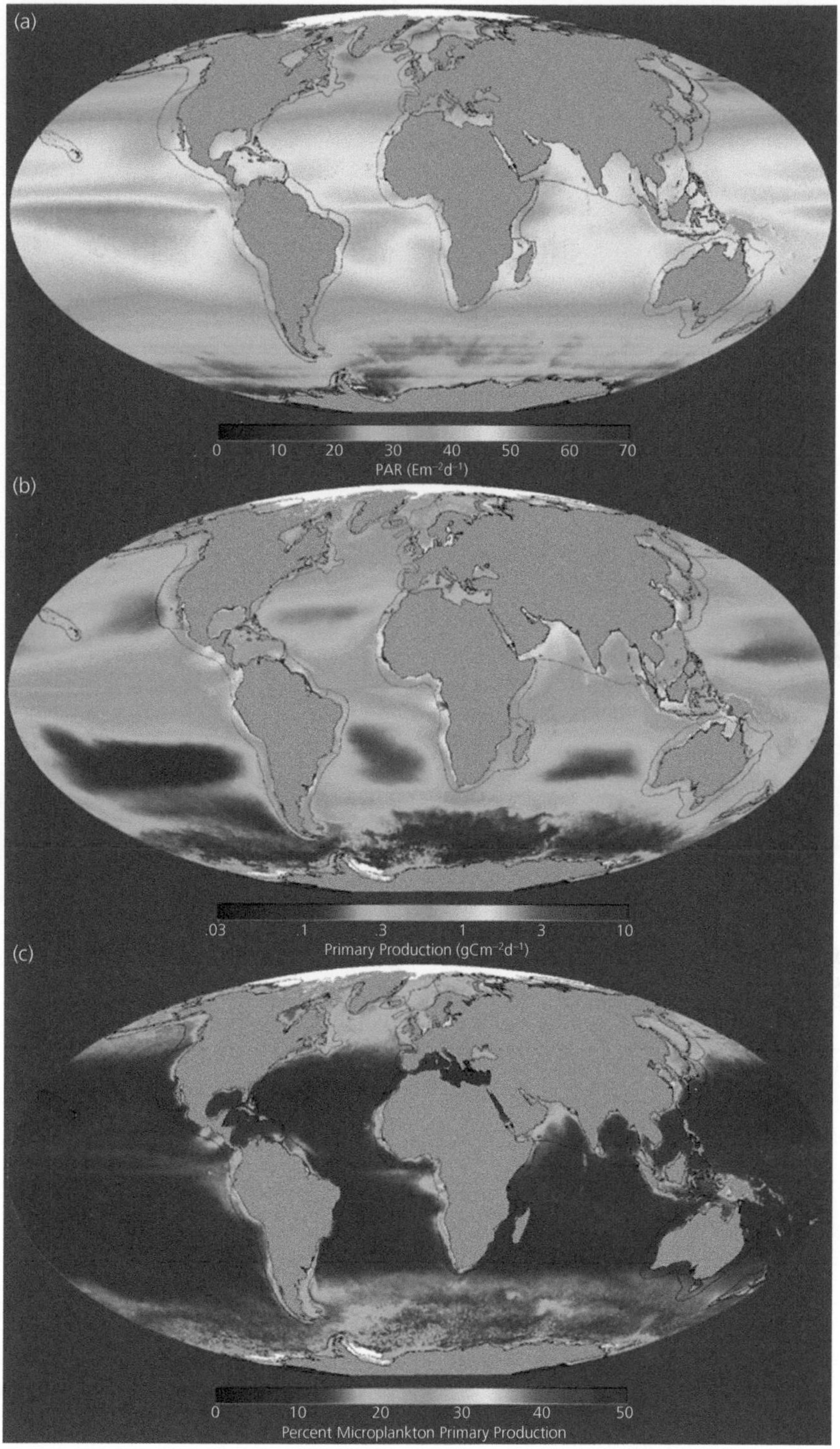

Figure 10.4 (a) Global estimates of photosynthetically active radiation (PAR, einsteins m^{-2} d^{-1}), (b) total primary production (gC m^{-2} d^{-1}), and (c) percent of primary production by microplankton (gC m^{-2} d^{-1}) in marine ecosystems. Courtesy Kim Hyde.

2001). Conversion factors have been developed to translate from one unit of measurement to another. As first approximations: 10 kcal ≈ 1 g carbon ≈ 10 g wet weight. As noted in Chapter 8, a comprehensive compilation for individual species is available at www.thomas.brey.de/science/virtualhandbook. Energy transfer among the lower planktonic groups may be more readily measured in units of nutrients. Phosphorus tends to limit production in many freshwater ecosystems and nitrogen in many parts of the world's oceans. To the extent that carbon, nitrogen, and phosphorus have a roughly constant ratio in organisms, nitrogen and phosphorus can be used as proxies for energy flow.

In instances where direct measurements of production and consumption are not available but estimates of biomass can be obtained, compilations of production to biomass (P/B) and consumption to biomass (Q/B) ratios can be employed. There are now estimates available from sources such as www.ecobase.org. It is also possible to derive estimates of these ratios for aquatic organisms from allometric principles (e.g. Ware 2000). The relationship between both P/B and Q/B ratios and body weight has been shown to follow the one-quarter scaling rule described in Chapter 8. The exponent of these allometric relationships is -0.25 while the intercept terms vary among taxa as a function of life history and other characteristics.

10.4 Linear network models

As noted in Chapter 1 models of energy flow in aquatic ecosystems have an extensive history. In his pioneering work *Elements of Physical Ecology*, Lotka (1925) provided a framework for the analysis of aquatic food webs. An early representation of a complex marine food web was constructed with Atlantic herring in the North Sea as the focal species (Hardy 1924). Clarke (1946) traced energy flow in a highly aggregated model for Georges Bank. The first detailed energy budget of a freshwater ecosystem was constructed by Lindeman (1942) for Lake Mendota in Wisconsin in what is now recognized as seminal contribution to aquatic ecology. This energy budget contained all the essential components, including pelagic and benthic energy pathways, and recycling of nutrients from all trophic components by bacteria.

10.4.1 Bottom-up calculations

We focus here on the production within trophic compartments (species, functional groups, etc.) instead of individual production, as in Chapter 8. The trophic compartments are basic building blocks and we are particularly interested in the flow of energy among them. The quantitative description of energy budgets begins by considering the inputs and outputs for each trophic compartment, i (Steele 1974). With the assumption of mass balance, the output should match the input after accounting for the trophic efficiency, e_i. The balance between consumption and production (P_i) of each trophic compartment can be expressed with a set of linear equations of the form:

$$P_i = e_i \left(\sum_j b_{ij} P_j + x_i \right) \tag{10.1}$$

where b_{ij} is the fraction of production P_j consumed by i and x_i allows for a constant rate of external input to or export from trophic compartment i. Energy flows from compartment j to i for $i = 1,n$.

Starting from primary production, this set of equations can be solved sequentially to estimate production of the top trophic level. In the case of loops in the food web and for mathematical convenience the terms of Equation 10.1 can be rearranged:

$$\frac{P_i}{e_i} - \sum_j b_{ij} P_j = x_i \tag{10.2}$$

and expressed in matrix notation:

$$\mathbf{M} \times \mathbf{P} = \mathbf{X} \tag{10.3}$$

where $\mathbf{M}$ is the matrix comprising $1/e_i$ on the diagonal and $-b_{ij}$ in the off-diagonal elements. The matrix $\mathbf{M}$ describes the transitions from columns j to rows i. The off-diagonal elements sum to one because the energy from each trophic compartment has to go somewhere (unless it is a terminal compartment). The column vector $\mathbf{P}$ contains the production of each compartment and $\mathbf{X}$ is the vector of external inputs.

If **M** and **X** are known, **P** can be calculated by matrix inversion:

$$\mathbf{P} = \mathbf{M}^{-1} \times \mathbf{X} \tag{10.4}$$

To illustrate the mechanics of the calculations we will focus on a simple example for the North Sea (see Figure 10.5). The matrix **M** for this example is shown in Table 10.1 and the bottom-up solutions are listed in Table 10.2. Because of the attenuation of production at each trophic level, the fisheries yield is only 0.42% of primary productivity. Even so, Steele found that relatively high trophic efficiencies were required to balance the food web.

10.4.2 Top-down calculations

An alternate, top-down approach is often used for food-web calculations when emphasis is on the higher trophic levels, and fish yields are the defining

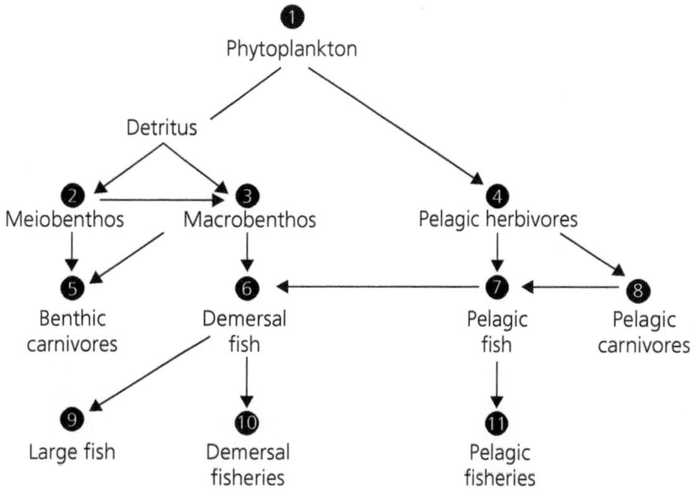

Figure 10.5 North Sea food web (Steele 1974). Numbers for each node in this diagram map to the numbered nodes in Tables 10.1–10.3.

Table 10.1 Matrix **M** for the North Sea food web shown in Figure 10.5. The trophic compartments or nodes are identified below in Table 10.2. The diagonal elements are $1/e_i$, the inverse of the trophic efficiency. The off-diagonal elements are $-b_{ij}$, the proportion of the production of category j (columns) going to category i (rows).

Node	1	2	3	4	5	6	7	8	9	10	11
1	1	0	0	0	0	0	0	0	0	0	0
2	−0.08	4.0	0	0	0	0	0	0	0	0	0
3	−0.12	−0.5	4.35	0	0	0	0	0	0	0	0
4	−0.80	0	0	5.0	0	0	0	0	0	0	0
5	0	−0.5	−0.67	0	7.69	0	0	0	0	0	0
6	0	0	−0.33	0	0	10.0	−0.5	0	0	0	0
7	0	0	0	−0.25	0	0	10.0	−1	0	0	0
8	0	0	0	−0.75	0	0	0	6.67	0	0	0
9	0	0	0	0	0	−0.24	0	0	1.0	0	0
10	0	0	0	0	0	−0.76	0	0	0	1.0	0
11	0	0	0	0	0	0	−0.5	0	0	0	1.0

Table 10.2 Trophic compartments for the North Sea food web shown in Figure 10.5. The trophic efficiency of primary production is assumed to be one, as is the efficiency of the terminal nodes. Production is calculated from Equation 10.4 assuming 100 units of primary production passing through the food-web matrix in Table 10.1. Consumption is calculated from Equation 10.6 with diet matrix Table 10.3.

Node	Compartment	Efficiency	Production	Consumption
1	Phytoplankton	1.00	100	100
2	Meiobenthos	0.25	2.00	8.00
3	Macrobenthos	0.23	2.99	13.00
4	Pelagic herbivores	0.20	16.00	80.00
5	Benthic carnivores	0.13	0.39	3.00
6	Demersal fish	0.10	0.13	1.30
7	Pelagic fish	0.10	0.58	5.80
8	Pelagic carnivores	0.15	1.80	12.00
9	Large fish	1.00	0.03	0.03
10	Demersal fisheries	1.00	0.10	0.10
11	Pelagic fisheries	1.00	0.29	0.29

input. Then the consumption in each compartment, C_i, becomes the state variable:

$$e_i C_i = \sum_j d_{ij} C_j + y_i \tag{10.5}$$

where e_i is again the ecological efficiency, d_{ij} is the diet composition of predator j, and y_i is the export or fisheries yield. Energy flow is now from compartment i to compartment j. If the ecological efficiencies and diet compositions are known from stomach-content data, the system of equations can be solved for the unknown consumptions, including the primary production necessary to support the observed fishery yields. Equation 10.5 can also be re-arranged and expressed in matrix notation as:

$$\mathbf{D} \times \mathbf{C} = \mathbf{Y} \tag{10.6}$$

where $\mathbf{D}$ is a square matrix with e_i on the main diagonal and diet composition on the off-diagonal (Table 10.3). The vector of consumption can again be obtained by matrix inversion. The consumptions estimated for the North Sea example are listed in Table 10.2. In practice, different parameters are known for each trophic box, sometimes as lower or upper bounds. Objective functions can be used to find the best fit (Vezina and Platt 1988). Elaborations of Equation 10.5 form the master equation of the widely used ECOPATH model (see Section 13.3), originally developed by Polovina (1984).

Recalling that $eC = P$, we see that the two approaches yield consistent results (cf. Table 10.2). The equivalence of the bottom-up and top-down approaches can be seen by defining diet composition of predator j as:

$$d_{ij} = \frac{b_{ij} P_i}{\sum_k b_{kj} P_k} \tag{10.7}$$

which is the flux from compartment i relative to the fluxes from all other compartments. In most applications, either the b_{ij} or d_{ij} are known, or some combination of the two.

Calculations of the energy or carbon fluxes through aquatic ecosystems provide a valuable check on estimates of the potential productivity associated with each component of the food web. These calculations can set limits on the expected fishery yields to humans as we discuss further in Chapter 13. Our knowledge of these food webs is provisional and the outcome of the calculations depends on the specification of food-web components and the links between them. The major limitation of network models is that the assumption of mass balance doesn't account for changes in food-web structure over time. One approach is to fit network models for different time stanzas to account for changing food-web components. This technique has been used to reconstruct food webs before the impact of human predation on fish and marine mammals (Steele and Shumacher 2000). But to get from one stanza to another, dynamic models are needed. We return to this issue in Section 10.7 and again in section 13.3.2.

10.5 Biogeochemical models

The simple network models described above typically do not trace the production pathways throughout aquatic food webs directly from nutrient supplies. In contrast, biogeochemical models do start with nutrient concentration but early applications focused on the lower trophic levels, often ending with the zooplankton component.

Table 10.3 Top-down matrix (**D**) for the North Sea food web (Figure 10.5). The nodes correspond to the compartments listed in Table 10.2. The diagonal elements are the trophic efficiencies (e_i). The non-diagonal elements represent the composition of prey (rows) in the diets of predators (columns). The non-diagonal elements of each row sum to −1; one because the diet proportions sum to one and negative because consumption is a loss to the prey species.

Node	1	2	3	4	5	6	7	8	9	10	11
1	1	−1	−0.92	−1	0	0	0	0	0	0	0
2	0	0.25	−0.08	0	−0.33	0	0	0	0	0	0
3	0	0	0.23	0	−0.67	−0.77	0	0	0	0	0
4	0	0	0	0.2	0	0	−0.69	−1	0	0	0
5	0	0	0	0	0.13	0	0	0	0	0	0
6	0	0	0	0	0	0.1	0	0	−1	−1	0
7	0	0	0	0	0	−0.23	0.1	0	0	0	−1
8	0	0	0	0	0	0	−0.31	0.15	0	0	0
9	0	0	0	0	0	0	0	0	1	0	0
10	0	0	0	0	0	0	0	0	0	1	0
11	0	0	0	0	0	0	0	0	0	0	1

More recently, enhanced computing power has opened avenues for a fuller representation of complete food webs starting with nutrient concentrations. The capability of these end-to-end (E2E) models is steadily increasing. Here we provide a brief introduction to the topic of biogeochemical models. Although a full treatment of E2E models is beyond the scope of this text, we provide an early example of such a model for the Georges Bank ecosystem.

10.5.1 Lower trophic level models

Nutrient (N)—Phytoplankton (P)—Zooplankton (Z) models have been used to investigate the relative importance of nutrients and grazing in phytoplankton bloom dynamics (Steele and Henderson 1981, 1992). Franks et al. (1986) investigated a model of the form:

$$\frac{dP}{dt} = \frac{V_m N P}{K_s + N} - mP - Z R_m \left(1 - e^{-\Lambda P}\right) \tag{10.8}$$

The three terms on the right-hand side represent nutrient uptake, mortality (m), and zooplankton grazing. Nutrient uptake is described by a hyperbolic relationship with maximum V_m and half-saturation constant K_s. Grazing has an Ivlev function with maximum R_m and grazing constant Λ (recall that we encountered the Ivlev function in Chapter 4 in our treatment of functional feeding responses). Nutrient uptake and zooplankton grazing are both saturating functions that provide density-dependent compensation to the model. The zooplankton equation,

$$\frac{dZ}{dt} = \gamma Z R_m \left(1 - e^{-\Lambda P}\right) - g' Z \tag{10.9}$$

includes the gain from grazing multiplied by growth efficiency, γ, and a linear loss term with mortality rate, g'. Finally the nutrient equation,

$$\frac{dN}{dt} = -\frac{V_m N P}{K_s + N} + mP + g'Z + (1-\gamma) Z R_m \left(1 - e^{-\Lambda P}\right) \tag{10.10}$$

includes the loss from nutrient uptake and gains from mortality and the fraction of grazing not assimilated by zooplankton. Since there is no upper trophic level in this simple model, all unused food returns to nutrients.

This system of equations can be integrated numerically to obtain the time dynamics of N, P, and Z. The fourth-order Runge–Kutta method is the most commonly used algorithm for numerical integration. Local stability analysis (see Box 6.2) reveals that the equilibrium values are unstable.

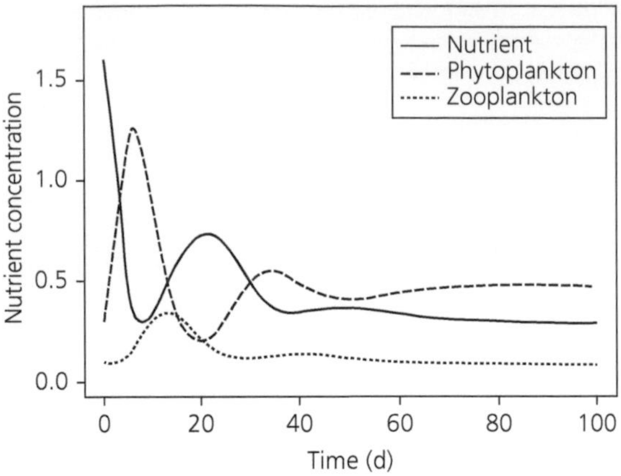

Figure 10.6 Biogeochemical ecosystem model incorporating nutrient–phytoplankton–zoooplankton dynamics (Miller 2004).

Because this model is a closed system, it results in cyclic dynamics with a degree of damping that depends on the initial conditions. Instability is introduced by positive feedback through the cycle N→P→Z→N.

Miller (2004) was able to simulate a damped spring bloom by making minor adjustment to this NPZ model (Figure 10.6). The conditions for stability include a lower zooplankton growth efficiency (γ) than used by Franks et al. (1986), 2% mixing per day of nutrients from below, and 50% fecal loss. With nutrient mixing and fecal loss, the system is no longer closed and the equilibrium becomes a stable node. This simple NPZ model can be extended to include an explicit compartment for detritus, where fecal losses accumulate and nutrients are regenerated. This one-dimensional model can be embedded in a circulation model to incorporate the processes of advection and diffusion. Finally, upper trophic levels can be added to account for grazing on zooplankton and the transfer of energy to fish and fisheries in the end-to-end models described below.

10.5.2 End-to-end models

The development of end-to-end (E2E) ecosystem models is a rapidly developing field. As noted in the introduction, reconciling the very different time and space scales for lower trophic level organisms relative to fish and other upper trophic level species presents special challenges. In an early development of an E2E model, Steele et al. (2007) developed a relatively simple network model tracing production processes on Georges Bank from nitrogen (nitrate and ammonium) to fish (Figure 10.7). Separate submodels for the lower and upper trophic levels were developed; the currency used in the lower web was nitrogen to accommodate the nutrient component while carbon was used for the upper web. The lower web encompasses the microbial food web. The nexus points for these subwebs involved the connection between meso-zooplankton (lower web) and carnivorous zooplankton (upper web) and between benthos (lower web) and carnivorous benthos (upper web) (Figure 10.7). The lower web was first solved using the basic methods described in Section 10.4.1 for a bottom-up system. These inputs were then used to set the initial conditions for the solution for the upper web. In this case, the top-down solution method (Section 10.4.2) was used. Diet composition data for three fish functional groups (Figure 10.7)—benthivores, planktivores, and piscivores—was used to develop the top-down estimates for the upper web.

This general approach allowed consideration of very different time and spatial scales for the upper and lower food webs, addressing one of the concerns identified at the start of this chapter. For the lower web, Georges Bank was partitioned into three seasonally varying spatial domains defined by hydrodynamic regimes: a well-mixed zone on the shallow crest of the bank, an outer zone that is thermally stratified during summer, and an intermediate transition zone. This permitted consideration of the very different seasonal mixing and nutrient regeneration regimes on the bank that profoundly affect production. Three within-year stanzas were employed (winter, summer, and fall–winter) to reflect differences in light and mixing levels. For the upper food web, spatial resolution was set at the bank scale. However, decadal scale analyses, starting in 1963, following the initiation of standardized research vessel surveys by the Northeast Fisheries Science Center were employed. In addition to abundance and demographic data of species caught by the gear, diet composition data and hydrographic information was collected. The temporal stanzas included 1963–72, 1973–82,

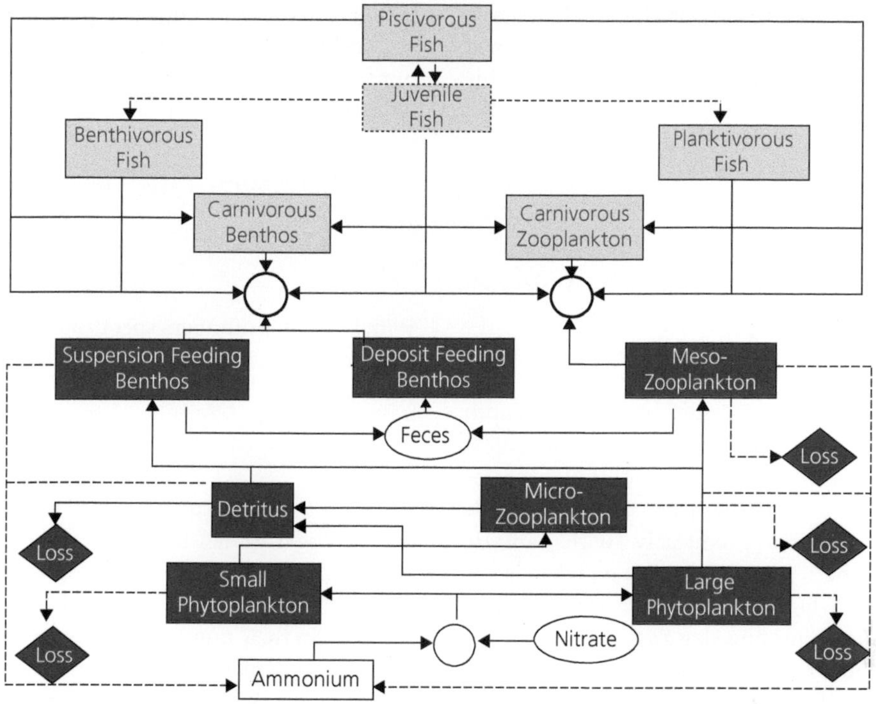

Figure 10.7 Georges Bank food-web model comprising lower and upper components. The lower food web contains the microbial loop and uses nitrogen as its currency and arrows represent the bottom-up flow of nutrients to plankton and benthos. The upper component encompasses the grazer/predator food web and arrows trace consumption by fish down to plankton and benthos where the upper and lower webs meet.

1983–92, and 1993–2002. During these four decades, important changes in nutrient regimes, fishing pressure, and environmental conditions related to the North Atlantic Oscillation (NAO) were observed (Steele et al. 2007). During the period 1960–71, the NAO was in a negative phase characterized by lower temperatures and nutrients (particularly nitrate) related to intrusions of the Labrador Current into the Gulf of Maine and on Georges Bank. During this first decadal period, relatively low levels of nitrate at depth were observed in the Gulf of Maine (Townsend et al. 2006). The estimated consumption by the total fish community was substantially lower during this stanza than in the succeeding decadal periods (Figure 10.8). The ability to explore the connection between nutrient dynamics and productivity in E2E models offers an important advantage in understanding mechanisms underlying bottom-up forcing in aquatic ecosystems. Because enhanced stratification can be expected in a warming climate, the energy required to transport nutrients from depth will be increased, reducing the new nutrients available in the illuminated surface waters. A number of studies

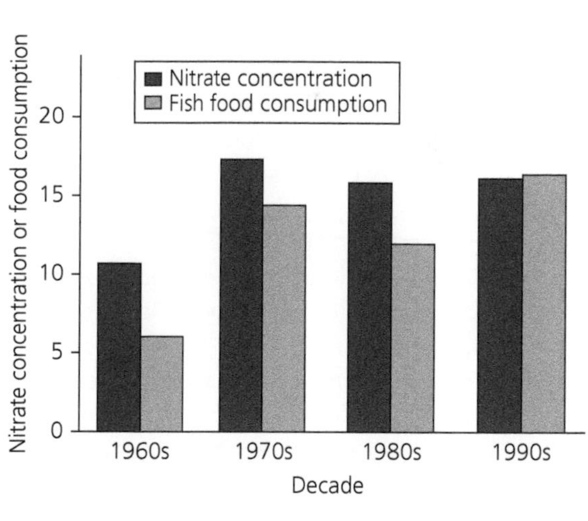

Figure 10.8 Estimates of nitrate concentration (micrograms $m^{-2}d^{-1}$; Townsendet al. 2006) in the Gulf of Maine and total fish consumption on Georges Bank (J. Collie, unpublished).

now predict a decline in new production in aquatic ecosystems under climate change and an increased dominance of regenerated production. In turn, this can lead to reduced productivity at higher trophic levels.

10.6 Biomass spectra

The classification of individuals by body size can be extended from communities (as we saw in Section 6.5) to include all taxa in the ecosystem (Kerr and Dickie 2001). Based on samples collected in the surface waters of the world's oceans and depth profiles from specific locations, Sheldon et al. (1972) constructed spectra of phytoplankton particles from 1 to 100 μm in diameter. Overall particle concentration varied predictably with latitude and depth. They found roughly equal relative concentrations when particles were grouped into logarithmically increasing size classes.

Based on these empirical results for phytoplankton, Sheldon et al. (1972) hypothesized that the size spectrum could be extended to all sizes of organism from bacteria to whales, and compiled preliminary evidence to support their hypotheses. If the standing stock of organisms is measured over a particular size range (e.g. phytoplankton), the size spectrum can be used to estimate the concentration of organisms in other size ranges. Sheldon and Kerr (1972) provide an amusing example, estimating the population density of monsters in Loch Ness.

Since the rate of population production scales inversely with body size (Figure 13 of Sheldon et al. 1972), the standing stock of each size class can be converted to production. In a follow-up paper, Sheldon et al. (1977) used these size relationships to estimate primary production in the Gulf of Maine and the North Sea from fish yields. In the opposite sense, they estimated fish production in the Peruvian Upwelling zone from estimates of primary production. Their estimates were consistent with independent measurements made at the time, and pioneered the use of biomass spectra for estimating ecosystem production.

Before considering more recent applications of biomass spectra, we need to review the quantitative theory. Continuing from Section 6.5, we assume a power relationship between population numbers and individual body weight: $N = k_1 w^{-b}$. The total number in the geometric weight interval w to aw is obtained by integration:

$$N_{tot}(w, aw) = \int_{w}^{aw} k_1 w^{-b} dw = \left| \frac{k_1 w^{1-b}}{1-b} \right|_{w}^{aw} = \frac{k_1 \left(a^{1-b} - 1\right) w^{1-b}}{1-b} \qquad (10.11)$$

With $b = 2$, this numbers spectrum has a slope of -1. Since biomass, $B = Nw$, the expected biomass spectrum is:

$$B_{tot}(w, aw) = \frac{k_1 \left(a^{1-b} - 1\right) w^{2-b}}{1-b} \qquad (10.12)$$

Again, if $b = 2$, the distribution of biomass in geometric (logarithmic) size intervals should be constant, as first proposed by Sheldon et al. (1972). In cases where the mass of organisms is measured, rather than their number, the normalized size spectrum is constructed by dividing $B_{tot}(w,aw)$ by w, which gives us back the numbers spectrum with expected slope -1.

The complete biomass spectrum, from algae to fish, has been measured by extensive sampling in a number of great lakes. The slopes of these spectra are very similar, ranging from -1.02 to -1.11. Sprules (2008) found remarkable similarity in the normalized size spectra of Lake Ontario and Lake Malawi, despite the large contrast in the geological, chemical, and biological characteristics of these lakes. Lake Ontario is a geologically young ($<$ 10000 yr) temperate lake that contains relatively few species, many of which are recent invaders. By contrast, Lake Malawi is a geologically old (10–20 million years) tropical lake in the African Rift Valley, which contains 500–1000 fish species, most of which are endemic (see Allison et al. 1996). Despite these intrinsic differences and the perturbation histories of the two lakes, their size spectra are statistically indistinguishable (Figure 10.9). The consistency of these empirical results, across time and among lakes, confirms that the size spectrum is a conservative property that emerges from highly size-structured aquatic food webs (Sprules 2008, Yurista et al. 2014). The inputs of solar radiation and nutrients determine the rate of primary

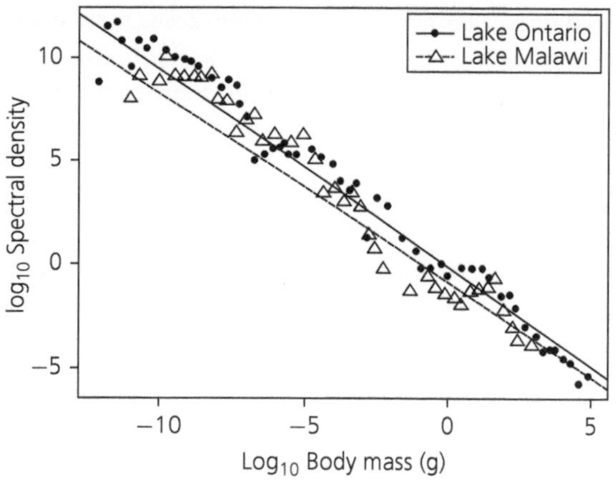

Figure 10.9 Normalized biomass size spectra from Lake Ontario and Lake Malawi. Spectra are annual averages. Adapted from Sprules (2008).

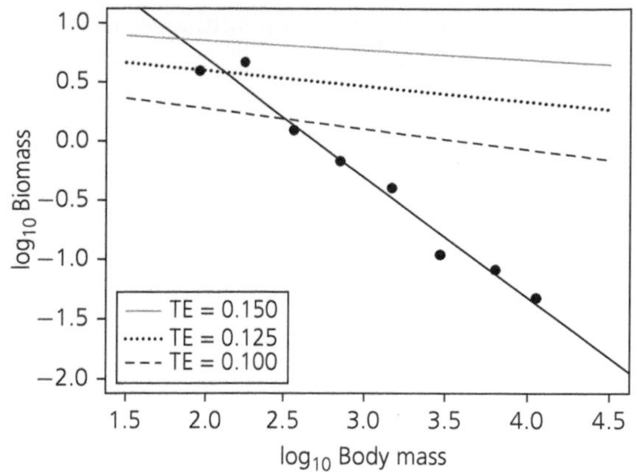

Figure 10.10 Predicted unexploited size spectra given primary production of 1956 g WWyear^{-1} and a predator–prey mass ratio of 390:1. Three size spectra corresponding to transfer efficiencies (TE) of 0.100, 0.125, and 0.150 are presented. Observed fish biomass at body mass for the North Sea fish community in 2001 (circles) and the fitted size spectrum (line, M^{-1}) are shown for comparison. (Adapted from Jennings and Blanchard 2004).

production, which propagates through the food web through the processes of consumption, growth, maturation, and reproduction, all of which depend on size.

The consistency of normalized size spectra is partly a consequence of the log-log transformation, which smooths out the irregularities although peaks are evident that correspond to the biomass of particular taxa (note the departure from simple linearity in Figure 10.9 in both Lake Ontario and Lake Malawi). This curvilinearity in size spectra is not simply a consequence of different sampling devices being used for different taxa. According to size-spectrum theory, the domes arise from ecological scaling at each trophic level and their spacing reflects predator–prey size ratios, which tend toward constant multiples in aquatic food webs (Kerr and Dickie 2001). A practical consequence is that we should expect spectra for particular taxa to be dome-shaped rather than strictly linear.

As first proposed by Sheldon et al. (1977) a biomass spectrum measured for one trophic group can be used to predict the biomass of other trophic groups. Starting with estimates of primary production, Jennings and Blanchard (2004) used this approach to estimate the biomass of fish in the North Sea prior to fisheries. The unexploited spectra were compared with actual biomass spectra measured from trawl surveys (Figure 10.10). The difference between the slopes of the predicted and measured spectra represents the deficit of large fish due to fishing.

10.7 Dynamic ecosystem models

It is possible to transform the linear equations of network models into a time-varying system of equations by writing:

$$\frac{dB_i}{dt} = \alpha_i B_i - \sum_j a_{ij} B_i B_j - m_i B_i \qquad (10.13)$$

which is reminiscent of the multispecies Lotka–Volterra equations (Equation 5.1) with biomass of each trophic compartment (B_i) replacing numbers as the state variable. Here α_i is the rate of increase, a_{ij} represents predation mortality from trophic group j, and m_i, is mortality from other natural sources such as disease. This set of linear differential equations can indicate the direction of change when the food web departs from a previous steady state, but because of the linear formulation, each trophic compartment is destined to increase or decrease without bound.

As we saw with two-species models in Chapter 4, density dependence is required in one or more of the equations for equilibrium solutions to exist. Walters et al. (1997) introduced the concept of foraging arenas whereby the rate of predation is an asymptotic function of predator abundance (see Chapter 4). This formulation was used to convert the linear equations of ECOPATH to the dynamic equations of ECOSIM, which we take up in Chapter 13.

10.8 Summary

The development of ecosystem models can be size-based, species-based, or trophocentric. In all cases, equilibrium mass-balance descriptions of ecosystems can be translated to dynamic models. Linear network models trace the flow of energy through food webs. Starting from the base of the food web, they can be solved *bottom up* to calculate how many predators can be supported for a given level of primary production. Conversely, the food web can be solved from the *top down* to calculate how much primary production is required to support fisheries yield, given the dietary requirements of top predators. The two approaches are complementary. The mass-balance equations can be made dynamic by converting them into a system of coupled differential equations. As we saw in Chapter 4, density dependence is required in one or more of the compartments to stabilize the system. Nutrient–Phytoplankton–Zooplankton models have been widely used to investigate the relative importance of nutrients and grazing on phytoplankton bloom dynamics. End-to-end trophic models have been constructed by coupling NPZ models with models of the upper trophic levels.

Additional reading

For more on descriptors of food webs and their characteristics see Pimm (1982). The multi-authored volume by Pascual and Dunne (2006) draws the connection between food webs and network theory. McCann (2012) approaches the problem from a dynamical systems perspective. Kerr and Dickie (2001) provide a definitive treatment of the biomass spectrum.

PART III

Harvesting Models and Strategies

CHAPTER 11

Harvesting at the Cohort and Population Levels

11.1 Introduction

In this chapter we explore alternative structures for single-species harvesting models. A rich and enormously important literature on this topic has emerged over the last half century or more, starting with the seminal work of Beverton and Holt (1957) and Ricker (1958). Influential precursors include Baranov (1918), Hjort et al. (1933), Thompson and Bell (1934) and Graham (1935), among others. In concert with essential theoretical developments, fisheries scientists have devised a broad suite of analytical approaches to demographic and population estimation to foster the translation from concept to implementation. These methods are designed to extract as much information as possible from both fishery-dependent and fishery-independent sources, often in very clever ways. Entire books are devoted to this topic (Quinn and Deriso 1999 provide encyclopedic treatment) and we cannot of course recapitulate this body of work in a single chapter. We instead focus on core elements of the theory and seek to complement other treatments by emphasizing topics sometimes accorded less attention in standard texts. We urge the reader to consult classics in the field including Ricker (1975), Gulland (1983) and Hilborn and Walters (1992) as well as recent books (e.g. King 2013, Ogle 2015). In keeping with our overall objective for this work, we adopt a perspective in which humans are viewed as integral parts of aquatic ecosystems.

Relative to terrestrial systems, harvesting in aquatic systems occurs in a comparatively opaque medium. Most often we cannot directly enumerate the population(s), nor do we directly observe the impacts of the fish catching process. Instead we must draw inferences from the characteristics and magnitude of the catch, and information from various fishery independent sampling programs. In many cases, the history of exploitation considerably pre-dates scientific study of the population and of the fishery. Even in instances where this is not the case, in general, fishing is not conducted in a controlled way designed to enhance our understanding of the population and demographic consequences of harvesting despite strong scientific reasons for doing so (Walters 1986). Fisheries development has historically proceeded in response to an interplay of food demand and food security, social and economic considerations, and conservation needs in which complex tradeoffs must be carefully evaluated and choices made.

One of the very few examples of strictly controlled exploitation derives from laboratory studies of small-bodied fishes. We again turn to the classic study of Silliman and Gutsell (1958) described earlier in Chapter 3. We have previously focused on the unexploited controls in this long-running experiment. We now turn to the treatments in which targeted removals of adult guppies were taken over the course of the experiment (Figure 11.1). For the first 40 weeks, the experimental populations were allowed to grow in the absence of exploitation. A controlled sequence of exploitation rates with target levels of 25, 10, 50, and 75 percent removals was then carried out over the next two and a half years, revealing the magnitude and shape of population declines and recoveries under changing exploitation rates (Figure 11.1a). If we focus on the mean biomass and yield levels

Fishery Ecosystem Dynamics. Michael J. Fogarty and Jeremy S. Collie, Oxford University Press (2020). © Michael J. Fogarty & Jeremy S. Collie 2020.
DOI: 10.1093/oso/9780198768937.003.0011

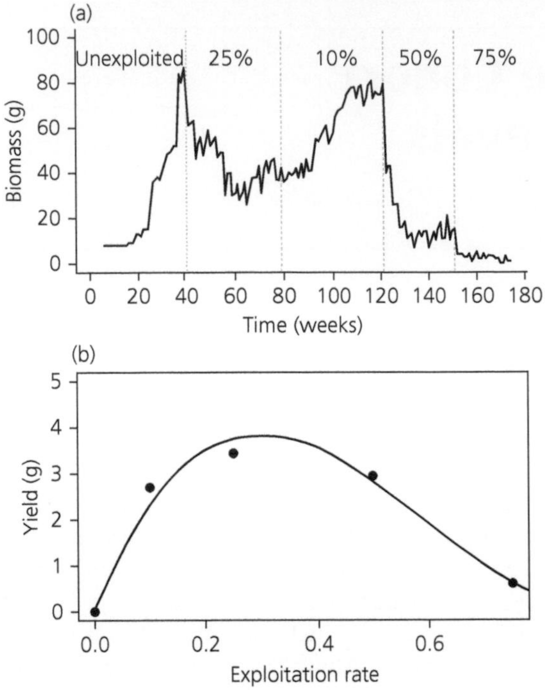

Figure 11.1 Experimental exploitation of guppies (*Poecilia reticulata*) at four levels of removal (Silliman and Gutsell 1958) showing (a) population trajectories throughout the course of the experiment and (b) the relationship between yield and target exploitation rate. Estimates of yield represent the mean over the last 6 weeks of each exploitation regime.

during the last six weeks of each experimental exploitation regime to reduce the transient effects of the sequence of harvest levels imposed, we obtain a dome-shaped relationship between yield and exploitation rate (Figure 11.1b). The yield is highest at some intermediate level of population biomass and exploitation rate. We will draw on insights from the outcomes of this experiment to motivate the development of models of population harvesting in real world applications later in this chapter.

11.1.1 Humans as predators

Fishers differ from natural predators with respect to both the intensity of predation and the size composition of their prey (e.g. Fowler 2009, Darimont et al. 2015). This difference is particularly acute in industrialized fisheries. In artisanal fisheries, the differences are often less pronounced. Exploitation rates by fishers are generally both higher than that of other predators and concentrated on larger size classes of prey. Indeed, conventional management practices typically focus on extraction of larger individuals at or beyond the size or age of maximum production. Collectively, these characteristics led Darimont et al. (2015) to label humans as "super predators." The selective removal of larger individuals by humans may induce genetic changes in exploited fishery resources (e.g. Conover and Munch 2002, Reznick and Ghalambor 2005, van Wilk et al. 2013), leading to reduced growth and earlier maturation. In addition, selective removal of older mature individuals can result in reduced reproductive output in instances where higher survival rates for progeny produced by larger and reproductively experienced females occurs (see Jakobsen et al. 2016, and contributions therein). In general, the reproductive value of smaller fish and invertebrates in marine ecosystems is lower than that of larger and older individuals. In contrast, natural predators generally remove smaller (younger) individuals from prey populations.

The fishing process involves both search and handling time in a way that invites comparison with natural predators. The search process involves the acquisition of local ecological knowledge by fishers. There are distinct parallels with other predators in this regard. It also involves the application of increasingly sophisticated remote-sensing technology (including both hydroacoustic and satellite imagery) to locate prey over large spatial scales. In this respect, fishing as a process transcends any analogy with natural predation and triggers concerns of a nearly unbridled capacity to locate and capture prey. To the extent that the fishing process involves onboard processing of the catch, handling time also emerges as a distinct issue in fisheries as with natural predators. This too has changed over time as onboard sorting and processing technology has increased in sophistication. Interestingly, although these issues have long been recognized (e.g. Hilborn and Walters 1992, Clark 1985), management-oriented models and applications almost invariably treat humans as Type-I predators in which species are harvested in direct proportion to their abundance. As we shall see in this chapter, the implications of this choice hold important consequences for our perception of how exploited aquatic populations respond to harvesting.

Throughout this and subsequent chapters, we will use fishing effort as a touchstone holding important implications for yield, fishing patterns and costs, options for input controls in management, and ecological impacts. It has become increasingly common to focus on fishing mortality rates without explicit consideration of the underlying levels of fishing effort, particularly for setting output controls such as quotas in management. The concept of the instantaneous rate of fishing mortality (F) is an immensely useful abstraction. It is not, however, a directly observable quantity. Rothschild (1977) notes that in the broadest sense, F is a function of fishing effort, population size, and time. In many standard fishery applications, F is assumed to be directly proportional to fishing effort and without explicit consideration of the effect of population size or density on the catch process. We will explore the implications of this assumption later in this chapter. Choices made by fishers concerning the amount and spatial distribution of fishing effort in relation to local resource abundance, and economic considerations, ultimately control critical aspects of yield, fishing costs and profits, habitat impacts, and incidental catch of non-target species (including protected and endangered species). Fishing effort is an indispensable quantity in any consideration of fishing from an ecosystem perspective and deserves renewed emphasis.

11.1.1.1 Humans as prudent predators?

Slobodkin (1972) introduced the concept of "prudent predation" for natural predators. Slobodkin noted that a prudent predator is one that most efficiently converts prey consumed into predator progeny and avoids depletion of its prey. The concept generated controversy over whether prudent predation would require some form of group selection (Maynard-Smith and Slatkin 1973). Slobodkin (1974) clarified that his focus on prudence in natural predators entailed consideration of prey characteristics rather than prey quantity per se. In particular, he emphasized that selection of prey with low reproductive value (typically small or senescent individuals; see Chapter 2) could obviate the need to consider group selection as a mechanism. There is a parallel set of considerations for human predators. In the absence of appropriate incentive/disincentive structures or regulations, individual fishers would presumably seek to maximize their catch, leading to the tragedy of the commons (Hardin 1968) through the collective action of all fishers. Resource economists have carefully considered the benefits of establishing individual property rights to avoid unbridled competition among fishers to attain the highest catch. Slobodkin's point concerning the reproductive value of individual prey is no less relevant. Conventional fisheries management strategies that focus on selective removal of larger, older prey can ultimately prove counterproductive by removing individuals with high reproductive value. Prudent predation by humans requires joint consideration of both the quantity and ecological/evolutionary quality of the prey to ensure resilience.

11.2 Harvesting at the cohort level

We begin with consideration of harvesting a single cohort over its lifespan, building directly on concepts developed in Chapter 8. We start with a specified number of individuals entering the population (recruitment) and then track the survival, growth, and maturation of members of the cohort over its lifespan. In the following, we simplify matters by expressing results on a per-recruit basis. We partition the losses due to natural causes and to harvesting at each age or size class. We are further interested in the numbers within the reproductive size or age classes. Summing over all age classes in the cohort and dividing by the number of recruits we determine both the yield-per-recruit and the egg production-per-recruit (or its proxy spawning biomass-per- recruit).

11.2.1 Yield-per-recruit

We first address the question of how to calculate the yield from a cohort of fish, given its growth and mortality characteristics. The rationale behind the per-recruit approach is that year-class size may vary from year to year in an unpredictable fashion or may be altogether unknown. Per-recruit analysis tells us nothing about the supply of recruits, but it does give guidance on how to manage the cohort for yield if year-class size can be determined or assumed. This approach enables fishery managers to compare the effects of alternative combinations of age at first capture and fishing intensity on yield from a cohort.

11.2.1.1 Continuous-time model

As we saw in Chapter 8, production of a cohort is determined by patterns of individual growth and mortality. The expression for yield over the lifetime of the cohort is:

$$Y = \int_{t=t_r}^{t_\infty} F_t N_t w_t \, dt \qquad (11.1)$$

where t_r is the age of recruitment to the fishery, F is the instantaneous rate of fishing mortality, N is the number in the cohort and w is the body weight. The product of the mean weight-at-age and the corresponding number for each age gives the cohort biomass ($B_t = N_t w_t$). Beverton and Holt (1957) provided an elegant solution to this problem for the case of isometric von Bertalanffy growth:

$$w_t = W_\infty \left(1 - e^{-k(t-t_0)}\right)^3 \qquad (11.2)$$

(see Section 8.3) and exponential decay in the number in the cohort at successive ages:

$$N_t = N_{t_r} e^{-Z(t-t_r)} \qquad (11.3)$$

where $Z = F + M$. The exponential term is a survival rate bounded between 0 and 1.

The distinguished Russian fisheries scientist Fydor I. Baranov had earlier solved Equation 11.1 using a simpler individual growth function (Baranov 1918) but unfortunately his work was not widely known in the west (Ricker 1975). Note that if the mean weight reaches an asymptote with increasing age and the number in the cohort decreases exponentially, the biomass of the cohort will typically exhibit a peak at an intermediate age (Figure 11.2). After this point, losses due to mortality exceed gains in production due to individual growth. If the cohort could be

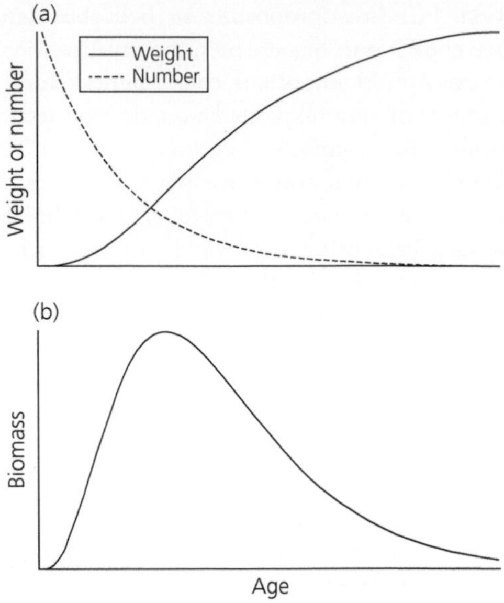

Figure 11.2 Changes in (a) number in a cohort subject to constant age-specific exploitation rates and average weight as a function of age and (b) biomass (product of number and weight-at-age) as a function of age.

harvested all at once, yield would be maximized at this point. The age at which the peak in the biomass curve occurs can be found by solving:

$$\frac{dB}{dt} = N\frac{dw}{dt} + w\frac{dN}{dt} = 0 \qquad (11.4)$$

using the chain rule to determine the age at peak biomass. We insert Equations 11.2 and 11.3 into Equation 11.4 to represent rates of growth and cohort decay and solve to obtain:

$$t_m = t_0 + \frac{1}{k}\log_e\left(1 + \frac{3k}{Z}\right) \qquad (11.5)$$

where t_m is the age at which cohort biomass is maximized (Quinn and Deriso 1999). In practice of course, it is not feasible to instantaneously harvest all individuals in a cohort at the age at which the optimum yield is attained. It is, however, entirely possible to do so in an aquaculture operation. In this case, we also must incorporate economic considerations (e.g. feed costs, discount rates, etc.) to determine the optimum age of harvest (e.g. Bjørndal 1988).

Finding an analytical solution to Equation 11.1 requires the imposition of certain constraints. In particular, we will confine our consideration to the case of "knife-edge" recruitment in which all members of the cohort become vulnerable to the fishery at the same age (t_r) and after which a constant rate of mortality is applied. The expression for yield-per-recruit is:

$$\frac{Y}{R} = F \cdot W_\infty \left[\frac{1}{Z} - \frac{3e^{-kt_r}}{Z+k} + \frac{3e^{-2kt_r}}{Z+2k} - \frac{e^{-3kt_r}}{Z+3k} \right] \tag{11.6}$$

(see Box 11.1). We can explore the implications for yield-per-recruit of varying the mortality rates and the age-at-recruitment to the fishery for combinations that result in the maximum yield per recruit. An illustration of jointly varying the fishing mortality rate and t_r is shown in Figure 11.3.

11.2.1.2 Discrete-time model

We can specify a generalized discrete-time analog of Equation 11.1 as:

$$Y = F \sum \bar{N}_a \bar{w}_a \tag{11.7}$$

where we now specify average values of cohort size and individual weight for each age class (here designated by the subscript a). In practice, data are

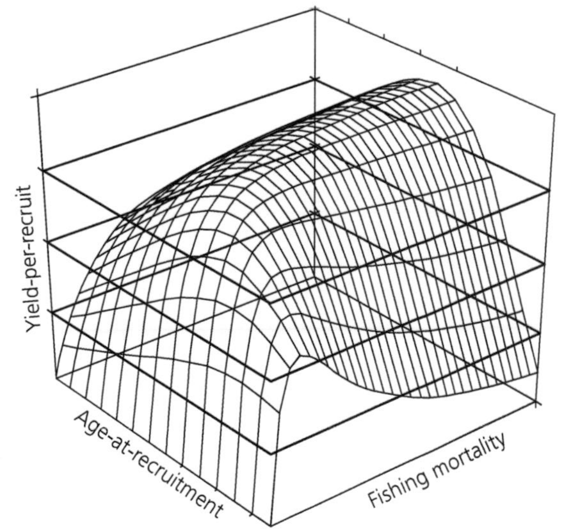

Figure 11.3 Yield-per-recruit as a function of fishing mortality and age-at-recruitment under the assumption of knife-edge recruitment. Horizontal slices through the surface define contours of equal yield.

Box 11.1 Beverton–Holt Yield-per-recruit

Beverton and Holt (1957) provided an analytical solution to the cohort yield problem by combining exponential mortality with the von Bertalanffy growth model. They noted that the number of fish surviving to recruit to the fishery is a function of the natural mortality rate, M:

$$R = N_0 e^{-Mt_r}$$

where R is the number of recruits to the fishery, N_0 is the initial number of fish at age t_0 (here set to $t_0 = 0$) and t_r is the age of recruitment to the fishery. Following recruitment to the fishery, we can specify the number surviving to age t as:

$$N_t = Re^{-(M+F)(t-t_r)}$$

Note that we are assuming a so-called 'knife-edge' selection pattern – all fish of a specified age become vulnerable to the fishery. The yield model is now:

$$Y = F \int w_x R e^{-(M+F)(t_r)} dx$$

where all terms are defined as before. As noted above, Beverton and Holt employed the von Bertalanffy growth model for weight at age:

$$w_t = W_\infty \left(1 - e^{-k(t-t_0)}\right)^3$$

and expanding the term in parenthesis, we have:

$$w_t = W_\infty \left(1 - 3e^{-k(t-t_0)} + 3e^{-2k(t-t_0)} - e^{-3k(t-t_0)}\right)$$

Substituting this expression into the yield model and solving gives:

$$\frac{Y}{R} = F \cdot W_\infty \left[\frac{1}{Z} - \frac{3e^{-kt_r}}{Z+k} + \frac{3e^{-2kt_r}}{Z+2k} - \frac{e^{-3kt_r}}{Z+3k} \right]$$

where certain (negligible) terms have been omitted. The model can be written more compactly as:

$$\frac{Y}{R} = F \cdot W_\infty \sum_{n=0}^{n=3} \Omega_n \left[\frac{e^{-nkt_r}}{Z+nk} \right]$$

where Ω_n is a dummy variable and $\Omega_0 = 1, \Omega_1 = -3$, $\Omega_2 = 3$ and $\Omega_3 = -1$. The Beverton–Holt method results in a closed-form solution for Y but requires knife-edge selection to the fishery and the power coefficient of the von Bertalanffy model to be exactly 3.

typically available for analysis on annual time scales. The mean number at each age can be determined from our exponential decay model and taking advantage of the standard formula for the first moment of a random process. To do this, we evaluate the expression for cohort decay over a one year time interval and solve. We will first consider the simplest case and take the total mortality rate to be constant over all ages. The mean number during the year for individuals of age a is given by:

$$\overline{N}_a = \int_0^1 N_a e^{-Z} da = -\frac{1}{Z}\left[e^{-z} - 1\right] N_a \qquad (11.8)$$

where N_a is the number alive at the start of age a. If we now substitute this expression into the model for yield and rearranging, we have for the yield for age class (a):

$$Y_a = \frac{F}{Z}\left[1 - e^{-Z}\right] N_a \overline{w}_a \qquad (11.9)$$

The total mortality rate can be decomposed into $Z, = F + M$ and we assume that in an infinitesimally small unit of time, a death can occur only due to fishing or natural causes. Equation 11.9 is known as the Baranov catch equation. Notice that the term in brackets is the complement of the survival fraction and therefore gives the fraction of the population dying during the time (age) interval. The ratio F/Z gives the proportion of total deaths that are due to fishing. The product of this ratio and the term in brackets is then the proportion of the population removed by fishing (the annual exploitation rate). Multiplying this product by the biomass at age gives the yield that age class. The total yield extracted from the cohort over its lifetime is the summation of age-specific yields over all age classes. Below, we will generalize these results to include age-specific mortality in the fishing process related to gear selectivity.

The discrete-time formulation affords much greater flexibility in specifying the yield equation than its continuous-time counterpart. In particular, we need not specify an individual growth function but rather can provide empirically derived estimates of the mean weight-at-age. In addition to the greater flexibility in the shape of the growth curve, this allows us to represent changes over time in the mean-weight-at-age of a cohort due to density or environmental factors. Empirical estimates of mean weight-at-age taken under different population levels and/or environmental conditions would give different results that can then be contrasted to explore the implications of population and/or environmental effects.

We are also no longer constrained to the assumption of knife-edge selection. We can partition the age-specific fishing mortality component into two parts, an overall fishing mortality rate (F) and a partial selection factor (p_a), which represents differential vulnerability-at-age to harvesting. This permits a more direct connection with the fish-catching process. Different gear types have different selectivity curves. The selection coefficients depend on the operating characteristics of the fishing gear. For towed fishing trawls (nets), the mesh size in the "cod end" of the net will have an important effect on the sizes of fish retained by the gear. However, the performance of the gear is also highly dependent on other structural features of net design. Similar considerations exist for other types of fishing gear. A logistic-type selection curve is typically used to represent towed fishing trawls:

$$p_l = \frac{e^{-(a-bl)}}{1 + e^{-(a-bl)}} \qquad (11.10)$$

where l is length and a and b are coefficients. An example of selectivity curves for several cod-end mesh sizes is provided in Figure 11.4a. After converting these estimates from length (or other measures of body size) to age, we can determine the selection coefficient p_a for use in Equation 11.12 below.

The probability of retention of fish of different sizes for gillnets is typically dome-shaped. A normal probability density function is often employed to represent the retention curve for fishing gear of this type:

$$p_l \propto \exp\left[\frac{-\left(l - \bar{l}\right)^2}{2\sigma^2}\right] \qquad (11.11)$$

where $\bar{l}$ is the mean size retained and sigma is its variance. Smaller individuals can pass through the net. Larger fish have a higher probability of not being enmeshed in the gear and the retention probability declines (Figure 11.4b).

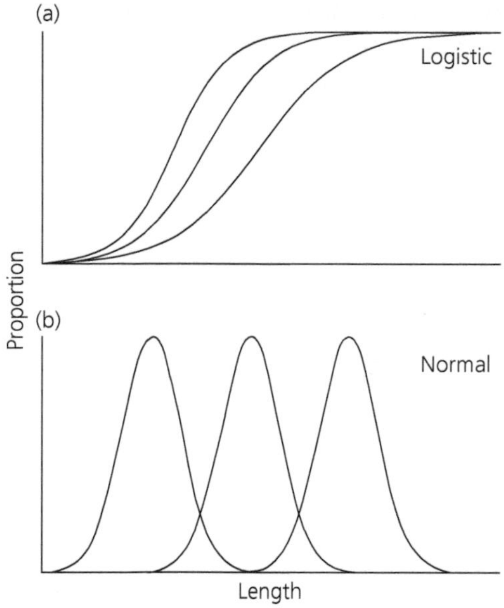

Figure 11.4 Probability of retention as a function of body size for three mesh sizes for (a) a trawl fishery and (b) a gill net fishery

Understanding catch processes for stationary fishing gears such as gill nets and traps also entails consideration of the soak-time of the gear. Individual units of gear can become saturated at longer immersion intervals and the catch may not be directly proportional to the soak time (e.g. Fogarty and Addison 1997). In contrast, the catch of large towed fishing gears is almost always taken to be a direct function of actual fishing time.

Traps are a major gear type employed in fisheries for a number of taxa including crustaceans, gastropods, cephalopods, and fish. While landings from traps may constitute a relatively small fraction of global fishery landings, they often represent a disproportionately high percentage of the economic value. In addition, except in deep-water fisheries where barotrauma may be high, animals caught in traps and certain hook fisheries can often be returned alive after being brought to the surface if they do not meet size limits or other regulations. Hence, in principle, there is greater control of the outcomes of the capture process in trap and hook fisheries, although individuals may be particularly vulnerable to natural predators following release.

The total yield over all age classes is given by:

$$Y = \sum_{a=a_{\min}}^{a=a_{\max}} \left(\frac{p_a F}{p_a F + M} \left(1 - e^{-(p_a F + M)}\right) \right) N_a \bar{w}_a \tag{11.12}$$

where a_{min} is the minimum age subject to capture and a_{max} is the maximum age. The yield-per-recruit is given by dividing through by the number of recruits entering the fishery (which can be specified as an arbitrary number). The term inside the large parentheses in Equation 11.12 is again the annual exploitation rate (u). We will provide an application of this approach as part of an illustration of the development of a full age-structured model for Icelandic cod in Section 11.5.

These plots give rise to what are referred to as biological reference points. When a well-defined peak in the yield curve exists, the level of fishing mortality resulting in the maximum yield-per-recruit is called F_{max}. Fishing at a level beyond F_{max} is referred to as *growth overfishing*—it forgoes potential yield from the cohort. F_{max} is the level of fishing mortality for a specified partial selection pattern that results in maximum yield per recruit. When natural mortality rates are high, the yield-per-recruit curve may not have a well-defined maximum (Figure 11.5a). In this case, the point on the yield-per-recruit curve whose tangent has a slope equal to one tenth of the slope of the curve at the origin ($F_{0.1}$) has been used as a reference point (Gulland and Boerema 1973). Although the choice of 10% is arbitrary, it has proven to be useful in practice and is widely used. The $F_{0.1}$ reference point is also useful for cases where a well-defined maximum does exist. In this case, $F_{0.1}$ still allows specification of a well-defined reference point. Given the shape of the typical yield-per-recruit curve (Figure 11.5a), $F_{0.1}$ can be substantially lower than F_{max} but entails only a small reduction in yield per recruit.

11.2.2 Egg production and spawning biomass-per-recruit

Yield-per-recruit analysis gives only a partial view of the implications of alternative harvesting strategies for the population. In particular, it does not consider the potential impacts on reproductive capacity.

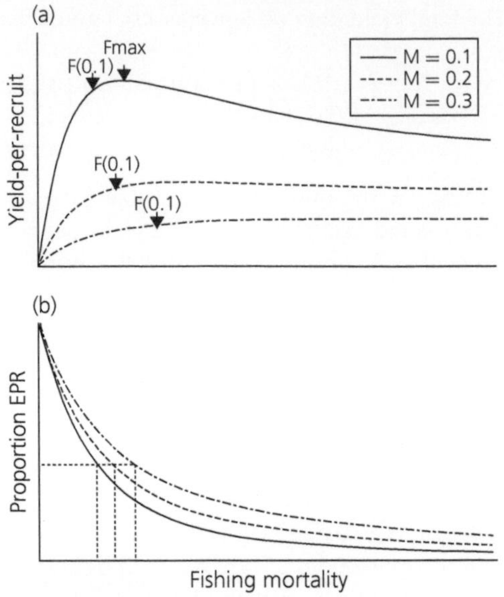

Figure 11.5 Yield-per-recruit (a) and egg-production-per recruit (proportion of maximum; b) as a function of fishing mortality under three natural mortality rates. A reference level (dotted lines) of 40% of maximum egg production per recruit (expressed as a proportion of the unfished state) is shown in panel (b).

The limitations of considering only the question of maximizing yield are clearly evident—a level of fishing mortality that optimizes yield from a cohort can conceivably result in substantial depletion of the reproductive potential of the population. We can expand our analysis using additional information on the reproductive biology of the stock to define the expected level of egg production ($\mathcal{E}$) resulting from a given level of recruitment. The egg production over the lifetime of the cohort is given by:

$$\mathcal{E} = \sum_{a=a_{\min}}^{a_{max}} m_a v_a f_a N_a \qquad (11.13)$$

where m_a is the proportion mature at age in the population, f_a is the mean fecundity–at-age of mature females, and v_a is the viability of eggs produced by females of age a (see Chapter 2). We can readily specify the relationship between age and maturity with a form of the logistic function:

$$m_a = \frac{1}{1 + e^{-\psi(a - a_{50\%})}} \qquad (11.14)$$

where m_a is the proportion mature at age a, $a_{50\%}$ is the age at which half the individuals are mature, and ψ is a slope coefficient.

We can calculate the egg-production-per-recruit (EPR) by again dividing through by the initial number of recruits. The maximum EPR will always occur when the fishing mortality rate is zero and it decays with increasing fishing mortality (Figure 11.5b). If spawning does not occur at the start of the time period, we decrement the population number-at-age to account for losses prior to spawning within the year (Gabriel et al. 1989). As we noted in Chapter 9, detailed information on fecundity as a function of age may not be available. The fall-back position is then to substitute the mean-weight-at-age for the terms $v_a f_a$ in Equation 11.13.

EPR analyses can be linked with recruitment models to determine limiting levels of fishing mortality required for population persistence. Replacement of the population requires that the inverse of the lifetime egg-production-per-recruit not exceed the slope of the egg-production recruitment curve at the origin (see Sissenwine and Shepherd 1987). If fishing mortality reduces lifetime EPR such that this threshold is exceeded, extinction of the population is predicted. Biological reference points for egg or spawning biomass per recruit are typically expressed as a percentage of the unfished level, often in the range 20–40%. In Figure 11.5b we show results for the 40% reference point.

11.3 Biomass dynamic models

The total biomass of a population in a given time period is of course simply the sum of its age (or size)-specific biomass levels. Here, we will consider biomass dynamic models (Hilborn and Walters 1992) that treat total population biomass without partitioning the whole by age or size classes, or other demographic characteristics. We will build directly on the results in Chapter 3 for models in which the currency of choice was population numbers (or density). The application of this general modeling strategy for exploited populations was pioneered by Graham (1935) and fully developed by Schaefer (1954, 1957), Pella and Tomlinson (1969), and Fox (1970). These models have a simplified structure in which recruitment, growth, and natural mortality are subsumed in a single function. They can provide a useful heuristic guide to the impacts of harvesting on exploited populations.

We now change the focus of our earlier ecological models by switching consideration from changes in population numbers to biomass. From the perspective of a harvester, it is the weight (and economic value) of the catch that is most important. In their simplest form, these models require only information on yield and fishing effort. Information on fishing effort is available for many fisheries and, as noted earlier, it provides a direct connection to the mechanics of harvesting. Quantification and standardization of fishing effort does require careful consideration of vessel and gear attributes (see Quinn and Deriso 1999 Chapter 1 for applicable methods). Human attributes related to fishing skill and experience also of course influence catch rates.

Biomass dynamic models are often referred to as "surplus production" models. The fundamental premise is that at high population levels, intraspecific competition for resources is correspondingly high and individual growth and/or reproductive output will be reduced. In addition, the mortality due to intraspecific interactions may be elevated at high population sizes. Recall that for the logistic model, the population growth rate is highest at one half the maximum population size and it is zero at the unexploited equilibrium level. Surplus production is the increased production that is generated by reducing intraspecific interactions. The term "surplus" is therefore used in a very specific (and limited) sense. In particular, it does not refer to production that would somehow remain unutilized if not harvested. Rather, under this construct, exploitation is the generating mechanism for increased production that would not otherwise exist.

11.3.1 Continuous-time models

The earliest biomass dynamic models were framed in continuous time as direct extensions of the logistic model (Schaefer 1954, 1957). We begin with models of the general form:

$$\frac{dB}{dt} = g(B)B - h(E,B) \tag{11.15}$$

where B is population biomass and E is standardized fishing effort. The right-hand side of Equation 11.15 therefore comprises one component specifying the per capita rate of change of population biomass $g(B)$ and another, $h(E,B)$, specifying a harvesting term incorporating fishing effort and population biomass. This latter term provides the expression for yield: $Y = h(E,B)$. The generalized function $g(B)$ represents the undifferentiated contributions of recruitment, natural mortality, and individual growth to per capita population growth. For simplicity and continuity we will use the same symbols as those in Chapter 3. Note, however, that the interpretation of these parameters now differs—they no longer represent the effects of birth and death rates alone but incorporate consideration of individual growth rates. In Section 11.4 we will introduce models in which separate growth, mortality, and recruitment functions are directly specified.

For the harvesting module, we will use a version of the well-known Cobb–Douglas production function employed in resource economics:

$$h(E,B) = Y = q \cdot E^{\varpi} B^{\gamma} \tag{11.16}$$

where q is the catchability coefficient (a measure of gear effectiveness per unit effort), ϖ and γ are shape coefficients. See Hannesson (1983) and Tsoa et al. (1985) for applications. When the coefficient ϖ is less than 1, we have an index of effort saturation or congestion. The coefficient γ relates to density-dependent vulnerability to capture. Given information on standardized fishing effort and population biomass, we can estimate the parameters of the harvesting module. The linearized form of the model is:

$$\log_e Y = \log_e q + \varpi \cdot \log_e E + \gamma \cdot \log_e B \tag{11.17}$$

and in this form is amenable to a multiple regression analysis. To simplify matters throughout the remainder of this chapter, we will set $\varpi = 1$ with the recognition that saturation and congestion of effort can potentially be important and should be directly examined when information on both fishing effort and biomass is available. To illustrate how a nonlinear harvest function can affect stability, we will examine the proposition that capture efficiency can be a function of population density or abundance such that $Y = qEB^{\gamma}$. If these conditions hold, the yield-per-unit-effort is:

$$\frac{Y}{E} = qB^{\gamma} \tag{11.18}$$

Hilborn and Walters (1992) designate three principal categories corresponding to different values of the exponent γ: hyperstability ($0 < \gamma < 1$); proportionality ($\gamma = 1$); and hyperdepletion ($\gamma > 1$); (see Figure 11.6). Hyperstability occurs when the yield-per-unit-effort remains relatively high as the overall population declines. Instances in which fishers are able to consistently locate higher concentrations of aggregated individuals give rise to this phenomenon. Although this pattern has typically been associated with pelagic schooling species (e.g. Csirke 1988), tests of Equations 11.17–18 have detected evidence of hyperstability in a number of demersal fish species (Harley et al. 2001). It has also been identified in species forming large spawning aggregations. Note that the concept of hyperstability is directly related to MacCall's basin model construct described in Chapter 7. Proportionality is associated with relatively even distribution patterns maintained by diffusive processes. Hyperdepletion occurs when a subset of the population is relatively immune from harvest (cf. Section 4.2.5 with respect to the role of refugia in predation processes). The issue of hyperstability is of particular concern because of the risk of sudden and unexpected population collapse (see Section 11.3.1.2).

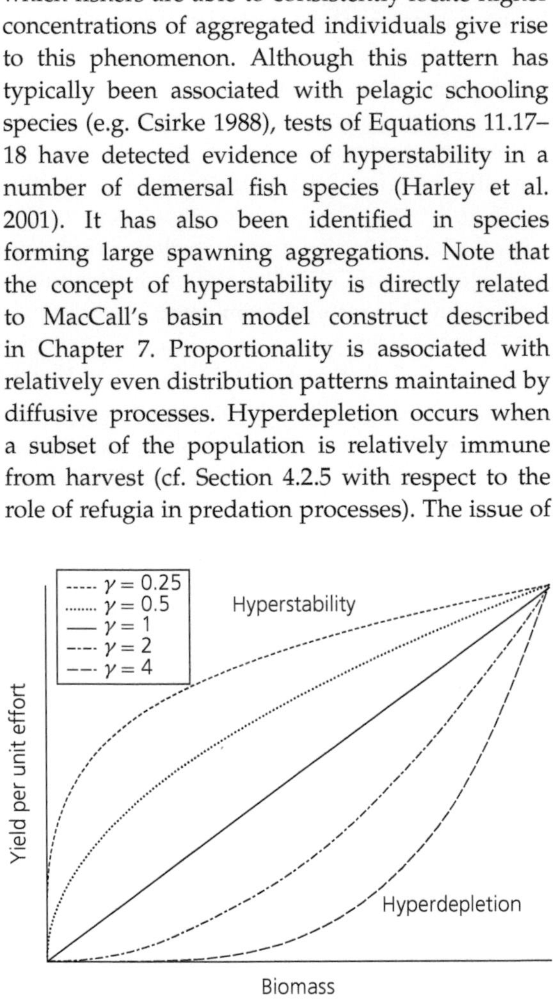

Figure 11.6 Yield per unit of effort as a function of population biomass. Standardized curves generated by the power function $Y/E = qB^{\gamma}$. For the case $\gamma = 1$, we have a proportional relationship between yield per unit effort and population biomass. For $\gamma < 1$, we have hyperstability Curves in which the Y/E remains comparatively high as biomass declines. For $\gamma > 1$, we have hyperdepletion in which the Y/E is comparatively low at low biomass levels.

Clark (1982, 1985, 2010) proposed a related concept in which the density of an exploited species also can be represented as a power function of population abundance or biomass. Clark further reasonably postulated that fishers would concentrate fishing effort in high density (and/or high profitability) locations. He identified four "concentration profiles" that can be mapped to the Hilborn and Walters classification scheme: Clark's Types I–III relate to hyperdepletion, proportionality, and hyperstability respectively. Clark further identified a Type IV category in which concentration remains constant over all levels of abundance.

11.3.1.1 Linear harvest functions

For ease of exposition, we will first consider models in which the internal dynamics of the population are governed by a logistic function and a linear harvest term. In an attempt to capture the interplay of different productivity levels and different harvesting practices, in Figure 11.7 we show logistic population models representing three productivity states in which an external force (e.g. environmental/climate factors) affects the intrinsic rate of increase of the population (the parabolas in Figure 11.7). We then superimpose harvest functions (dashed lines) and examine the points of intersection of the population and harvest functions.

We will begin with a very simple case, but one with very important ramification for stability of the resource—attempts to continuously remove a constant biomass as harvest. This is a linear harvest function with a zero slope (i.e. target removals do not vary as a function of population biomass). The model is then:

$$\frac{dB}{dt} = (\alpha - \beta B)B - Y \tag{11.19}$$

where Y is a fixed, time invariant quantity. At equilibrium we have:

$$Y = (\alpha - \beta B)B \tag{11.20}$$

and applying the quadratic formula we see there are potentially two equilibrium points at biomass levels of:

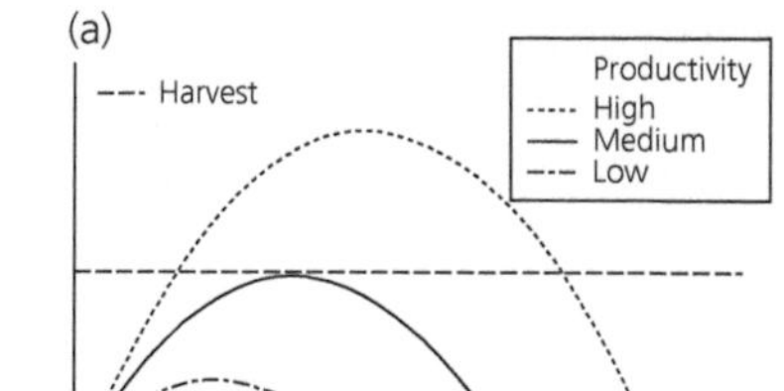

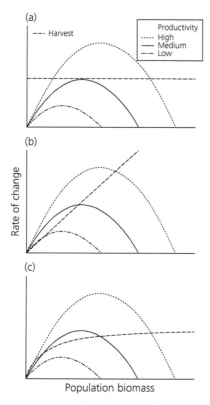

Figure 11.7 Relationship between the population rate of change and population biomass for logistic production functions for three intrinsic rates of change showing points of intersection between the logistic functions and harvest functions (dashed lines) for (a) constant yield, (b) proportional (linear) harvest rate, and (c) non-linear harvest rate.

$$\frac{-\alpha \pm \sqrt{\alpha^2 - 4\beta Y}}{2\beta} \tag{11.21}$$

(note that β will enter this expression as a negative number). To visualize this, consider our three environmentally dependent logistic functions, and note that the horizontal harvest function intersects the highest productivity logistic model at two points (Figure 11.7a). The intersection point at the higher population biomass level is globally stable. In contrast the lower intersection point is an unstable equilibrium. Any perturbation that reduces the population biomass below the lower intersection point will result in a collapse. A time-invariant constant harvest policy is therefore a very dangerous strategy entailing high risk to the population. The intersection point between the intermediate (middle) logistic function and the flat harvest function in Figure 11.7a is set at the peak of the parabola representing the internal dynamics. This is a highly precarious (and untenable) situation. Environmental perturbations will ultimately result in extinction if reduced productivity states persist. Finally, we see that for the lowest productivity state, there is no intersection at all between the logistic model and the harvest line and the stock will collapse.

Here we wish to convey the implications of not recognizing or responding to changes in the productivity state of the resource. Suppose managers were operating under the presumption of no changes in productivity when a decline in productivity had in fact occurred. Any resulting stock collapse would be entirely unanticipated under what was assumed to be a sustainable harvest policy.

In Figure 11.7b we again represent three environmental states but now consider a fixed harvest *rate* strategy represented by a straight line through the origin with slope qE. This corresponds to the proportionality designation of Hilborn and Walters and the Type II concentration profile of Clark. Where the logistic curves and the harvest line intersect in Figure 11.7b, we have stable equilibria. The model is now:

$$\frac{dB}{dt} = (\alpha - \beta B)\, B - qEB \tag{11.22}$$

This is the case considered by Schaefer in his highly influential paper. The equilibrium biomass (B^*) for the Schaefer model is given by:

$$B^* = \left(\frac{\alpha - qE}{\beta}\right) \tag{11.23}$$

and the equilibrium yield–effort relationship is:

$$Y^* = qE\left(\frac{\alpha - qE}{\beta}\right) \tag{11.24}$$

The effort level resulting in the maximum sustainable yield (also called maximum sustainable yield, MSY) is:

$$E_{\text{MSY}} = \left(\frac{\alpha}{2q}\right) \tag{11.25}$$

and the maximum sustainable yield is:

$$Y^*_{\text{MSY}} = \frac{\alpha^2}{4\beta} \tag{11.26}$$

MSY is now enshrined in national and international legislation and provides the foundation for fisheries management throughout the world. MSY can be broadly defined as the largest average yield that can be sustained over time and is not tied to a specific model. In practice, many proxy measures are used for MSY, some of which are altogether model-free. The MSY concept has long been challenged over concerns that this level may not necessarily be sustainable or desirable from broader ecological, social, and economic perspectives (e.g. Larkin 1977, Gulland 1969, Sissenwine 1978, Clark 1990); 2010. In a characteristically trenchant observation, Ricker (1963) noted that "... any really close approach to the point of optimum yield will usually be too dangerous to be practical." For this reason, precautionary buffers are now often applied to biological reference points to reduce risk to the population.

An application of the Schaefer model to the western rock lobster *Panulirus cygnus* (Penn et al. 2015) is provided in Figure 11.8. This fishery was managed by input controls (effort limitation) from its inception in 1945 to 2014 when management switched to output controls (individual transferable quotas).

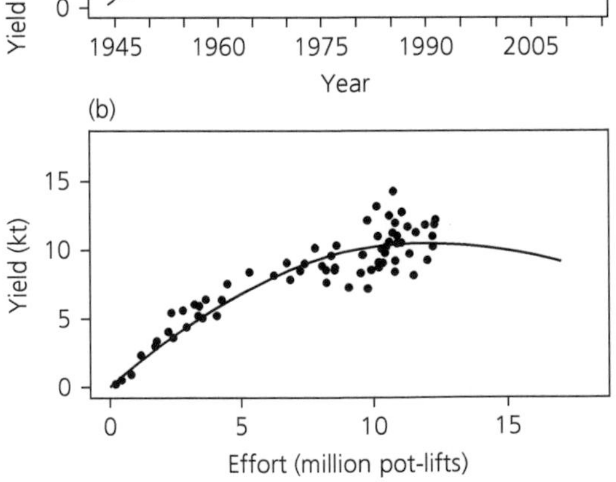

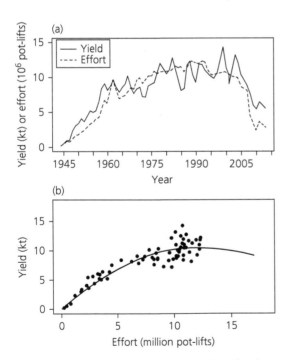

Figure 11.8 (a) Yield (kt) and fishing effort (million pot-lifts) of western rock lobster (*Panulirus cygnus*). (b) Relationship between yield and fishing effort. Data from Penn et al. (2015)

Fishing effort was carefully controlled to prevent over-exploitation. As a testament to the effectiveness of management of this population, this fishery was the first to be awarded certification for sustainable management by the Marine Stewardship Council. In Figure 11.8a, we show trends in yield and fishing effort (millions of pot-lifts) over time; the relationship between yield and fishing effort is shown in Figure 11.8b. In this case, fishing effort was deliberately controlled before potential declines in yield due to overfishing were experienced. See Penn et al. (2015) for an account of the management history of this fishery in response to changing social, economic and ecological conditions.

11.3.1.2 Non-linear harvesting

We now consider the case of the nonlinear harvest function given by Equation 11.16. Under the high productivity state, we have a single stable intersection point (Figure 11.7c). There are two intersection points for the medium productivity state shown. We have a stable equilibrium at high biomass and an unstable point at lower biomass. The population will collapse if biomass drops below the lower intersection point. If the population shifts to a lower productivity state, no intersection points exist and a population collapse is predicted.

Below, we explore the implications of non-linear harvest rates and show that they can result in backward bending yield curves for the case of hyperstability (see Figure 11.9). The full model is:

$$\frac{dB}{dt} = (\alpha - \beta B)\, B - qEB^{\gamma} \tag{11.27}$$

and to determine the inflection point at which the biomass curve bends backward (Figure 11.9b), we take the relationship between fishing effort and biomass at equilibrium:

$$E = \frac{1}{q}\left(\frac{\alpha - \beta B}{B^{\gamma - 1}}\right) \tag{11.28}$$

and setting $dE/dB = 0$ we find that the inflection will occur when $B = \alpha(1-\gamma)/\beta(2-\gamma)$. We now have two equilibrium biomass and yield levels for a given level of fishing effort comprising stable (solid lines in Figures 11.9b and c) and unstable (dashed lines) points.

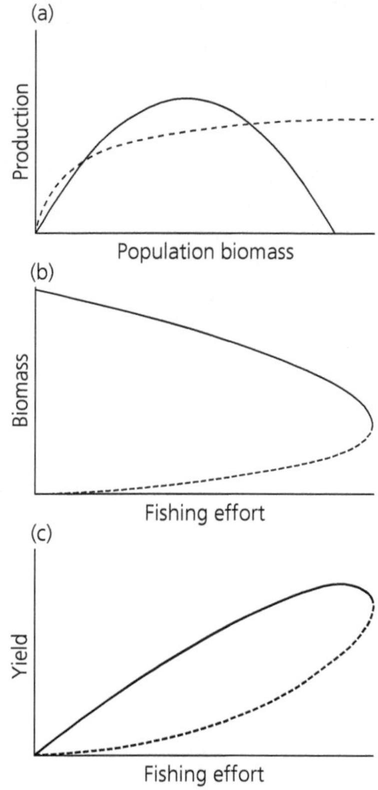

Figure 11.9 (a) Production model with non-linear harvest function, (b) relationship between population biomass and fishing effort, and (c) yield and fishing effort.

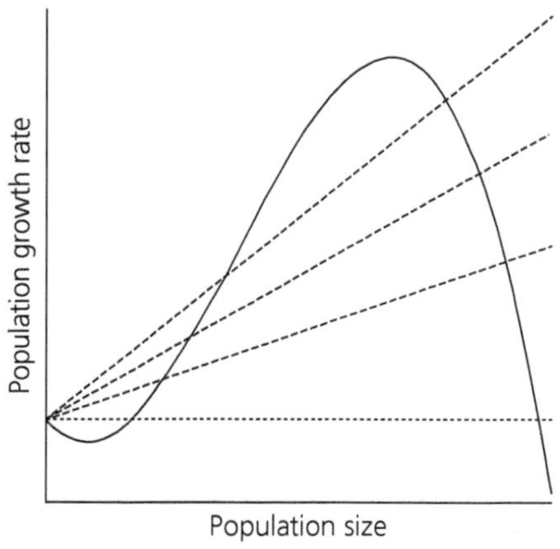

Figure 11.10 Depensatory population model showing intersection points at three levels of fishing effort (dashed lines).

11.3.1.3 Depensatory production

The population production functions considered above were all compensatory in form. We now turn to population models embodying depensatory dynamics (see Chapter 3) and examine the implications of harvesting populations exhibiting depensation. For simplicity we will again explore the case of a linear harvest function through the origin. As we saw in Chapter 3, the production function has an inflection point at low population size. Here we will illustrate the case of critical depensation in which production can be negative at low population size and we superimpose three levels of harvest intensity (Figure 11.10). We employ a model permitting a simple interpretation of the effect of depensation:

$$\frac{dB}{dt} = (\alpha - \beta B)\left(\frac{B}{B_{crit}} - 1\right) B - qEB \qquad (11.29)$$

where B_{crit} is the critical population threshold below which production is negative. Notice that when $B = B_{crit}$ the term in large brackets is zero. For the case $dB/dt = 0$, the relationship between fishing effort and biomass is:

$$E = \frac{1}{q}\left[(\alpha - \beta B)\left(\frac{B}{B_{crit}} - 1\right)\right] \qquad (11.30)$$

and the resulting expression can be solved to obtain the intersection points as in Section 11.3.1.2. We again obtain a backward-bending yield curve similar to Figure 11.9c but generated by very different mechanisms.

11.3.2 Discrete-time models

Although fishery production models were originally developed as differential equations, it is now quite common to employ difference-equation versions. Difference equations may be more appropriate for organisms in seasonal environments. As we have seen in previous chapters, they also can exhibit a much richer array of dynamical behavior than low-dimensional differential equations.

11.3.2.1 Biomass dynamic models

Quite often, discrete-time biomass dynamic models are specified as an Euler approximation to their continuous-time analogs. The basic model structure is then:

$$B_{t+1} = B_t + g(B_t)B_t - u_t B_t \qquad (11.31)$$

Box 11.2 Management Reference Points for the Discrete Pella–Tomlinson Model

In Chapter 3, we introduced the theta-logistic model as a generalized production function (Gilpin and Ayala 1973). Pella and Tomlinson (1969) had earlier introduced a generalized production model for fishery applications. The original Pella–Tomlinson model was framed in continuous time. Here, we will examine management reference points for the corresponding discrete-time formulation shown in Equation 11.33. For the case where $\theta > 1$ we have:

The equilibrium yield–exploitation rate relationship is $Y^* = uB^*$ where B^* is given in Equation 11.34. We then have:

$$Y^* = u\left(\frac{\alpha - u}{\beta}\right)^{\frac{1}{\theta-1}}$$

and the fishing effort level resulting in maximum yield is then:

$$u^*_{MSY} = \alpha\left(1 - \frac{1}{\theta}\right)$$

and the maximum sustainable yield is:

$$Y^*_{MSY} = \alpha\left(\frac{\alpha}{\theta\beta}\right)^{\frac{1}{\theta-1}} - \beta\left(\frac{\alpha}{\theta\beta}\right)^{\frac{\theta}{\theta-1}}$$

For $\theta = 2$ we recover the reference points for the Schaefer Model. For the case $\theta < 1$, the signs of the coefficients α and β are reversed in each of the above expressions. Notice that θ cannot equal 1. However, as $\theta \rightarrow 1$, we can write:

$$B_{t+1} = B_t + (\alpha - \beta \log_e B_t)\, B_t - u_t B_t$$

which is a discrete time version of the Fox (1970) production model. The equilibrium biomass is then:

$$B^* = \exp\left(\frac{\alpha - u}{\beta}\right)$$

The equilibrium relationship between yield and exploitation rate is given by:

$$Y^* = u\left[\exp\left(\frac{\alpha - u}{\beta}\right)\right]$$

and the exploitation rate resulting in maximum sustainable yield is simply $u^* = \beta$. Substituting this result into the equilibrium yield-exploitation rate relationship gives the maximum sustainable yield:

$$MSY^* = \beta\left(\exp\left(\frac{\alpha}{\beta} - 1\right)\right)$$

where $g(B_t)$ is a function of biomass and u_t is the exploitation rate. We use the exploitation rate (fraction of the population removed by harvesting) in this context because it is bounded between 0 and 1, ensuring that removals cannot exceed biomass. We can specify the annual surplus production (ASP) as:

$$\text{ASP} = B_{t+1} - (1 - u_t)B_t \qquad (11.32)$$

which is simply the change in population biomass between successive points in time plus the yield removed during the time interval. This quantity does not depend on the specific form of the production function and is therefore broadly applicable. Walters et al. (2008) strongly recommended that the annual surplus production (ASP) be routinely monitored as a key diagnostic quantity in assessing population status. Inserting the generalized production model of Pella and Tomlinson in Equation 11.31 we have:

$$B_{t+1} = B_t + \left(\alpha - \beta B_t^{\theta-1}\right)B_t - u_t B_t \qquad (11.33)$$

where all terms are defined as before. The equilibrium biomass is now:

$$B^* = \left[\frac{1}{\beta}\left(\alpha - u^*\right)\right]^{\frac{1}{\theta-1}} \qquad (11.34)$$

and expressions for the biomass, exploitation rate, and yield at MSY in this formulation (see Box 11.2) are comparable to the results for the continuous-time case.

11.4 Delay–difference models

We next describe the development of discrete-time population harvesting models with a simplified age structure. The delay–difference models provide a bridge between the nonage-structured models considered in the last section and the full age-structured models to be considered in the next section. We can build on the simple delay–difference model introduced in Chapter 3 (see Equation 3.37) which dealt with survival and recruitment but not individual growth. We now incorporate specific submodels for growth. Deriso (1980) pioneered the development of this approach, which was then was further generalized by Schnute (1985). The population biomass at time $t+1$ is, by definition:

$$B_{t+1} = \sum_{a=r}^{\infty} w_a N_{a,t+1} \quad (11.35)$$

where w_a is the mean weight at age a and $N_{a,t}$ is the number at age a at time t. We can partition this expression into terms for recruits and post recruits:

$$B_{t+1} = \sum_{a=r+1}^{\infty} w_a N_{a,t+1} + w_r N_{r,t+1} \quad (11.36)$$

where w_r is the mean weight of recruits at age r. If we assume a monomolecular relationship between age and weight (see Chapter 8), the relationship between the weight at successive ages is:

$$w_a = W_\infty (1-\rho) + \rho\, w_{a-1} \quad (11.37)$$

where $\rho = \exp(-k)$. If we now express w_{a-1} as a function of w_{a-2} and substitute into Equation 11.37 we obtain, after simplifying:

$$w_a = (1+\rho)\, w_{a-1} - \rho\, w_{a-2} \quad (11.38)$$

resulting in a second-order difference equation for the weight-at-age. In the Deriso model as described below an initial condition of $w_{r-1} = 0$ is invoked.

Next, survivorship (s_t) can be expressed in at least two principal ways. If fishing and natural mortality operate concurrently throughout the year, we can build on our earlier treatment and note that $s_t = \exp(Z_t)$ where Z_t is the instantaneous rate of total mortality.

Alternatively, if fishing mortality alone occurs during a relatively brief period at the beginning or end of the time period and natural mortality occurs in the remainder, we have:

$$s_t = s' (1 - u_t) \quad (11.39)$$

where s' is the "natural" survival rate (assumed to be constant) and u_t is the annual exploitation rate. Here the exploitation rate again takes the form $u = 1-\exp(-qE)$. Substituting these components into Equation 11.36 we have:

$$B_{t+1} = \sum_{a=r+1}^{\infty} [(1+\rho)\, w_{a-1} - \rho\, w_{a-2}]\, s_t N_{a-1,t} + w_r N_{r,t+1} \quad (11.40)$$

which can be written:

$$B_{t+1} = \sum_{a=r+1}^{\infty} (1+\rho) s_t w_{a-1} N_{a-1,t} - \sum_{a=r+2}^{\infty} \rho\, s_t s_{t-1} w_{a-2} N_{a-2,t-1} + w_r N_{r,t+1} \quad (11.41)$$

and translating to biomass throughout we have:

$$B_{t+1} = (1+\rho) s_t B_t - \rho s_t\, s_{t-1} B_{t-1} + f_B\left(B_{t+1-r}\right) \quad (11.42)$$

where $f_B\left(B_{t+1-r}\right)$ is a function relating adult biomass to recruitment. This equation defines how biomass propagates over time. At equilibrium, $B^* = B_{t+1} = B_t$, and solving we have:

$$B^* = \frac{f_B(B^*)}{(1-\rho s^*)(1-s^*)} \quad (11.43)$$

(Quinn and Deriso 1999).

An application of this model to yellowtail flounder on Georges Bank is shown in Figure 11.11. Here, we have used biomass estimates derived from stock assessments (NEFSC 2008) and independent growth estimates to depict results using process-error and observation-error estimation techniques. The process-error approach assumes that all deviations from the model are due to various forms of model mis-specification (including structural uncertainty, exclusion of important factors, random variation etc.). The observation-error approach assumes that the underlying model is known exactly and deviations are due to measurement error (including random and systematic errors).

11.4.1 Complex dynamics

Here we ask how exploitation might alter the dynamic landscape of simple harvesting models expressed as difference equations. We have seen that for unexploited populations, complex dynamics can readily emerge in discrete-time models. Does harvesting alter this conclusion?

We first provide an example drawn from a simple biomass dynamic model. In the absence of harvesting, an increase in the intrinsic rate of increase (α) of the population will result in the classical

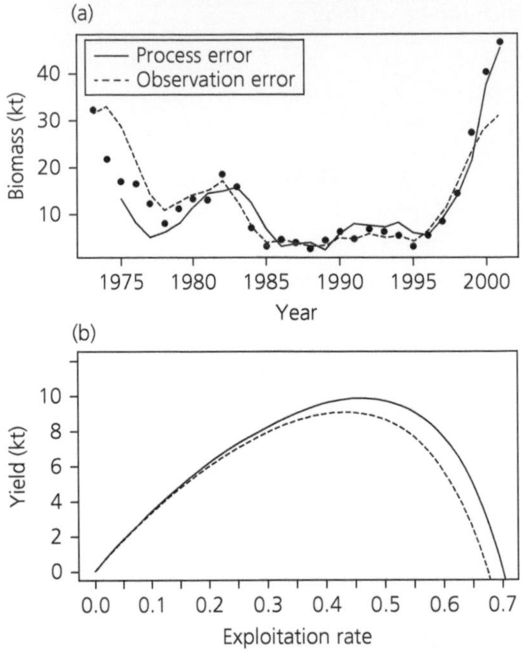

Figure 11.11 Delay–difference model for the Georges Bank yellowtail flounder (*Limanda ferruginea*) population. (a) Time-series fits and (b) equilibrium yield curves.

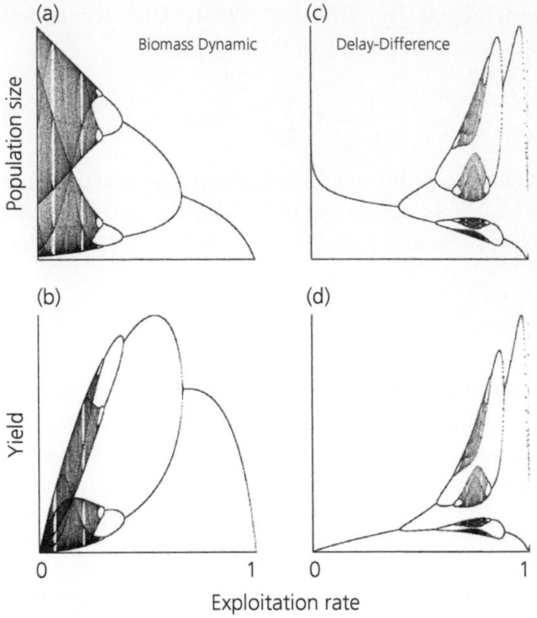

Figure 11.12 Bifurcation diagrams of population biomass for (a) a biomass dynamic model and (b) a delay–difference population model as a function of exploitation rate on adults. Bifurcation diagrams of yield for (c) a biomass dynamic model and (d) a delay–difference population model as a function of exploitation rate on adults.

period-doubling route to chaos (see Chapter 3). However, for an exploited population this dynamic can be fundamentally altered. For models without demographic structure, an increase in the exploitation fraction can dampen or eliminate the expression of complex dynamics. A bifurcation diagram for the case in which the population size fluctuates chaotically in the absence of exploitation but stabilizes with increasing fishing effort is shown in Figure 11.12a. The corresponding bifurcation diagram for yield with increasing exploitation is depicted in Figure 11.12b. In this case, the expression of complex dynamics is a function of the magnitude of the realized rate of population increase ($\alpha - u$). An increase in exploitation rate can result in a transition from complex to stable dynamics as fishing effort increases. In this case, harvesting reduces the population rate of increase. Demographic structure is of course not represented in the biomass dynamic model. One way to view this is that we are seeing the implications of a non-selective harvesting strategy with respect to size or age on the dynamical response to exploitation.

We note that for a delay–difference model with a simple age structure in which harvesting is confined to adults, an increase in fishing effort can result in destabilization of the system dynamics (Figure 11.12c, d; see Basson and Fogarty 1997, Fogarty et al. 2016). Here, when a robust age structure is maintained under low harvesting rates, even with relatively high rates of intrinsic increase, the population is stable. However with increasing exploitation and a degradation in the age structure, complex dynamics are revealed (Fogarty et al. 2016). This situation changes if both juveniles and adults are subject to harvest—in effect approaching the non-selective harvesting strategy in the biomass dynamics model. In this case harvesting results in a reduction of the slope of the recruitment curve at the origin and we see a manifestation of the results shown in the non age-structured model.

11.5 Full age-structured models

We can connect the information in this and previous chapters to specify a full age-structured population model. Information from stock-recruitment analyses can be combined with information from a yield and spawning biomass per recruit analysis to develop a full age-structured model (Sissenwine and Shepherd 1987). We require information on the parameters of the stock-recruitment relationship and the results of yield and egg production (or spawning biomass) per recruit analyses. For example, the Ricker model can be expressed:

$$\log_e\left(\frac{R}{\mathcal{E}}\right) = \log_e a - b\mathcal{E} \tag{11.44}$$

where R is recruitment, $\mathcal{E}$ is egg production, and a and b are parameters (see Section 9.3.1). Solving for $\mathcal{E}$ we have:

$$\mathcal{E} = \frac{1}{b}\log_e\left[a\left(\frac{\mathcal{E}}{R}\right)\right] \tag{11.45}$$

Notice that the expression inside the brackets includes $\mathcal{E}/R$ (egg production-per-recruit—EPR). We can therefore determine the level of EPR for each level of fishing mortality and use this to determine the predicted total egg production for that level of fishing mortality. Similarly, for the Beverton–Holt model, we have:

$$R = \frac{a\mathcal{E}}{1 + b\mathcal{E}} \tag{11.46}$$

and solving for $\mathcal{E}$ we now have:

$$\mathcal{E} = \frac{1}{b}\left[a\left(\frac{\mathcal{E}}{R}\right) - 1\right] \tag{11.47}$$

which again explicitly contains $\mathcal{E}/R$. Similar expressions can be readily derived for other stock-recruitment models.

Once the total egg production corresponding to a particular level of fishing mortality is determined, the corresponding recruitment can be obtained by the simple identity $R = \mathcal{E}/(\mathcal{E}/R)$ and the equilibrium yield is then obtained by the identity $Y = (Y/R)R$.

We illustrate these steps with data for Icelandic cod using yield and spawning biomass per recruit analyses as a proxy for EPR (Figure 11.13a) and derived estimates of a Ricker stock- recruitment relationship for this population (Figure 11.13b). Combining the information from the stock-recruitment relationship and the yield and spawning biomass-per-recruit analyses as described above, we obtain the total yield curve (Figure 11.13c). We have superimposed the actual (non-equilibrium) yield trajectory for this population on the equilibrium yield curve.

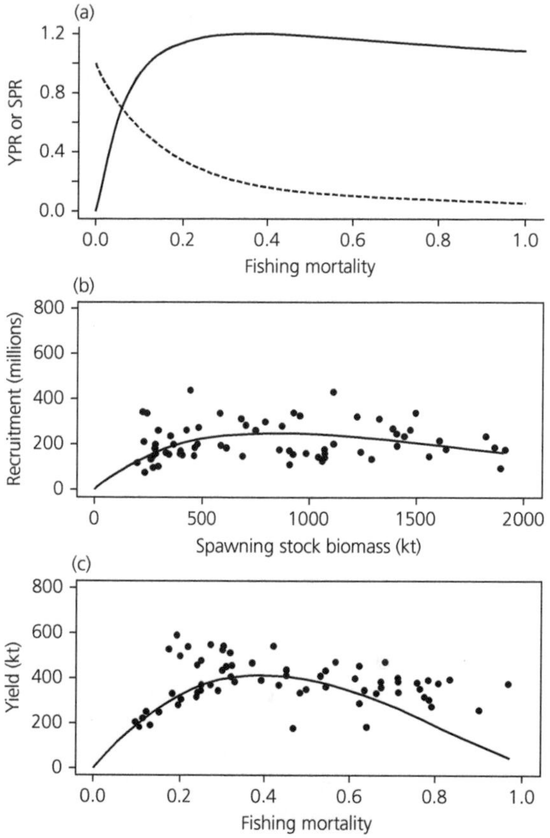

Figure 11.13 Age-structured population model for Icelandic cod (*Gadus morhua*) based on (a) yield and spawning-biomass-per-recruit, and (b) stock-recruitment relationship, to derive (c) yield-fishing mortality relationship. Note that the curve in panel (c) is not a fit to the observed data points (dots) but rather predicted from the yield and SPR curve (panel a) and the stock-recruitment function (solid line panel b).

11.6 Harvesting in randomly varying environments

In this chapter, we have thus far focused on deterministic models with no explicit consideration of random environmental variability. In our treatment of recruitment processes in Chapter 9 we saw that processes governing the growth and survival of individual aquatic organisms may be strongly affected by random environmental events, particularly during the early life-history stages. Stochastic models can provide more realistic representations of recruitment processes. With adoption of the stochastic approach, we focus on probability distributions of outcomes rather than fixed-point predictions. Stochastic harvesting models have been developed in both discrete and continuous time. A full analytical treatment of stochastic differential equations for harvested populations is beyond the scope of this introductory text (for an overview see Sissenwine et al. 1988). In general, stochastic production models indicate more conservative estimates of maximum sustainable yield and associated reference points than their deterministic counterparts. In the following, we provide an example of numerical studies of a stochastic harvesting model in discrete time. We focus on the probability of quasi-extinction with increasing exploitation rates. The potential interplay between population variability and harvesting pressure can substantially enhance these risks.

11.6.1 Discrete-time models

Stochastic difference equations evaluated by numerical methods can afford considerable flexibility for analysis. Detailed analytical treatments are also possible in some cases; see Bousquet et al. (2008) and Bordet and Rivest (2014).

Here we will consider a Ricker-logistic type production function:

$$B_{t+1} = B_t e^{[(\alpha+\varepsilon_t)-\beta B_t]} - u_t B_t \qquad (11.48)$$

where ε_t is a normally distributed random error term with a mean of 0 and constant variance. Notice that the stochastic element resides in the exponential term and we therefore have a multiplicative error structure. As before, the yield is given by the product of the exploitation rate μ and biomass.

We are particularly interested in the effect of random variability on population status and the probability of quasi-extinction under different

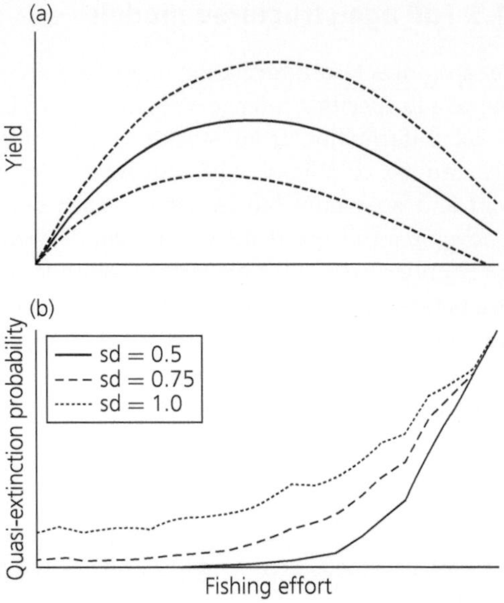

Figure 11.14 (a) Yield as a function of fishing effort for the stochastic Ricker-logistic model (mean +/− 1 standard deviation) and (b) probability of quasi-extinction (population driven to below 20% of unexploited biomass) as a function of fishing effort (effort scaled 0–1) at standard deviation levels ranging from 0.5 to 1.0.

levels of environmentally induced variability. Here, the quasi-extinction probability is the chance that the mean population level will fall below 20% of the unfished population state, although other choices for this threshold level could of course be made. In Figure 11.14 we show the results of stochastic simulations in which we vary the standard deviation of the random normal probability distributions of ε_t at three fixed levels. Increasing the levels of variability from sd = 0.5 to 1.0 markedly increases the probability of quasi-extinction in these simulations.

11.6.2 Low-frequency variation and climate change

In this section, we address the issue of directional change in the environment due to climate change. The full magnitude of change now unfolds over decades and represents a form of low-frequency variation. Future climate predicted under increasing anthropogenic emissions of greenhouse gases will strongly alter the physical structure of the oceans, with direct implications for marine ecosystems and human societies. These changes need to be considered in the context of impacts resulting from other human activities including fishing, pollution, and habitat loss due to coastal development. Climate change can interact with other human-induced changes to alter the fundamental production characteristics of marine ecosystems. Climate change will potentially exacerbate the stress on some living marine resources imposed by harvesting and other anthropogenic activities in many regions, but will potentially enhance production for other populations in some areas. Considerable emphasis is now being placed on understanding the causes and consequences of climate change in temperate marine systems (e.g. Harvell et al. 2002, Helmuth et al. 2002, Barnett et al. 2005, Drinkwater 2005, Sutton and Hodson 2005) to prepare for anticipated alterations in ecosystem structure and function.

Climate change scenarios have been employed in fishery production models to explore the consequence of alteration of temperature regimes. For example, Fogarty et al. (2008) examined the implications of increasing temperature on projected yields of cod in the Gulf of Maine under climate change scenarios corresponding to specified categories defined by the International Panel on Climate Change (IPCC). The IPCC scenarios chosen represent alternative futures based on projections of population and economic growth, technological development, and energy use patterns bounded by higher and lower emission scenarios. The IPCC A1 fi scenario envisions a world in which fossil fuels continue to dominate energy-use patterns leading to steadily increasing greenhouse gas emissions, approaching 30 Gt/yr by the end of the twenty-first century. The IPCC B1 alternative involves a reduced emission scenario leading to output by the end of the century of just under 10 Gt/yr. Fogarty et al. (2008) considered an ensemble of three general circulation models. The dynamics of Gulf of Maine cod were represented by a full age-structured model as described in Section 11.5. However, in this case, the effect of projected bottom temperature on individual growth and recruitment was incorporated. The effect on yield reflects the interplay of decreased recruitment and increased individual growth with increased water temperature. Increased water

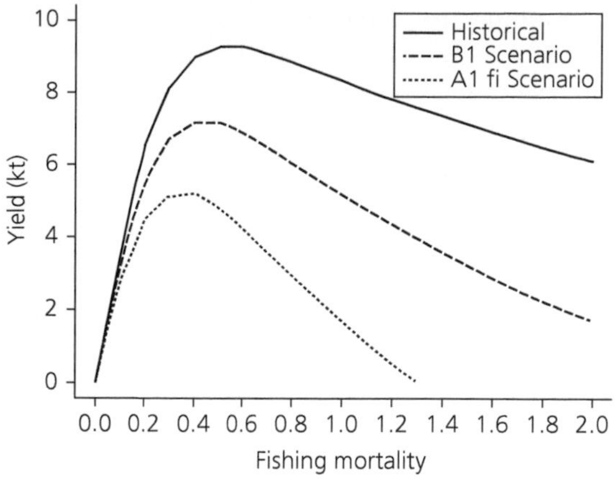

Figure 11.15 Relationship between yield and fishing mortality for Gulf of Maine cod (*Gadus morhua*) under (a) average bottom temperature during 1982–2003 (solid line), projected yield under the IPCC B1 scenario (dashed lines) and A1 f1 scenario (dotted lines). See text for description of scenarios (Fogarty et al. 2008).

temperatures of 1 °C and 2 °C relative to the baseline levels, corresponding approximately to projected increases under the B1 and A1 fi emission scenarios, result in declines in projected yields and in the fishing mortality rate resulting in maximum yield (Figure 11.15). A one-degree increase would result in an approximate 21% decline in maximum yield relative to a baseline historical standard while a 2 °C temperature increase is predicted to result in a 43% decline in maximum yield. At the higher temperature level, the stock is predicted to approach extinction at fishing mortality rates in excess of $F = 1.3$, a level which is sustainable (though far from optimal) under a one-degree increase in temperature. These results qualitatively point to a decrease in yield and resilience to fishing under increasing temperatures, particularly under the projected levels under the high-emission scenario. A key lesson is that we must focus on the interactive effects of climate and fishing intensity and not each individually.

11.7 Summary

In this chapter we have attempted to complement traditional treatments of single-species harvesting models by exploring dynamical behaviors that go beyond globally stable outcomes. The possibility of multiple equilibria (both stable and unstable) emerges quite readily in models with non-linear harvesting functions. In practice, most fisheries management protocols at least implicitly assume that harvested populations have well-behaved stable equilibrium properties. If this is not the case, then sudden changes (including collapse) can occur and be totally unanticipated. The problem of hyperstability is likely much more pervasive than commonly assumed. We argue for more direct consideration of fishing effort as a driver with far-reaching consequences for harvested populations, habitat, and incidental catch of non-target species. Fishing mortality estimates derived assuming a linear relationship between F and fishing effort can be misleading.

Biomass dynamic models provide a useful heuristic guide to the impacts of harvesting on an exploited population. Biomass dynamic model may be the only choice when the demographic structure of a population is unknown. Ludwig and Walters (1985) showed that biomass-dynamic models can provide more robust estimates of management parameters than more complex models (where parameter uncertainty can become an important issue). Delay–difference models provide a bridge between biomass dynamic models and full age-structured models. They require the estimation of additional parameters, but since the parameters have clear life-history interpretations, some of them can be estimated independently and fixed to reduce the number of parameters to be estimated. Dynamic pool models combine yield-per-recruit and egg-per-recruit with a stock-recruitment model to obtain an equilibrium yield curve. These single-species models are used to estimate biological reference points with which to assess stock status.

As in previous chapters, we examine the possibility of complex dynamical behaviors in discrete-time models and show that for certain parts of the parameter space, difference-equation models for harvested populations can provide markedly different population levels and yield than their continuous-time counterparts. Contrasts between the behavior of discrete-time biomass dynamic models and delay–difference models are also marked. In biomass dynamic models, exploitation can be stabilizing while it can be destabilizing in delay–difference models depending on whether harvesting occurs only on post-recruit individuals or whether a broader age/size spectrum is harvested.

It is important to recognize that harvesting occurs in randomly varying environments and that consideration of stochastic forcing argues for precautionary management. The probability of population collapse increases with increasing exploitation rates and environmental variability. Finally, we note that climate change offers important challenges to effective management. It often entails a type of low-frequency forcing that can be particularly problematic if it goes unrecognized and its effects on population production are not taken into account in management. Climate change and harvesting pressure can act synergistically to destabilize exploited populations.

Additional reading

The fundamentals of single-species harvesting theory have been very-well laid out in classical fisheries texts. Beverton and Holt (1957), recently republished, remains a treasure trove of information and is well worth consulting. Not simply of historical interest, it covers many of the topics in this chapter in depth and much more. Ricker (1958, 1975) is a standard as is Gulland (1983). The more recent texts by Hilborn and Walters (1992) and Quinn and Deriso (1999) are invaluable resources. King (2013) provides an introductory survey of fishery assessment models.

CHAPTER 12

Harvesting at the Community Level

12.1 Introduction

The harvesting models considered in Chapter 11 apply to the dynamics of single species, but these species interact with others in the community and ecosystem. Mangel and Levin (2005) argue for the central importance of community ecology as a foundation for ecosystem-based fisheries management. Although additional ecological considerations will be essential (see Chapter 13), adoption of community ecology as a fundamental framework for EBFM does hold certain advantages for the development of tactical fisheries management advice. Models of exploited communities are typically of intermediate complexity and can readily be tailored to address specific management issues. Further, we can ask whether it is possible to capitalize on emergent ecosystem properties in devising management strategies in ways that might simplify the problem. Emergent properties are ones that cannot be deduced by examining the parts of a system in isolation. For example, the collective biomass or production of a defined community may be more stable and predictable than that of the species constituting the assemblage. Interactions among species may result in covariance patterns that are stabilizing at the community level. We can then ask if there may be an advantage to focus management at higher levels of ecological organization, such as the community, rather than the population/stock level. What risks might be entailed by such a strategy? We will explore these questions for species groups subject to technical and/or biological interactions.

In this chapter we examine the implications of technical and biological interactions for the community dynamics of exploited aquatic communities. *Technical interactions* occur because species are caught together by the same fishing gears. *Biological interactions* occur because the species interact through competition and predation. In mixed-species fisheries, understanding the dynamical response of co-occurring species to exploitation becomes critically important. How a species will respond to harvesting is determined by factors such as its susceptibility to capture (catchability) and its life history characteristics. Susceptibility is a function of morphological and behavioral attributes of the species sought, and the behavior of fishers in response to economic incentives. The differential response of species and species groups in turn is a critical determinant of how an aquatic community as a whole will respond to exploitation.

12.2 Technical interactions in mixed-species fisheries

Fisheries in which many species are caught simultaneously are common. In these fisheries, a single input (fishing effort) results in multiple outputs (mixed-species catches) in a joint production system. The catch composition is a function of species distribution patterns, selectivity characteristic of the gear, choices made by fishers, and behavior of the species caught. There are necessarily some constraints on the compositional control and fishing mortality rates exerted on the species mix derived from this common input. It follows that our ability to effectively target levels of fishing mortality for individual species has limits. Detailed studies of this problem suggest that it is both pervasive

Fishery Ecosystem Dynamics. Michael J. Fogarty and Jeremy S. Collie, Oxford University Press (2020). © Michael J. Fogarty & Jeremy S. Collie 2020.
DOI: 10.1093/oso/9780198768937.003.0012

and problematic in fisheries resource management (e.g. Squires 1987; Scheld and Anderson 2016). Successful fishers are skilled naturalists who can unquestionably exert some level of compositional control of their catch, particularly when appropriate incentives are present. However, it is inevitable that there will be collateral impacts on the assemblages as a whole in mixed fisheries that must be taken into account in management.

In the following, we will outline two major approaches that have been used to address the issue of technical interactions: cohort models and biomass dynamic models in which technical interactions are important and biological interactions among species in the assemblage are assumed to be minimal or non-existent. Both are natural extensions of the harvesting models described in Chapter 11.

12.2.1 Mixed-species cohort models

With appropriate modifications, the single-species cohort models described in Chapter 11 can be adapted to address aspects of the mixed-species fisheries problem. Although the development of continuous time models is feasible in this setting, discrete-time models again offer considerably enhanced flexibility and we will focus on this approach. Murawski (1984) pioneered the development of mixed-species yield-per-recruit models of this type. His generalized treatment considered not only multiple species but multiple fleets (see also Pikitch 1987). Because recruitment levels differ among species, it is necessary to scale the results in a way that accounts for differential recruitment patterns. Murawski (1984) applied species-specific recruitment multipliers based on research vessel surveys to address this issue. If estimates of total recruitment are available, these can be used directly to generate total yields for each species and for the assemblage as a whole. Because recruitment varies over time for each species, a dynamic approach in which yield and egg production (or its proxy, spawning biomass) estimates are updated periodically is desirable. The yield for species i subject to a single fleet is given by a modification of the single-species case:

$$Y_i = \sum_{a=a_{rec}}^{a=a_{\max}} \left(\frac{p_{ia} q_i E}{p_{ia} q_i E + M_i} \left[1 - e^{-(p_{ia} q_i E + M_i)} \right] \right) \times N_{ia} \overline{w}_{ia} \tag{12.1}$$

where we have indexed the inputs by species (i) age (a) and where for each species, p_{ia} is an age-specific fishery selectivity factor, q_i is a catchability coefficient, E is standardized fishing effort, N_{ia} is the number of individuals alive at age a, and $\bar{w}_{ia}$ is the mean weight at age a. The summation over age classes is from the age at recruitment to the fishery (a_{rec}) to the maximum age (a_{max}). Because there are no interactions among species in this formulation, we can simply evaluate each species independently and sum the results to get the total yield.

Similarly, we evaluate the egg-production per recruit by a simple modification of Equation 11.13 indexed by species:

$$\mathcal{E}_i = \sum_{a_{rec}}^{a_{\max}} m_{ia} v_{ia} f_{ia} N_{ia} \tag{12.2}$$

where for each species m_{ia} *is* the proportion mature at age in the population, f_{ia} is the fecundity at age and v_{ia} is a measure of age-specific egg viability. Again, for the common case where fecundity data are not available, we use spawning biomass as a proxy.

Given that different species have different economic values in the marketplace, maximizing revenue or profits from a fishery is not the same as maximizing yield in weight. Price differentials and related considerations are major determinants of the harvesting strategies fishers choose. Gross revenue is the product of yield and price per unit mass. Net revenue is the difference between gross revenue and the cost of fishing. Costs can be partitioned into fixed and variable costs. Fixed costs include capital investments in vessels, fishing gear, and other foundational necessities for fishing operations. Variable costs include on-going operational requirements for fuel, bait, and other needs that depend on the expenditure of fishing effort. The variable cost net revenue for a mixed-species, single-fleet fishery can be expressed as:

$$\pi = \sum_{i=1}^{n} (q_i P_i B_i - C_e) E \tag{12.3}$$

where P_i is the price per unit weight for species i, C_e is the variable cost per unit of fishing effort.

Jacobson and Cadrin (2008) provide an application of a mixed-species cohort model for groundfish species in the northeastern United States. Here we show results for a reduced subset of the species assemblage considered by Jacobson and Cadrin. We focus on three of the mainstays of the mixed-species New England ground-fishery: cod, haddock, and yellowtail flounder. Although relatively strong biological interactions occur among some species in the northeast groundfish assemblage as a whole (Tsou and Collie 2001; Curti 2012), interactions among post-recruit individuals of the three species evaluated here appear to be relatively weak. Results are provided using average recruitment levels for the three species as reported in Jacobson (2009). Several main points emerge. The $E_{0.1}$ yield occurs at very different levels of fishing effort for each of the three species considered (Figure 12.1a). It is not possible to achieve target levels of exploitation simultaneously for the three species for a given effort level. The threshold level of fishing effort at which the spawning biomass was reduced to 40% of the unfished state varied considerably among the three species (Figure 12.1b), indicating differential risk to the populations under a common level of fishing effort. Finally, unlike the total yield curve, the total profit (here variable cost net revenue) function was dome-shaped and the maximum profit occurred at a lower level of fishing effort than the biological reference points for the individual species (Figure 12.1c). A more conservative effort level therefore is also economically desirable.

12.2.2 Mixed species biomass dynamic models

In this section we extend the biomass dynamic models introduced in Chapter 11 to account for harvesting of an assemblage of co-occurring species subject to technical interactions. The general form for a production model for an individual species in the assemblage is:

$$\frac{dB_i}{dt} = \left(\alpha_i - \beta_i B_i^{\theta-1}\right) B_i - q_i E B_i^{\gamma_i} \qquad (12.4)$$

where q_i is a species-specific catchability within a particular fishery (and gear type) with fishing effort E. As with the single-species models described in Chapter 11, we wish to afford flexibility to account for non-symmetric production functions and non-linear harvest functions. With respect to the latter, it is recognized that in situations where we lack fishing effort information and/or supplemental biomass data, it may be necessary to substitute the observed yield for the harvest term.

An example for a hypothetical three-species fishery in which each species is governed by a logistic function ($\theta = 2$) is provided in Figure 12.2. The total yield for a given level of fishing effort is the sum of the individual species yields at that level. Note, however, that now the total production curve is no longer a simple quadratic function. Further, the choice of target levels of fishing mortality is complicated by the fact that each species has its own level of fishing effort that results in maximum yield. Notice that in this example, if we were to harvest at the fishing effort level that maximizes yield for either Species 2 or Species 3, it would result in the extinction of Species 1 (Figure 12.2a). As shown in Figure 12.2b, the total revenue curve and optimal fishing strategies can be altered by the relative prices for each species. Depending on the price differential among species, fishers may have an economic incentive to target (to the extent possible) the most valuable species and possibly discard lower valued species.

12.2.3 Identifying vulnerable species

The preceding sections make clear the need to understand the relative vulnerabilities to fishing of the species constituting mixed-species assemblages. In general, it is not possible to simultaneously attain target levels of exploitation for each individual species. Species with lower productivity levels will be overexploited in mixed-species fisheries unless additional protective measures are implemented. A key metric to understanding vulnerability is the intrinsic rate of increase of the population (Musick 1999). In turn, this metric is a function of demographic characteristics related to reproductive biology (age and size at maturation, fecundity), individual growth (maximum size, rate of growth), and other factors. Fortunately, substantial information is available on exploited

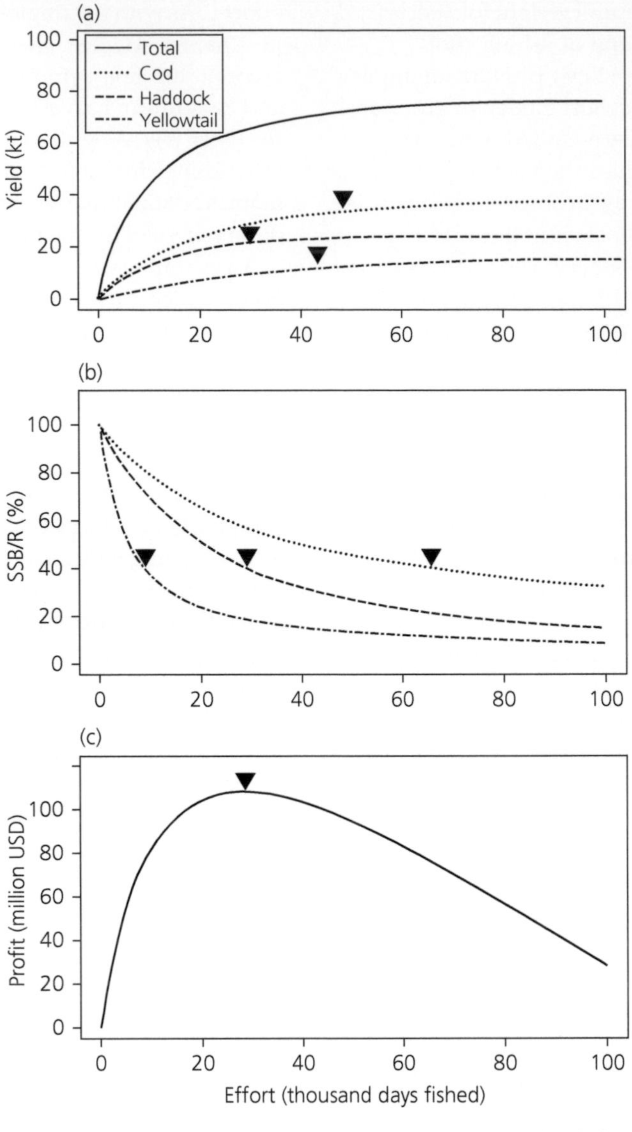

Figure 12.1 Multispecies technical interaction model for (a) yield (b) spawning biomass, and (c) total profit for cod (*Gadus morhua*), haddock (*Melanogrammus aeglefinus*), and yellowtail flounder (*Limanda ferruginea*) off the Northeastern U.S. (based on Jacobson 2009). Arrows represent (a) $E_{0.1}$ levels for each species, (b) and $SPR_{40\%}$, and (c) peak profit. $E_{0.1}$ is an analogue of $F_{0.1}$ (see Section 11.2.1)

aquatic species to provide guidance on likely vulnerability to harvest. Fenchel (1974) developed an allometric function describing the relationship between the intrinsic rate of increase and body mass for species ranging from unicellular organisms to large-bodied vertebrates. Fenchel reported a common exponent of −0.25 for species groups and different intercept terms for different taxa. Extensive and readily available compilations of fishery, population, and life-history information for exploited aquatic species (e.g. FishBase, RAM II) provide fertile ground for quantifying risk in this context. Myers et al. (1999) provide estimates of the maximum reproductive rate (a proxy for the intrinsic rate of increase) for over 230 species of exploited fishes.

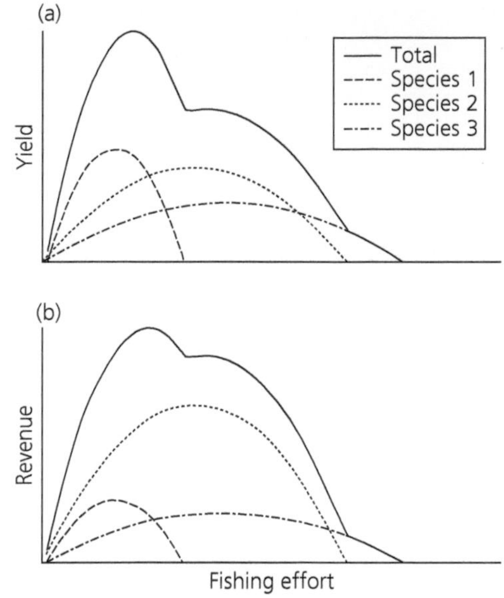

Figure 12.2 (a) Individual and total species yield as a function of fishing effort in an assemblage of co-occurring species with no biological interactions and (b) corresponding total revenue curves with price differential in which the market value of species 2 is 3 times that of species 1 and the market value of species 3 is 1.5 times that of species 1.

12.2.3.1 Productivity-susceptibility analysis

Joint productivity and susceptibility attributes have been used to evaluate relative risk to exploited aquatic species (see Hobday et al. 2007 and references therein). Potentially large numbers of variables related to species productivity and susceptibility are combined into categorical vulnerability scores for each species and attribute. Productivity attributes include variables such as the intrinsic rate of increase, maximum size, longevity, age-at-maturity, fecundity, natural mortality, and individual growth rate (von Bertanffy k). Variables included in the susceptibility scores include factors related to catchability (species concentration indices, species distribution–effort overlap; body morphology in relation to gear selectivity factors; behavior of target species including schooling and migration/movement patterns); and revenue and/or profitability. The vulnerability score is given by:

$$V_i = \sqrt{(P - P_0)^2 + (S - S_0)^2} \tag{12.5}$$

where P_0 and S_0 specify the origin of this 2-dimensional coordinate system. The vulnerability score is the Euclidean distance of the productivity (P) and susceptibility (S) scores. Patrick et al. (2010) developed vulnerability scores for 133 marine species in U.S. waters using three categories for productivity scores ranging from 1 (high productivity) to 3 (low productivity). Susceptibility scores ranged from 1 (low susceptibility) to 3 (high susceptibility). Figure 12.3 depicts scores for five species arrayed along these axes. Species A, B, and C fall along the median bisector in the vulnerability plane, ranging from high to low vulnerability (risk). Species D and E have equal risk probabilities with species D characterized by high productivity and susceptibility while species E has low productivity and susceptibility.

12.2.3.2 Eventual threat index

Burgess et al. (2013) developed a predictive threat index based on measures of catch-per-unit effort, the intrinsic rate of increase, and population size for mixed species fisheries prosecuted by one or more fleets. First, a vulnerability score for each species is determined. Using our earlier notation, the index can be written as:

$$V_{ik,t} = \frac{Y_{i,t}}{r_i N_{i,t} E_{k,t}} \tag{12.6}$$

where the subscripts indicate species (i) and fleets (k), Here, a fleet (or métier) is defined as a fishery that captures an assemblage of species populations in time and space with a particular gear type, and r_i is the maximum population growth rate for a population growing according to the logistic equation. Notice that we can isolate Y_i/N_i in this expression and substitute $q_{ik}E_k$ for this term. We then see that the vulnerability term can be written $V_{ik} = q_{kj}/r_i$. Species with the highest catchability and the lowest intrinsic rate of increase will therefore be the most vulnerable. In this case, q represents susceptibility and r provides an index of productivity.

To develop a predictive index, Burgess et al. then identified "key" species to provide a reference level. Key species are ones that are highly sought because of high value and with high catchability and low productivity, and that will trigger a fishery closure if depleted. A threat index for any species is defined

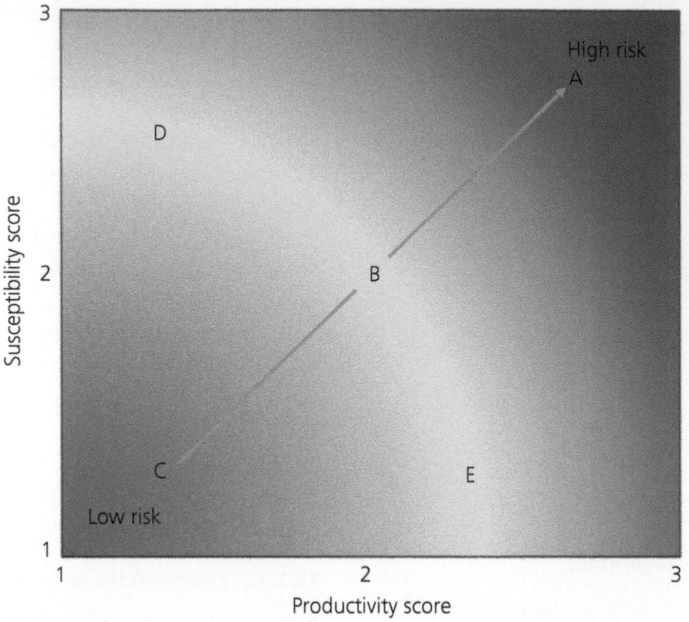

Figure 12.3 Productivity-susceptibility plane for exploited species. Productivity ranges from high (1) to low (3). Susceptibility ranges from low (1) to high (3) based on a categorical evaluation of sets of individual species productivity and susceptibility metrics (see text). Adapted from Hobday et al. (2007) and Levin et al. (2008).

relative to the key species in any fishery. It consists of the ratio of the vulnerability score for each species captured in fleet k to that of the key species for each fleet, weighted by the catch of the key species in fleet k relative to the total catch of this species over all m fleets in which it is caught:

$$T_{i,t} = \sum_k \left[\left(\frac{V_{ik,t}}{V_{k,t}^{key}} \right) \cdot \left(\frac{Y_{kk,t}^{key}}{\sum_m Y_{km,t}^{key}} \right) \right] \quad (12.7)$$

For populations exhibiting logistic growth, recall that extinction is predicted if the fishing mortality rate exceeds the intrinsic rate of increase of the population (see Section 11.3.1.1). The production will be highest when the population is one-half of the unexploited equilibrium state. Under these assumptions, a population is threatened with extinction when $T_{i,t} \geq 2$. Burgess et al. (2013) further proposed cut points for high ($1 < T_{i,t} < 2$), moderate ($0.5 < T_{i,t} \leq 1$) and low ($T_{i,t} \leq 0.5$) risk of overfishing. They showed that this metric can be successfully used to predict future risk in an application to the Western and Central Pacific tuna and billfish fishery. This fishery captures eight species, two of which are targeted because of high value and designated as key species. The eventual threat index successfully predicted the future status of these populations, in some cases decades in advance.

12.3 Aggregate biomass dynamic models

Simple biomass dynamic models are used to assess individual stocks around the world (Prager 1994). These approaches are also applicable to aggregate-species groups in multispecies fisheries. Although this method has been employed to remedy data limitations and/or address system complexity (Ralston and Polovina 1982; Sugihara 1984), a more general rationale has centered on its ability to implicitly account for interspecific interactions (Brown et al. 1976). The trajectory of the whole is taken to integrate the effects of fishing and species interactions on the parts. We simply apply the single-species biomass dynamic framework to the aggregate species group using the equations specified in Section 11.3. The approach has been applied to large numbers of species (e.g. Brown et al. 1976; Mueter and Megrey 2006) or subsets of a larger species complex (e.g. Sparholt and Cook (2010); Fogarty et al. 2012). For additional examples, see the contributions in Bundy et al. (2012).

In many tropical and subtropical fisheries, individual species cannot be readily separated and/or identified in the catch. There then may be little recourse except to employ models for aggregate species groups. The Gulf of Thailand mixed species trawl fishery is a case in point. Pope (1977) developed one of the earliest multispecies production models for this fishery. Chotiyaputta et al. (2002) presented an updated analysis for the Gulf for the period 1967–1995 using an aggregate-species Gompertz–Fox ($\theta \to 1$) production model (Figure 12.4). Over 120 species are caught in the Gulf of Thailand trawl fishery, a substantial fraction of which are not identified to species in landings statistics.

In these aggregated analyses, we must again confront the potential risk to individual species when a broad range of life-history strategies are represented in the aggregate group. Where feasible, the approaches to defining vulnerability and threat described in Section 12.2.3 should be employed to understand risk. It is imperative that a precautionary approach be taken to avoid quasi-extinction (see Chapter 11) of any of species in the assemblage.

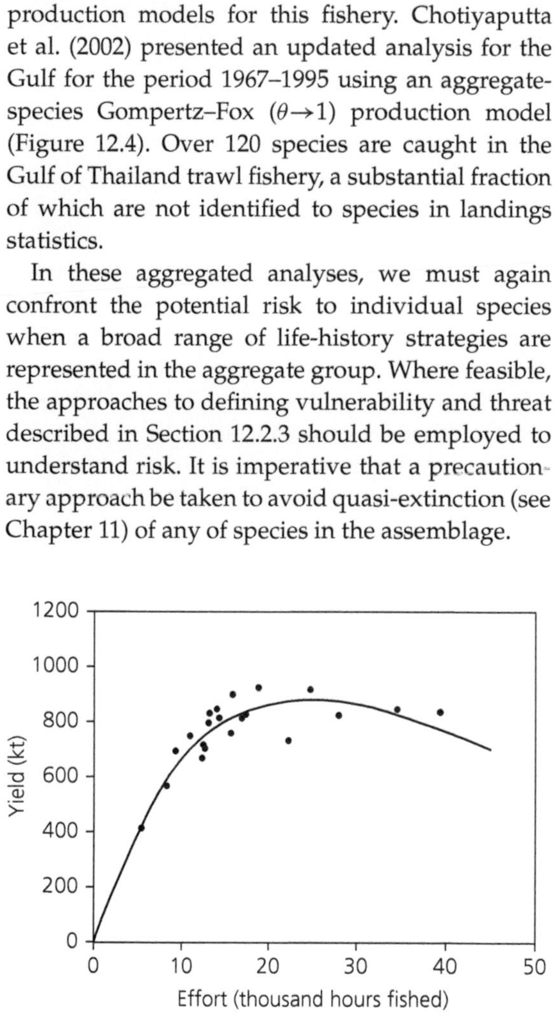

Figure 12.4 Fitted production curve of yield on fishing effort for the total aggregate landings of fish species from the Gulf of Thailand from 1966 to 1995 (adapted from Chotiyaputta et al. 2002).

12.4 Multispecies biomass dynamic models

We next turn to models in which we explicitly specify interactions between pairs of species. Important early developments in the application of simple multispecies models for exploited fish communities (e.g. Walter and Hoagman 1971; Pope 1977; May et al. 1979) adopted this approach.

12.4.1 Continuous-time models

In the following, we examine the implications of the specific form of the functions depicting interspecific interactions and show that a broad spectrum of dynamical behaviors is possible, including the emergence of alternative stable states. We begin with simple linear interaction terms and then turn to models with nonlinear interactions.

12.4.1.1 Linear interaction terms

We can extend the basic biomass dynamic framework to encompass interspecific interactions in simple models of exploited aquatic communities. The model can be specified as:

$$\frac{dB_i}{dt} = \left(\alpha_i - \beta_i B_i^{\theta-1}\right) B_i + \sum_{j \neq i}^{n} a_{ij} B_j B_i - q_i E_i B_i^{\gamma_i} \quad (12.8)$$

where a_{ij} is the effect of species j on species i, and all other terms are defined as before. The interaction terms are negative for pairs of competitors and for the effect of a predator on a prey species but positive for the effect of a prey species on a predator. Notice that in this formulation, predators don't depend solely on the prey species included in the model. Rather, they implicitly can also derive energy from non-specified "other prey". For the purposes of exposition, here we will set $\theta = 2$ and $\gamma = 1$. The equilibrium biomass for species i is found by setting the rate of change of all species to zero:

$$B_i^* = \frac{1}{\beta_i} \left[\alpha_i - q_i E_i + \sum_{j \neq i}^{n} a_{ij} B_j \right] \quad (12.9)$$

where now it is clear that the equilibrium biomass is affected not only by fishing mortality but by interspecific interactions. Equating the surplus production to yield at equilibrium, we have:

$$Y_i = (\alpha_i - \beta_i B_i) B_i + \sum_{j \neq i}^{n} a_{ij} B_j B_i \qquad (12.10)$$

The level of fishing effort resulting in maximum sustainable yield for species i (MSY_i) is obtained by differentiating Equation 12.10 with respect to B_i and solving:

$$E_{\mathrm{MSY}_i} = \frac{1}{2q_i} \left[\alpha_i + \sum_{j \neq i}^{n} a_{ij} B_j \right] \qquad (12.11)$$

and the MSY of species i is:

$$\mathrm{MSY}_i = \frac{1}{4\beta_i} \left[\alpha_i + \sum_{j \neq i}^{n} a_{ij} B_j \right]^2 \qquad (12.12)$$

The above expressions indicate clearly that when calculating MSY and E_{MSY} for a particular species, we must account for the biomass levels of interacting species. The equilibrium levels are not fixed quantities but rather change as fishing pressure and the biomass of interacting species change. In the single-species context, the species interactions were implicit in the α and β parameters and were therefore assumed to be constant with time. However, it is clear from the above equations that as the abundance of one species changes due to harvesting or other changes in the ecosystem, the productivity and potential yield of other species will also change. It is further possible to determine the total yield of the fish assemblage and corresponding reference points by solving Eq. 12.10 for all species simultaneously (Gislason 1999, Collie et al. 2003).

A graphical representation of this situation is shown in Figure 12.5 for a two-species predator-prey system. For the case of a quadratic production model we see that the prey yield surface is a symmetrical function of the fishing effort or mortality on both the predator and the prey. The model predicts that for a given level of fishing pressure on the prey, the prey yield increases with increasing fishing effort on the predator.

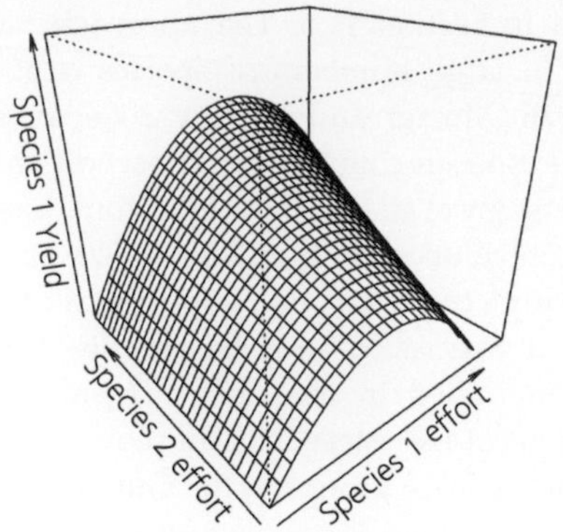

Figure 12.5 Prey (Species 1) yield surface as a function of fishing pressure on the predator (Species 2) and the prey for a multispecies Schaefer-type harvest model with linear interspecific interaction and harvest functions.

12.4.1.2 Non-linear predation terms

The multispecies production models considered in Section 12.4 have globally stable equilibrium points that depend on the joint fishing mortality rates on each species. Recall from Section 4.2.4 that inclusion of non-linear predation terms can lead to multiple equilibria. In this case, changes in the predation rate can cause the predator–prey system to flip rapidly between alternative stable states. In this section we show that in systems with multiple equilibria, changes in fishing mortality rates can flip the system from one equilibrium to the other, resulting in human-induced regime shifts Collie et al. (2004).

In the following, we will introduce a nonlinear functional feeding response term to the multispecies production model. For simplicity, we will focus on predation and assume interspecific competition to be negligible. The model is then:

$$\frac{dB_i}{dt} = \left(\alpha_i - \beta_i B_i^{\theta-1}\right) B_i + \frac{a_{ij} B_j B_i^{\omega}}{1 + \sum_{i \neq j} d_{ij} B_i^{\omega}} - q_i E_i B_i^{\gamma} \qquad (12.13)$$

where a_{ij} is the consumption rate of predator j on species i, d_{ij} is a coefficient of overall prey suitability and ω is a shape parameter. When species j is a predator of species i, a_{ij} is negative. The effect of a prey species on a predator is positive. For the case $\omega = 1$ and $d_{ij} = 0$ for all j, we have a linear functional feeding response for the predator. When d_{ij} is non-zero and $\omega = 1$ we have a Type-II functional response; and for non-zero d_{ij} and $\omega = 2$ we have a Type-III response. The predator expends a common search time during which multiple prey types can be encountered, each of which has differential vulnerability to the predator. The time taken to capture and subdue a prey item now affects the time available to consume alternative prey items.

For a Type-III functional response, the system can have multiple stable equilibrium points (Figure 12.6a). Collie and Spencer (1993) demonstrated this point using a Steele-Henderson model (Equation 3.25). with an added fishing term. In the case of low fishing and predator mortality we have a single globally stable equilibrium point at high prey biomass. At high levels of combined fishing and predation, a single intersection with the prey production curve exists and a stable equilibrium is predicted at low prey biomass. At intermediate levels of combined mortality, two stable equilibrium points are possible (Figure 12.6a). The resulting yield curve for this case shows that as fishing effort increases we have a single upper equilibrium yield curve and

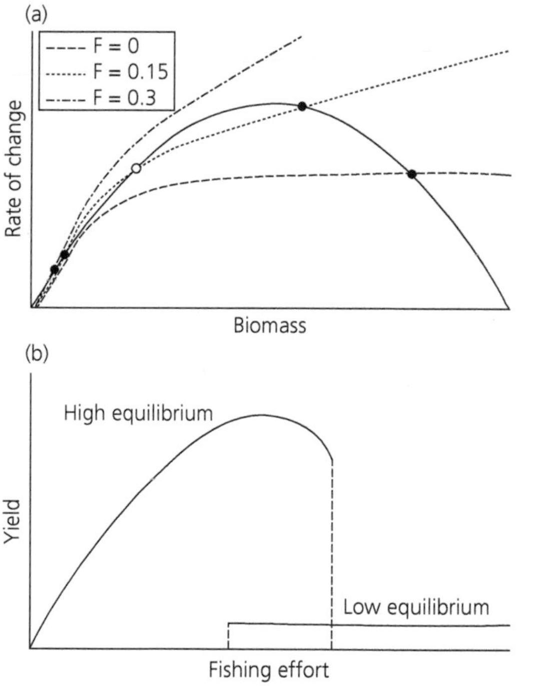

Figure 12.6 (a) Prey production curve (solid line) with superimposed curves representing combined human and natural predation at three levels of exploitation on the prey and a Type-III functional response. For the intermediate level of fishing pressure (dotted line) there are three equilibrium points (an upper stable, an intermediate unstable, and a lower stable point) and (b) upper and lower yield curves. Closed and open circles indicate stable and unstable equilibrium points.

then progress through a region where high or low equilibrium yields are possible and then to a low yield region (Figure 12.6b). At the threshold indicated by the vertical broken line, a small increase in fishing mortality could flip the system from the high to low equilibrium point.

12.4.2 Discrete-time models

As we saw in Chapter 6, multispecies models can also be written as coupled difference equations, which may be more appropriate for species in seasonal environments. As model complexity increases, discrete-time models may also be more tractable than their continuous-time analogs. Within certain ranges of the parameter space, multispecies difference equations give results similar to the corresponding differential equations. However, as we have observed consistently throughout, they can also exhibit quite complex dynamics and instabilities. In the following, we will consider discrete multispecies production and delay-difference models applied at the individual species and functional group levels.

12.4.2.1 Multispecies production models

It is common in application to employ a Euler approximation with a time step of one year to the differential equation (Equation 12.13) to specify the corresponding discrete-time model:

$$B_{i,t+1} = B_{i,t} + \left(\alpha_i - \beta_i B_{i,t}^{\theta-1}\right) B_{i,t} + \frac{a_{ij} B_{j,t} {B_{i,t}}^{\varpi}}{1 + \sum_j d_{ij} {B_{i,t}}^{\varpi}} - u_{i,t} B_{i,t} \tag{12.14}$$

where we now use the annual exploitation rate in the harvesting module and all terms are defined as before.

Uchiyama et al. (2016) applied discrete-time multispecies models of this general form to seven groundfish species in the Bering Sea. Three species were treated individually: walleye pollock (*Gadus chalcogrammus*), Pacific cod (*Gadus microcephalus*), and arrowtooth flounder (*Atheresthes stomias*). The remaining four species: yellowfin sole (*Limanda aspera*), northern rock sole (*Lepidopsetta polxystra*), flathead sole (*Hippoglossoides elassodon*) and Alaska plaice (*Pleuronectes quadrituberculatus*) were combined into a flatfish complex. Model inputs included biomass estimates derived from research vessel surveys and catches. Uchiyama et al. (2016) provided a detailed analysis in which both multispecies biomass dynamic and delay-difference models (see Section 12.4.2.2) were applied. The model for walleye pollock included additional demographic structure. Type-III functional feeding responses were employed throughout. Diet composition information and multispecies estimation models provide evidence for the importance of predator–prey interactions in this system. Uchiyama et al. were then able to directly compute losses due to predation. Here, we will provide results for a simplified model employing total pollock biomass and a Type-I functional response. Uchiyama et al. suggested that a simpler model structure employing Type-I functional responses would likely provide a more tractable setting for parameter estimation. For illustration, here we show the results using a process-error formulation and maximum likelihood estimation. The simple multispecies production model can track the main features of the population trajectories of these species (Figure 12.7). It must be noted, however, that some of the parameter estimates, including some predation terms, are characterized by high uncertainty (Uchiyama et al. 2016). In simulation studies, Oken and Essington (2015) showed that a combination of observation error and environmental variability can easily obscure underlying predator–prey dynamics and complicate estimation when applying multispecies production models to observed time series. Situations in which predation was concentrated on juvenile prey were found to be more amenable to analysis using these simple models in the simulations conducted by Oken and Essington (2015).

12.4.2.2 Multispecies delay–difference models

We next consider an extension of the delay–difference model structure to consider interspecific interactions affec with an emphasis on predation. The adoption of the delay–difference format allows us to account for age structure and to separate the effects of predation on pre-recruit and fully-recruited individuals. The simplest possible form of the model is:

$$B_{i,t+1} = \vartheta_{i,t} B_{i,t} + w_{ir} \sum_j h\left(B_{i,t+1-r} B_{j,t+1-r}\right) \qquad (12.15)$$

where $\vartheta_{i,t}$ is the combined individual growth-survival term for post-recruit individuals of species i, w_{ir} is the weight of a recruit (age r) of species i and $h\left(B_{i,t+1-r} B_{j,t+1-r}\right)$ is a recruitment function incorporating predation on pre-recruits. In our treatment of single-species delay-difference models in Chapter 11, individual growth was represented using a monomolecular growth function. Here we begin with formulations for growth that entail a simpler time delay structure for the multispecies model. The post-recruit survival term reflects three components—mortality due to fishing, predation and other sources of natural mortality. The other natural mortality component is typically assigned a constant value. If the individual growth rate is constant, we can specify the growth-survival component in a particularly convenient form:

$$\vartheta_{i,t} = \exp\left[g_i - q_i E_t - M1_i + \sum_j a_{ij} B_j \right] \qquad (12.16)$$

where g_i is the instantaneous growth rate of species i, $M1_i$ is the natural mortality rate of post-recruits from all natural sources other than predation, and the final term in Equation 12.16 represents the effect of predation on post-recruit individuals of species i. The sign of the interaction term a_{ij} is negative for the effect of a predator on a prey species and positive for the effect of a prey species on a predator. Alternatively, if mean weight at age for all age classes is known and growth is monomolecular, the combined growth-survival term for post-recruits can be specified:

$$\vartheta_{i,t} = s_{i,t} \left[\frac{W_{i,\infty}(1 - \rho_i) + \rho_i}{\bar{w}_{i,t}} \right] \qquad (12.17)$$

where the survival ($s_{i,t}$) term incorporates predation, non-predation natural mortality, and losses due to fishing, $W_{i,\infty}$ is the asymptotic weight of species i and $\bar{w}_{i,t}$ is the mean weight of post-recruits (Hilborn and Walters (1992). If the average mean weight for the post-recruit segment of the population is not available, we can extend Equation 11.42

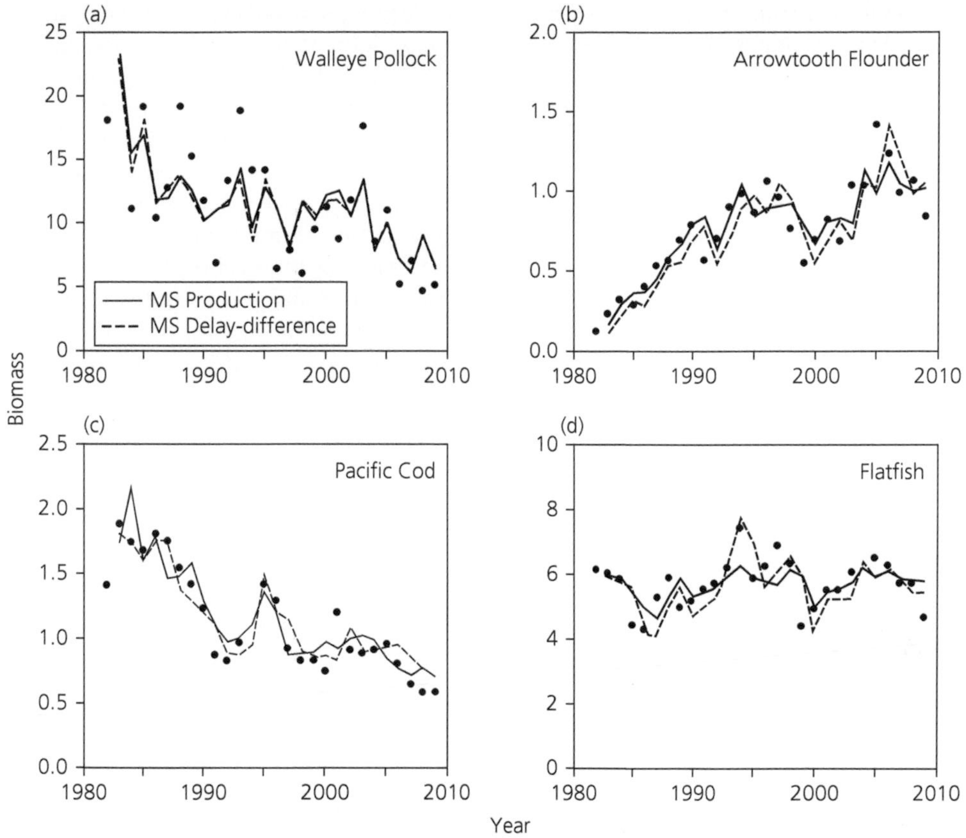

Figure 12.7 Estimated biomass (closed circles) based on research vessel surveys and fitted multispecies production (solid lines) and delay–difference models (dashed lines) for (a) walleye pollock (*Theragra chalcogramma*), (b) arrowtooth flounder (*Atheresthes stomias*), (c) Pacific cod (*Gadus macrocephalus*), and (d) small-mouthed flounder complex in the Bering Sea (data courtesy of Tadayasu Uchiyama). Process-error model fit by maximum likelihood.

for the single-species case to include losses due to predation in the survival and recruitment terms. (Uchiyama et al. 2016). An example of the multispecies recruitment function as an extension of the Ricker model (Hilborn and Walters 1992) can be expressed:

$$R_{i,t} = a_i B_{i,t-r} e^{-b_i B_{t-r} + \sum_j a_{ij} B_{j,t-r}} \tag{12.18}$$

where all terms are defined as before. We provide an application of this simplified delay–difference structure to the Bering Sea case study in Figure 12.7. The fit of the multispecies delay–difference model was significantly improved relative to the simple production model for Pacific cod and the small-mouthed flounder complex; nearly identical for walleye pollock; and significantly worse for arrowtooth flounder where a very strong autoregressive effect was seen in the delay–difference model, resulting in a one-year offset in the model predictions (see also Uchiyama et al. 2016).

12.4.2.3 Functional group models

For situations involving many individual species, the number of parameters to be estimated can become formidable. Further, as we noted in Chapter 5, most multispecies models involving pairwise Interactions implicitly assume that predators and competitors operate independently within their respective assemblages and that their effects on focal species are additive. This may, of course, not be true but dealing with non-independence may require the introduction of many more parameters.

In addition, it may be difficult to specify the form of the necessary functions or to estimate the necessary parameters.

Here, we explore an approach that occupies the middle ground between the aggregate models for all species combined described in Section 12.3 and the more detailed models described in Section 12.4 on the other. Existing approaches have tended to define functional groups as taxonomically similar species or in terms of trophic guilds (e.g. Gaichas et al. 2012). For a reasonable number of functional groups, the estimation problem may be tractable and the issue of higher-order interactions can be subsumed within the aggregate biomass of the functional group. Bell et al (2014a) reported that interactions within trophic functional groups in general were stronger than between-group interactions.

In the Bering Sea example described above, the flatfish complex was essentially treated as a functional group. A broader application of the principle entails using models of the form specified in Equations 12.14 or 12.15 with functional groups replacing individual species in the analysis. The classic study of the Hawaiian archipelago deepwater fishery by Ralston and Polovina (1982) identified four functional groups based on a cluster analysis of fish catch composition. Ralston and Polovina then fit production models to the functional groups. The models for the functional groups provided a better fit to the data than production models for each individual species. This is in fact a common outcome in the use of aggregate production models and has been cited as one factor supporting their use. Ralston and Polovina found little evidence of interactions among functional groups and ultimately fit a single production model to all species combined. The lack of interaction was attributed to the fact that the functional groups each occupied distinct depth zones along the bathymetry gradient.

Bohaboy (2010) modeled 10 commercially and ecologically important fish species in the Georges Bank fish community classified into both taxonomically defined and guild-based functional groups. Bohaboy's (2010) guild-based functional groups comprised piscivores [spiny dogfish (*Squalus acanthias*) and winter skate (*Leucoraja ocellata*)]; shrimp-eaters [cod (*Gadus morhua*), silver hake (*Merluccius bilinearis*), little skate (*Leucoraja erinacea*)]; demersal benthivores [haddock (*Melanogrammus aeglefinus*), yellowtail flounder (*Limanda ferruginea*), and

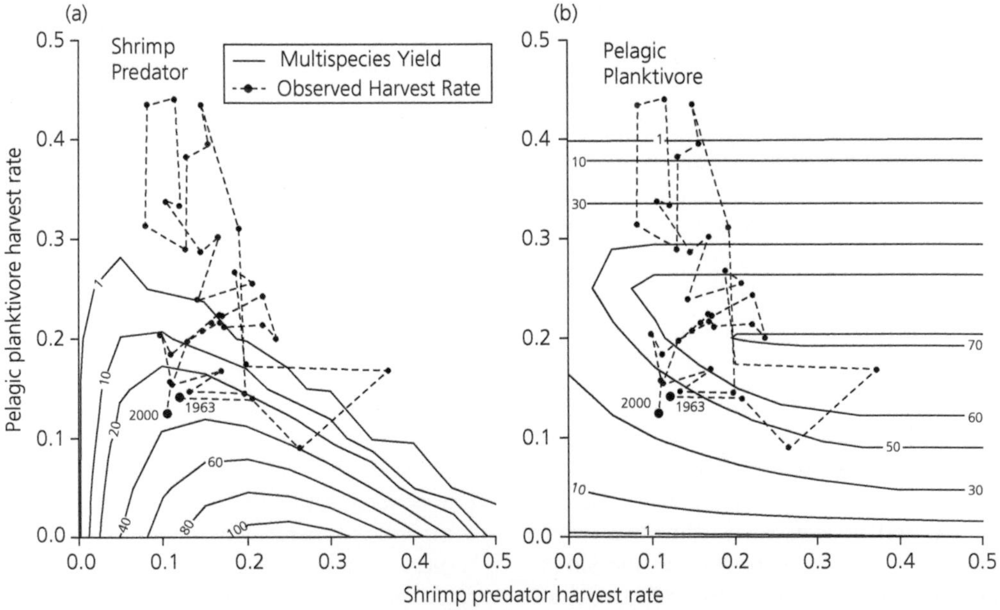

Figure 12.8 Contours of equilibrium yield (kt) of (a) shrimp predators and (b) pelagic planktivores for pair-wise combinations of harvest rates for each group with observed harvest rates (circles). Harvest rates were held constant at ½ of the single-species F_{MSY} for piscivores and benthivores. (Courtesy of Erin Bohaboy).

winter flounder (*Pseudopleuronectes americanus*)]; and pelagic planktivores [herring (*Clupea harengus*) and mackerel (*Scomber scombrus*)]. The shrimp-eater group fed extensively on macrozooplankton but larger cod and silver hake are also highly piscivorous and preyed on the planktivore group. The multispecies functional group model fit significantly better than models without interactions (Bohaboy 2010).

As a result of these interactions, the equilibrium yield of a given trophic group may depend on the harvest rate of other trophic groups. In this example, the yields of shrimp predators and pelagic planktivores are linked by their predator–prey interaction. The shrimp predator yield declines with increasing harvest of its prey; as the planktivore harvest rate increases, the shrimp predator yield would be maximized at a lower harvest rate (Figure 12.8). Conversely, the planktivore yield increases with the harvest rate of its predator. For intermediate levels of the shrimp predator harvest rate, planktivore yield is maximized at lower harvest rates because the fishery no longer competes with the shrimp predator group. At higher levels of shrimp predator harvest rate, this group becomes depleted and no longer affects the planktivore harvest (see Figure 12.8).

Multispecies models often imply higher harvest rates than their single-species equivalents. In this example, predator yield in the absence of planktivore harvest, or conversely, prey yield in the absence of predator harvest, is maximized at higher harvest rates in the multispecies model than the corresponding harvest rates with no interactions. However, lower combinations of harvest rates would be required with fisheries on both species groups. The observed harvest rates suggest that the pelagic planktivores were historically overharvested.

12.5 Complex dynamics

In earlier chapters, we showed that complex dynamics can emerge even in single-species difference equation models of exploited species. In contrast, continuous-time models do not, except in certain specialized circumstances, exhibit these types of complex dynamics in low dimensional models of less than three species. In Chapter 6, we demonstrated that simple food-web models with as few as three species can exhibit chaotic dynamics. For example, models of intraguild predation with very simple interaction terms can generate very complex population behaviors (e.g. Tanabe and Namba 2005). We have noted that intraguild predation appears to be pervasive in aquatic food webs (Irigoien and Roos 2011). Because of broad ontogenetic shifts in the diets of most fish species, analysis of stage-structured multispecies models often entails consideration of patterns of omnivory over the life cycle. Intraguild predation is a particular form of omnivory in which top predators also prey extensively on basal resource species. To explore how harvesting might affect the expression of complex dynamics in IGP systems, we revisit the Tanabe–Namba model introduced in Chapter 6 (see Equation 6.6) and consider the effects of harvesting of the top predator. Expressed in terms of biomass, the model is now:

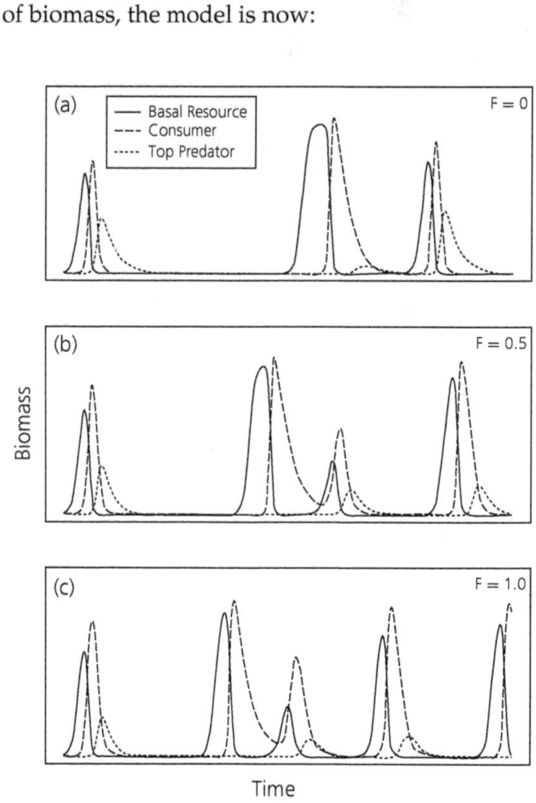

Figure 12.9 Effects of exploitation at three levels ($qE = F = 0$, 0.5, and 1.0) on a three-species intraguild predation model with a basal resource species, an intermediate consumer, and a top predator, which feeds on the two lower trophic levels.

$$\frac{dB_1}{dt} = [\alpha_1 - \beta_1 B_1 - a_{12}B_2 - a_{13}B_3]B_1$$
$$\frac{dB_2}{dt} = [a_{21}B_1 - d_2 - a_{23}B_3]B_2 \qquad (12.19)$$
$$\frac{dB_3}{dt} = \left[a_{31}B_1 + a_{32}B_2 - d_3 - qE\right]B_3$$

Numerical integration of this system of equations for three levels of exploitation reveals that population irruptions occur with increasing frequency with escalation in the harvest levels on the top predator (Figure 12.9). Release of predation pressure on the basal resource and the intermediate consumer in this three-species system affects a critical top-down control mechanism is this system. If only the top predator is deemed desirable for the market, as it is depleted by harvesting and the lower trophic level species become increasingly dominant, the economic returns will steadily diminish.

12.6 Harvesting in random environments

As we saw in Chapter 11, consideration of stochastic variation can alter our expectations for population trajectories and yield in time in important ways. With increasing environmental variability, the chance of quasi-extinction increases as populations are driven to low levels by chance events. Outcomes are defined by probability distributions rather than point predictions. These lessons hold for multispecies systems with the added consideration that chance events affecting the individual species can reverberate throughout the community through direct and indirect pathways. The nature of stochastic change in multispecies systems is further complicated by the possibility of environmentally driven regime shifts. We have seen that nonlinear functional feeding responses can lead to the possibility of alternate stable states in multispecies assemblages. Here we will further explore the role of stochasticity in effecting sustained shifts in population levels with a particular focus on the implications of autocorrelated random variability.

Steele and Henderson (1984) showed how multidecadal shifts in abundance can emerge as a result of environmental perturbations. Collie and Spencer (1994) expanded on this treatment for a two-species predator–prey system. The predator–prey model is a pair of coupled, first-order differential equations describing the rate of change of prey (B_1) and predator (B_2) biomass:

$$\frac{dB_1}{dt} = (\alpha - \beta B_1)B_1 - \frac{cB_2B_1^2}{D^2 + B_1^2} - q_1E_1B_1$$

and (12.20)

$$\frac{dB_2}{dt} = B_2\left(\frac{gB_1^2}{D^2 + B_1^2} - MB_2\right) - q_2E_2B_2$$

The prey equation is based on the logistic model with intrinsic growth rate, α, and density-dependent parameter, β. The predator is assumed to have a Type-III functional response with maximum consumption rate c and half-saturation constant D (see Equation 3.25). Predator production depends on prey abundance with growth rate, g, and a quadratic closure (mortality) term, M. Collie and Spencer explored the case of exploitation of the prey only, although the predators could be harvested as well without changing the main lesson of this analysis. The time dynamics of the predator–prey system were simulated by numerically integrating Equation 12.20 with stochastic random variability introduced into the predator mortality rate, such that: $M_t = M_0 + \nu_t$ and $\nu_t = \rho\nu_{t-1} + \varepsilon_t$. Parameter ρ is the first-order autocorrelation coefficient and ε is a standard normal random deviate (Steele and Henderson 1984).

In the absence of environmental variation, stable equilibria were attained for both species with the parameters chosen (Figure 12.10a). When simulated with autocorrelated random variability, however, the predator–prey system flips between high and low equilibria on multidecadal time scales (Figure 12.10b). The prey fluctuates much more than the predator, as small changes in predator abundance cause booms and busts in the prey population. The two stable equilibria act as domains of attraction between which the populations alternate. Predator outbreaks cause the prey to shift to the low equilibrium level; low predator abundance allows the prey to increase to the upper stable equilibrium. In the deterministic case, once a population is driven to a lower

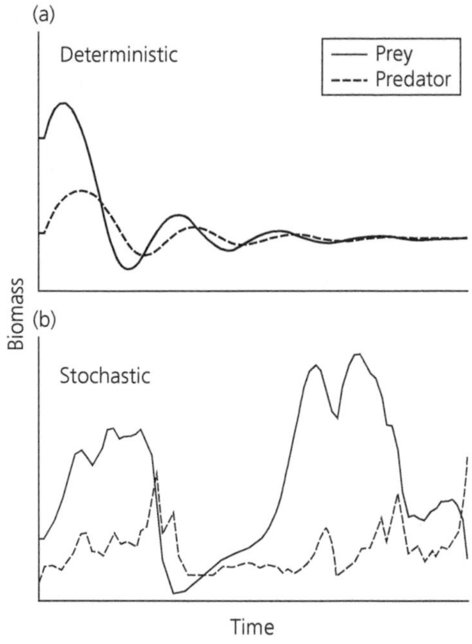

Figure 12.10 Population trajectories for a predator–prey system with a Type-III functional feeding response for the predator over a 100-year time period. For the deterministic case (a), both species coexist in a stable equilibrium for the baseline parameter set. For the model with autocorrelated stochastic forcing on the predator mortality term, the prey population alternates between two main levels on multidecadal timescales.

equilibrium point, it may be unable to escape this lower level without a subsidy from another population. In the stochastic case, it is possible that chance events can provide a means of escaping the lower domain of attraction. The interplay of autocorrelated environmental variability and harvesting pressure can give rise to dramatic changes in population biomass.

12.7 Size- and age-structured multispecies models

Multispecies analogs of size- and age-structured models have been developed for many fishery ecosystems. Like their single-species counterparts, these models can be partitioned into two principal categories. The first comprises models designed to provide an historical reconstruction of population sizes and demographic characteristics of an assemblage of interacting species using variants on the theme of catch-at-age or size models. These estimation models include terms to partition natural mortality into two components, "other" natural mortality and predation mortality, typically designated as *M*1 and *M*2 respectively. In the following, we will refer to these as assessment models. We will also consider age- or size-structured process models which entail specification of the entire life cycle of species in the designated communities and the nature of their interactions.

Some of the overall data requirements for age- or size-structured models of both types are depicted in Figure 12.11 along with those required for multispecies production and delay–difference models for comparison. The information needed to apply these approaches involves many additional data types and sources relative to the simpler multispecies models. By themselves, multispecies assessment models based on catch or size-at-age analysis do not provide direct information with which to designate multispecies management reference points or to provide projections of future population levels and yields. Nor do they explicitly consider competitive interactions among species within the community.

The second category of multispecies models does include feedback through recruitment processes and most often incorporates externally derived model parameters. These process models also provide modules for individual growth and share with multispecies assessment models a detailed treatment of predation processes. They are often used in a simulation context and can be used to create virtual worlds in which the efficacy of alternative options for management are systematically tested in a Management Strategy Evaluation (see Chapter 15).

The genesis of both the multispecies estimation and process models can be traced to the influential work of Andersen and Ursin (1977). The original Andersen–Ursin model was a complex full ecosystem simulation model with externally derived model parameters. The Andersen–Ursin model provided the nucleus for the predation component of the estimation and process models described below. We accordingly address this common element first.

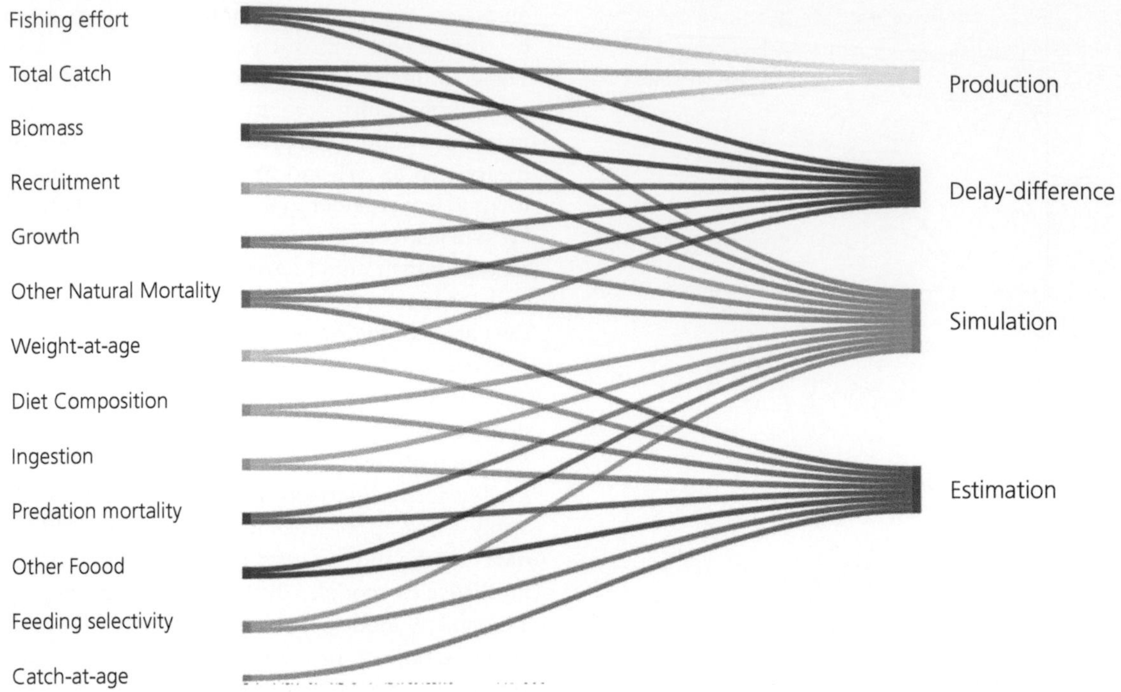

Figure 12.11 Principal data inputs for multispecies production, delay–difference, simulation, and estimation models.

12.7.1 Predation module

We begin by partitioning the natural mortality rate (M) into two components: sources of natural mortality due to all factors other than predation ($M1$) and predation mortality ($M2$), such that $M = M1 + M2$. The other mortality component includes starvation, disease, environmental conditions, and predation mortality by species other than those included in the model. In order to specify the predation mortality rates further, we decompose the predation process into size selectivity, species preference, and the consumption rate. The amount (weight) of available food for a predator in group j from prey category i in a given year is:

$$\varphi_{ij} = \psi_{ij}\bar{N}_i w_i \tag{12.21}$$

where ψ_{ij} is an index of prey suitability (ranging from 0 to 1), $\bar{N}_i$ is the mean number of prey individuals in category i, and w_i is their mean weight. Summing over all m prey categories, the total amount of available prey for a predator in category j is:

$$\varphi_{\cdot j} = \sum_{i=1}^{m} \varphi_{ij} \tag{12.22}$$

where the dot notation indicates summation over that index. We can now express the predation mortality rate as:

$$M2_{ij} = \frac{N_j I_j}{N_i w_i} \times \frac{\varphi_{ij}}{\varphi_{\cdot j}} \tag{12.23}$$

where I_j is the rate of food ingestion for a predator of group j. Note that $\varphi_{ij}/\varphi_{\cdot j}$ occurs as a ratio such that φ_{ij} could be multiplied by an arbitrary constant. It is common to normalize ψ_{ij} to sum to one for a given predator species and size/age (Sparre 1991). It can be seen that this definition of predation mortality implies a Type-II functional response. This equation can be recast as:

$$M2_{ij} = \frac{N_j I_j \psi_{ij}}{\varphi_{\cdot j}} \tag{12.24}$$

Note that the term $\varphi_{\cdot j}$ can include an additional term to account for prey not explicitly considered in the multispecies complex under analysis (other food). This expression simply indicates that the instantaneous predation mortality rate depends on the food intake per predator, the number of predators and the fraction of a particular prey category to the total available prey, and the biomass of the prey. The total predation rate exerted on prey category i (encompassing both species and size/age class) is obtained by summing over all n predator categories:

$$M2_{i.} = \sum_{j=1}^{n} M2_{ij} \tag{12.25}$$

where the subscript j indicates predator category (species and size/age class).

It is now necessary to specify the index of prey suitability, ψ_{ij} in greater detail. An extensive body of empirical information indicates that there is an important size selectivity component in the predation process. A lognormal function is often used to describe the relationship between predator and prey size classes:

$$\vartheta_{ij} = \exp\left(\frac{-\left[\log_e(w_j/w_i) - \eta_j\right]^2}{2\sigma_j{}^2} \right) \tag{12.26}$$

where η_j is the preferred predator/prey weight ratio on a logarithmic scale, and $\sigma_j{}^2$ is the variance in predator size preference (Ursin 1982). We now have $\psi_{ij} = \rho_{ij}\vartheta_{ij}$ where ρ_{ij} indicates the component of species preference or vulnerability that is independent of size.

12.8 Multispecies assessment models

Both age- and size-structured multispecies assessment models have been developed as direct extensions of corresponding single-species approaches. Helgason and Gislason (1979) and Pope (1979) independently derived the age-structured version of multispecies virtual population analysis by merging the Andersen–Ursin predation module with classical single-species sequential population analysis. Size-based multispecies assessment models were subsequently developed by Pope and Jiming (1987), which opened the possibility of application of the general technique to communities in which age-structured information is not routinely available. In the following, we will focus on age-structured assessments to illustrate the main features of the general approach.

12.8.1 Multispecies virtual population analysis

Here we outline the steps in a generic multispecies assessment analysis; the following sections provide more details on statistical parameter estimation. Population abundance is propagated forward in time with:

$$N_{i+1} = N_i e^{-(p_i F + M1_i + M2_i)} \tag{12.27}$$

If we know the consumption rates of the predators from one of the methods described in Chapter 8, and their prey preferences, we can calculate the suitability coefficients and hence $M2_i$. Fish are assumed to consume their prey in proportion to its availability in the environment and we can therefore write:

$$\frac{d_{ij}}{d_{.j}} = \frac{\varphi_{ij}}{\varphi_{.j}} \tag{12.28}$$

where d_{ij} is the fraction of prey type i in diet of predator j. This proportionality provides a statistical basis for the estimation of $M2$. Given an estimate of $M1_i$ (we often assume that other natural mortality is constant over ages) and N_i for the youngest age group, the fishing mortality rate is obtained by solving the catch equation:

$$C_i = \frac{p_i F}{p_i F + M1_i + M2_i} \left(1 - e^{-Z_i}\right) N_i \tag{12.29}$$

where C_i is the catch in numbers , Z_i is the sum of fishing mortality, predation mortality, and other sources of natural mortality for prey category i. We can now use Equation 12.29 to estimate N_{i+1} for each age class of each species. Because available food is calculated from the mean number of individuals over the year (Equation 12.21), which is not immediately known at the start of the year, it is necessary to iterate these steps before moving on to the next year and age. It is also necessary to iterate the calculations over years because $M2$ depends on the abundance of predators in older age classes. In this sequence, the fate of a cohort can be tracked from year to year, accounting for the losses to predation mortality and fisheries catch.

This approach and set of equations formed the basis of Multispecies Virtual Population Analysis (MSVPA, see review in Sparre 1991). As with single-species VPA, the equations were solved backward in time, reconstructing each cohort starting with the oldest age. MSVPA has been applied in a number of ecosystems, including the North Sea, Baltic Sea, eastern Bering Sea, Georges Bank, and the mid-Atlantic Bight. The time-varying estimates of predation mortality ($M2$) from MSVPA have been used as inputs to single-species sequential population analyses.

12.8.2 Multispecies statistical catch-at-age analysis

Just as statistical catch-at-age models have largely replaced cohort analysis for single-species stock assessments, MSVPA is giving way to multispecies statistical catch-at-age analysis (MSSCA). MSSCA uses the same equations to estimate prey suitability and predation rates as MSVPA (Section 12.7.1) and essentially the same input data. The input data include catch-at-age (C_a), diet composition, estimates of prey abundance other than species contained in the model, weight-at-age, ingestion rates (I), and residual natural mortality. Unlike MSVPA, which is a deterministic reconstruction of cohort abundance, statistical models allow for error in the input data and the estimation of parameters with maximum-likelihood methods. The set of parameters to be estimated includes annual fishing mortality rates and fishing selectivity by age, cohort abundance by age in the first year and at age one in each subsequent year (recruitment), survey catchability, and age selectivity and food-selection parameters.

Curti et al. (2013) used a three-species model to test the performance of a MSSCA. In this submodel of the Georges Bank fish community, Atlantic cod preys on silver hake and Atlantic herring; silver hake preys on herring; cod and silver hake are both cannibalistic. As in other studies, the multispecies model was compared with the equivalent single-species models without predation. Both models closely fit the commercial catch data. Fits to the survey data were also good, but not quite as tight, because of the interannual sampling variability inherent in trawl surveys.

Total abundance and fishing mortality were compared with the corresponding stock assessment of each species (Figure 12.12). The multispecies and single-species models gave almost identical results for cod because it is the top predator species. For the prey species, the multispecies model predicts high levels of abundance to account for the additional losses due to predation. Conversely, the estimates of fishing mortality are lower in the multispecies model, because more of the total mortality is attributed to predation and less to fishing.

Predation mortality of silver hake and herring was high and variable over time (Figure 12.12). Predation mortality is generally higher for the younger ages. An exception is herring, for which $M2$ at age 2 was equal to or exceeded $M2$ at age 1 because age-2 herring is closer to the preferred size of cod and silver hake. The temporal pattern in predation mortality can be explained by the interplay of predator and prey abundances (Figure 12.12). During the 1980s and early 1990s, predation mortality declined on all three species, reflecting declines in predator abundance, especially cod. Predation mortality on silver hake increased in the 1990s and 2000s as silver hake abundance leveled off and increased slightly. Herring abundance increased as its predation mortality was falling, which implies that it escaped predator control.

Through numerous Monte Carlo simulations, Curti et al. (2013) showed that the parameters of the multispecies model can be estimated with acceptable precision from known data with levels of error similar to those in this three-species example. Following this pilot study a full nine-species model was fit to the Georges Bank fish community. Multispecies statistical catch-at-age models have also been developed for the North Sea, the Baltic Sea, and the eastern Bering Sea. A spatially explicit version, GADGET, has been applied to the waters around Iceland and to the Barents Sea fish community.

The objective function has three components: one for the commercial catch, one for the research survey, and one for the fit to diet data. The first two are assumed to be lognormally distributed; a multinomial distribution was used for the diet data since they are proportions. In this likelihood function, the "hat" represents quantities estimated with the model and a small constant is typically added to prevent taking the log of zero:

$$LL = \sum_{t,sp,a}\left(\log_e C - \log_e \hat{C}\right)^2 + \sum_{t,sp,a}\left(\log_e N - \log_e \hat{N}\right)^2 + \sum_{t,sp,a} 50 \cdot d \cdot \log_e \hat{d} \tag{12.30}$$

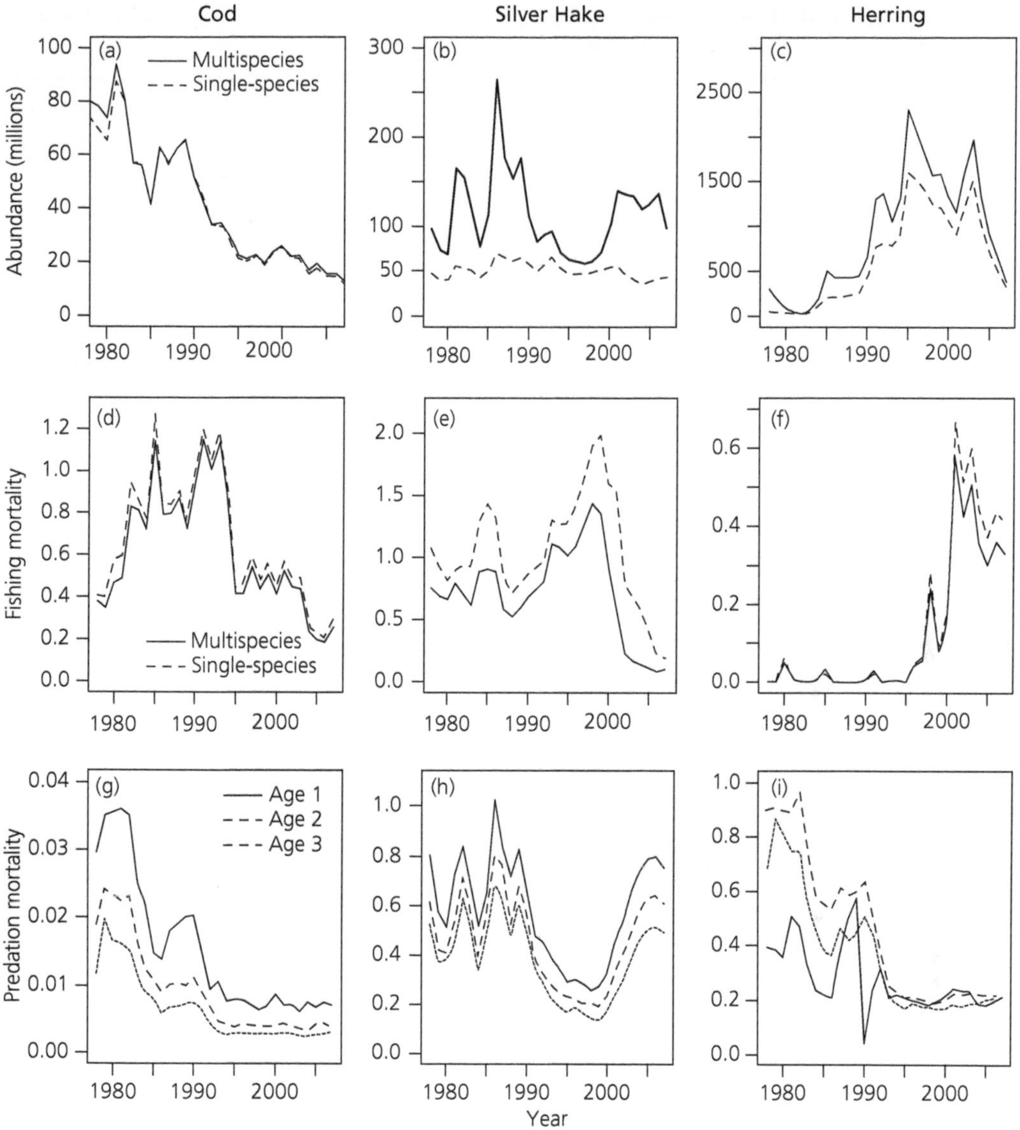

Figure 12.12 Estimates of (a–c) total abundance, (d–f) fully recruited fishing mortality, and (g–i) predation on age 1-3 fish for cod (*Gadus morhua*), silver hake (*Merluccius bilinearis*) and herring (*Clupea harengus*) on Georges Bank. Abundance and fishing mortality shown based on multispecies (solid lines) and single-species (dashed lines) analyses. (Data courtesy of Kiersten Curti).

Uncertainty in the parameter estimates can be quantified and propagated in model projections. The output variables include the suitability coefficients (ψ), predation mortality rates ($M2$), and fishing mortality (F) and stock numbers (N) by species and age. One advantage of age-structured multispecies models is that the output tables of N and F by age and year are in a format that is familiar to stock-assessment scientists.

12.9 Multispecies process models

A number of multispecies process models have been developed and applied to exploited aquatic ecosystems. Both age- and size-structured models have been constructed to examine multispecies reference points and as vehicles for simulation studies. Size-structured models offer a somewhat broader scope for application to exploited aquatic communities because age information may not be routinely available in many species. To illustrate some general principles and outcomes, we will focus on one implementation of a length-based model: LeMANS (length-based multispecies analysis by numerical simulation; for a full description see Hall et al. 2006). Gaichas et al. (2017) describe a structurally similar model for use in simulation studies.

In LeMANS, growth in length is governed by a deterministic von Bertalanffy equation (Equation 8.18) for each species. Length is discretized into size intervals of equal width. The time taken for each species i to grow through size interval j is obtained by solving Equation 8.18 for time:

$$t_{ij} = \frac{1}{k_i}\log_e\left[\frac{L_{\infty,i} - L_{lower}}{L_{\infty,i} - L_{upper}}\right] \qquad (12.31)$$

where k_i is the growth coefficient. The proportion of individuals leaving the size interval in a time interval, solely because of growth, is $1/t_{ij}$. Recruitment is specified with a Ricker model for each species (Equation 9.27) in which the number of eggs spawned is assumed proportional to the mature biomass. The density-independent parameter of the Ricker model, a_i is scaled to asymptotic length ($L_{\infty,i}$) while the density-dependent parameter, b_i, is scaled relative to the maximum observed spawning-stock biomass. Hall et al. (2006) developed predictive models for the parameters of the Ricker function to permit application in situations where direct stock and recruitment information is not available. In subsequent extensions to LeMANS, Thorpe et al. (2016, 2017) employ a linear segmented (hockey-stick) recruitment function.

Fishing mortality is modeled as a logistic function of length, residual natural mortality is specified with a beta function which permits a U-shaped pattern of natural mortality as a function of size, and predation mortality is calculated with the steps described in Section 12.7.1. The size-specific ration of each predator is calculated from the weight increment between size classes, divided by a size-specific growth efficiency. Predation mortality is calculated by Equation 12.24 with a mean predator-to-prey weight ratio of $\eta = 3.5$ and variance $\sigma^2 = 1$.

Implementation of LeMANS requires seven life-history parameters for each species, a matrix specifying which predators eat which prey species, and a vector indicating which species are fished. LeMANS has been parameterized for the Georges Bank and North Sea fish communities (Rochet et al. 2011).

Worm et al. (2009) examined multispecies harvesting strategies on Georges Bank using. LeMANS to illustrate the effect of maximizing aggregate catch on several community metrics (Figure 12.13). The exploitation rate that would maximize multispecies yield would reduce total biomass by ~60% relative to unfished levels, decrease the weighted mean size of fish in the community by ~30% and cause eight

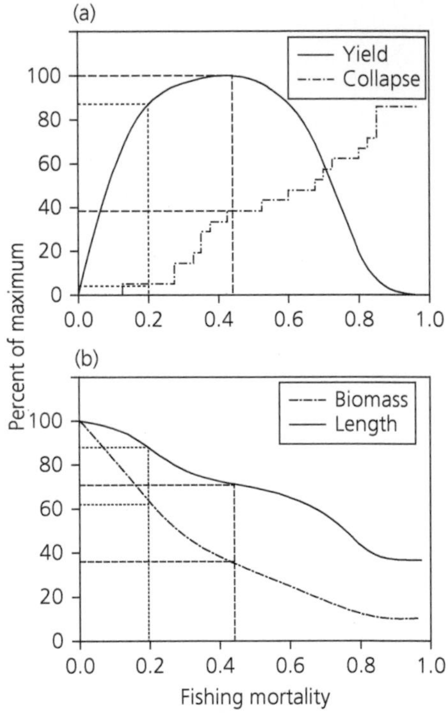

Figure 12.13 Effects of increasing exploitation rate on the Georges Bank fish community, as modeled with LeMANS for (a) yield and proportion of collapsed stocks and (b) equilibrium biomass and mean body length as a function of fishing mortality. Collapsed species are those for which stock biomass has declined to less than 10% of their unfished biomass. Dashed lines show position multispecies maximum sustainable yield. Dotted lines show the effect of adopting a more conservative fishing policy of F = 0.2. Adapted from Worm et al. (2009).

species to collapse. Reducing the exploitation rate to 0.2, would produce ~90% of the maximum catch while reducing total biomass by approximately 40%, mean length by ~10%, and resulting in the collapse of one species.

12.10 Multispecies biological reference points

The specification of biological reference points, and control rules based on them, has been instrumental in halting overfishing and rebuilding overfished stocks (Worm et al. 2009). The recognition that biological reference points may change due to climate or multispecies interactions can challenge the authority of fishery management agencies. Fishermen wonder why they should try hard to meet a target or avoid an overfishing threshold that is likely to change anyway. Viewed in a multispecies context, biological reference points for individual species depend on the abundance of interacting species. This dependence is stronger for prey species subject to variable predation rates than it is for predator species whose growth rates are buffered by the availability of alternative prey (Collie and Gislason 2001). There is particular concern about harvest levels on forage species, defined as small pelagic species, which are important in the diets of piscivorous fish, seabirds, and marine mammals (Smith et al. 2011).

There are two contrasting approaches for accommodating multispecies considerations into single-species reference points and control rules. Risk-averse reference points can be adopted for fish communities for which there is limited understanding or data on species interactions. This approach was taken by the Commission for the Conservation of Antarctic Marine Living Resources in the Southern Ocean, where quantitative estimates of species interactions remain uncertain, despite a good qualitative understanding of the food web. But risk-averse harvest control rules require foregoing potential fish yields, which in the case of small pelagic species could represent important protein sources. In fish communities with quantitative estimates of species interactions, the biological reference points for single species can be adjusted in response to changes in the abundance of interacting species. Reference points based on the concept of MSY lend themselves to this type of adjustment because they would compensate for changes in productivity (Collie and Gislason 2001). By contrast, proxy reference points, such as those based on "per-recruit" calculations, do not lend themselves to formulaic adjustments in response to changes in predation mortality.

Biological reference points are useful for evaluating the current status of a fishery relative to long-term potential yields. They can tell us which direction to go, but are seldom useful as targets. In a multispecies context we need to consider trade-offs involved in harvesting interacting species, as shown in Section 12.9. Although it is possible to calculate the maximum multispecies yield from a fish community (MMSY), it is necessary to consider the differential impacts of harvesting strategies on individual species. Even if we convert total yield to total revenue, the costs and benefits of harvesting predator and prey species accrue to different fleets, and in the case of transboundary fisheries, to different nation states, which may have different social, economic, and political priorities.

Managing a multispecies fishery involves identifying combinations of fishing mortality rates that maintain each species within sustainable levels. Rather than a single reference point for each species, we can envision a range of sustainable fishing mortality rates as a multidimensional volume. To illustrate this volume in three dimensions, we return to the cod–herring sprat example in the Baltic Sea. The equilibrium yield of each species was calculated for each factorial combination of fishing mortality rates (Figure 12.14). When viewing yield isopleths for three species in two dimensions, it is necessary to freeze the abundance of the third species. We fixed the fishing mortality rate of the third species at its highest and lowest value to create the greatest contrast. In this example, all combinations of F are expected to result in sustainable yields. Cod yield (not shown) is fairly insensitive to herring and sprat F because cod growth is not prey limited in this model. Conversely, herring and sprat yields both increase with increasing cod F—with less prey consumed by cod, more is available for the fishery. This effect is more pronounced for sprat, which has higher predation mortality rates. There is also an indirect interaction between herring and sprat because they both occur in the cod functional

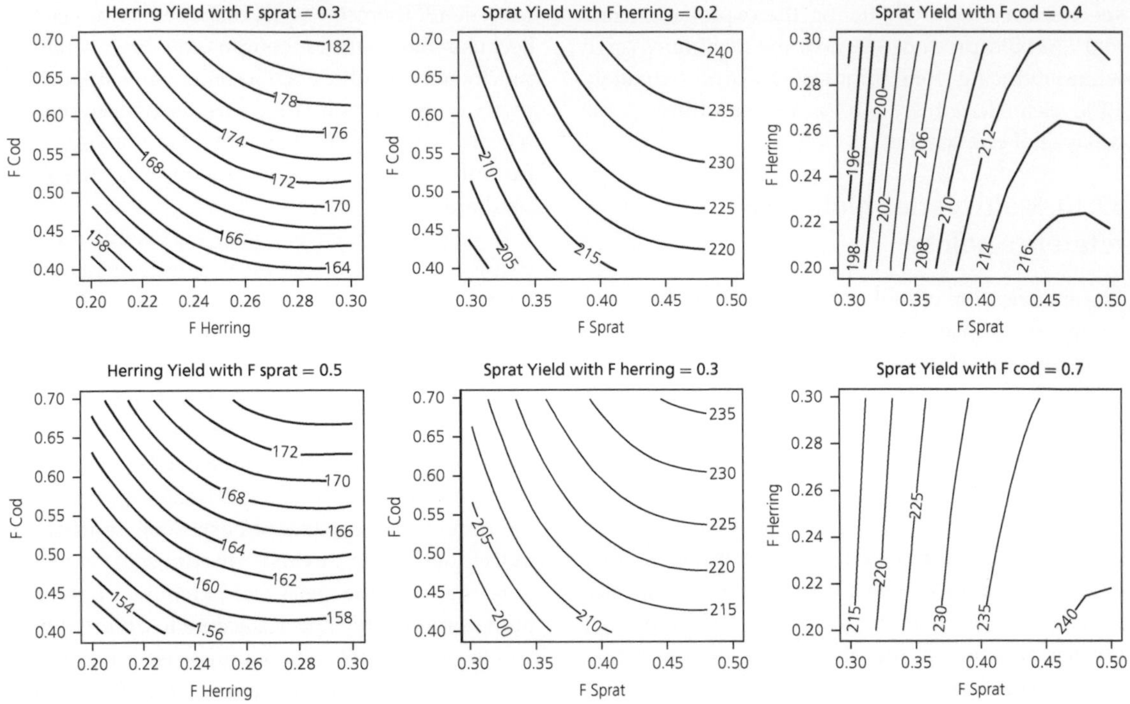

Figure 12.14 Equilibrium yields (kt) of herring (*Clupea harengus*) and sprat (*Sprattus sprattus*) in the Baltic Sea for combinations of fishing mortality (F) on cod (*Gadus morhua*), herring, and sprat. Yields were calculated with the Stochastic Multispecies (SMS) Model (Lewy and Vinther 2004). Model results were provided by Morten Vinther, DTU.

response (Eq. 12.24). Herring yields are higher at low sprat *F* because predation mortality on herring is lower when there is more sprat. This indirect interaction between herring and sprat is more pronounced at low cod *F*, when predation mortality is higher. Sprat yield would be maximized with a high *F* on its predator cod and a low *F* on its competitor herring. The combination of fishing mortality rates that is chosen will depend on the relative abundance of interacting species and the priority placed on the yield of one species over others. The chosen combination will seldom be maximal or even optimal. Instead of the elusive concept of multispecies MSY, it would be more pragmatic to drop *Maximum* and rebrand MSY as *Multispecies* Sustainable Yield (Collie et al. 2014).

12.11 Summary

Harvesting of multispecies communities is one of the first components of an ecosystem approach to be implemented by fishery management agencies. The mixed-species problem remains one of the most pressing challenges. Technical solutions can help to avoid some species while targeting others, but a comprehensive solution requires creating the right economic incentives. Accounting for trophic interactions means that biological reference points depend on the abundance of other taxa. Most single-species reference points are difficult or impossible to use in a multispecies context. Multispecies maximum sustainable yield should not be considered as a target; instead we should identify combinations of fishing mortality rate that result in multispecies sustainable yield. Socioeconomic and nature conservation considerations are necessary to resolve tradeoffs among various users.

Multispecies production models have played a central role in understanding biological interactions. Sustainable harvest of prey species is contingent on predator abundance. Continuous-time models are useful for examining the equilibrium and stability properties; discrete-time models can be fit statistically to time series of catch and biomass data. Parameter estimation is also facilitated by aggregating species into functional groups. Predator and prey species need to be differentiated in separate groups, but competitor species can be aggregated, provided that they share similar life-history characteristics. The introduction of simple non-linear terms into production models can result in multiple equilibria and regime shifts, which qualitatively resemble the abrupt shifts observed in some fish stocks. Production models may be the method of choice for fish communities for which routine size and age measurement are not collected.

Where they can be supported with data, size- and age-structured models are needed to account for size-dependent processes, especially predation and fishing. Predation mortality is high and varies over time; the dynamics of each species can be explained by the interplay of fishing and predation mortality. To date, size-based models have mostly been used as simulation tools to investigate the consequences of harvesting strategies on the fish community. By contrast, age-structured multi-species models are being fit to time series of catch, abundance, and diet data with statistical methods in a growing number of ecosystems. Statistical parameter estimation is necessary for multispecies models that are to be used for tactical decision-making. The output of these models can be readily used to identify sustainable combinations of fishing mortality rates for the species in the community.

Additional reading

In their classic text, Hilborn and Walters (1992 Chapters 3 and 14) provide an introduction to multispecies analysis. Although their appraisal at the time was somewhat pessimistic, the need to go beyond single-species approaches was made clear. An important early treatment of multispecies modeling for fishery ecosystems can be found in the contributions to Mercer (1982). The multiauthored symposium volume of Daan and Sissenwine (1991) contains a number of examples of application of community-level modeling in support of fisheries management. Jennings et al. (2001 Chapter 8) offer a concise overview of multispecies analysis, covering many of the topics found this chapter. The 16th Lowell Wakefield Fisheries Symposium volume (Alaska Sea Grant College Program 1999) has descriptions of a number of applied community-level analyses in support of fisheries management.

CHAPTER 13

Harvesting at the Ecosystem Level

13.1 Introduction

We have now progressed through a sequence of harvesting strategies and related considerations at increasingly higher levels of ecological organization. In this chapter we focus on the broader dimensions of harvesting aquatic ecosystems. We describe ways of estimating potential yield at the ecosystem level and the analytical tools available for analyzing fishery ecosystems. Consideration of both the direct and indirect effects of exploitation is paramount in evaluating the ecosystem effects of exploitation. Accordingly, we will consider not only the effects of fishing on species directly targeted by the fishery, but impacts on other dimensions of the ecosystem including habitat and protected resources subject to incidental catches. We focus in particular on searching for tractable solutions to coping with ecological complexity. Models selected for analysis will necessarily depend on the objectives chosen by managers. Whole ecosystem models can be invaluable in helping to frame and evaluate broad strategic objectives. Other models focused on selected parts of the ecosystem or species of concern requiring management may be best suited for addressing tactical management concerns.

A very broad spectrum of models is now available for analysis of exploited aquatic ecosystems. Here, we can take up only a small subset of these and we refer the reader to earlier reviews by Hollowed et al. (2000) and Whipple et al. (2000). Plaganyi (2007) provides a very insightful review of the properties, strengths, and weaknesses of different classes of ecosystem models used in a fisheries context.

In Chapters 8 and 10, we suggested that energy is the logical currency for considering both potential harvest and the natural constraints that bound the harvesting problem. In this chapter we return to this theme and discuss harvesting and potential ecosystem yield from this perspective. The energy entering at the base of the food web sets the stage for production throughout the food web, although there are many environmental, ecological, and anthropogenic forcing factors that shape the final levels of production in any given ecosystem. A very extensive literature has now developed on application of mass-balance ecosystem models and their extension to dynamic models through the widely applied Ecopath with Ecosim (EwE) framework. Although other complementary approaches are available, we will focus on EwE because of its ubiquity, ready availability (https://ecopath.org) and wide user community. A recently developed R package (Rpath) is also now available Lucey et al. (2020). We then approach the issue of harvesting at the ecosystem level through the lens of biomass spectrum theory in which size categories rather than taxonomic designation are the state variables.

With respect to the broader dimensions of the harvesting problem, estimating the effects of harvesting on habitat and non-target species using more narrowly focused models is then taken up. Our call in Chapter 11 to return to the basics of the fishing process was motivated in part by exactly these considerations. Impacts on habitat and non-target organisms are related to the magnitude and spatial distribution of fishing effort and are not directly captured in concepts such as fishing mortality rates on target species. In earlier chapters we saw that populations and communities can exhibit alternative stable states. Here we extend these considerations to encompass alternative ecosystem states.

Fishery Ecosystem Dynamics. Michael J. Fogarty and Jeremy S. Collie, Oxford University Press (2020). © Michael J. Fogarty & Jeremy S. Collie 2020.
DOI: 10.1093/oso/9780198768937.003.0013

13.2 Fishery ecosystem production

The simplest and perhaps most broadly employed methods to estimate the potential production of fishery ecosystems focus on the relationship between chlorophyll concentration or primary production and observed catch levels. This approach has been applied in both freshwater and marine ecosystems. Most often, these methods pair average levels of chlorophyll *a* concentration or primary production with average catches over specified time periods for individual water bodies. In effect, these observations are treated as spatial replicates in attempting to discern the relationship between measures of chlorophyll or primary production and fishery yield. Deines et al. (2015) provided a recent global synthesis of observations of this type for freshwater ecosystems and demonstrated a linear relationship between fishery yield and chlorophyll *a* concentration. An extensive literature demonstrating relationships between chlorophyll or primary production and yields has now accrued for marine fishery ecosystems (see recent compilations and reviews by Conti and Scardi 2010, Chassot et al. 2010, Mcowen et al. 2015, Friedland et al. 2012).

The methods described above depend on the assumption that fishery production depends in large measure on "bottom-up" forcing resulting in direct relationships between incoming energy at the base of the food web and fish yield (Section 10.3). Chassot et al. (2010) concluded that fisheries yield was constrained by primary production in large marine ecosystems around the world. In a comparison of the relative strength of bottom-up and top-down forcing in 47 large marine ecosystems, Mcowen et al. (2015), however, classified 20 as dominated by bottom-up forcing while 16 also reflected important top-down controls; 11 systems were classified as indeterminate. Bottom-up forcing was predominant in systems characterized by higher productivity and fishing pressure. In contrast, top-down forcing was associated with lower productivity and fishing pressure. A strict dichotomy between bottom-up and top-down forcing is of course overly simplistic (Mcowen et al. 2015) and most systems experience a blend of these types of controls. Nonetheless, this attempt to partition system types by dominant drivers of fishery yield provides important insights into expected yields, which can provide first-order estimates for planning and management.

Linear models employing physico-chemical measurements as proxies for primary production have also been used to provide predictions of expected yield in aquatic ecosystems. The morphoedaphic index (MEI) developed for freshwater ecosystems (Ryder 1965, 1982) is an early and widely known example. Ryder's (1965) classic study examined observed fish yields in 23 high-latitude North American lakes in relation to a suite of sixteen variables. The final model expressed yield as a function of the ratio of total dissolved solid concentration and water depth (calculated as total volume divided by surface area). Subsequent analyses treated these elements individually in a multiple regression framework and included factors such as temperature, ion concentration, and other variables (see Pitcher 2016).

13.2.1 Simple food chain models

The development of predictive models in which a fishery food web is collapsed to a linear food chain has an extensive history. The most well-known early example of this general approach was developed by Ryther (1969)[1]. The basic model can be written:

$$Y = u\left(PP \cdot TE^{TL-1}\right) \tag{13.1}$$

where Y is the expected yield in weight, u is the fractional exploitation rate, PP is primary production expressed in units of biomass, TE is the transfer efficiency for the flow of energy between successive trophic levels, and TL is the mean trophic level at which catch is extracted. Primary producers are classified as trophic level 1 by convention. The expression in parentheses is the fish production at a specified trophic level. The transfer efficiency ranges from 0 to 1 but is typically relatively low (the canonical value often used in many early studies was 0.1 or a 10% transfer efficiency).

Ryther (1969) estimated the potential world fish catch to be on the order of 100 million mt. Ryther's work was the first to apply a partitioning of fishery

[1] In fact, Ryther's (1969) analysis was just one of a number of similar approaches (e.g. Graham and Edwards 1962; Schaefer 1965; Ricker 1969; Gulland 1970), some of which used both food-chain models and different extrapolation approaches to estimate potential global fish production and yield.

production among different oceanic domains including coastal, offshore, upwelling, and open ocean systems. Ryther further applied different estimates of food-chain length in these different domains to reflect fundamental differences in ecosystem structure and patterns of energy flow. At the time of Ryther's projection, the global marine fish catch was on the order of 55 million mt. The annual landings from marine capture fisheries are now very close to Ryther's estimate of potential yield although Pauly (1996) suggested that Ryther's estimate might have been "right" for the wrong reasons as a result of countervailing errors of some parameters.

Efforts to link primary production and fish catches at global scales are complicated by differences in energy pathways at the base of the food web (Carr et al. 2006). Note that in Equation 13.1, a single functional group of primary producers is assumed. Ware (2000) generalized the Ryther model to allow consideration of the contribution of nano-, pico- and net phytoplankton (principally diatoms) to total annual primary production. The picoplankton component is an important element of the microbial loop (recall Section 10.3) and energy from this component must pass through two or more steps to reach the mesozooplankton at the nexus of many aquatic food webs. Ware's model is:

$$Y = u\left(M_l PP \cdot TE^{TL-1}\right) \tag{13.2}$$

where M_l is the fraction of total annual primary production plus the fraction of heterotrophic microzooplankton production derived from the microbial food loop and potentially available to mesozooplankton (see Ware 2000 for details). Ware further allowed for specification of the fraction of phytoplankton production retained within a given ecosystem (R_f); the product $R_f M_l$ is the mesozooplankton ecotrophic efficiency (the fraction of total primary production available to the mesozooplankton in the system). In an application to the North Sea, Ware (2000) estimated the mesozooplankton ecotrophic efficiency to be on the order of 0.4 assuming a 95% retention rate. The pathway through the microbial food web therefore results in a substantial diminution of the total primary production ultimately available to higher trophic levels.

Stock et al. (2017) modified the Ryther approach in two important ways: (1) by sequentially altering the baseline model relating yield to net primary production to consider additional lower trophic level and environmental inputs; and (2) by explicitly including observed catches in the analysis and solving for parameters considered to be less well known. Stock et al. assembled information for large marine ecosystems around the globe on primary production, mesozooplankton production, temperature, particle export to the benthos, mean trophic level of the aggregate catch, and catch (Figure 13.1). See Stock et al. (2017) for further information on data sources and processing.

Stock et al. used the Ryther model (Equation 13.1) as a baseline (their Model 1). In this case, given information on primary production, catch, and the equivalent trophic level of the yield (TL_{eq}), the exploitation rate (u) and transfer efficiency (TE) were estimated as free parameters. The Ryther and Ware models described above do not use observed yield directly in developing estimates of expected yield but rather trace the flow of energy through successive steps in the food chain to determine the amount of production reaching the mean trophic level of the catch. An assumed exploitation rate is then applied to determine expected yield.

Stock et al. next replaced the net primary production term with the detrital flux to the benthos (see Friedland et al. 2012) and mesozooplankton production to give their Model 2:

$$Y = u\left(f_{det} \cdot TE^{TL_{eq}-1} + P_{mz} \cdot TE^{TL_{eq}-2.1}\right) \tag{13.3}$$

where f_{det} is detrital flux, P_{mz} is mesozooplankton production and all other terms are defined as before. The mesozooplankton trophic level was set at 2.1 and again estimates of the exploitation rate and the transfer efficiency were made. Model 3 of Stock et al. involved a temperature adjustment to the transfer efficiency term to Model 2 for warm-water ecosystems: $TE_{warm} = f_T TE$ if $\underline{T_{100}} < \underline{T_{100,\,warm}}$ where $\underline{T_{100}}$ is the average temperature at 100m depth in each LME. The adjustment was intended to account for higher metabolic demands at higher temperatures, resulting in reduced transfer efficiencies. The dimensionless factor f_T and the threshold temperature $\underline{T_{100,\,warm}}$ are estimated in the model, now giving 4 estimated

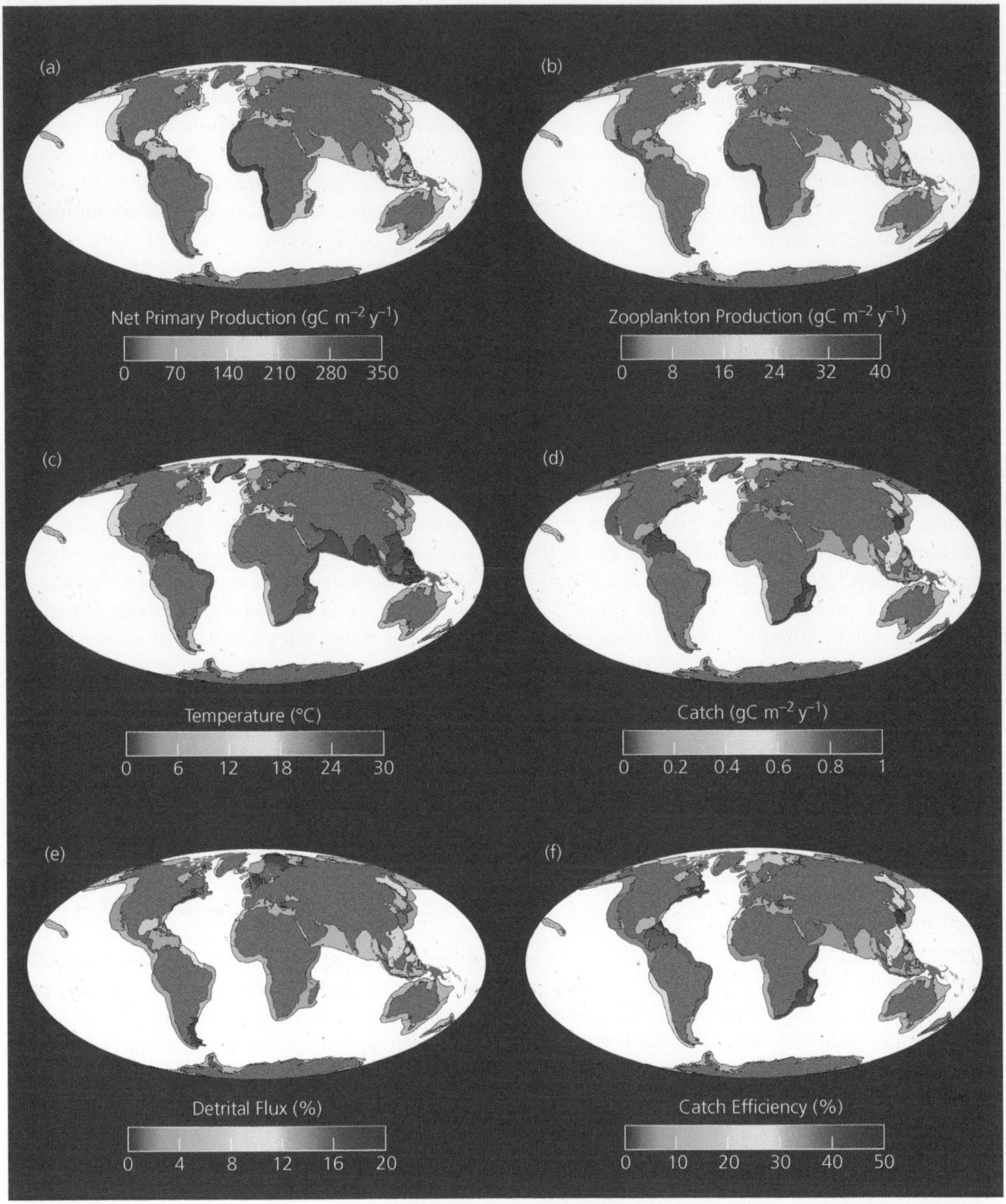

Figure 13.1 (a) Net primary production (gC m^{-2} y^{-1}), (b) mesozooplankton production (gC m^{-2} y^{-1}), (c) temperature, (°C), (d) average catch (landings and discards gC m^{-2} y^{-1}), (e) detrital flux (percent of net primary production, and (f) catch efficiency (ratio of catch to net primary production) by LME. (Data courtesy of Charles Stock, NOAA; map courtesy of Kim Hyde, NOAA.)

parameters. Model 4 of Stock et al. builds on Model 3 but estimated separate transfer efficiencies for the benthic and pelagic components of the model under the hypothesis that energetic costs of foraging were higher in the three-dimensional pelagic realm relative to the two-dimensional world of the benthos.

Stock et al. (2017) showed that by accounting for the fate and pathways of production at lower trophic levels in general and temperature effects on transfer efficiencies, a significant improvement in the correspondence of observed and predicted catch predicted was possible relative to the baseline model incorporating only NPP as a driver. The correlations between the observed and predicted catch for Models 1–4 were 0.47, 0.64, 0.75, and 0.79 respectively. The change in the Akaike Information Criterion (ΔAIC) scores between successive models indicated that the improvements in model fit after adjusting for the number of parameters estimated were statistically significant. In Figure 13.2, we show the observed and predicted catch for Model 4. The highest yields are for colder-water ecosystems with higher f_{det}/P_{mz} ratios. Many of the temperate-boreal shallow bank ecosystems historically supporting highly productive fisheries fall into this category. In contrast, warmer water ecosystems with lower f_{det}/P_{mz} ratios are associated with lower productivity fisheries (Stock et al. 2017). Examination of Figure 13.2 shows that the best fitting model (4) still somewhat underestimated the positive effect of colder temperature and higher benthic-pelagic ratios.

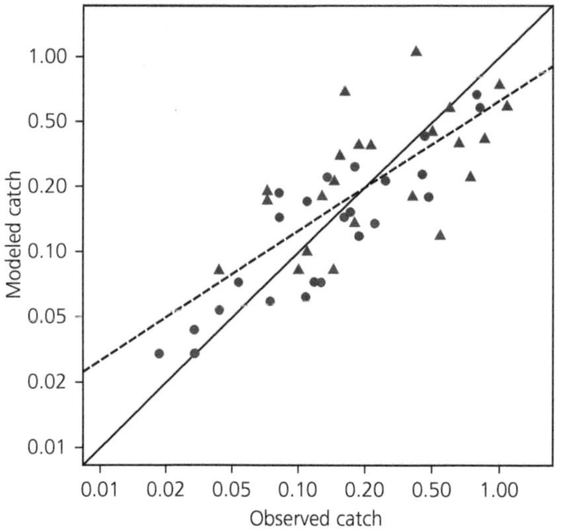

Figure 13.2 Relationship between (a) observed catch (gC $m^{-2}yr^{-2}$) and predicted catches (dashed line) for Model 4 of Stock et al. (2017). Each symbol represents a different LME. Observations categorized into four compartments: (1) $T < 20°C$ and $f_{det}/P_{mz} > 1$ (blue diamonds); (2) $T < 20°C$ and $f_{det}/P_{mz} < 1$ (blue circles); (3); $T > 20°C$ and $f_{det}/P_{mz} < 1$ (red triangles); and (4) $T > 20°C$ and $f_{det}/P_{mz} < 1$ (red circles). (Data courtesy of Charles Stock.)

13.3 Network models for exploited ecosystems

In this section, we expand the energy budget concepts introduced in Chapter 8 for individual organisms and extended to the ecosystem level in Chapter 10. We introduced many of the fundamental considerations for system-level energy budgets in Section 10.3. Here we build on that foundation to construct mass-balance models of exploited aquatic ecosystems.

13.3.1 Mass-balance models

In a seminal contribution, Polovina (1984) described a formal analytical approach to balancing an energy budget for French Frigate Shoals in the Northwest Hawaiian Islands. The network diagram for this system is shown in Figure 13.3. Previous approaches relied principally on ad hoc methods for balancing energy budgets. Christensen and Pauly (1992) subsequently expanded the approach and developed a widely used software package to facilitate the application of the method which has since grown beyond the static mass-balance approach to encompass dynamic ecosystem analysis (Ecosim; see Section 13.3.2), spatially explicit models (Ecospace), and other modifications. A catalog of over 400 Ecopath with Ecosim models now exists in the EcoBase repository (see http://ecobase.ecopath.org).

In a balanced system, consumption must meet the requirements for production, respiration, and egestion at the system level. The energy balance model for each compartment within the system can be expressed:

$$Q_i = P_i + R_i + U_i \tag{13.4}$$

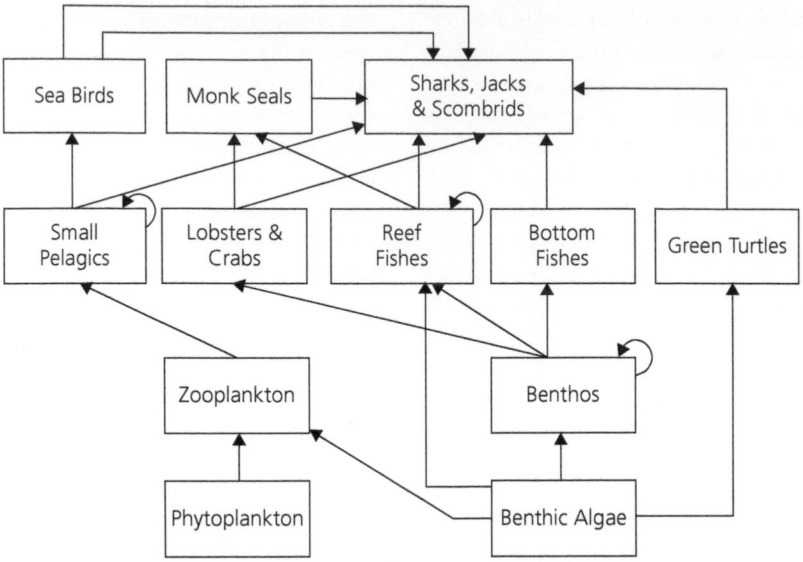

Figure 13.3 Ecopath food web structure for French Frigate Shoals used in the development of the original Ecopath model (after Polovina 1984).

where, Q_i is consumption (ingestion)[2], P_i is production, R_i is respiration, and U_i is egestion (unassimilated food) of the ith trophic compartment. In most existing applications, lower trophic level compartments comprise a functional group of organisms. Where possible, individual species are typically specified for upper trophic levels, particularly economically important species or species of special concern. We will use the generic term compartment to represent functional groups or species depending on the context. The production for the ith compartment in a closed steady-state exploited system is given by:

$$P_i = Y_i + M2_i B_i + (1 - EE_i)\, P_i \qquad (13.5)$$

where B_i is biomass, Y_i is yield ($Y_i = F_i B_i$), $M2_i$ is the predation rate per unit time (typically one year), and EE_i is the ecotrophic efficiency (the proportion of production consumed within the system or harvested); its complement is the proportion of mortality not accounted for in Equation 13.5 or Other Mortality (MO_i). The EE cannot exceed one and checks of this term in initial fitting of the model provide important diagnostics for evaluating imbalances in the initial solutions. If net migration (emigration−immigration) is not negligible, a term accounting for this loss (typically designated E_i) can be added to the right-hand side of Equation 13.5. EwE also allows for the addition of a biomass accumulation term ($BA_i = B_{i,t} - B_{i,t-1}$) to account for situations in which the system is not in a steady state.

Predation mortality on species i is given by:

$$M2_i = \frac{1}{B_i} \sum_j d_{ij} Q_j \qquad (13.6)$$

where d_{ij} is the proportion of prey i in the diet of predator j. In many instances, direct observations of production and consumption may not be readily available. However, biomass and production to biomass (P/B) and ingestion to biomass ratios (Q/B) may be available through a number of avenues including meta-analyses of compiled estimates (available in EcoBase and FishBase), allometric relationships (e.g. Peters1986; Ware 2000), and other predictive models (e.g. Palomares and Pauly 1998). The Ecopath framework accommodates this situation by specifying biomass and P/B and Q/B ratios in the mass-balance equation. When net migration (emigration−immigration) is negligible

[2] In Chapter 8, we designated ingestion or consumption as I. We note the equivalence but here provide the original notation of Christensen and Pauly (1992) familiar to the wide network of EwE users. Similarly, in Equation 13.8 we designate "other" mortality as MO in accordance with EwE usage.

and biomass accumulation is zero, the balance equation can be expressed:

$$B_i\left(\frac{P_i}{B_i}\right)EE_i - Y_i = \sum_j \left(\frac{Q_j}{B_j}\right)d_{ij}B_j \qquad (13.7)$$

and the product of biomass and the production-to-biomass ratio is production (and similarly for consumption). Dividing the production-to-biomass ratio by the consumption-to-biomass ratio gives P/Q, the growth efficiency.

13.3.2 Ecosim

Walters et al. (1997) translated the static view provided by the Ecopath equations to a dynamic setting in the simulation module Ecosim. The Ecopath mass balance model (Equation 13.7) can be recast as a system of differential equations:

$$\frac{dB_i}{dt} = g_i\sum_j Q_{ji} - \sum_j Q_{ij} - MO_iB_i - Y_i \qquad (13.8)$$

where g_i is the growth efficiency, Q_{ji} is the consumption rate of group j by group i and Q_{ij} is the loss due to predation of group i by group j MO_i is the instantaneous rate of natural mortality due to all factors other than modeled predation, and Y_i is the yield of group i (see Walters and Martell 2004 for an overview). Groups can be further partitioned into stanzas to represent ontogenetic shifts in diet composition or differential vulnerability to predation and fishing. Ecosim also allows for the specification of multiple fleets contributing to the overall yield. Again it is possible to account for emigration and immigration terms. Estimates from the corresponding balanced Ecopath model for the system are used to set initial conditions for numerical integration of this system of equations.

Ecosim invokes the foraging arena concept of Walters and Kitchell (2001; see Chapter 4) to account for the vulnerability of a prey species to a predator such that V_{ij} is the biomass of the species/group i that is vulnerable to predator/group j at a given time and $B_i - V_{ij}$ is the biomass safe from predation at that time. The rate of change of V_{ij} is a function of the exchange between the vulnerable and safe biomass pools and the loss due to predation:

$$\frac{dV_{ij}}{dt} = v_{ij}\left(B_i - V_{ij}\right) - v'_{ij}V_{ij} - a_{ij}V_{ij}B_j \qquad (13.9)$$

where v_{ij} and v'_{ij} are instantaneous rates of exchange between the safe and the vulnerable pools. The first term on the right-hand side of Equation 13.9 is the flux rate into the vulnerable pool and the second term is the flux rate from the vulnerable to the safe pool. The parameter a_{ij} here represents an encounter rate between predators and prey. Setting Equation 13.9 to zero and solving gives the equilibrium state (which should hold over short time steps):

$$V_{ij} = \left(\frac{v_{ij}B_i}{v_{ij} + v'_{ij} + a_{ij}B_j}\right) \qquad (13.10)$$

and, noting that $Q_{ij} = a_{ij}B_j$ (Walters and Martell 2004), the flow of consumption from species i to j is then:

$$Q_{ij} = \left(\frac{a_{ij}v_{ij}B_iB_j}{v_{ij} + v'_{ij} + a_{ij}B_j}\right) \qquad (13.11)$$

where all terms are defined as before. These steps apply to all trophic levels except the first. For primary producers, production is taken to be a saturating function as biomass increases, reflecting light limitation with crowding, nutrient limitations, etc. Notice that the numerator of Equation 13.11 invokes a mass-action dynamic as in the classical linear functional feeding response. The terms in the denominator, however, reflect a predator-dependent functional response (See Section 4.2.5) in which interference among predators affects the resulting predation mortality rates. Strong levels of predator interference limit the overall impact of the predators on prey. Conversely, if predator interference is weak, top-down controls are imposed on the prey.

For the purposes of illustration, an example of the types of simulation outputs for an Ecosim model for French Frigate Shoals using Polovina's initial formulation is provided in Figure 13.4. Here we examine a scenario in which a fishery for small pelagic fishes on French Frigate Shoals is initiated from its previously unexploited state and explore its ramifications for other parts of the ecosystem. In many aquatic ecosystems small planktivorous fish play a central role in the transfer of energy from lower to upper trophic levels. An

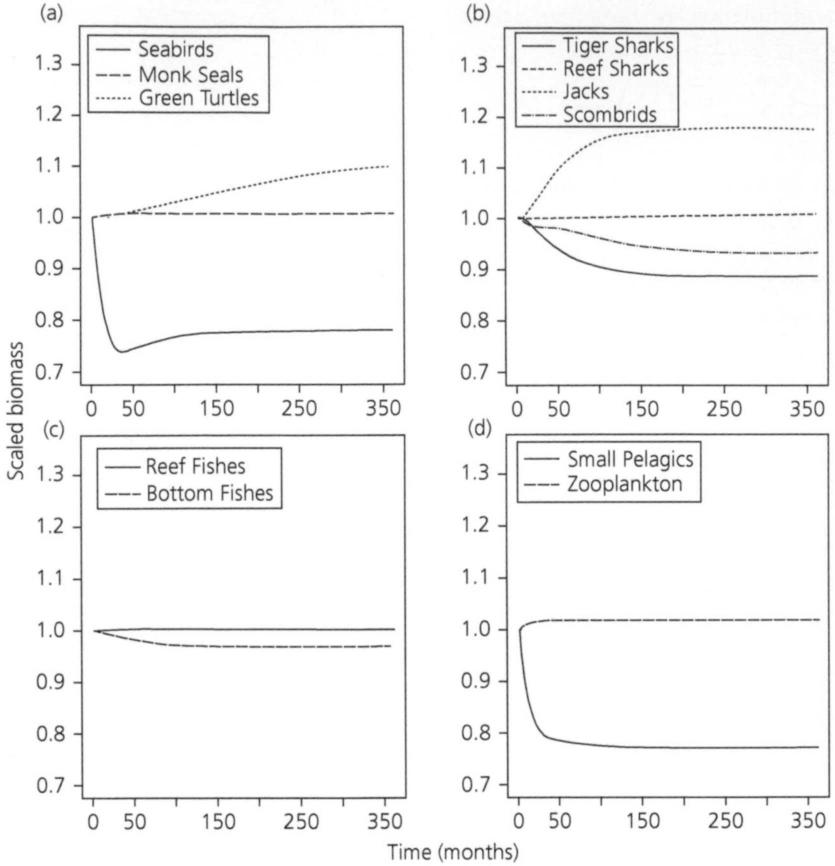

Figure 13.4 Predicted population trajectories on French Frigate Shoals using the Rpath model showing response to the establishment of a fishery on small pelagic fishes at an exploitation rate of 0.3 for (a) "charismatic megafauna", (b) upper trophic level predators, (c) reef and bottom-dwelling fishes, and (d) zooplankton and small pelagic fish (analysis courtesy of Sean Lucey).

examination of Polovina's flow energy diagram (Figure 13.3) provides valuable insights into where we might begin to see effects of the pelagic fishery manifest. A pronounced decline in seabird biomass relative to the non-exploited state is observed with an exploitation rate of 0.3 on the small pelagics, an important prey item for the birds, while no appreciable effect is seen in monk seal biomass (Figure 13.4a). Tiger sharks and scombrids show a modest decline while reef sharks are unaffected (Figure 13.4b). Bottom fishes show a minor decline and reef fishes are unaffected (Figure 13.4c). In Figure 13.4d the impact of the fishery on the target pelagic fishes is shown; interestingly, the impacts of one of their major prey items (zooplankton) is minor where we might have expected somewhat greater increase with a release from predation by one of their predator groups. Observed increases in green turtles (Figure 13.4a) and jacks (Figure 13.4b) appear to reflect indirect effects in the system. Effects on all other lower trophic elements (not shown) were negligible.

13.4 Size spectra

Jennings et al. (2008) combined theoretical developments in macro-ecology, life-history theory, and food-web ecology to estimate the global biomass of marine fish. Primary production was computed from satellite remote sensing measurements of chlorophyll *a* based on the approach of Longhurst et al. (1995).

Macro-ecological theory was used to convert from production to biomass. Production (P) is a power function of body mass (W):

$$P = e^{25.22-\frac{E}{kT}} W^{0.76} \tag{13.12}$$

where E is the activation energy of metabolism (0.63 eV), k is Boltzmann's constant (8.62×10^{-5} eV K^{-1}) and T the temperature in degrees Kelvin (Brown et al. 2004). Dividing both sides of Equation 13.12 by W, production per unit mass is:

$$\frac{P}{W} = e^{25.22-\frac{E}{kT}} W^{-0.24} \tag{13.13}$$

such that P/W decreases with W and increases with T. Total biomass (B) within each weight class was estimated as:

$$B_{tot} = \frac{P_{tot}}{P/W} \tag{13.14}$$

The primary production and median size of phytoplankton were used as starting points to construct a size spectrum in each spatial grid cell with slope:

$$\frac{\log \varepsilon}{\log \mu} + 0.25 \tag{13.15}$$

where ε is the trophic transfer efficiency, μ the predator-to-prey mass ratio, and logs are base 10. Assuming $\varepsilon = 0.125$ and $\log\mu = 3$, gives a slope of -0.051 for the biomass spectrum, which is close to expected values (see Section 10.6 and Edwards et al. 2017).

Based on these methods, the global biomass of marine animals larger than 10^{-5} g was estimated as 2600 Mt with a corresponding production of 10 000 Mt yr^{-1}. The contribution of fish to total biomass was estimated by tracking population numbers from hatch to maximum size, with the key assumption that 50% of all individuals in size class 1.1g are fish. Total fish biomass and production were estimated as 899 Mt and 791 Mt yr^{-1}, respectively. This estimate of total fish production is more than three times higher than that of Ryther (1969), mostly because higher levels of primary production are obtained from remote sensing data. Production of marine animals could be estimated for 90% of the total ocean area. Polar regions were excluded because remote sensing is limited at high latitudes.

Jennings and Collingridge (2015) refined the size-based approach by using primary production estimates from global climate models and by estimating the fraction of phytoplankton exported or consumed at higher trophic levels as a function of cell size. Where the phytoplankton is dominated by small cells (nano- and picoplankton), there are more trophic steps to mesozooplankton and less export from surface waters to the benthos. The biomass of individuals with body sizes ranging from 1 g to 1000 kg was predicted to be highest at temperate latitudes and in coastal upwelling zones (Figure 13.5a). By contrast, predicted biomass was lowest in the mid-ocean gyres and the Mediterranean Sea (Figure 13.5a). Consumer production was highest in temperate and upwelling zones, by virtue of high biomass and also in equatorial regions because production-to-biomass ratios are temperature dependent (Figure 13.5c). Eighty percent of the predicted consumer biomass is concentrated in less than one third of the total area of the oceans.

Median global biomass of consumers weighing between 1 g and 1000 kg was estimated at 4.9 billion tons (Bt) with a 90% uncertainty interval of 0.3 to 26.1 Bt. Previous estimates of 0.9 Bt based on size-based methods (Jennings et al. 2008) and 2 Bt from a box model (Wilson et al. 2009) were lower than the median but still within the uncertainty interval. The median is still lower than a global estimate of 11–15 Bt based on acoustic data from mid ocean ridges (Irigoien et al. 2014).

Unexploited baseline predictions for the simple macro-ecological model were used to calibrate a more complex size- and trait-based model in order to estimate potential fishery yields. The two models were calibrated by minimizing the difference in their size-spectrum slopes. The size- and trait-based model is similar to LeMANS (Section 12.9) in tracking the size distribution of multiple species. This application was modified to include the effects of varying primary production and temperature on ecological rates.

Fishing impacts were assessed with a range of overall fishing mortality rates and a minimum size of either 8 cm or 20 cm, with the smaller size simulating fisheries for forage species in upwelling zones. Global maximum multispecies equilibrium yield (MMEY) for consumers in all body size classes ranged from 130 to 512 Mt yr^{-1}, depending on selectivity. Much of this yield is from small species

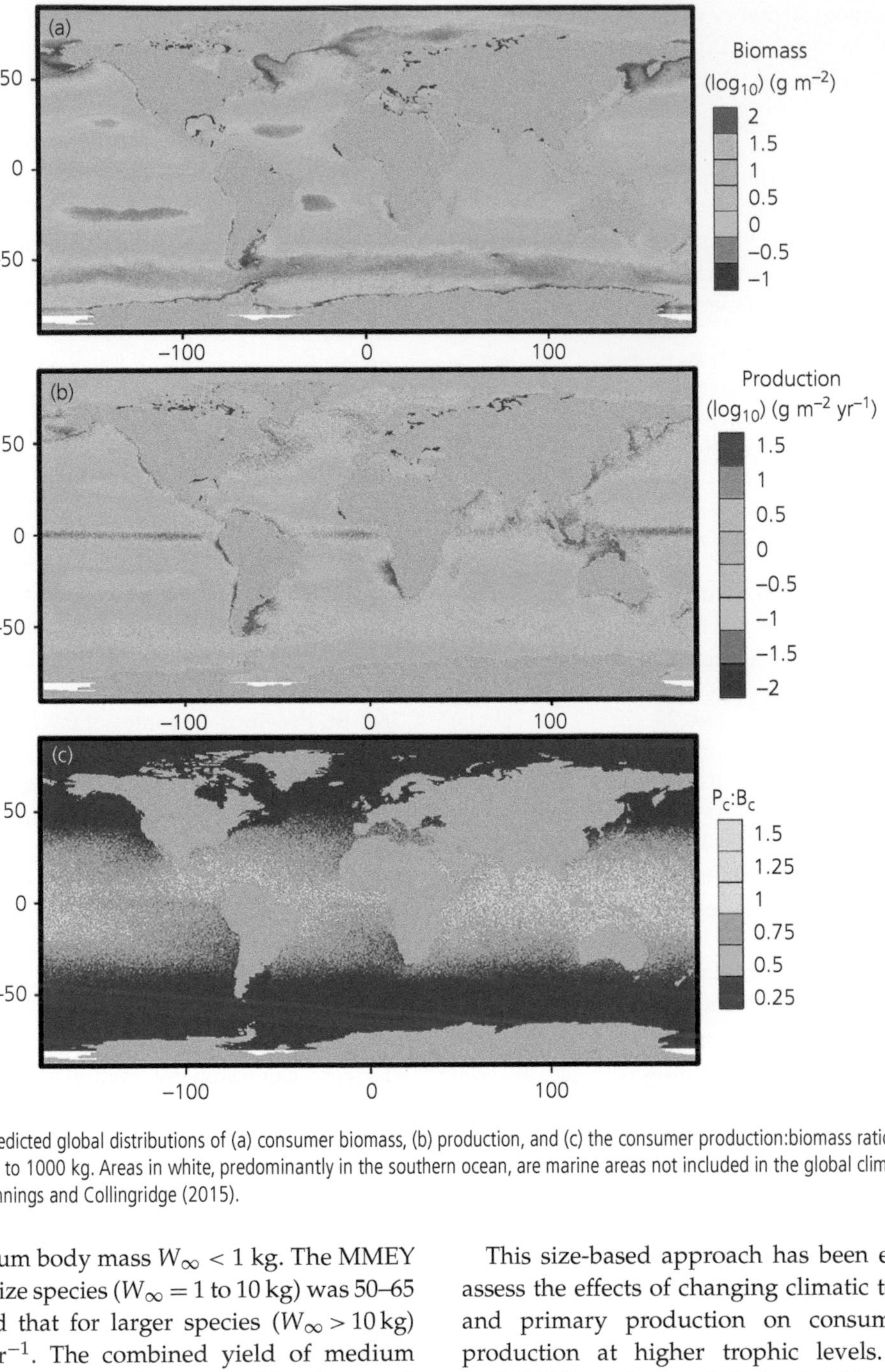

Figure 13.5 Predicted global distributions of (a) consumer biomass, (b) production, and (c) the consumer production:biomass ratio for individuals of body mass 1 g to 1000 kg. Areas in white, predominantly in the southern ocean, are marine areas not included in the global climate model domain. From Jennings and Collingridge (2015).

with maximum body mass $W_\infty < 1$ kg. The MMEY of medium size species ($W_\infty = 1$ to 10 kg) was 50–65 Mt yr^{-1} and that for larger species ($W_\infty > 10$ kg) 19–26 Mt yr^{-1}. The combined yield of medium and large species spans the reported landings and discards for 2015–2016 of 80 Mt yr^{-1} (FAO 2018). Increasing the global fisheries yield would require targeting small individuals and species, including mesopelagic fish and krill.

This size-based approach has been extended to assess the effects of changing climatic temperature and primary production on consumption and production at higher trophic levels. Blanchard et al. (2012) used a coupled physical-biogeochemical model that has been parameterized for 11 large regional shelf areas, comprising 28 Large Marine Ecosystems. The physical-biogeochemical model was forced with trace gases set either to their 1980

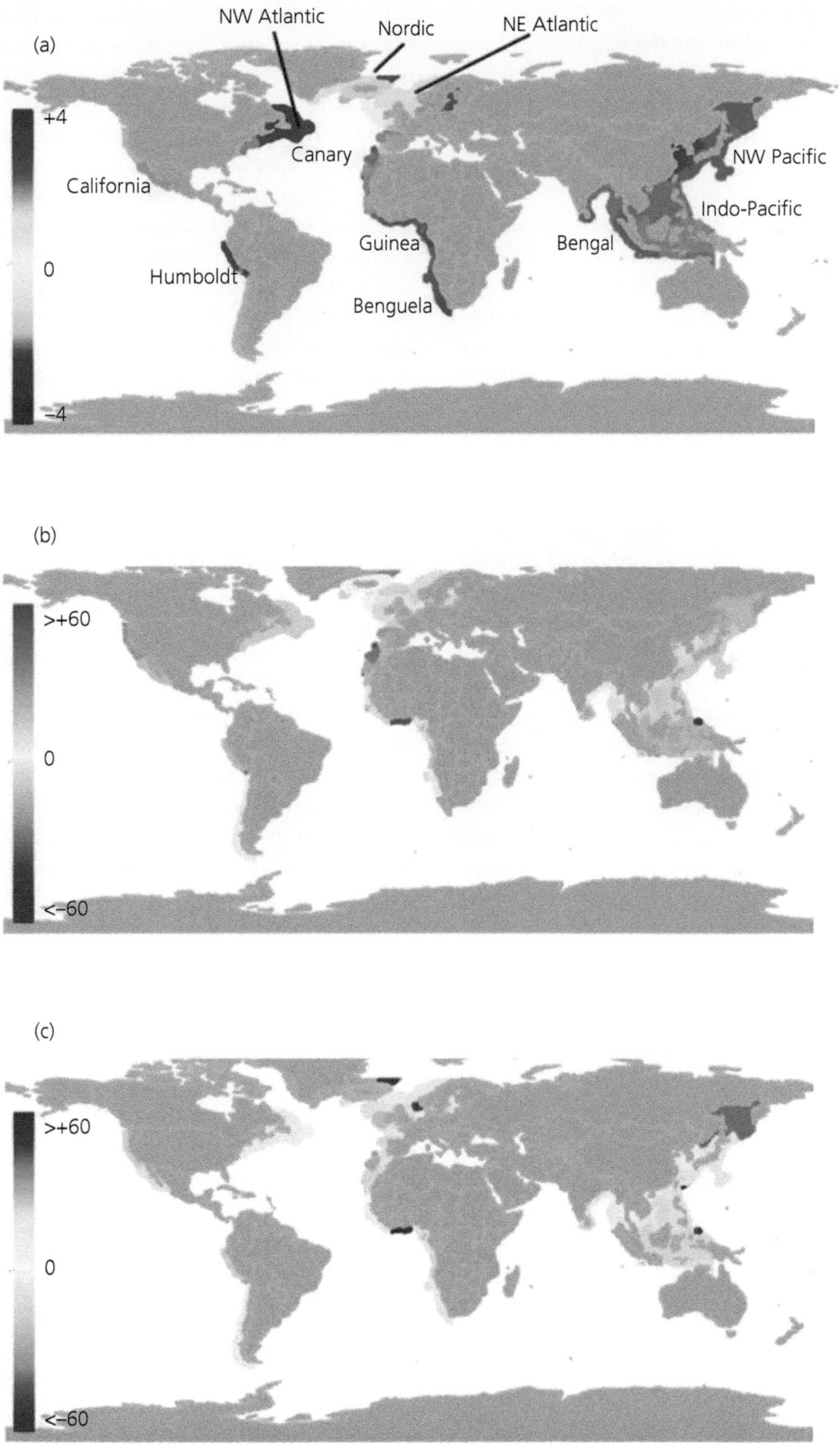

Figure 13.6 Mean-predicted relative changes for 2050 under the climate-change scenario. Maps of change in (a) mixed layer temperature (°C); and percentage changes in: (b) density of phytoplankton and (c) biomass density of pelagic predators. Adapted from Blanchard et al. (2012).

levels or the 2050 values from the IPCC "business-as-usual" scenario to simulate the effects of climate change. Outputs from the physical-biogeochemical model were used as inputs to a size-based model of pelagic predators and benthic detritivores, modified to include a temperature effect on the feeding and intrinsic mortality rate of organisms.

With climate change, potential fish production was predicted to decline by 30–60% in tropical shelf and upwelling areas (especially in the Indo-Pacific, Northern Humboldt, and Canary Current) and to increase by 28–89% on some high-latitude shelf seas (Figure 13.6). Predicted changes in pelagic predator biomass generally mirrored the changes in phytoplankton density, more so than the changes in temperature. For example, fish biomass is predicted to increase in the Nordic Sea and Guinea Current, but decrease in the Canary Current, even though all three areas have roughly similar projected temperature increases. These results confirm that potential fisheries production is primarily governed by available primary production.

The pelagic size range from 1.25 g to 100 kg was labeled as fish because fish typically dominate this size range. All sizes within this range were subject to the same fishing mortality rate ($F = 0.2\ yr^{-1}$ or $F = 0.8\ yr^{-1}$) to assess the combined effects of fishing and climate change on fish biomass, production, and potential yield. With low fishing pressure ($F = 0.2$) climate effects dominated the deviations from the baseline size spectrum, whereas fishing dominated when mortality rates were high ($F = 0.8$). Depending on whether phytoplankton density is predicted to increase or decrease, climate change may either amplify or compensate the effect of fishing on the community size spectrum.

13.5 Habitat impacts and carrying capacity

The productivity and carrying capacity of a fish species is determined in large part by the set of physical and biological conditions necessary for it to complete its life cycle. As these conditions vary for eggs, larvae, and juveniles, essential fish habitat is defined for each stage. Beyond the direct mortality of catching fish, fishing gears can impact fish habitat (Auster and Langton 1999). These impacts are of most concern for fishing gears towed over the seafloor, namely bottom trawls and dredges, which are widely used on continental shelves around the world to harvest about one quarter of the catch of marine fish species (FAO 2018).

The effects of bottom fishing on benthic habitats have been well studied with trawling experiments and spatial comparisons between trawled and untrawled areas. From meta-analyses of these studies we know that a single pass of trawl gear through an undisturbed habitat can kill 20 to 50% of organisms in the trawl track (Collie et al. 2000; Kaiser et al. 2006). The percent mortality depends on the fishing gear, depth, sediment type, and the sensitivity of different taxa. The mortality caused by bottom-fishing gear depends on the area in contact with the seafloor and the depth of penetration into the sediments (Hiddink et al. 2017). Per unit area, dredges cause more disturbance than trawls. The impacts of trawling disturbance are greatest in deeper habitats with either mud or gravel substrates (Collie et al. 2005). Biogenic habitats with emergent epifauna are particularly sensitive to bottom-trawling disturbance.

As a general paradigm, benthic communities in habitats that experience frequent natural disturbance from tides, waves, and storms are less sensitive to the additional mortality caused by bottom trawling than deeper habitats, which lie below the penetration depth (~50m) of most storm waves. To the extent that this paradigm is valid, the sensitivity of benthic habitats to bottom trawling can be predicted and mapped based on physical variables including depth, tidal currents, and sediment type (Kostylev et al. 2005a). Although this disturbance paradigm is appealing, Hall (1999) pointed out that, as long as the trawl-induced mortality is additive, it does not replace natural disturbance.

The sensitivity of benthic fauna to bottom fishing disturbance depends on their size, rate of increase, and hardness. In the meta-analysis of Collie et al. (2000), sea anemones and crustacea were most vulnerable, whereas oligochaetes and seastars were least vulnerable. It follows that in heavily trawled areas, the species composition will shift to smaller species with higher turnover rates, such as small annelids. However, the increased ratio of production-to-biomass does not compensate for the reduction in biomass (Jennings et al. 2002). While bottom trawling causes some less sensitive species to increase in *relative* abundance, on average all taxa decrease in *absolute* abundance.

The disturbance paradigm applies in reverse to recovery. Recovery rates have been measured following experimental trawling studies and from comparative studies (Hiddink et al. 2017). Numerical abundance in disturbed areas recovers more quickly than biomass, which also requires individual growth of longer-lived taxa. The median time for recovery from 50% to 95% of unimpacted biomass ranges from 1.9 to 6.4 years, on sediments ranging from mud to gravel (Hiddink et al. 2017). There are too few studies of biogenic habitats to include them in meta-analyses, but recovery of fragile taxa, such as glass sponges and corals, could take decades or centuries (NRC 2002).

Like fishing mortality, trawl-induced mortality is modeled as an exponential process. The first pass through an undisturbed habitat does the most damage; subsequent passes cause the same *relative* mortality but a diminishing *absolute* mortality of the remaining fauna. Spatial management regulations that freeze or reduce the spatial footprint therefore reduce trawl-induced mortality compared with regulations that distribute the same amount of effort over a wider area (Jennings et al. 2012). The spatial distribution of trawling is very patchy; small areas of the continental shelf are heavily fished (1–10 times per year), while larger areas are lightly fished (< 1 time per year) or unfished (Amoroso et al. 2018). It is the intersections of heavily fished areas with vulnerable habitats that are of most concern—these are the best candidates for restricting bottom-trawling with closed areas.

13.5.1 Effects on productivity and yield

In addition to the direct mortality of benthic fauna, there is concern that bottom trawling may indirectly affect the productivity of demersal fish species that depend on benthic habitats for food and shelter (Auster and Langton 1999). Historical data suggest that continental shelf ecosystems may have supported high fish yields before the widespread advent of bottom trawling (Bolster 2008) but such high yields could simply reflect the removal of the standing stock as the fisheries developed. It is difficult to distinguish the direct effects of fishing mortality from potential indirect effects of bottom trawling on fish productivity.

A large-scale fishing experiment was conducted off the northwest Australian shelf to test several hypotheses about factors regulating the dynamics of the demersal fish community (Sainsbury et al. 1991). Motivated by a desire to rebuild depleted fish stocks following the declaration of the 200-mile fishery zone, this experiment was a rare example of active adaptive management. Sainsbury et al. developed delay–difference models to describe the dynamics of four species groups: *Lethrinus* and *Lutjanus* (*L&L*) and *Saurida* and *Nemipterus* (*S&N*). Model variants were created for each of four hypotheses: (1) the dynamics of each species group were regulated independently by intraspecific processes and fishing pressure, (2) an interspecific mechanism whereby *L&L* exert a negative influence on the population growth rate of *S&N*, (3) an interspecific mechanism whereby *S&N* exert a negative influence on the growth rate of *L&L*, and (4) *L&L* were limited by the percent cover of large benthos, which was destroyed by bottom trawling.

The multispecies delay–difference models were parameterized with observations made up to 1985 and used to project the consequences and likely benefits of adaptive management. Bayes theorem was used to estimate the relative support for each hypothesis:

$$P_y(m_i) = \frac{P_0(m_i)L(O_y|m_i)}{\sum_j P_0(m_j)L(O_y|m_j)} \tag{13.16}$$

where P_y is the probability in year y of model m_i being correct, P_0 is the initial probability (prior), and L the likelihood of obtaining the observations O up to year y if model i were correct.

Prior to the experiment, there was little support for Hypotheses 1 and 3, but support was almost equally split between Hypotheses 2 and 4 (Table 13.1). The experiment consisted of sequentially closing two large areas to bottom trawling, while leaving a third area open. After five years, the percent cover of large and small benthos

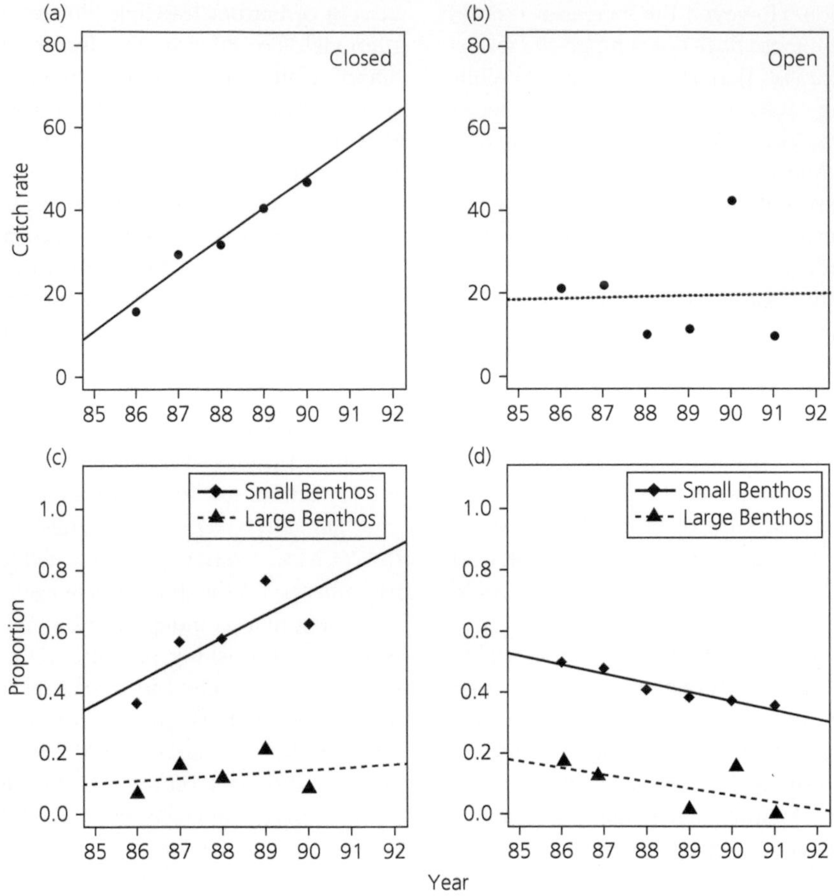

Figure 13.7 (a) Total catch rate of *Lethrinus* plus *Lutjanus* (kg/30 min trawl) in the zone closed to trawling in 1985 and (b) the zone left open to trawling, based on annual research data. (c) Proportion of the seabed with large and small benthos in the closed area (left) and (d) open area. Modified from Sainsbury et al. (1997).

(sponges) increased in the closed area (Figure 13.7), providing increased support for the habitat hypothesis (Table 13.1). Sainsbury et al. (1997) concluded that the species composition of this multispecies fish community depended to a large extent on the epibenthic habitat and that high-value yields would be possible if the habitat modification could be avoided or greatly reduced with alternative gears, such as fish traps. Since Sainsbury's seminal experiments, there have been few field studies of the indirect effects of bottom fishing on benthic productivity, but the habitat-dependent fishery models have been expanded.

To account for the time dynamics of habitat, Collie et al. (2017) extended this approach into a coupled model of habitat (H), benthic prey (B), and fish (F)[3]:

$$\begin{aligned}\frac{dH}{dt} &= r_H(1-H) - q_H EH \\ \frac{dB}{dt} &= r_B(B_{\max}-B) - a_F BF - q_B EB \\ \frac{dF}{dt} &= \gamma a_F BF - PF[a_{PH}H + a_P(1-H)] - \mu_F F - q_F EF\end{aligned} \tag{13.17}$$

Fishing effort (E) affects H, B, and F with relative catchabilities (vulnerabilities) determined by q_H, q_B,

[3] Here we retain the original notation of the authors; different notation for some coefficients playing similar roles was used in earlier equations in this book.

Table 13.1 The probability placed on each hypothesis in 1985, prior to the northwest shelf experiment and in 1990, after the first five years of the experiment. Species groups are *Lethrinus & Lutjanus* (L&L) and *Saurida & Nemipterus* (S&N). Modified from Sainsbury et al. (1997).

Hypothesis	Prior to experiment	After five years of experiment
1. Intraspecific control	0.01	0.02
2. Interspecific control (L&L < S&N)	0.52	0.33
3. Interspecific control (S&N < L&L)	0.01	0.03
4. Habitat dependence	0.46	0.62

and q_F. Benthos is preyed on by fish with an attack rate a_F. In turn, B is converted to F with efficiency γ. Habitat is assumed to provide structural shelter from fish predators (P), such that the attack rate within H (a_{PH}) is one tenth the rate outside H (a_P). Finally, natural mortality of fish from other causes occurs at rate μ_F.

A number of scenarios were run to investigate the effect of these processes on the equilibrium yield of fish, particularly MSY, E_{MSY}, and E_{ext}, the maximum effort level beyond which the fish stock would collapse (Figure 13.8). As the vulnerability of habitat to fishing increases, equilibrium yield decreases, with little change to E_{MSY}, which depends primarily on B. The decline in yield with increasing q_H, is rapid at first and then saturates when H is depleted. If predators are depleted by fishing, the protection afforded by habitat is no longer important. This scenario could mimic a real situation in which H and P are both depleted by fishing, allowing a higher yield of the target species F.

Increasing the vulnerability of benthic prey to fishing (q_B) reduced MSY slightly while reducing E_{MSY} and markedly reducing E_{ext} (Figure 13.8). This effect can be mitigated if a fraction of the benthos killed by fishing is available to the fish as carrion. If the model is altered slightly such that benthic prey production depends on habitat (as opposed to the fish predation risk), the equilibrium yield curve for fish is pushed down and to the left with increased vulnerability of H to fishing pressure. Finally the recovery of a depleted fish stock is much slower when habitat has also been reduced to 0.25 of its unfished level.

These heuristic scenarios illustrate the processes whereby the effects of bottom trawling on structural habitat and benthic prey species could indirectly affect fish productivity. Empirical studies of these processes are increasing, but only a few have been able to measure simultaneously the effects of bottom trawling on benthic prey, fish feeding, condition, and growth (Johnson et al. 2015, Hiddink et al. 2016). High-intensity trawling on vulnerable habitats could reduce fish productivity, thereby altering the biological reference points. On sand habitats the indirect effects of bottom fishing are likely to be small relative to the direct fishing mortality. The patchiness of fishing effort, coupled with the foraging behavior of demersal fish, may mitigate these indirect effects of bottom fishing.

13.5.2 By-catch and impacts on protected species

Most fishing gears catch multiple species and most species are caught by multiple gears—hence the term *technical interactions*. Catch can be divided into three main components: a) landings: the proportion retained because it has economic value (target and non-target species); b) by-catch: the portion discarded dead at sea; c) release: the portion released alive (and survives). *By-catch* is the part of the catch that is discarded at sea, dead or injured to the extent that death is the result (Hall 1996). *Prohibited species* are any species that must by law be returned to the sea. *High grading* is the discard of marketable species in order to retain the same species at a larger size and price, or to retain another species of higher value.

Alverson et al. (1994) estimated global discards of 27 million tons (range 17.9 to 39.5 Mt). This means that about one quarter of global fisheries catch (~100 Mt) was discarded at that time. The highest discard levels were in the northeast Pacific Ocean. Shrimp trawl fisheries have the highest discard rates. The estimated value of discarding has been estimated in the millions. By-catch is therefore considered a waste of food and potential economic benefits; it remains a central challenge of fisheries management. More recent estimates suggest a decline in discarding, presumably reflecting

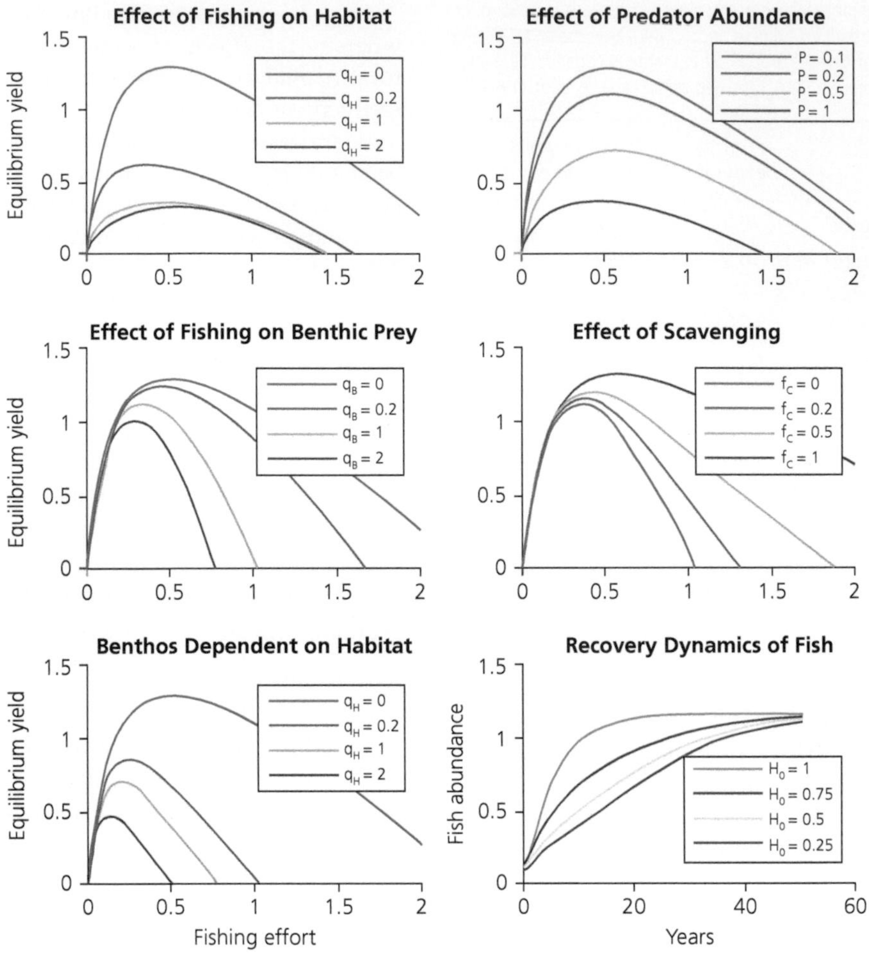

Figure 13.8 Equilibrium yield curves for fish calculated with the coupled habitat-benthos-fish model of Collie et al. (2016). Reprinted with permission (© J. Wiley and Sons Ltd.)

vigorous management efforts to reduce this practice. More current estimates are of somewhat less than 10 million Mt (e.g. Pérez Roda et al. 2019).

The FAO Code of Conduct for Responsible Fisheries (1995) stated precisely that discarding should be discouraged. U.S. National Standard 9 states that "conservation and management measures shall, to the extent practicable, (a) minimize by-catch and (b) to the extent that by-catch cannot be avoided, minimize mortality of such by-catch. Declaration 15 of the Kyoto Plan of Action (1995) stated that "they would promote fisheries through research and development and use of selective, environmentally safe and cost-effective gear and techniques."

The regulation of by-catch has been considered as a "two-lever" system. The basic by-catch equations can be expressed as: Total by-catch = effort × by-catch per unit effort. A general solution to reducing the second term of this equation is to develop and implement more selective fishing gears. With many gear types, by-catch can be reduced without substantially reducing the catch of target species (Hall and Mainprize 2005). Extrapolating these experimental gear studies to the fishery scale suggested that total by-catch could be substantially reduced with much smaller reductions in global catch. However, implementation faces many challenges because more selective gears tend to increase costs and reduce efficiency.

13.5.2.1 Protected species

Incidental takes of protected species in fishing gear are potentially a major source of mortality for many marine mammal, sea turtle, and seabird populations. Mandated modifications to fishing gear and spatio-temporal management strategies play critical roles in attempts to reduce by-catch and mortality of protected species. Models are routinely employed to evaluate management options and to select those with the highest probability of success. For example, the loggerhead sea turtle *Caretta caretta* on the eastern seaboard of the United States is currently listed as threatened under the Endangered Species Act of 1973. Major threats to the species include by-catch in fishing gear and disturbance to nesting beaches. Conservation programs in place to address these problems include protection of nesting sites and use of escapement devices in fishing trawls. Modeling efforts sharply focused on the target species and specific human impacts are often employed to evaluate the potential conservation benefits of tactical management options in these cases.

In an early example, Crowder et al. (1994) used a stage-structured matrix model of loggerhead turtle population dynamics to evaluate the efficacy of turtle excluder devices (TEDs) in fishing gear to minimize incidental takes. Crowder et al. developed a five-stage model composed of eggs/hatchlings, small juveniles, large juveniles, subadults, and adults, building on the work of Crouse et al. (1987). Crowder et al. employed an elasticity analysis to evaluate the relative sensitivity of model parameters. Recall that scaling of the elasticity estimates allows direct comparisons of relative sensitity of each parameter to perturbations (including potential management options). The elasticities reveal that the highest relative sensitivities are for the elements on the main diagonal (probability of surviving and staying in the same stage during a given time step; see Figure 13.9). Elasticities for fertility are lower than for the "survive and stay" values for the juvenile and adult stages. Protection of nesting beaches is intended to increase survival of eggs and hatchlings. TEDs are designed to enhance the survival probability of the juvenile and adults stages vulnerable to capture in fishing gear. In the absence of any additional protective measures, the loggerhead population is projected to decline given the baseline parameters used in the analysis. An increase in the survival rate of the egg/hatchling stage alone is insufficient to allow positive population growth. Crowder et al. found that a 50% decrease in the mortality rate of the large juvenile stage while keeping all other matrix elements at their baseline levels would allow positive population growth. However, 50% decreases in mortality of the small juvenile, subadults, and adults individually were insufficient to permit postive growth (although a decrease of 90% would permit positive population growth). In practice, of course, the deployment of TEDs potentially provides benefits to all life stages vulnerable to capture in the trawl. A conservation program resulting in a reduction in trawl-induced mortality of the large juveniles, subadults, and adults of 20% or higher would permit positive population growth (Crowder et al. 1994) with higher reductions leading to faster recovery. However, the long lifespan and delayed maturation of this species means that the time horizon for recovery spans decades. We note that Botsford et al. (2019) caution against using stage-structured models in place of age- or size-structured models where possible. Population growth rates are overestimated in stage-structured models. Here our focus has been on qualitative insights into the potential utility of alternative management options

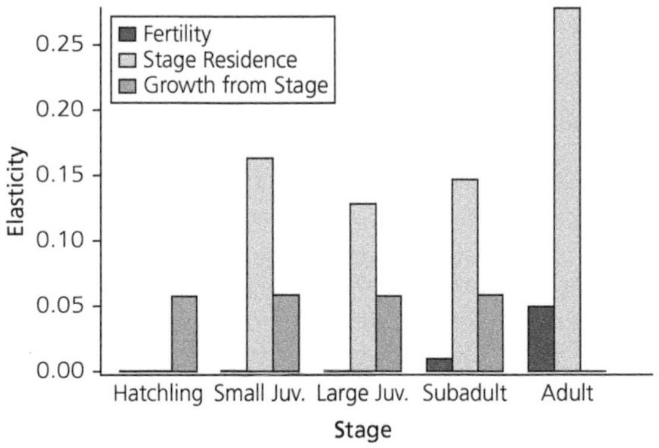

Figure 13.9 Estimates of elasticity (relative sensitivity) by life stage for the matrix model of Crowder et al. (1994) for fertility, the probability of surviving and staying within a life stage within a specified time period, and surviving and growing into the next life stage.

and priorities gleaned from the stage-structured analysis of Crowder et al. (1994). Further research into the validity of this use would be most valuable, particularly because the information requirements to support an age-structured analysis may not be met for many threatened or endangered species.

13.6 Alternative ecosystem states

We have seen that populations and communities can potentially undergo rapid transitions to alternative states if perturbed. Fundamental shifts in ecosystem structure and function are also possible. As with changes at the population and community levels, these shifts in state may not be readily reversible. Human interventions can reverberate through ecosystems in ways that go far beyond the proximal effect on target species. Scheffer (2009) provides a comprehensive treatment of this issue, documenting a broad array of shifts to alternative states in terrestrial and aquatic ecosystems under perturbation (see also Petraitis 2013).

Here we will provide an illustration using shifts in coral reef ecosystems arising from the exploitation of herbivorous fish and other disturbances. Although shifts to alternative ecosystem states are context-dependent, they have been reported in a number of coral reef ecosystems. Hughes (1994) documented a rapid transition from a coral-dominated ecosystem to one in which a reduction in grazing pressure on macroalgae resulted in a switch to algal dominance in Jamaica. Corals and macroalgae directly compete for space on the reef. Grazing on macroalgae by a guild of herbivores (including fish and sea urchins) keeps the algae in check. The joint effect of the decimation of sea urchin populations by disease and declines of herbivorous fish under exploitation reduced grazing on macroalgae and led to a switch from a coral-dominated reef to one in which corals were overgrown by macroalgae. We can gain insight into the underlying processes by returning to some core principles described in Chapter 5. The effects of different levels of herbivory on the outcome of the coral–algal competition are illustrated in Figure 13.10. Here we construct isoclines for macroalgae under three levels of herbivory and for corals. Superimposed on these isoclines is the observed trajectory of coral and macroalgal cover on the reef based on Hughes (1994). Notice that under high levels of herbivory, algal cover is kept under control by grazers and corals dominate. As herbivore control is weakened, the coral–algae isoclines cross and both corals and algae coexist. As herbivory further decreases there is no longer an intersection of the isoclines and algae dominates (Figure 13.10).

A number of models of phase transitions in coral reef ecosystems have been developed, including ones that specifically take fishing into account in dynamic models (for a recent overview, see Blackwood et al. (2012; 2018) and references therein). Here we show one such model for a coral reef ecosystem focusing on macroalgae (M)[4], corals (C), algal turf (T), and parrotfish (P) (Blackwood et al. 2012). Parrotfish, which graze on macroalgae, are exploited. The system of equations is:

$$\frac{dM}{dt} = aMC - \frac{g(P)M}{M+T} + \gamma MT$$

$$\frac{dC}{dt} = rTC - dC - aMC$$

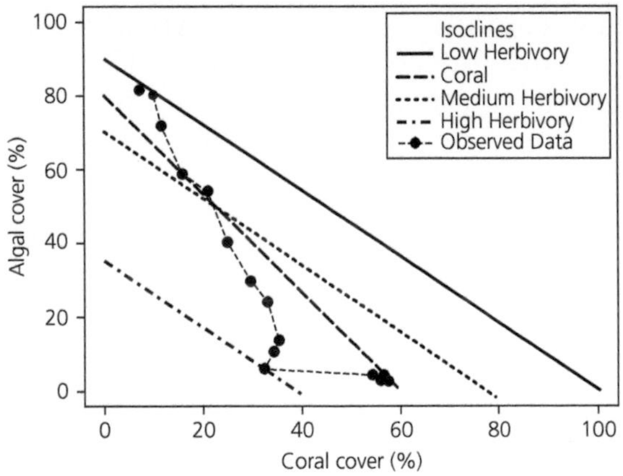

Figure 13.10 Competition isoclines for macroalgae cover under three levels of herbivory: high (solid bold line), medium (dotted line) and high (broken line), and coral cover (bold dashed line). Observed trajectory of algal and coral cover (closed circles) based on Hughes (1994); average values over all depth zones sampled.

[4] Again, we retain the original notation of the authors (see Blackwood et al. 2012); letters used to represent specific state variables should not be confused with coefficients used elsewhere in this book.

$$\frac{dT}{dt} = \frac{g(P)M}{M+T} - \gamma MT - rTC + dC$$

$$\frac{dP}{dt} = sP\left[1 - \frac{P}{\beta K(C)}\right] - FP \qquad (13.18)$$

where a is the rate at which macroalgae directly overgrow corals, γ is the rate at which macroalgae overgrows algal turfs, r is the rate of coral overgrowth of algal turfs, d is the natural mortality rate of corals, s is the intrinsic growth rate of parrotfish, β is the maximum carrying capacity of parrotfish, $K(C)$ is a term that limits the carrying capacity as a function of coral cover, and F is fishing mortality. Grazing intensity $g(P)$ is taken to be a dynamic function of parrotfish abundance and assumed to be proportional to P/β. To simplify the model, substrate coverage is assumed to be complete and scaled such that $M+C+T = 1$. It is therefore possible to eliminate algal turf from further direct consideration. We adopted the non-dimensional form of this model proposed by Fattahpour et al. (2019) for computational ease.

In the absence of exploitation of parrotfish, grazing of macroalgae is sufficiently intense to allow coral to dominate (Figure 13.11a). An increase of fishing mortality on parrotfish to 0.15 results in stable coexistence of corals and macroalgae (Figure 13.11b). With a further increase of fishing mortality to 0.3, grazing intensity by parrotfish is sufficiently low to result in macroalgal dominance (Figure 13.11c). These results support the qualitative conclusions of the isocline analysis shown in Figure 13.10 and stability analyses of this system of equations.

Another prominent example of alternative aquatic ecosystem states under perturbation by humans is a switch in vegetation types in shallow lakes with increasing nutrient loading and cultural eutrophication (see Scheffer 2009 and references therein). With low nutrient loading submerged macrophytes dominate, water clarity is high, and piscivores are important components of the fish community. Macrophytes reduce sediment resuspension, compete for nutrients with algae, and provide a refuge for planktonic organisms. Under increasing nutrient loading, there is a regime shift to a system dominated by algae, reduced water clarity, and a switch to a dominance of planktivorous fish.

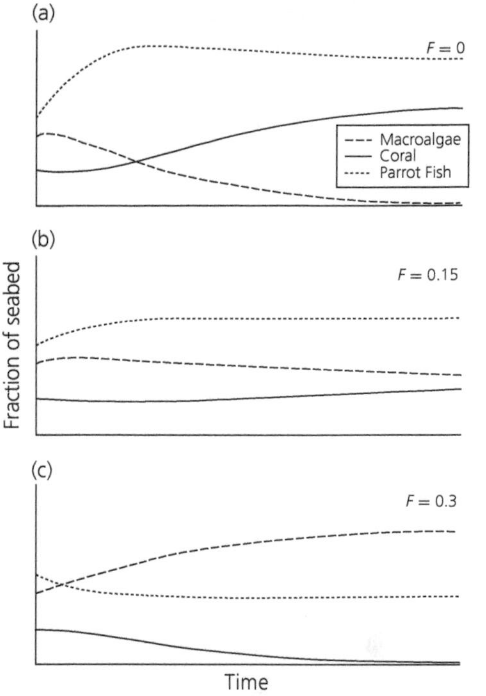

Figure 13.11 Fraction of seabed covered by macroalgae, coral, and the fraction of carrying capacity of parrot fish based on the Blackwood et al. (2012) model for three levels of fishing mortality (a) unexploited, (b) $F = 0.15$, and (c) $F = 0.3$ as modified by Fattahpour et al. (2019).

13.7 Conceptual and qualitative models

We introduced the use of qualitative models in Chapter 6 as one way to confront the issues of increasing complexity in models of community dynamics. The issue of complexity is of course amplified in any consideration of full ecosystem models. Variations on Levin's loop analysis have found a number of applications to aquatic ecosystem analysis (e.g. Yodzis 1989, Dambacher et al. 2003, 2009).

Loop analysis can be extended to more complex food webs to determine the direct and indirect effects of one species on other species or groups. Rochet et al. (2013) considered a temperate food web with planktonic and benthic pathways (Figure 13.12a). In this simplified case, the effect of pelagic piscivores on plankton is $(-)(-) = (+)$, which is a simple example of a trophic cascade. Following through the food web, the effect of

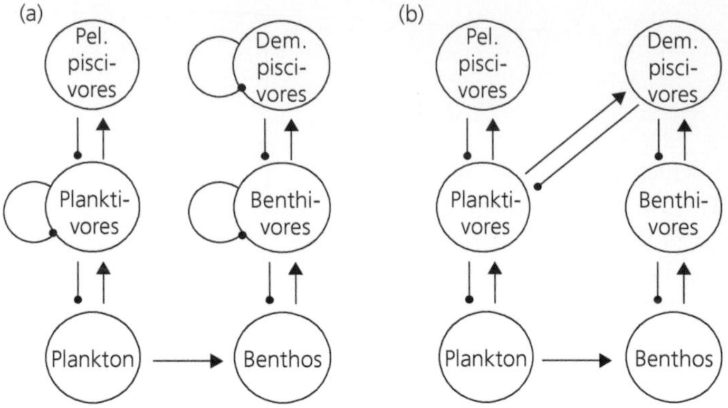

Figure 13.12 A food web with a planktonic pathway leading to Pelagic piscivores and a benthic pathway leading to Demersal piscivores. Redrawn from Rochet et al. (2013).

pelagic piscivores on demersal piscivores is (−)(−)(+)(+)(+) = (+). This indirect effect is seen to be positive because pelagic piscivores reduce grazing on plankton, making more of the planktonic production available to the benthic pathway. With more linkages, a given pair of species may be linked through several different interaction chains. We can calculate the signs of the indirect effects through each possible pathway. For example, Figure 13.12b recognizes that most fish species have catholic diets; demersal carnivores eat planktivores and benthivores. Now, plankton affects demersal carnivores through planktivores (+)(+) = (+) and through the benthic pathway (+)(+)(+) = (+). In this case, both effects are positive; plankton is the base of the food web and has a positive effect on all consumers.

In other cases, the signs of the interactions through differing pathways may be opposite. Pelagic piscivores have a positive effect on demersal piscivores through the plankton but have a (−)(+) = (−) effect via planktivores to demersal piscivores because both piscivore groups compete for planktivores (13.12b). The overall interaction is indeterminate unless additional information is available. In systems with a small number of nodes, knowledge of the relative strengths of the interactions can be sufficient to predict the overall interaction. Otherwise, the overall signs of the interactions can be expressed as probabilities based on the numbers of positive and negative paths (see Dambacher et al. 2009 for a clear exposition in a fisheries context). In our example (Figure 13.12b), if the feedback from exploitation competition were stronger than the feedback through the plankton and benthos, the interaction between pelagic and demersal piscivores would likely be negative. Finally, if all the interactions can be measured, quantitative analysis of the community matrix is possible (Case 2000).

13.8 Summary

Energy entering at the base of the food web sets constraints on total harvest from the ecosystem. Empirical relationships have been established between fishery yield and measures (or proxies) of production at lower trophic levels. These relationships can be useful for comparing among ecosystems and, in data-limited situations, obtaining initial estimates of potential yield from aquatic ecosystems. However, a more mechanistic understanding is required to trace production up the food chain. Network models for exploited aquatic ecosystems are perhaps the most widely applied models in fisheries ecosystems, in large measure due to the implementation of the EwE framework in both freshwater and marine ecosystems throughout the world. The extension of the mass-balance approach to a dynamic setting in EwE has greatly expanded the range of issues that can be addressed. Linear food-web models group species taxonomically, but require knowledge about how much production is passed on from each trophic compartment and where it goes. Size-based models require less input data, but do not explicitly separate fish from other consumers. The two approaches make consistent predictions about the regional distribution and magnitude of potential fish yield, but these predictions come with wide uncertainty intervals. These models suggest that current fishery yields are close to the ecological maximum that is attainable without tapping small fish in remote areas (e.g. mid-ocean ridges, polar regions). These models can be used to understand the implication of climate change. In addition to direct effects on target species, fisheries can affect the structure and function of ecosystems. The damage cause by towing trawls and dredges across the seafloor is the most pervasive effect of fishing on fish habitats. High intensity trawling on vulnerable habitats could reduce fish productivity, and hence sustainable yields. It is essential to understand the magnitude and effective spatial footprint of fishing effort. Fishing activities also result in the incidental catch of non-target organisms, including species of particular concern. Recognition of the direct effect of fishing on target species and collateral effects on habitat and non-target species is critically important in devising management strategies at the ecosystem level. Finally, we note that dramatic shifts in ecosystem structure and function can occur as a result of anthropogenic impacts, including fishing and cultural eutrophication, and that these changes may not be readily reversible.

Additional reading

Jennings et al. (2001) provide excellent coverage of the broad range of issues that emerge in any consideration of human impacts on marine ecosystems and their implications for management (see also Hall 1999). Walters and Martell (2004) describe core elements of the EwE framework in greater detail than has been possible here and also provide important perspectives on the key issues in ecosystem-based fisheries management. Christensen and Maclean (2011) review the broad range of social and ecological considerations underlying adoption of an ecosystem approach to fisheries management.

CHAPTER 14

Empirical Dynamic Modeling

14.1 Introduction

The complexity of aquatic ecosystems presents special challenges for both researchers and managers. To this point, we have focused on ecological processes and corresponding mechanistic models at the individual, community, and ecosystem levels. These processes are described with models in which explicit functional relationships among components of an ecosystem are specified *a priori*. These models are intended to represent the ecosystem (or parts thereof) in some simplified way that nonetheless captures its essential dynamics. They are based on certain governing processes within and between species and other aspects of ecosystem structure and function.

In general, however, we cannot fully specify a known set of governing equations for aquatic ecosystems. Approaches to addressing this uncertainty range from the application of multimodel inference to various forms of qualitative and non-parametric approaches. Multimodel inference entails the specification of a suite of alternative models and the application of different strategies to synthesize the results of these models. This may involve taking weighted or unweighted averages of the model outputs. Qualitative modelling can take the form of loop analysis based on graph theory as described in earlier chapters. An alternative (and complementary) approach is to 'let the data speak' and to construct flexible empirical models based on observed patterns and relationships among ecosystem components. This approach in many cases offers greater predictive power than the use of traditional models (e.g. Perretti et al. 2013). Peters (1991) argued forcefully for the use of empirical models in support of the development of a predictive ecology to meet the needs of environmental management and other concerns.

In this chapter, we focus on a class of empirical predictive models that can accommodate the diverse spectrum of dynamical behaviors we have identified in our treatment of differential and difference equation models in earlier chapters (Figure 14.1). Traditional ecological and fisheries models embody hypotheses concerning the functional forms of specific processes. The dynamical properties of the model(s) follow from these choices. For example, we have seen how cannibalism can generate periodic and aperiodic dynamics and how nonlinear functional feeding responses can lead to alternative stable states in predator-prey systems. However, as we progressively add layers of detail, increasing model complexity can potentially result in higher levels of uncertainty and a reduction in predictive capability with the requirement to estimate more parameters (see Figure 1.6). There is also an attendant increase in the data requirements for these models that will be constraining in already data-limited situations.

The non-parametric empirical models considered here do not involve the *a priori* designation of particular functional forms for underlying processes. They are specifically aimed at developing forecasts for nonlinear dynamical systems. In many instances, short-term forecasts are important in providing tactical management advice (e.g. setting annual quota allocations). If these empirical models can provide greater forecast skill than traditional models, they can be extremely valuable.

The time series methods at the heart of the approach deal specifically with time-varying

Fishery Ecosystem Dynamics. Michael J. Fogarty and Jeremy S. Collie, Oxford University Press (2020). © Michael J. Fogarty & Jeremy S. Collie 2020.
DOI: 10.1093/oso/9780198768937.003.0014

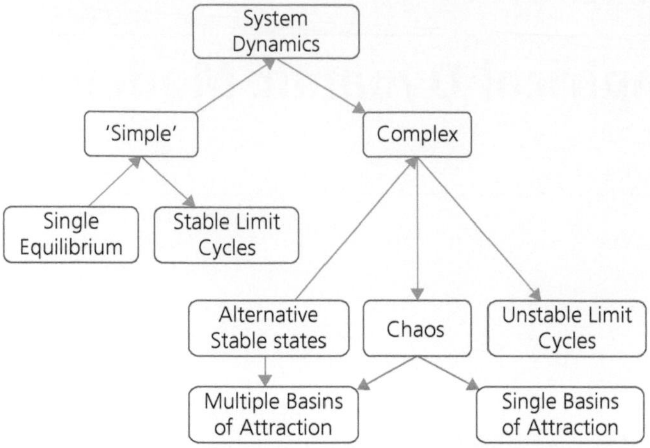

Figure 14.1 Dynamical system roadmap for topics covered in this book (adapted from Fogarty et al. 2016a). Types of models typically employed in management of exploited aquatic species are often (but not exclusively) limited to those addressing simple dynamics (upper left region of this diagram). Empirical dynamic modeling is specifically designed to address models and systems exhibiting complex dynamical behavior (right portion of the diagram).

changes in system properties and therefore treat a critically important issue in fisheries analysis. Changes at the population, community, and ecosystem levels in response to natural and anthropogenic forcing are prevalent. The resulting effects on production must be taken into account in devising and implementing effective management strategies. This issue is an important focal point in fisheries analysis and the approach described here can complement these ongoing efforts.

A central question to be addressed in this chapter is the extent to which we can extract information on the dynamical properties of a system[1] from a series of observations of one or more variables over time. To illustrate some key principles with a very simple example, consider the time series of 30 observations depicted in Figure 14.2a. We see what appear to be random fluctuations in population size. If however we lag the observations by one time step and plot $N(T+1)$ against $N(T)$, an underlying order is revealed when viewed in this two-dimensional state space (closed circles in Figure 14.2b). In fact, the time series in Figure 14.2a was generated by simulating the Ricker-logistic model introduced in

[1] The approaches to be described in this chapter have been applied to individual populations, communities, and ecosystems. We use the generic term system to encompass this spectrum of levels of ecological organization.

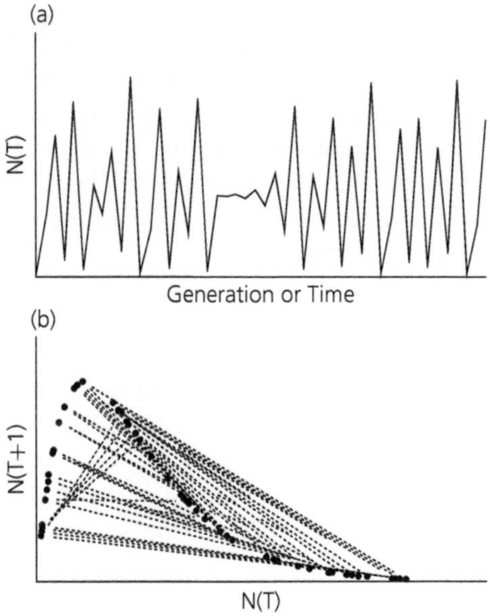

Figure 14.2 (a) Time series generated by a Ricker-logistic model exhibiting complex dynamics and (b) two-dimensional state-space representation of this time series constructed by lagging the observations in this simulated series by one time unit (here, one generation, T). The closed circles show the resulting points in state-space, revealing the underlying order in the process. Dotted lines connect successive points in state-space.

Chapter 3. The cobweb diagram in Figure 3.10c showed the generating mechanism for the irregular time series in Figure 3.10d. In Figure 14.2 we have essentially reversed this process to recover the relationship between population sizes at successive points in time.

Our focus here is on the rapidly developing field of Empirical Dynamic Modeling (EDM), designed specifically for use in problems of nonlinear dynamics. Applications of EDM in fisheries analysis have been steadily increasing. Our hope is to bring this method to the attention of a broader audience of students and fisheries scientists. In earlier chapters we set the stage by providing examples of the types of complex behavior possible in commonly used ecological and fisheries models. Our task, as indicated above, is to now reverse-engineer the problem. If we accept the premise that we do not know the underlying model with certainty, can we still extract critical insights concerning nonlinear dynamics from time series observations such as population size, catch, or other metrics and provide forecasts of future states of the system? EDM attempts to address just this question.

The term nonlinear dynamics is used here in a very specific way. We have seen that models with nonlinear functional forms are common in ecological and fisheries applications. Some of these models can also generate very complex dynamics in which small changes in an input variable or parameter can result in large changes in a response variable. We refer to these as nonlinear dynamical systems. In contrast, for linear dynamical systems, we see a proportional response to a change in an input variable. Nonlinear dynamical systems cannot be decomposed into simple additive elements. Here, the key issue is not simply the functional form of a model but its dynamical properties. Some models with nonlinear functional forms can only generate stable dynamics. Others, for certain areas of the parameter space, can give rise to very complex dynamics as we have seen throughout this book. For example, recall the bifurcation diagrams for the discrete logistic and Ricker-logistic models in Figure 3.11 showing the transition from stable dynamics through a sequence of period-doubling bifurcations to chaos as we change the intrinsic rate of increase.[2]

Empirical Dynamic Modeling (EDM) builds on the foundation established by Sugihara and May (1990) in their development of nonparametric models for forecasting and analysis of nonlinear dynamical systems (see also Sugihara 1994). A critical consideration entails identifying and quantifying state-dependence in a time series of observations. State-dependence refers to situations in which relationships among ecosystem elements change with changes in the state of the system. This essential characteristic sets nonlinear dynamical systems apart from linear additive ones. Chang et al. (2017) and Munch et al. (2019) provide excellent recent overviews of EDM. For a visual tour through the main elements of the approach see: https://deepeco.ucsd.edu/video-animations/

14.2 Core elements of the approach

The EDM approach first entails determination of the dimensionality of a system represented by a time series of observations and then quantifying the degree of state-dependence using locally weighted nonparametric estimators. The degree of state-dependence is taken as a measure of nonlinearity in the system. EDM employs out-of-sample forecast skill for model selection in a comparison against a linear null model. Given a sufficiently large number of observations, the time series is divided into two and a model is fit to the first half, called the library set, which is essentially a training set. The second half—the prediction set—is held in reserve to check predictions based on the library set. For shorter series, a cross-validation procedure (often randomly leaving off one data point at a time) is used. The missing point is predicted with a model generated by the remaining points. If forecast skill is significantly higher for the nonparametric EDM estimator relative to the null model, the system is deemed to be state-dependent. The approach has now been expanded to encompass multivariable observations and to

[2] We have also seen distinct differences in dynamical behavior between continuous-time and discrete-time forms of low dimensional models involving one or two species. Although discrete-time models in fisheries are often presented as approximations to continuous-time models (as represented in the use of an Euler approximation to a differential equation), they have potentially very different dynamical properties, as we have repeatedly seen in earlier chapters.

provide pragmatic tests for causality (Sugihara et al. 2012). The general approach has found application in ecology, epidemiology and medicine, and economics, among other fields.

The dimensionality of a system refers to the number of interacting processes that generate the deterministic component of a time-evolving system. We will take the estimated dimensionality of a system as an index of its complexity. Quantifying the dimensionality and nonlinearity of a system can allow us to distinguish between patterns in nature with a detectable nonlinear deterministic component from those arising from linear stochastic processes or containing significant observational error (Sugihara and May 1990). To estimate dimensionality and nonlinearity, we will employ a process of state-space reconstruction as described below.

14.2.1 State-space reconstruction

The time-evolution of nearby points in state space emerges as a critical consideration in understanding and classifying nonlinear dynamical behavior. Although complex nonlinear systems may appear random, they do in fact have important deterministic features and exhibit bounded fluctuations in state space. Because regions of state space of a nonlinear system are expected to be revisited in arbitrarily close proximity, methods based on estimating the recurrence of observations in state space have also been used to detect nonlinear dynamics (e.g. Kaplan and Glass 1995, pp. 315–18; Kantz and Schrieiber 2003, pp 44–5). Royer and Fromentin (2006) provide an application of recurrence mapping to extensive catch histories of Atlantic bluefin tuna in the Mediterranean. As we shall see, we can also take advantage of this property in the form of projections based on the trajectories of points in close proximity in state space.

We begin with the recognition that observations of one or more elements of an ecosystem also implicitly embody information on other components that affect the observed variable(s). These additional variables may not be directly observed. Can we extract information on the structure of the system based on the idea that an observed series also encodes information on the system as a whole? Takens (1981) showed that for nonlinear dynamical systems it is, in principle, possible to decode some of this information by translating an observed time series to a system of higher order by constructing a vector of lagged observations of the original series. In effect, the broader dimensions of the system may be captured by a more limited set of observations on even one variable. To do this, we construct a time-delayed coordinate system using the lagged variables as proxies for the unobserved variables. For a univariate series, this takes the form:

$$X_t = g\left(X_{t-\tau}, X_{t-2\tau}, X_{t-3\tau}, \ldots X_{t-(E-1)\tau}\right) \qquad (14.1)$$

where X is the variable of interest, $g\{\cdot\}$ is an unspecified function, τ is a specified lag period and E is the number of past values needed to capture the dynamics of the system (the embedding dimension). The embedding dimension represents an index of the complexity of the system recovered from the time series of observations. For a nonlinear system, viewing the time-delayed series in state space reveals a geometric shape that is directly related to the true underlying attractor, giving a "shadow" attractor which retains the dynamical properties of the original provided that a sufficient number of lags have been incorporated.

In practice, of course, we do not know the true dimensionality of real ecosystems. Whitney (1936) specified a theorem stating that a system of dimensionality D can be embedded in E dimensions where $D \leq E < 2D + 1$. It follows that E is not necessarily equal to the true dimensionality of the system. In practice, estimates of E can be relatively low, suggesting that a restricted number of critical variables strongly influence the dynamics of some systems in ways that can simplify analysis. From a mechanistic perspective, this also suggests that it might be possible to capture the principal features of a system using relatively simple models when dimensionality is low.

Recall that in Chapter 6 we introduced the concept of multidimensional attractors in relation to a three-species model of intraguild predation (Tanabe and Namba 2005). For certain ranges of the parameter space, this deterministic model can produce highly variable trajectories for the three species. However, constructing a three-dimensional representation of this community reveals the underlying order governing the process in the form of an attractor manifold (see Figure 6.11d). If we now take any one of the species trajectories and construct a time-delayed coordinate representation with delays of 0, 1, and 2 time steps, we can view the resulting shadow attractors in relation to the true manifold (Figure 14.3). The shadow attractors constructed using information for each of the individual species differ in detail but not fundamental properties (Figure 14.3a–c). Each bears a clear relationship to the true underlying attractor (Figure 14.3d) and can be shown to preserve the fundamental mathematical properties of the original attractor manifold. Construction of the shadow attractor need not be constrained to time delayed coordinates of just one of the variables as in the simple example in Figure 14.3. When observations on multiple variables are available, we can allow the axes to represent different time delays of more than one of the variables. Although we can only visualize the state space in three dimensions, the analytical procedure is, of course, not constrained in this way. We seek the minimum number of embeddings required to unfold the attractor such that the trajectories in state space do not cross.

In the example above, we know the true dimensionality of the IGP system. For application to real world data, we need an objective way to determine the minimum dimensionality of the full system. Higher dimensional embeddings can in principle fully resolve the dynamics, but they have greater uncertainty due to the possibility of contamination of nearby points (Sugihara and May 1990) and therefore the principle of parsimony is invoked. To determine the "best" embedding dimension we will take advantage of simplex projection, a nearest-neighbor forecasting algorithm (Sugihara and May 1990). We sequentially identify points and

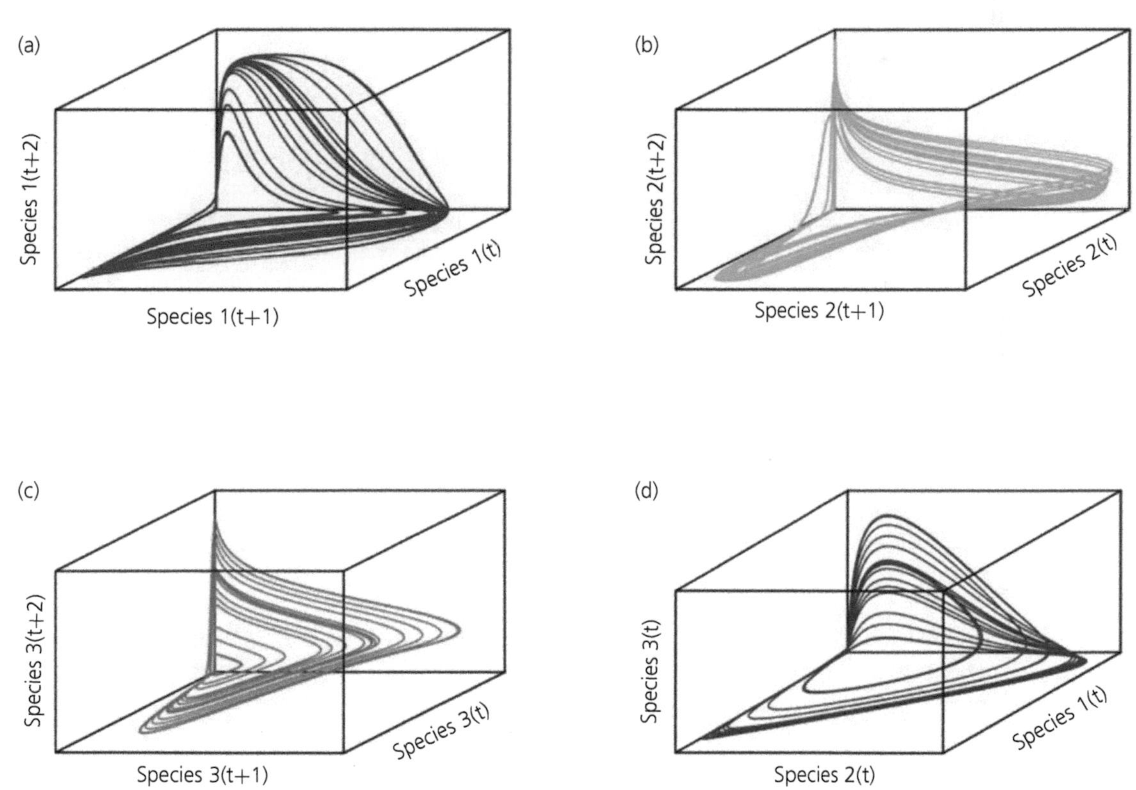

Figure 14.3 Shadow attractors for the three-species intraguild predation model shown earlier in Figure 6.1. Panels a–c show the shadow attractors using lagged coordinate systems of species 1–3 respectively. The true underlying attractor with axes representing each of the three species is shown in panel d. After Fogarty et al. (2016a).

systematically search for similar regions in the state space from which to make forecasts. These nearby points represent similar past events on the shadow attractor.

To illustrate the method with actual data, we provide an example using Pacific herring (*Clupea pallasii*) from the central coast of British Columbia (Perretti et al. 2015). The time series of abundance for this population is shown in Figure 14.4a. For a two-dimensional embedding, we require three nearest neighbors to form the simplex. Suppose we wish to project ahead one time step for the point indicated by the red triangle in the lower right quadrant of Figure 14.4b. We first identify the three nearest neighbors to this point (blue triangles). We wish to make a prediction for the cyan square in the upper left quadrant. Its three nearest neighbors are indicated by the dark blue squares. A weighted average of the nearest neighbors of the predictee gives the prediction (red diamond) for this point.

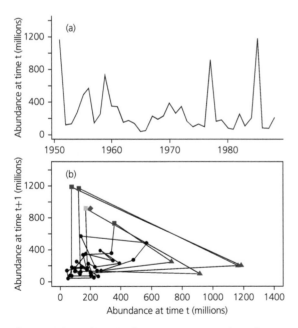

Figure 14.4 Two-dimensional state-space reconstruction and simplex projection prediction of Central Coast British Columbia Pacific herring (*Clupea pallasii*) abundance (adapted from Perretti et al. 2015). The three blue triangles closest to the red triangle are the $E+1$ nearest neighbors used to predict the red point. The blue squares are the three nearest neighbors to the observed value (cyan square). The red diamond is the prediction for this point based on the weighted average of its nearest neighbors.

This process is repeated for all points in the time series and for different embedding dimensions. In practice, the number of embedding dimensions examined is generally restricted to less than 10 given the typical length of ecological and fisheries time series. The embedding dimension with the highest correlation between observed and predicted values and/or the lowest mean absolute error is selected as the best estimate. In the following, we will outline the sequence of steps to be followed for the case in which we have observations for a single variable. The process is, however, readily generalized to the multivariate case (Deyle and Sugihara 2011; Sugihara et al. 2012). The account given here follows Deyle et al. (2013).

We build on the fundamental premise that nearby points in state space can be followed to provide forecasts of future states. For chaotic systems, these points will diverge exponentially with time (with important implications for the time window for which reliable forecasts can be made). However, for short forecast windows, they will retain a connection with the recent past and can be used to construct predictions. Our objective is to predict future values of the state variable starting with time t^*. To apply the approach, we require an uninterrupted series of observations taken at equally spaced points in time. The process of reconstructing the attractor requires a sufficient number of observations to allow us to specify its underlying form. In practice this typically requires a minimum of 35–40 points (Sugihara et al. 2012). More observations, and correspondingly higher data density, permit higher resolution of the attractor. Strategies of enhancing the time series length by directly employing closely related observations (e.g. spatial replicates) in the analysis have been shown to be effective (Hsieh et al. 2008; Glaser et al. 2011; Clark et al. 2015). We further require that the time series be stationary prior to analysis. For nonstationary series, this can be achieved by differencing the time series. For example, first-order differencing entails subtracting the previous value of a series from the current value. If necessary, higher-order differences can also be taken by subtracting points with longer time lags. In EDM it is also standard practice to transform the data to standard normal deviates prior to analysis.

In the following, we define the Euclidean distance between two vectors x and y in an E-dimensional space as:

$$d(x,y) = \|x - y\| = [(x_1 - y_1)^2 + (x_2 - y_2)^2 + \ldots (x_E - y_E)^2]^{1/2} \quad (14.2)$$

where $||\cdot||$ indicates a vector norm given by the term in brackets. Our identification of nearest neighbors depends critically on the embedding dimension of the system. The set of $E+\ 1$ nearest neighbors define the minimum number of points needed to uniquely define the location of x_t in E-dimensional space. These points define a simplex. For a two-dimensional system, we require a minimum of three points to define the vertices of the simplex (a triangle). Higher-dimensional embeddings result in more complex simplex structures. For a three-dimensional system, we require four points and the simplex is a tetrahedron, and so on. We sequentially increase the dimensionality and determine the "best" embedding dimension based on the forecast skill of each model. For a time series of length n, we will have $n - E + 1$ vectors.

To generate a forecast for a variable X, we determine the $E + 1$ nearest neighbor to the target vector x_{t^*} starting with the first nearest neighbor with time index t_1. To project p time steps into the future from t^* we have:

$$X_{\tau^*+p} \mid \boldsymbol{M}_x = \frac{\sum_{i=1}^{E} w_i X_{t_i^+ + p}}{\sum_{j=1}^{E} w_j} \quad (14.3)$$

where we have conditioned on the manifold ($\boldsymbol{M}_x$) used in the reconstruction. The weighting term (w_t) is given by the exponential kernal:

$$w_i = \exp\left(-\frac{\left\|\underline{x}_{t^*} - \underline{x}_{t_i^+}\right\|}{\left\|\underline{x}_{t^*} - \underline{x}_{t_1^+}\right\|}\right) \quad (14.4)$$

where t_i^+ is the time indicator for the ith nearest neighbor to the target point (t^*).

We sequentially vary the embedding dimension and then calculate the resulting mean forecast based on the average of the points defining the simplex. The choice of embedding dimension that gives the highest forecast skill is selected as "best" for this simplex projection. We define forecast skill, or ρ, as the correlation coefficient between the observed data and model predictions for each point in the time series. As noted above, we can also employ additional metrics such as the mean absolute error of the predictions.

Although the approach outlined above is based on a number of familiar principles, the book-keeping operations involved in the sequence of calculations required may seem somewhat daunting. Fortunately, an R package (rEDM) is available to translate the precepts of EDM into action (Ye et al. 2015). In Figure 14.5 we show results of an analysis for the embedding dimension for the Ricker-logistic model discussed in Section 14.1. We see that the highest forecast skill is for an embedding dimension of 1, consistent with the known model structure (Figure 14.5a). With higher-order embeddings, the forecast skill decays. We can also evaluate forecast skill as we progressively broaden the prediction window with increases in p. Forecast skill as a function of the time to prediction with an embedding dimension of 1 is shown in Figure 14.5b. For this example, the forecast skill remained relatively high for up to 4 time periods before rapidly declining (Figure 14.5b).

14.2.2 State-dependence

As noted above, state-dependence refers to situations in which the effect of one variable on another is conditioned on the state of the dynamical system. In an early suggestive example, Skud (1982) identified cases in which the sign of a correlation between landings (taken as an index of abundance) and temperature of Atlantic herring (*Clupea harengus*) and Atlantic mackerel (*Scomber scombrus*) in the Northwest Atlantic changed over different stanzas of time depending on whether herring or mackerel were more abundant. Skud (1982) reported similar results for California sardine (*Sardinops sagax*) and anchovy (*Engraulis mordax*). In these cases, a simple correlative analysis of the relationship between landings of either species alone with temperature would be inconclusive.

In another early example, Brander (2005) combined information for six Atlantic cod (*Gadus morhua*) stocks in the Northeast Atlantic and partitioned the spawning stock and recruitment

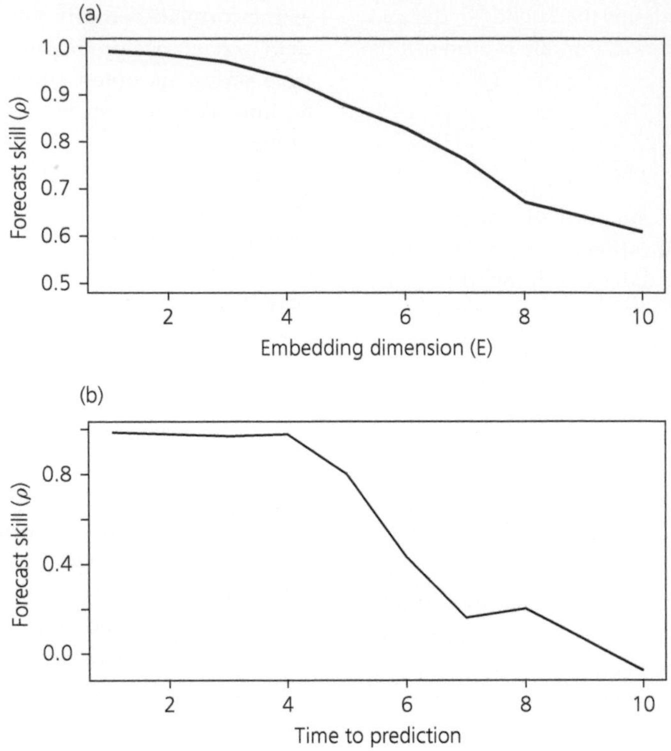

Figure 14.5 Simplex estimates of (a) embedding dimension and (b) forecast skill as a function of forecast window for the Ricker-logistic model used to generate the time series in Figure 14.1.

values of each into high, medium, and low categories. A similar categorization scheme was applied to values of the North Atlantic Oscillation (NAO) Index, a measure of the sea level pressure differential at two locations in the North Atlantic with dominant high and low pressure attributes (typically Lisbon and Reykjavik respectively). The NAO is a dominant climatic indicator for this ocean basin and adjacent land masses. Brander found that under negative NAO conditions (indicating a dominance of the low pressure cell), the probability of low recruitment at low spawning stock sizes was significantly higher than when positive NAO conditions (dominance of the high pressure cell) prevailed.

An extensive literature involving attempts to identify environmental correlates of state variables such as abundance or landings in fisheries research is available. Myers (1998) noted that most of these correlations do not stand the test of time. Reported correlations among state variables, (landings, abundance, environmental metrics, etc.) often fail with the accumulation of additional observations over time. Although there are many possible explanations for this outcome, one is that state-dependence as shown in the early examples documented by Skud (1982) and Brander (2005) are far more prevalent than is commonly recognized.

To quantify state-dependence, Sugihara and May (1990) developed the sequentially weighted global linear map (S-map) forecasting method. The application of locally weighted regression or other estimators is quite common in analysis of fisheries time analysis—e.g. use of the LOWESS (LOcally WEighted Scatter-plot Smoother) procedure. In these applications, higher weights are assigned to points close in time to the target observation. In EDM, the distinction is that higher weights are assigned to nearby points in state space. A forecast from a target time point t^* to p time steps ahead is reconstructed from the lagged coordinate system using a linear model:

$$\widehat{X}_{\tau^*+p} = C_0 + \sum_{j=0}^{E-1} \mathbf{C}_i x_{j,t^*} \tag{14.5}$$

Unlike the simplex forecasting methods, S-map uses all vectors in the state-space rather than just nearest neighbors. The model **C** is given by the system of equations:

$$\mathbf{B} = \mathbf{DC} \tag{14.6}$$

where **B** is a vector of dimension n of weighted future values X_t^+ for each observed time t^*:

$$B_i = w\left(\|\underline{x}_{t_i^+} - \underline{x}_{t_i^*}\|\right) X_{t_i^+ + p} \tag{14.7}$$

and **D** is an $n \times d$ dimensional matrix with elements:

$$D_{ij} = w\left(\|\underline{x}_{t_i^+} - \underline{x}_{t^*}\|\right) x_{j,t_i^+} \tag{14.8}$$

Because **D** is a non-square matrix, the solution (which requires inverting **D**) necessitates the use of singular value decomposition. The weighting function w is defined as

$$w(d) = e{-}\theta d/\overline{d} \tag{14.9}$$

where the parameter θ determines the degree of nonlinearity. The expression is normalized by distance between $\underline{x}_{t^*}$ in the state space and the observed points $x(t)$:

$$\overline{d} = \frac{1}{n} \sum_{j=1}^{n} \|\underline{x}_{t_j^+} - \underline{x}_{t^*}\| \tag{14.10}$$

The system is deemed linear if $\theta = 0$ and nonlinear for $\theta > 0$. When $\theta = 0$, all points in the state space are given equal weight and we have a global linear model. The resulting vector autoregressive model is taken to be the null model. In selecting a model, we then test to see if the forecast skill of the nonlinear models for different values of θ is significantly greater than the null model. If none meet this criterion, the null linear model is accepted.

We note that, in principle, it is possible to search over all combinations of the embedding dimension E and values of the parameter θ to determine the choice that provides the greatest out-of-sample forecast skill. Sugihara and May (1990) noted however that a more robust approach is to first determine the dimensionality of the system using the simplex method and to then estimate the degree of nonlinearity (state dependency) given this embedding dimension.

To illustrate application of the S-map procedure, we again turn to rEDM in an analysis of our Ricker-logistic example. We use the embedding dimension estimated in the previous section and progressively increase the value of θ from 0 to 10 and search for the value giving the highest forecast skill. In this case, we find that forecast skill initially rapidly increases with increasing θ and then levels out (Figure 14.6); after $\theta = 8$, forecast skill did not increase.

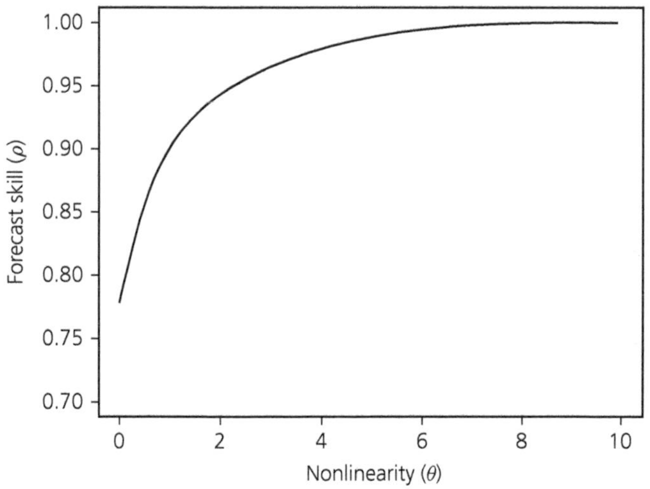

Figure 14.6 Application of the S-map forecasting procedure to determine the best estimate of the parameter θ indicating the degree of non-linearity and providing an index of state dependence. Models for sequentially increasing values of θ are fit and forecast skill (ρ) is assessed using the correlation between the observed and predicted observations to select the best value of θ.

14.3 Multivariate analysis

The power of Taken's theorem resides in the recognition that a time series of observations on a single variable can, in principle, capture important information on the system as a whole. In many instances, we may have observations on number of key elements of aquatic ecosystems (multiple species, environmental factors, human interventions, etc.). Given that observations on real ecosystems may be subject to observation error, stochastic forcing, and restricted time series length, taking advantage of multiple variables simultaneously is potentially quite valuable in understanding system dynamics. In fact, even in simulated ecosystems in which we can control all aspects of the process, we find that incorporating information on multiple state variables in EDM can enhance our ability to predict and to gain further understanding of mechanistic linkages among the elements of the virtual system. To make this possible, Deyle and Sugihara (2011) developed embedding theorems for multivariate systems that include Taken's theorem as a special case. Sugihara et al. (2012) built on this foundation to provide a generalized structure for EDM to take advantage of observations on suites of ecosystem variables.

To illustrate this point, we will generate observations from a multispecies model and explore how increasing the number of species included can improve forecast skill. In this example, we assume that observation error and environmental stochasticity are negligible. Deyle et al. (2016) describe a five-species system comprising a basal resource species, two intermediate consumer species that competed for a common prey (the basal resource), and two predators that each preyed on one of the intermediate consumers. See the supplemental materials section of Deyle et al. (2016) for details of the model structure (an expansion and modification of the Powell–Hastings model described in Chapter 6). We consider a three-species subset of the whole comprising the basal resource species and the two competing intermediate species. The simulated population trajectories of these three species over 50 time steps[3] is shown in Figure 14.7a. We next ask whether we can predict the trajectory of the resource species using information on the abundance of Consumer Species 1. Here, we use the abundance of the consumer as the library set and predict the abundance of the resource species. The correlation between the observed and predicted resource species using this approach is $\rho = 0.77$ (see also Figure 14.7b). We next add the abundance of the second consumer species to the library set. The correlation between the observed and predicted

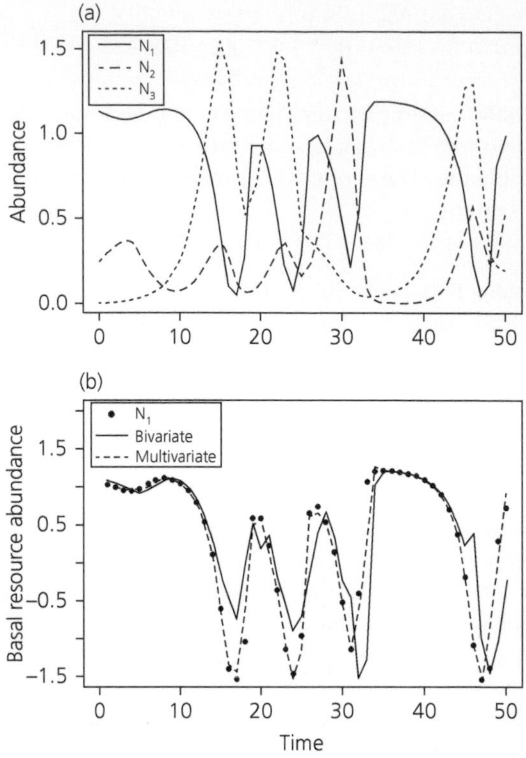

Figure 14.7 (a) Simulated abundance of a basal resource species (N_1), a consumer resource species (N_2) and a competing consumer resource species (N_3). (b) Prediction of the basal resource species (N_1) based on abundance of consumer resource species (N_2) in a bivariate analysis (solid line) and a multivariate analysis using both consumer species to predict the abundance of the basal resource species.

[3] We note that in simulation studies we can generate arbitrarily long time series. Here and in the use of subsequent simulations, we have elected to portray results with time scales consistent with the general range of observed series available for analysis (~30–50 years).

resource species abundance is now $\rho = 0.99$ (Figure 14.7b). In this example we clearly see the value of incorporating as much available information on the system as possible (subject to evaluation of data quality and other considerations) in a process of cross-prediction in multivariate systems[4]. In Section 14.3.1 below, we describe the extension of these concepts to provide a way to assess causality in nonlinear systems (Sugihara et al. 2012).

14.3.1 Causality and convergent cross-mapping

The Nobel Laureate in Economics, David Granger proposed a test for causality in linear stochastic systems (Granger 1969). Granger causality (or G-causality) is inferred when the forecast skill of a model is degraded by the elimination of one or more explanatory (independent) variables. A model in which forecast skill is enhanced by the inclusion of one or more independent variables (after appropriate adjustment in the model degrees of freedom and other considerations) is said to indicate Granger causality. Although this approach does not address the entire range of philosophical issues that emerge in any full treatment of causal inference, it does provide a pragmatic way forward that is broadly applicable. The focus is not on classical hypothesis testing, but rather on predictability.

Sugihara et al. (2012) proposed a practical approach specifically designed to assess causality in nonlinear deterministic systems. In the cross-prediction example described in the previous section, Deyle et al. (2016a) specified the exact causal relationships in the simulation model. In applications to empirical data, we may have supporting ancillary information to infer the possibility of causal relationships among species (e.g. diet composition data documenting predator–prey interactions). However, this is not universally the case, nor are ancillary observations necessarily definitive when they are available. It is possible that cross-prediction can pick up on statistical associations that are not causal (e.g. reflecting the effect of a common driver rather than direct interactions).

To help guard against this possibility, Sugihara et al. (2012) proposed a process labelled convergent cross-mapping. The approach builds on the recognition that as more observations accrue, the definition of any underlying attractor reflecting causal relationships will be enhanced and predictability will increase. In practical terms, this means that given a set of observations on a system we can draw random samples of progressively increasing library size. The sampled library sizes range from a minimum number of points equal to the embedding dimension to a maximum equal to the length of the time series in its entirety. The test for convergence involves two elements. The first is a test for a trend in forecast skill with increasing library size and the second tests whether the forecast skill at the maximum library size is significantly greater than that of the minimum. The hypothesis of causality is accepted if both conditions hold. Sugihara et al. (2012) provide details on the test statistics used (see also Chang et al. 2017 for a clear account).

Our expectation is that if a causal relationship exists increasing the library size will result in a detectable increase in predictability, converging on the maximum level of predictability. By taking a specified number of random samples at each library size, we have a way of determining if any increase in predictive skill is statistically significant. Rather than using the library sets to predict future values of a target variable, we set the time window of prediction to zero to give a contemporary estimate for the target. Accordingly we use the term convergent cross-mapping for this process.

To see how this process plays out in practice, we return to our example involving competition between two species in Section 5.2.4, in which we first encountered the phenomenon of mirage correlations (Figure 5.10b). There we saw that no consistent relationship is evident in the simulated time series for the two species despite the fact that the model specifies a causal relationship. If we apply the convergent cross mapping approach to these data, we do see a clear increase in forecast skill with increasing library size for both species (Figure 14.8). To interpret Figure 14.8, we note that the

[4] In earlier literature, this was referred to as co-prediction. Liu et al. (2012) provide an analysis using empirical observations on a multispecies system on Georges Bank.

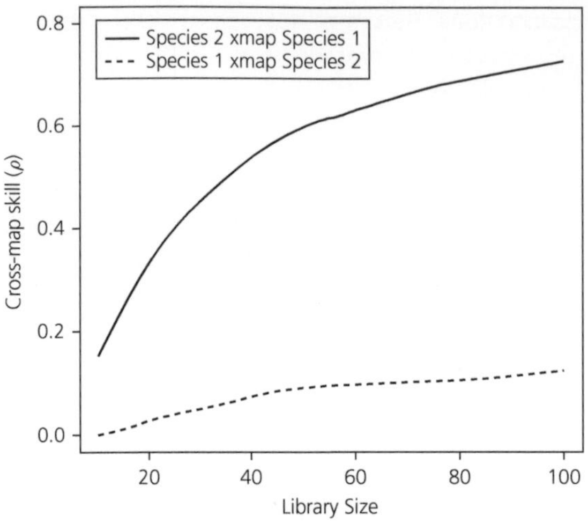

Figure 14.8 Cross-map forecast skill as a function of library size for the 2 species discrete time competition model depicted in Figure 5.16b and exhibiting an ephemeral (mirage) correlation pattern. The application of cross-mapping demonstrates convergence and indicates bi-directional causality (each species affects the other).

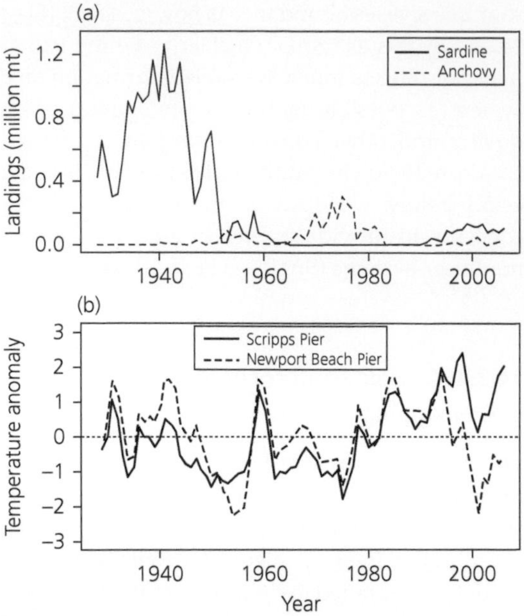

Figure 14.9 (a) Landings time series for sardine (*Sardinops sagax*) and anchovy (*Engraulis mordax*) in California (after Sugihara et al. 2012) and (b) standardized anomalies of annual sea surface temperature records for Scripps Pier and Newport Beach Pier.

designation Species 1 xmap Species 2 maps the effect that Species 2 has on Species 1. If there is a causal relationship, the effect of Species 2 is encoded in the past history of Species 1 and therefore past values of Species 1 can be used to predict the current value of Species 2. Notice that the cross-mapping is in the opposite direction of the causal effect (Species 2 has a causal effect on Species 1 and we use information on Species 1 to predict Species 2). In our mirage correlation example there is bidirectional causality, consistent with the underlying model generating the process. The results also reflect the fact than in our original simulations, we specified that the effect of Species 1 on Species 2 was substantially larger than the converse.

In an application to fisheries data, Sugihara et al. (2012) employed convergent cross-mapping to examine the relationships among Pacific sardine, anchovy, and temperature. Sardine and anchovy populations are known to exhibit regime-like behavior across ocean basins, suggesting the importance of environmental drivers operating on large spatial and temporal scales (Chavez et al. 2003). This pattern is reflected in California landings of these two species (Figure 14.9a). Direct competitive interactions between sardine and anchovies have also been hypothesized (see Deyle et al. (2013) for a review of the alternative hypotheses). In Figure 14.10 we show the cross-mapping results for (a) sardine and anchovy landings, (b) sardine landings and Scripps Pier water temperature, and (c) anchovy landings and Newport Pier water temperature. The sardine–anchovy cross-map indicates no direct causal relationship between these species (Figure 14.10a). The cross-map skill is uniformly low and does not provide any indication of convergence. In contrast, the cross-map exercise for both anchovy and sardine indicates a significant causal effect of water temperature on each species (Figure 14.10b,c). As expected, landings of sardine and anchovy show no causal effect on water temperature. The cross-map skill for temperature effects on sardine and anchovy landings is modest and it is clear that other factors are operative, including of course the effects of the fishery itself and management initiatives that have been implemented over time.

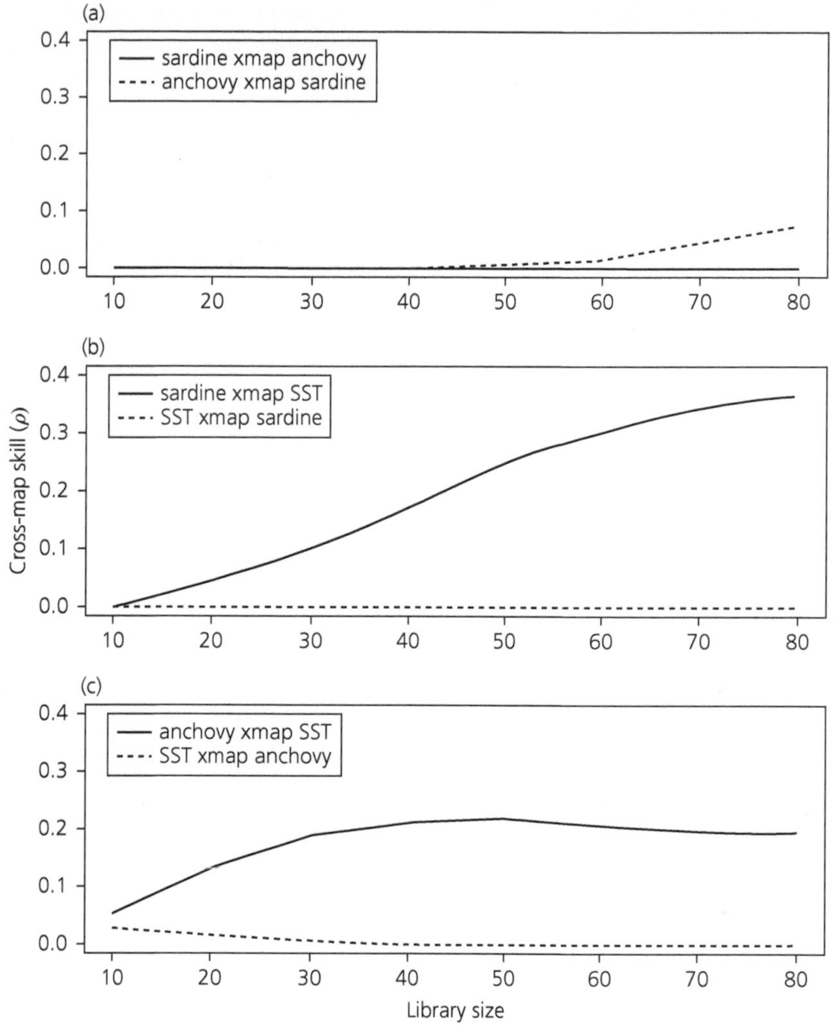

Figure 14.10 Application of convergent cross-mapping to test for causality of (a) interactions between sardine (*Sardinops sagax*) and anchovy (*Engraulis mordax*) as reflected in fishery landings, (b) the relationship between sardine landings and water temperature at Scripps Pier and (c) anchovy landings and water temperature at Newport Beach Pier.

The temperature connection has direct relevance to sardine management in the California Current. A pioneering temperature-dependent harvesting strategy was implemented in 1999 for sardine management in which higher exploitation rates were permitted under warmer conditions. Exploitation was reduced during periods of cooler temperatures based on analyses conducted by Jacobson and McCall (1995), who used a general additive model to evaluate the influence of sea surface temperature on an index of recruitment success of sardine. A significant relationship was found between recruitment success and a three-year running average of SST measured at the Scripps Institute of Oceanography pier. Based on this finding, the Pacific Fishery Management Council modified the sardine management plan to explicitly account for SST. Exploitation rates were set between 5 and 15% depending on temperature, with higher exploitation rates permitted with warmer temperatures. Upper and lower threshold biomass levels were also specified beyond which harvest was prohibited. A subsequent analysis using a different statistical approach applied to a longer time series suggested that the relationship between Scripps Pier SST and recruitment success of sardine no longer held (McClatchie et al. 2010) and the temperature-based control rule was subsequently rescinded by the Pacific Fishery Management Council. The analyses conducted by Sugihara et al. (2012) using sardine landings, and those of Deyle et al. (2013) using CalCOFI indices of larval abundance as state variables support the importance of incorporating temperature in sardine management. The analyses of Sugihara et al. (2012) and Deyle et al. (2013) further point to the need to consider state-dependent responses to temperature, opening a more nuanced and dynamic view of incorporating environmental control rules in fisheries management. Deyle et al. (2013) were further able to evaluate potential implications of climate change for sardine. Based on subsequent evaluation of the methods employed (Jacobson and McClatchie 2013) and consideration of alternative temperature series, a temperature-dependent control rule was re-instated in 2013.

14.4 Assessing species interaction strength

The development of convergent cross-mapping allows further empirical insights into the nature of species interactions. Our treatment of interspecific interactions in previous chapters has been predicated on the assumption that interaction coefficients are constant. Deyle et al. (2016) provided a way to quantify state-dependent changes in interaction strength. In Chapter 6, we employed the Jacobian of the community matrix to evaluate stability properties of simple models of community dynamics (see Box 6.2). In this case, we linearized the model around the equilibrium points. Recall that the Jacobian is a matrix of all pairwise partial coefficients of each species with respect to all other species in the community. The elements of the Jacobian matrix are the interaction coefficients. Deyle et al. adapted key elements of this basic approach to permit sequential calculation of the Jacobian at any point to estimate interaction strengths throughout the manifold. It is therefore possible to evaluate state-dependence of the interaction coefficients rather than assuming that they are constant as in most mechanistic multispecies models we have encountered to this point. As we saw earlier, the S-map procedure entails sequentially implementing a locally weighted linear regression scheme throughout the manifold. Rather than linearizing around a single fixed point as in the traditional community matrix analysis considered in Chapter 6, we now have a much richer representation of interaction strengths in state space.

Each row in the Jacobian matrix for a given point on the attractor specifies a linearization surface at that point. For example, given an n-species system in discrete time, we have for Species 1:

$$\left[\frac{\partial N_{1,t+1}}{\partial N_{1,t}}, \frac{\partial N_{1,t+1}}{\partial N_{2,t}}, \dots \frac{\partial N_{1,t+1}}{\partial N_{n,t}}\right] \tag{14.11}$$

where each element of the row represents the net local effect of each species on Species 1 at time $t+1$. In practice, the coefficients estimated using the S-map procedure for the multivariate embeddings approximate the elements of the Jacobian matrix at each successive point along the attractor (Deyle et al. 2016).

To examine the question of state-dependence in species interaction strength using this approach, we return to the multispecies simulation example described in Section 14.3 based on Deyle et al. (2016). Here we will focus on consumer species N_2 as the focal species and examine the effect of the basal resource species (N_1), competing consumer species (N_3), and a predator species (N_4) on this target species. The partial derivatives $\partial N_2/\partial N_1$, $\partial N_2/\partial N_3$, and $\partial N_2/\partial N_4$ can be interpreted as measuring the strength of bottom-up forcing, competition, and predation (top-down controls) respectively. The dominant message that emerges is that the interaction strengths are intermittent and variable rather than constant for each of the interactions considered in these simulations (Figure 14.11). The effect of the basal resource species (N_1) on species N_2 is positive as expected and the competitive and predation effects on are of course negative. The magnitude of the bottom-up effect is considerably stronger in this example than the competitive and predation effects.

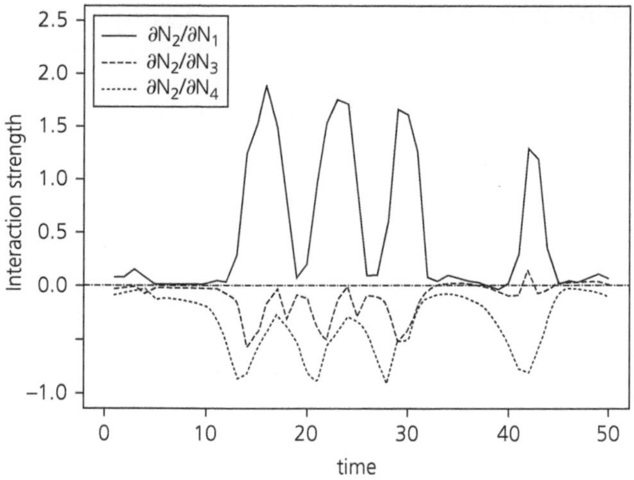

Figure 14.11 Estimation of time-varying interaction strength using the multispecies simulation model of Deyle et al. (2016) showing the effect of the basal resource species (N_1) on consumer species (N_2) (solid line); the competitive effect of consumer species (N_3) on species (N_2), and the effect of predator species (N_4) on species (N_2).

14.5 Forecasting

One of the underlying rationales of applying EDM in this context is of course to generate forecasts that may permit reliable predictions of state variables relevant to management. Forecasts that could be used to guide short-term management decisions would be extremely useful. As we have seen, for nonlinear dynamical systems, the time horizon for effective prediction can be limited (see Glaser et al. 2014). Nonetheless, even short-term forecasts can be valuable. Perretti et al. (2013) showed that EDM produced better short-term forecasts than the known mechanistic models when standard statistical approaches were used to fit simulated data. Improved forecasts were also obtained using EDM when applied to experimental data relative to mechanistic models applied to these data. In an important evaluation of the predictability of recruitment, Munch et al. (2018) showed in a global meta-analysis of 185 fish populations that EDM provided significantly lower prediction error than traditional stock-recruitment models. Nonlinear-nonparametric models using information on recruitment alone also could explain a substantial fraction of recruitment variation, averaging approximately 40% on average. Predictability increased as a function of the number of generations for which observations were available. While a significant effect of adult population size on recruitment was detected using convergent cross-mapping in about 60% of the populations examined, the fraction of the variability in recruitment explained was relatively low, consistent with the inferences drawn in Chapter 9. Perretti et al. (2015) had earlier presented evidence of predictability of recruitment using EDM in an analysis of over 500 populations. In this case, univariate analyses of recruitment were undertaken to expand the number of species/stocks that could be examined (estimates of adult population size and environmental drivers were not available for all). Deyle et al (2018) demonstrated that recruitment of two menhaden species (*Brevoortia spp.*) was predictable from adult stock size and other ecological factors. Convergent cross-mapping revealed that Atlantic menhaden (*Brevoortia tyrannus*) recruitment could be predicted from lagged indices of adult abundance. In this case, sea surface temperature was found to be related to recruitment but that the temperature effect was encoded in the adult biomass index and predictions could be made using the latter alone. In contrast, direct incorporation of an environmental driver (sea level pressure) was necessary to achieve reasonable forecast skill for Gulf menhaden (*Brevoortia patronus*).

Collectively, these findings concerning the predictability of recruitment offer important tools for management. Recruitment forecasts are essential for setting management measures (e.g. quotas) for forthcoming fishing seasons. In many instances, forecasts are currently made under the assumption that recruitment is predominately a stochastic process. Often, time series of observed recruitment levels are used to construct empirical probability distributions of recruitment. Forecasts based on random draws from this distribution are then used to generate probabilistic estimates of expected yield in concert with other pieces of information under specified extraction rates. If direct forecasts of recruitment can be made using EDM or other methods, more refined estimated of expected yields should be possible on timescales relevant to management decisions.

Ye et al. (2015) applied multivariate state space reconstruction to sockeye salmon stocks in the Fraser River (British Columbia) to generate forecasts of near-term abundance to guide management. These stocks exhibit pronounced cyclic dominance and have undergone wide fluctuations in abundance (see Chapter 3). Because of the ecological, economic, and cultural significance of these stocks, forecasts of Fraser River sockeye salmon are extremely important. A general decline in returns and dramatic and unexpected shifts over the last several decades has elicited substantial concern over the efficacy of management (Ye et al. 2015). In 2009, forecasts using traditional stock-recruitment models produced a forecast an order of magnitude higher than the actual returns. To evaluate an alternative approach, Ye et al. employed convergent cross-mapping in their analysis of Fraser River sockeye stocks, using sea surface temperature as a covariate. The fitted nonparametric model with salmon recruitment of the Seymour stock as a function of adult population size (lagged by 4 years) and water temperature is shown in Figure 14.12. Recruitment is highest at intermediate levels of spawning stock size and temperature. Forecasts of sockeye returns using EDM were significantly better than those of conventional stock-recruitment models previously used for assessment and management for 7 of the 9 sockeye stocks tested in this analysis.

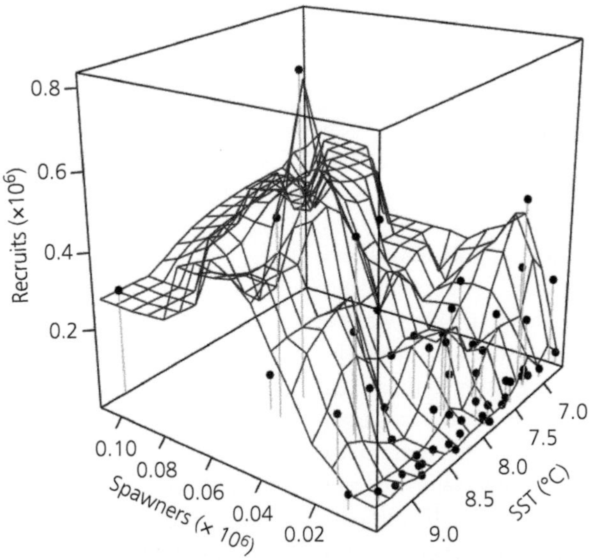

Figure 14.12 Nonparametric response surface of sockeye salmon (*Oncorhynchus nerka*) recruitment for the Seymour stock of the Fraser River complex as a function of adult abundance and water temperature. Closed circles show observation points. Mesh surface shows the nonparametric relationship between recruits, adults, and water temperature. (Courtesy Hao Ye).

14.6 Complexity in social-ecological data

Glaser et al. (2014) compared the dynamics governing fisheries at two scales of observation: the natural resource subsystem (represented by survey-derived indices of abundance of unfished and fished species) and the encompassing fishery social-ecological system (represented by fishery landings) in two North American marine ecosystems. Landings reflect the interplay of ecological dynamics, economic and marketing factors, and management interventions. We might hypothesize that fished taxa and landings would exhibit higher dimensionality and be more be likely to exhibit nonlinear dynamics than the abundance of unfished taxa because of the layered complexity of the coupled social-ecological system (Mullon et al. 2005, Chapin et al. 2006, Rouyer et al. 2008, Anderson et al. 2008).

Glaser et al (2014) analyzed over 200 time series of abundance and landings from the California Current Large Marine Ecosystem and Georges Bank off New England. The two regions differ markedly with respect to exploitation histories, species composition and community structure, and physical drivers. Abundance estimates derived from standardized resource surveys were available for both regions. In the California Current system, larval data from long-standing CalCOFI plankton surveys were available while on Georges Bank sampling was conducted using bottom-trawl surveys capturing juvenile and adult organisms. Approximately 70% of time series were considered to be predictable in one-step-ahead forecasts, but prediction skill declined exponentially over a five-year window. The ability to make long-term fishery predictions is therefore limited, as we might anticipate in nonlinear systems.

Human exploitation of marine populations was shown to affect both the estimated dimensionality and the probability that dynamics were nonlinear. The estimated dimensionality was highest for catch. But also, contrary to expectation, higher in unfished than fished populations (Figure 14.13). Nonlinear dynamics were significantly more likely to be found in landings and abundance estimates of fished taxa than in abundance estimates of unfished taxa

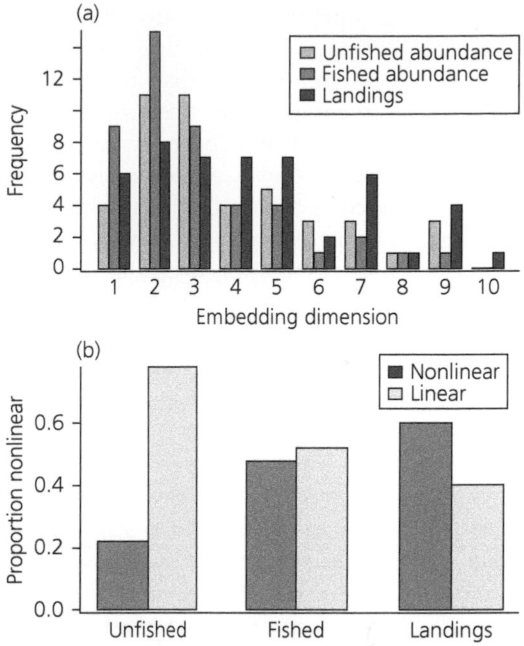

Figure 14.13 (a) Frequency of occurrence estimated embedding dimensions ranging from 1 to 10 for unfished and fished abundance data and for landings from the California Current and Georges Bank regions combined ($n = 135$ and (b) prevalence of linear and nonlinear dynamics for time series ($n = 135$) of abundance for unfished and fished populations, and for landings data from the California Current and Georges Bank regions combined. Adapted from Glaser et al. (2014).

(Figure 14.13). Accounting for differences in life-history characteristics of the species considered did not alter these general conclusions. These results support the hypothesis that human intervention does fundamentally alter the dynamical properties of exploited marine ecosystems in ways that reduce predictability and other attributes of the ecosystem.

14.7 Summary

The core elements of the EDM approach entail determining the effective dimensionality and the degree of nonlinearity of a system. Dimensionality is determined through the process of state space reconstruction. We first determine an embedding dimension using lagged coordinates of the observed variable(s). The true dimensionality of any given ecosystem (comprising all of its physical and biological elements) is of course not known. In practice, we often find that the estimated embedding dimension is relatively low (<10) in systems that exhibit nonlinear dynamics. This suggests that, in these cases, a modest number of ecosystem components dominate the system dynamics. To determine if the system is nonlinear, we employ a form of locally weighted regression in state space (S-maps) and compare forecast skill and other metrics against a null (linear) model with no local weighting. Evidence of significant local weighting indicates the importance of state-dependent dynamics in which the relationships among elements of the system can change in different regions of the state space.

Although EDM can be applied to a single observation series in an attempt to detect the influence of other unobserved variables and to develop forecasts, the extension of the method to include multiple ecosystem variables can provide both improved predictability and insights into the nature of interactions among system components. This opens avenues to understanding the importance of key mechanisms operating in the system. These can include environmental and climate influences, interspecific interactions, and other factors. EDM employs a process of convergent cross-mapping (CCM) to evaluate the significance of these drivers and whether causal connections can be inferred. We again focus on forecast skill – in this case on the question of whether inclusion of additional factors improves predictability. We take random samples from observation series and progressively increase the sample size to see whether forecast skill significantly increases (converges) as a function of library size.

As with all other analytical techniques, limitations of the method must be recognized. The performance of EDM can be adversely affected by high observation error, missing data, non-stationarity, and short time series (Chang et al. 2017 and Munch et al. 2019). EDM shares these constraints with other approaches, including parametric time series analysis. Non-stationarity can potentially be addressed by differencing the data, and it may be possible to leverage observations of related variables to effectively increase sample size. Causal inference using CCM is sensitive to the effect of strong unidirectional forcing which can induce synchrony in the system as a whole (see Sugihara et al. 2012). Ye et al. (2015) developed an extension to CCM involving careful evaluation of expected time delays in response to the driver that can remedy this situation.

EDM can, in principle, address the diverse spectrum of issues we have considered in previous chapters in a way that complements traditional approaches. As we have noted, the differential and difference equation models as the heart of previous chapters can be viewed as testable hypotheses about how aquatic systems 'work', and the structural forms of the relationships among component parts. In a multivariable setting, EDM can provide different insights into important mechanisms that are not tied to specific functional forms connecting different ecosystem components. The focus on forecast skill rather than classical hypothesis testing as a measure of success casts the issue in quite pragmatic terms. We use the approach if it pays off in terms of predictability relative to alternative methods and not otherwise.

Additional reading

Kaplan and Glass (1995) provide a very readable and informative introduction to nonlinear dynamics and methods available for their analysis. Nicolis and Prigogine (1989) give an important account of complexity theory in general.

CHAPTER 15

Toward Ecosystem-Based Fisheries Management

15.1 Introduction

In the preceding chapters we have assembled key concepts and analytical tools relevant to the development of strategies and models in support of Ecosystem-Based Fisheries Management (EBFM). EBFM will necessarily unfold differently in different places to meet specified objectives and requirements. It will be tailored to the available understanding, scientific infrastructure, and data/analytical resources available for different aquatic ecosystems. Throughout this book we have portrayed a diverse set of ecological processes and associated models operating at the individual, population, community, and ecosystem levels. These elements can be viewed as modules that can be assembled in different ways to develop models to address specified management objectives, contingent on the scientific information and resources available.

General principles for fisheries management in an ecosystem context have been advanced by the Food and Agriculture Organization of the United Nations, identifying the need to:

...balance diverse societal objectives, by taking into account the knowledge and uncertainties about biotic, abiotic and human components of ecosystems and their interactions and applying an integrated approach to fisheries within ecologically meaningful boundaries.

(Garcia et al. 2003). The 1992 UN Convention on Biological Diversity (CBD) calls for "ecosystem and natural habitats management" designed to:

... meet human requirements to use natural resources, whilst maintaining the biological richness and ecological processes necessary to sustain the composition, structure and function of the habitats or ecosystems concerned.

These general principles are mirrored in a growing number of national and regional initiatives defining policies and providing roadmaps toward implementation of EBFM. However, despite the longstanding and widespread recognition of the need to integrate ecosystem principles into aquatic resource management (e.g. Botsford et al. 1997; Pikitch et al. 2004; Link 2010), there has been considerable inertia in implementation (e.g. Pitcher et al. 2009; Skern-Mauritzen et al. 2016). This delay reflects a current lack of binding legal and regulatory mandates for implementation of EBFM on a specified timetable. Patrick and Link (2015) argue that the concept of optimum yield provides a sufficiently broad framework to accommodate the fundamental principles of EBFM. In the United States, the Magnuson–Stevens Fishery Management and Conservation Act as amended in 1996 defines optimum yield as the "*... maximum sustainable yield from the fishery, as reduced by any relevant economic, social, or ecological factor*" and calls for the protection of marine ecosystems and rebuilding of overfished stocks. Other nations have adopted variants on the theme of optimum yield and extension to the concept of EBFM may be possible in other jurisdictions (Patrick and Link 2015).

Important progress has been made in establishing flexible strategies for the transition to EBFM. For example, organizing principles for the development

Fishery Ecosystem Dynamics. Michael J. Fogarty and Jeremy S. Collie, Oxford University Press (2020). © Michael J. Fogarty & Jeremy S. Collie 2020.
DOI: 10.1093/oso/9780198768937.003.0015

of marine Integrated Ecosystem Assessments have been established (Levin et al. 2009a) to provide a strategic and analytical framework for Ecosystem-Based Management. The approach outlines an iterative, multistage process comprising: (1) *Scoping*: consult with stakeholders to identify goals and objectives; (2) *Indicator development*: identify metrics and targets to assess the state of the ecosystem; (3) *Risk analysis*: evaluate threats to the system as measured by the indicators and the probability that the system will be driven to an undesirable state by anthropogenic or natural stressors; (4) *Management strategy evaluation*: test the performance of alternative management options and decision rules in a virtual world to identify promising management strategies and to eliminate ones with low probability of success; and (5) *Monitoring and evaluation*: continue monitoring and assessment of indicators to determine the efficacy of management actions. If management strategies are not achieving desired outcomes, new management options are entertained and tested by simulation. Building on the IEA framework, Levin et al. (2018) provided a generalized template for the development of next generation Fishery Ecosystem Plans (FEPs) to guide the development and implementation of EBFM in a way that is responsive to the needs and objectives of management authorities in different jurisdictions.

A steady evolution in the maturity of modeling approaches designed to meet the needs for EBFM is also clearly evident (e.g. Hollowed et al. 2000; Whipple et al. 2000; Plagányi 2007; Christensen et al. 2009; Plagányi et al. 2014). Multispecies and ecosystem models have been developed for many aquatic ecosystems throughout the world. For example, the EcoPath with EcoSim (EwE) modeling framework has now been applied to over four hundred marine and freshwater ecosystems (Colléter et al. 2015). End-to-end models such as Atlantis (Fulton et al. 2011) have also now been developed and applied to over three dozen aquatic systems, principally in marine ecosystems but also including Lake Victoria (Audzijonyte et al. 2019).

Despite inertia in implementation, effective foundations for the development of EBFM have been established. The state of aquatic ecosystem science and modeling is sufficiently advanced in many parts of the world to provide analytical support for integration of ecosystem principles into fisheries management now.

In this chapter we will develop themes centered on the development of place-based management strategies; managing for resilience by preserving biodiversity and ecosystem structure and function; and the need for a fully integrated approach recognizing the inter-relationships among the physical, ecological, and human components of the system.

15.2 Place-based management

A central objective in defining EBFM as a place-based approach is to foster a holistic approach to management in a defined geographical area. Conventional fisheries management encompasses conservation and yield of target resource species, habitat protection, and protection of non-target species (including threatened and endangered species). Under EBFM, these issues, now addressed in a quasi-independent fashion, would be treated as parts of a fully integrated whole.

15.2.1 Delineating the ecosystem

In order to undertake EBFM, we need to clearly delimit the spatial domains of concern for management. Because fish are among the most speciose of the vertebrate taxa and constitute a major element in global biodiversity, consideration of their representation in marine and freshwater ecosystems has played a major role in defining aquatic ecoregions. Biogeographers have long engaged in efforts to identify regions defined by consistent species compositions and associated physical characteristics in both marine and freshwater ecosystems. Global syntheses are now available for both freshwater (Abell et al. 2008) and marine ecosystems (Spalding et al. 2007; see Figure 15.1). These ecoregions could provide an important starting point for the specification of relevant spatial domains for EBFM. These considerations, of course, differ markedly for freshwater and marine systems. For freshwater systems, specific management measures are often tailored to individual water bodies delineated by river margins, lake boundaries, or watersheds. Distinct water bodies in relatively close proximity or under similar jurisdictions do,

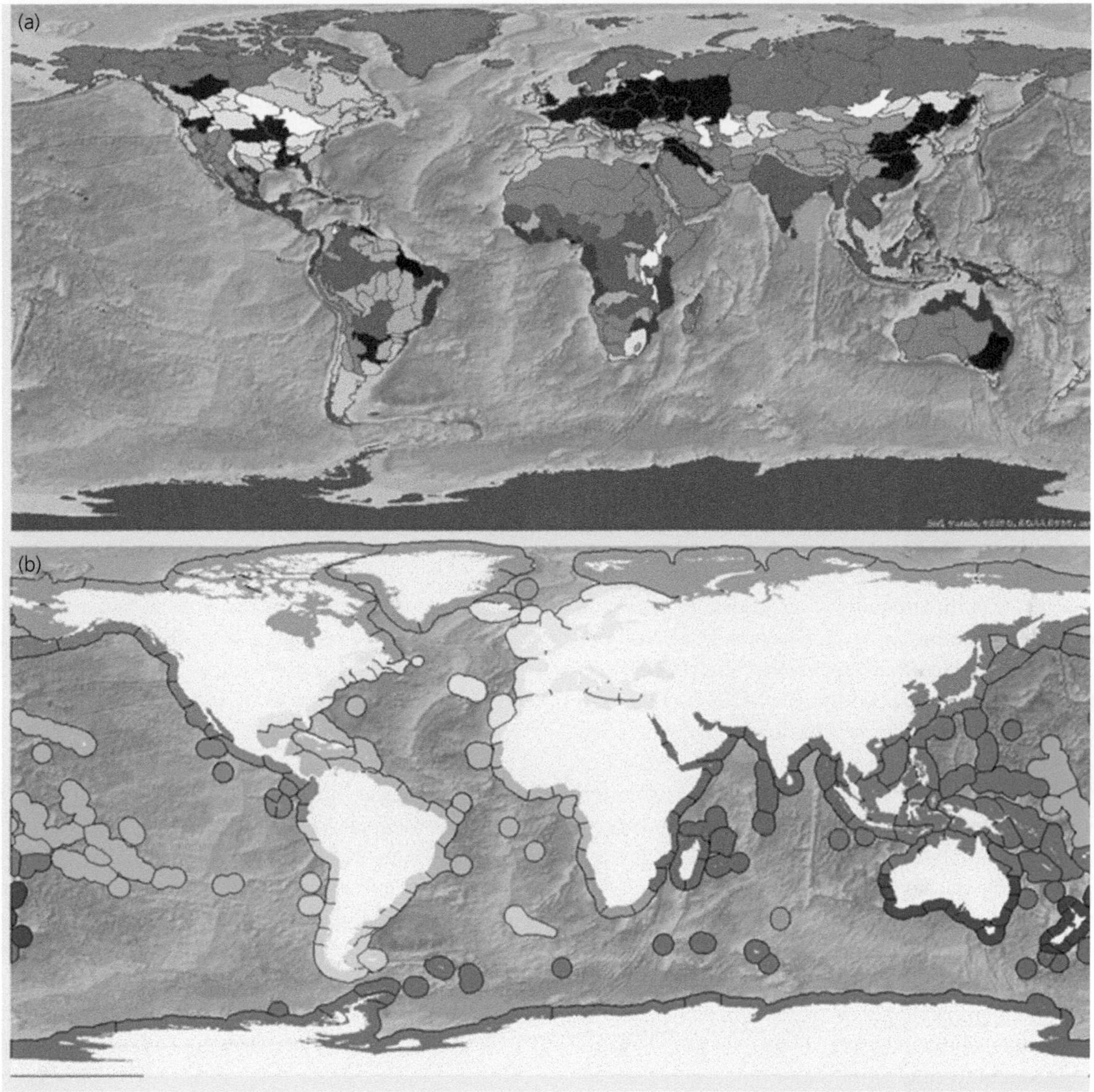

Figure 15.1 (a) Freshwater ecoregions of the world (Abell et al. 2008) and (b) marine ecoregions of the world (Spalding et al. 2007). Shape files courtesy of The Nature Conservancy.

however, often share basic management approaches and precepts. Barriers to dispersal in freshwater ecosystems result in high levels of endemism, a major consideration in conservation management. In contrast, levels of interconnectedness are much higher for marine ecosystems linked through large-scale oceanographic processes and features. These considerations ultimately will determine the relevant spatial scales for management in marine and freshwater ecosystems.

In an analysis of freshwater ecosystems, Abell et al. (2008) defined a total of four hundred and twenty-six ecoregions throughout the world to define biodiversity conservation units (Figure 15.1a). Because information on fish faunal assemblages was among the most common and comprehensive descriptors of water bodies, this classification scheme relied heavily on this metric (see www.feow.org). Spalding et al. (2007) employed a nested hierarchical approach to identify marine ecoregions for coastal and continental shelf areas. At the highest level of organization, twelve marine realms (defined as very large coastal, benthic, or pelagic regions) were identified. Nested within these realms are sixty-two marine provinces identified as large marine areas with distinct biotas. Finally, two hundred and thirty-two ecoregions were identified as subsets of these marine provinces, with clearly distinct species compositions relative to adjacent regions (Figure 15.1b; see www.meow.org).

The Large Marine Ecosystems (LME) concept defined by Sherman and Alexander (1986) provides an alternative starting point for identifying spatial domains for management in marine systems. We encountered the use of LMEs as spatial strata for analysis in Chapter 13 (see Figure 13.1). LMEs are "relatively large regions on the order of 200,000 km^2 characterized by distinct bathymetry, hydrography, productivity, and trophically dependent populations" (www.lme.noaa.gov). Sixty-six LMEs have been identified based on these considerations. The LME approach differs from the sharp focus on biogeography in the Spalding et al. (2007) analysis to encompass physiographic, hydrographic, and food web considerations. The Sea Around Us Project of the University of British Columbia has partitioned global marine catch, fishing effort, and other fishery metrics by LME (see www.seaaroundus.org). These products have greatly facilitated LME-scale analysis relevant to EBFM. EwE models have been developed for each of the sixty-six LMEs (Christensen et al. 2009), providing an important tool for incorporating ecosystem considerations into fisheries management on a global scale. Large marine ecosystems have now been used as fundamental spatial units in a number of analyses including the relationship between primary production and fish yield (Chassot et al. 2010; Friedland et al. 2012; Stock et al. 2017); quantification of ecosystem overfishing levels (Coll et al. 2008; Link and Watson 2019); and fishery production potential (Fogarty et al. 2016b).

15.2.2 Spatial management strategies

Various forms of spatial management have long been applied as fishery management tools in concert with other regulatory measures (direct control of catch or fishing effort, size limits, etc.). A variety of strategies have been employed, including: seasonal area closures to protect spawning and nursery areas; rotational harvest schemes in which areas are sequentially opened and closed to fishing; and the specification of long-term closures imposing partial or complete strictures on extraction of resources. Given the spatial orientation of EBFM as defined here, consideration of spatial management tools to meet an array of ecosystem-based management objectives meets an important need.

In the following, we will focus on long term-closures (designated as Aquatic Protected Areas, APAs). The use of APAs as part of an overall management strategy is currently much more common in marine ecosystems. To date, nearly 12000 Marine Protected Areas have been designated (http://www.mpatlas.org), but they cover only 4.8 per cent of the marine surface area of the globe. The global coverage of freshwater protected areas has been more difficult to ascertain because the relative contribution of freshwater and land-based biodiversity is not always differentiated in current reporting mechanisms for terrestrial protected areas (Juffe-Bignoli et al. 2016). There is particular interest in the contribution of protected areas to the overall preservation of biodiversity in freshwater systems. It is estimated that freshwater ecosystems account for as much as 12 per cent of total biodiversity and approximately one third of the vertebrate biodiversity of the combined freshwater–terrestrial total (Juffe-Bignoli et al. 2016). For a recent overview of the status and development of freshwater protected areas, see Finlayson et al. (2018).

Meta-analyses of the documented performance of marine APAs consistently demonstrate positive effects on average biomass, density, species richness, and individual body size within the protected areas (Micheli et al. 2004; Lester and Halpern 2008). However, differential impacts at the species level are also evident, with some species exhibiting neutral or negative responses with the implementation of reserves. This suggests that the direct and indirect effects of interspecific interactions and other factors must be recognized and anticipated. For APAs to have positive effects on yield, moderate levels of dispersal must occur at one or more life stages from the protected area(s) to areas open to fishing. Short-distance dispersal can benefit conservation goals (for example, protection of critical biogenic habitat) but provide little or no benefits for yield. Conversely, long-distance dispersal can dissipate the overall benefits of protected areas. Relatively few direct estimates of actual dispersal distances at the egg and larval stages are available for aquatic organisms. Estimates of planktonic egg and larval stage durations are more readily available and may permit inference concerning potential dispersal ranges. Species with longer stage durations may exhibit greater dispersal distances depending on the exact nature of hydrographic transport mechanisms (Fogarty et al. 2000; Fogarty and Botsford 2007). Dispersal distances for juvenile and adult aquatic organisms can be estimated from tagging studies but these can be costly and, in most instances, are undertaken for species with high economic value. Dispersal remains a major source of uncertainty in evaluating the potential performance of an APA. It is also clear that the diversity of individual species dispersal patterns ensures that any protected area configuration cannot be optimal for all species.

The outcome of the establishment of an APA with respect to yield also depends on the response of fishers with respect to the redistribution of fishing effort from closed to open areas. Adaptive behaviors by fishers that concentrate at the boundaries of the APA ("fishing the line") strongly affects its performance with respect to yield (e.g. Walters 2000, Walters and Martell 2004; Kellner et al. 2007).

Fulton et al. (2019) and Botsford et al. (2019) provide important recent reviews of strategic and tactical models for APAs, drawing on the experience in marine systems. To date, the preponderance of modeling efforts designed to guide the design of aquatic reserves or to assess their efficacy have focused on yield and single-species conservation requirements in relation to other management tools. Given the focus of this book, we are particularly interested in the potential utility of APAs in meeting a broader suite of objectives for EBFM, including: preservation of biodiversity and ecosystem structure; habitat protection; conservation of target and non-target species; and yield (see also Hilborn et al. 2004; Browman and Stergiou 2004; Halpern et al. 2010).

Because APAs will be nested within larger management units for EBFM, this topic fits naturally into this chapter. We will focus on very simple model structures that can be expressed within a common framework to address core considerations for EBFM. We note that quite sophisticated treatments of critical issues such as dispersal patterns, the role of demographic structure, and options for the establishment of APA networks have been undertaken (see Fulton et al. 2015; Botsford et al. 2019; and references therein). Ultimately, any consideration of tactical management options will of course require evaluation at this richer level of detail. Actual implementation will require models embodying still greater realism. The most advanced treatments in this regard have been developed within the Ecospace modeling framework (Walters et al. 1999; Walters 2000; Walters and Martell 2004).

15.2.2.1 Single-species models

To set the stage, we introduce simple strategic models for APAs for single species with no demographic structure and implicit dispersal mechanisms. These models are intended to provide heuristic insights into the potential efficacy of protected areas, with particular emphasis on no-take APAs (commonly referred to as reserves). Our goal is to show how extensions of some of the simple population models we have encountered in earlier chapters can be modified to consider elementary models for aquatic reserves. We will then broaden our focus to encompass issues related to biodiversity, habitat, and interspecific interactions.

We begin with models for single species with areas open and closed to fishing in which dispersal is assumed to be widespread throughout the spatial domain of interest. Discrete-time logistic or Ricker-logistic models have been used extensively in these constructs (e.g. Mangel 1998; Lauck et al. 1998). In the following, we will use biomass[1] as the model currency. We further assume that exploitation (u) precedes reproduction in each time step

[1] Most of the strategic models for aquatic reserves have used population number rather than biomass. Here, we use a simple translation to biomass and again note that now recruitment, natural mortality, and individual growth are subsumed in the production function.

(e.g. Mangel 1998). If a specified fraction (p_R) of the total spatial domain is held in reserve, the total population biomass can be expressed as the sum of biomass within areas open and closed to fishing:

$$p_R B_t + (1-u)(1-p_R) B_t = (1-u+p_R u) B_t \quad (15.1)$$

In the following, we will assume that the population is governed by the discrete logistic model. We then have:

$$B_{t+1} = \vartheta B_t + r\vartheta B_t \left(1 - \frac{\vartheta B_t}{K}\right) \quad (15.2)$$

where $\vartheta = (1-u+p_R u)$. At equilibrium, $B_{t+1} = B_t = B^*$. Solving Equation 15.2 for B^* we have:

$$B^* = \frac{K}{1-u(1-p_R)}\left(1 - \frac{u(1-p_R)}{r[1-u(1-p_R)]}\right) \quad (15.3)$$

and the equilibrium yield is given by:

$$Y^* = u(1-p_R) B^* \quad (15.4)$$

An illustration of the shape of the equilibrium yield curve as a function of exploitation rate for several values of the proportion of the area held in reserve is shown in Figure 15.2. Under the simplifying assumptions of this model, with an increasing proportion of the area held in reserve, we see a shift in the peak of the production function in relation to the exploitation rate. Under this construct, the reserve can, in effect, supply a safety net against high exploitation in the harvested part of the domain.

Notice that the term $u(1-p_R)$ appears in the expressions for both equilibrium biomass (Equation 15.3) and yield (Equation 15.4). Mangel (1998) showed that $u(1-p_R)$ is a conservation invariant (I) for which any combination of u and p_R will give equivalent results for fixed I. The equilibrium biomass expressed in terms of the invariant is:

$$B^* = \frac{K}{1-I}\left(1 - \frac{I}{r(1-I)}\right) \quad (15.5)$$

For a management objective designed to maintain biomass at or above a designated fraction of carrying capacity we can explore combinations of the maximum exploitation rate permissible and proportion of the spatial domain held in reserve to obtain this result (see Mangel 1998, 2000 for full treatment of this issue). In Figure 15.3 we illustrate results for several levels of the intrinsic rate of increase of the population. We see that for slower population growth rates, more of the area must be held in reserve to meet the objective as exploitation rates outside the reserve increase. Conversely, more productive populations can withstand higher exploitation rates with lower fractions of the spatial domain held in reserve.

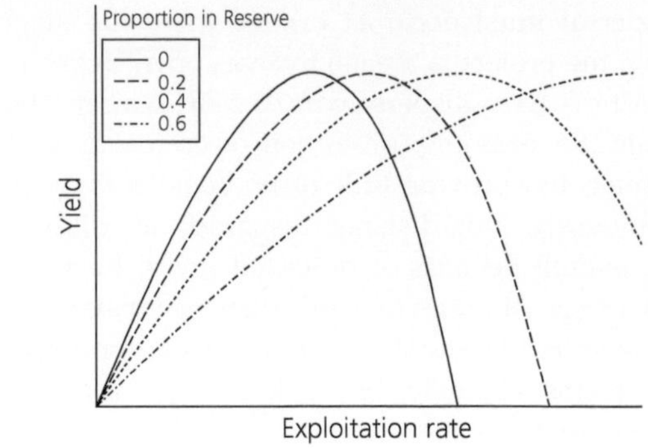

Figure 15.2 Equilibrium yield curves for a discrete logistic population model for an Aquatic Protected Area as a function of exploitation rate in the area open to fishing and the proportion of the spatial domain held in reserve.

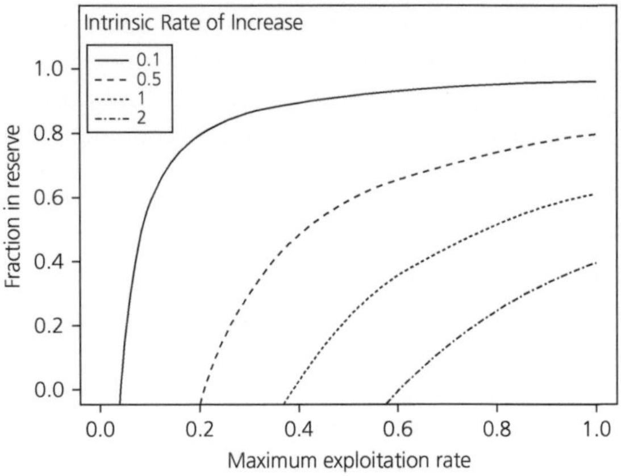

Figure 15.3 Fraction of spatial domain held in reserve as a function of maximum exploitation rate and the intrinsic rate of increase required to maintain population biomass at 60% of carrying capacity (after Mangel 1998).

Hastings and Botsford (1999) provided results for a generalized delay–difference model with sedentary adults and also showing the potential equivalence of no-take protected areas and control of exploitation rates in achieving management objectives (see review in Chapter 11 of Botsford et al. 2019). Given this equivalence, an active debate remains concerning the relative merits of achieving single-species yield and conservation goals by using APAs versus direct control of exploitation rates (see Hilborn 2014 for a review of this issue). The clearest benefit of employing APAs for achieving yield targets under single-species management in these simple models is for over-exploited stocks in which direct control of exploitation rates has not, for one reason or another, been successful. In this case, spatial closures can provide higher yields and greater resilience to exploitation through spillover from the restricted area.

In general, the degree to which it is possible to closely control exploitation rates is a central issue. Lauck et al. (1998) noted that control of exploitation rates relative to target levels is subject to uncertainty related to implementation error, which can be considerable. Lauck showed that reserves could be used to offset uncertainty in achieving the target exploitation rates and the probability of maintaining the population above a specified threshold level. More generally, Mangel (2000) showed the importance of APAs as critical buffers in the face of uncertainty.

15.2.2.2 Habitat and biodiversity

Designation of boundaries for protected areas in aquatic ecosystems naturally holds implications for the entire suite of species and associated habitat(s) within the APA. Any full appraisal of the potential merits of APAs must ultimately adopt this broader perspective. Fishing activities that adversely affect habitats can also impact system-wide carrying capacity, biodiversity, and community composition of aquatic living resources.

Levin et al. (2009b) extended the discrete-logistic APA model to address the effects of harvesting on habitat destruction and expected levels of biodiversity for the U.S. West Coast continental shelf. Specification of a species–area relationship (see Chapter 7) for this region provided the vehicle for explicitly incorporating biodiversity into the analysis. Levin et al. further included *de facto* refuge areas defined by untrawlable bottom in their analysis. For simplicity, in the following we will assume that the entire area can, in principle, be harvested. This results in scalar changes in estimated yield and species richness relative to the analysis of Levin et al. but preserves their fundamental conclusions. The total biomass is again the sum of biomass levels inside and outside of the reserve. We adopt the framework in Equations 15.1–15.4. The biomass in the area open to fishing, however, is now given by:

$$B_{O,t+1} = B_{O,t} + rB_{O,t}\left(1 - \frac{B_{O,t}}{d_h K}\right) \qquad (15.6)$$

where B_O is the biomass in the open area. The coefficient d_h is a multiplier ranging from 0–1, indicating the adverse impact of trawling on carrying capacity (K) of the habitat (lower values of d_h indicate greater reduction in carrying capacity). The effective area of the region supporting biological diversity is also affected by trawling activity:

$$A_e = p_R A + d_h (1 - p_R) A \qquad (15.7)$$

where A is the total (nominal) area of the spatial domain considered. In the area subject to harvesting, the nominal area is decremented to account for habitat destruction by trawls. The number of species (S) supported by the effective area on the shelf is computed using the species–area relationship $S = cA_e{}^z$ where c and z are coefficients.

In Figure 15.4, we illustrate the tradeoff between yield and biodiversity as measured by species richness at three levels of exploitation ($u = 0.1$, 0.15, 0.2). Each of the curves shows the relationship between yield and species richness for values of p_R ranging from 0 to 1 [the y-intercept gives the value for species richness when $p_R = 1$ (no harvest) and the terminus of each of the curves occurs at $p_R = 0$]. In all cases, the lowest species richness occurs with $p_R = 0$ where all protection against habitat damage by trawls is eliminated and the effective habitat area is accordingly diminished. As the exploitation rate progressively increases, the curves initially show a positive relationship (albeit weak, for $u = 0.1$) between yield and species richness as p_R increases from 0 to 1 but then reach an inflection point after which the relationship becomes negative. The initial positive relationship between yield and richness is a result of elimination of overfishing as with the establishment of the protected areas. However, as the

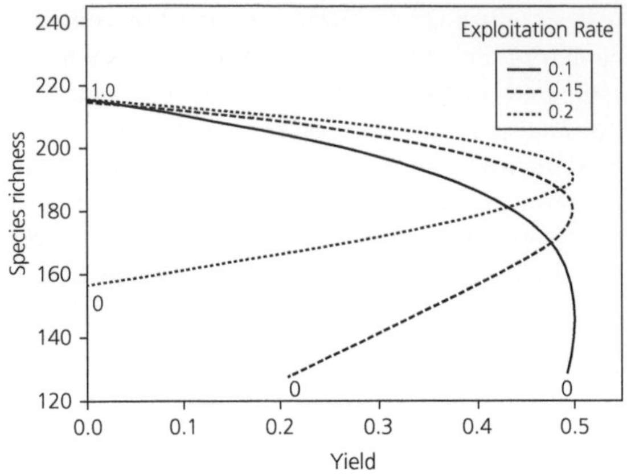

Figure 15.4 Tradeoff between species richness and yield at three levels of exploitation as the proportion of the area held in reserve increases from 0 to 1 based on the Levin et al. (2009b) model under the assumption that the entire spatial domain is trawlable.

proportion of the area held in reserve increases beyond the inflection point, negative impacts on yield are manifest. We then see the tradeoff between yield and species richness with increases in the extent of the protected area.

15.2.2.3 Interspecific interactions

In contrast to the now voluminous literature on single-species APA models, fewer examples of multispecies models have been developed and explored. Here, we will restrict ourselves to a summary of results of existing models and analyses. We noted earlier that in empirical evaluations some species show negative responses to the establishment of APAs and that interspecific interactions may account for some of these outcomes. Extension to the multispecies case introduces a very wide range of possible model structures (see Micheli et al. 2004; Baskett 2006; Baskett et al. 2007; Kellner et al. 2010; and Takashina et al. 2012).

Not surprisingly, extension to the multispecies case also reveals a very broad array of possible outcomes. An overarching finding suggests that consideration of the efficacy of APA models for multispecies systems requires larger areas to be held in reserve than for most single-species models. Recall that this mirrors the general result for maintenance of biodiversity described above. In addition, multispecies APA models generally agree with single-species models, in that APAs are most consistently effective when exploitation rates outside the area held in reserve are high.

Trophic models can demonstrate positive benefits for both prey and predators or adverse effects on the prey depending on whether the predator is a specialist or a generalist (Kellner et al. 2010) and whether the prey benefits from a refuge in size or space (Baskett 2006). Protection of a specialist predator in an APA can result in a trophic cascade leading to declines in its focal prey (e.g. Micheli et al. 2004; Takashina et al. 2012). Kellner et al. (2010) showed that availability of alternative prey substantially increases the area of the state-space in which predators and prey can co-exist in a protected area. The nature of the functional feeding response of the predator is also critically important. Similarly, models embodying a cultivation-depensation mechanism (Walters and Kitchell 2001) in which larger/older predators prey on the competitors of their juveniles can exhibit alternative stable states (Walters et al. 1999; Walters 2000; Baskett et al. 2007).

15.2.2.4 Priority areas for APAs

Given the array of possible yield and conservation outcomes related to different life history and dispersal patterns of individual species within any given APA, the nature and intensity of interspecific interactions, and behavioral responses of fishers to the establishment of APAs, can we provide selection criteria for designating priority areas for protection from a whole-system perspective? Ultimately, this question can only be answered in relation to specific objectives set by managers. It is clear, however, that no single APA can be optimal for all species within its borders.

One possibility is to concentrate on potential system-wide benefits of habitat protection. Habitats with high structural complexity would be strong candidates for priority designation. Both biogenic structures and geomorphological features contribute to structural complexity and often support high diversity assemblages of organisms, food resources, and shelter sites that can provide a refuge from predators (Kovalenko et al. 2012). These high-relief features are particularly vulnerable to disturbance by mobile bottom-tending fishing gears such as trawls and dredges. Recall that the adaptive management experiment conducted by Sainsbury (1987) on the continental shelf of northwestern Australia found strong support for the importance of habitat in defining species assemblages (see Section 13.5.1). The Collie et al. (2014) modeling study in that section also suggests adverse impacts of yield and recovery times related to loss of biogenic habitat due to fishing.

Cabral et al. (2016) report that marine protected areas delineated on the basis of habitat quality and extent provide the greatest benefit relative to alternative siting criteria in simulation tests for metapopulations. We suggest that this approach may also provide the greatest return on investment from a broader ecosystem perspective and should be thoroughly evaluated. Other considerations for high-priority designation would be for areas in which the probability of incidental catch of threatened or endangered species is high and where effective control of exploitation rates is limited.

15.3 Maintaining ecosystem structure and function

A principal objective of EBFM is to minimize disruption to ecosystem structure and function to ensure the sustainable flow of ecosystem services. Although some alteration of aspects of ecosystem structure cannot be avoided under harvesting, alternative management strategies can be evaluated with the objective of constraining adverse impacts of this type. The use of APAs can potentially provide one avenue to maintaining important elements of ecosystem structure and function within the boundaries of no-take reserves. Wing and Jack (2013) reported that a reserve network in Fiordland, New Zealand conserved biodiversity, stabilized communities, and preserved food web structure.

Conventional fisheries management has led to alterations of ecosystem structure as a result of selective removal of species occupying different trophic levels. Targeting of high-value, upper trophic level species has led to disruption of aquatic food webs, leading to cascading effects throughout the ecosystem (e.g. Frank et al. 2005; Daskalov 2002). Fishing inevitably truncates community size structure as larger individuals are removed and large species become less common. As large fish tend to be piscivorous, their depletion reduces predation pressure, thus altering predator–prey interactions in the food web (Hall 1999). Selective depletion of particular species can cause a loss of biodiversity if the community becomes dominated by a few non-target species; a less-diverse community undermines resilience to human and environmental perturbations (Lotze et al. 2006).

This phenomenon has led to consideration of strategies of maintaining ecosystem structure or "balance." In the most general sense, this would entail an integrated management strategy in which exploitation rates throughout defined elements of the ecosystem would be implemented in a coordinated fashion rather than treating each in isolation.

15.3.1 Concepts of balance in fishery ecosystems

An enduring belief that nature exists in a state of balance in the absence of human impacts, although now largely discredited among scientists, remains pervasive in other spheres of society (Kricher 2009). In Chapter 1, we saw that reconstructions of fish population abundance on millennial time scales based on scale deposition rates show dramatic patterns of variability and change prior to the advent of intensive human intervention and impacts on aquatic ecosystems (Figure 1.2). It is of course widely recognized that human activities such as harvesting can fundamentally alter the structure of populations, communities, and ecosystems. The question of how to minimize adverse impacts of human activities to prevent large-scale disruption in ecological processes has assumed increasing importance with the confluence of global climate change, pollution, exploitation, and other stressors. Here we review issues related to concepts of balance in fishery ecosystems. The importance of maintaining trophic balance in fish communities has long been recognized in freshwater fisheries management. More recently, strategies attempting to minimize impacts on demographic structures, sex ratios, genetic variability, trophodynamics, and spatial patterns have elicited considerable interest while also raising important concerns.

15.3.1.1 Trophic balance

Swingle (1950) introduced the concept of trophic balance in fish populations in his pioneering work on pond management. Monocultures of mid-trophic level species in these ponds exhibited extreme density-dependent growth patterns (stunting) such that harvestable body sizes were not attained and natural reproduction was sharply curtailed. Introduction of predators resulted in cropping of the prey species and reduction of intraspecific competition in the prey. Swingle sought to define an optimum ratio of prey to predators that resulted in consistently high yields from the system as a whole. To this end, he defined and monitored several metrics including: the ratio of total prey biomass to predator biomass; the ratio of prey biomass within size ranges vulnerable to predation-to-predator biomass; and the percentage of harvestable-size fish in the population. Cropping of the prey by predators resulted in increased production and overall yield in these multispecies systems. The value of being able to conduct controlled and replicated experiments is clearly evident in these studies.

Although Swingle's work essentially entailed creating artificial ecosystems in experimental ponds, its relevance can be extended to natural systems in understanding the importance of prey to predator ratios. Large-scale distortion of trophic structure through selective and differential removal of trophic groups can hold long-term consequences for the resilience of aquatic ecosystems.

Swingle also experimented with manipulation of nutrient levels in the experimental ponds and stocking of broader multispecies assemblages of prey and predators. These experiments shed light on the importance of overall patterns of energy utilization for production and yield (see also Tang 1970). Higher-diversity assemblages provided pathways for fuller utilization of available energy and the possibility of higher yields from the assemblages as a whole. This result reinforces the value of maintaining biodiversity in exploited ecosystems.

Pauly et al. (2000) introduced the Fishing-in-Balance (FIB) index to quantify the effects of selective removal of lower trophic level fishes on upper trophic levels. A focus on harvesting lower trophic level fish is expected to generate increased overall catch because of the higher productivity at these levels. The FIB index examines whether the increase in yield is commensurate with decline in the mean trophic level of the catch. If catches do not increase at the expected rate, an imbalance induced by the fishery is inferred. More recently, the FIB has also been used to detect geographical expansion of fisheries to new fishing grounds which also alters the patterns of yield.

15.3.1.2 Balanced harvest

Consideration of the effects of fishing at the ecosystem level has led for calls to adjust selectivity regulations to balance the impact of all fisheries in an area with the relative production of the species and sizes of organisms in the ecosystem (e.g. Zhou et al. 2010; Garcia et al. 2012). In its most expansive form, balanced harvesting would entail exploiting all species in an aquatic ecosystem in proportion to their production. This interpretation has elicited important concerns over whether, in this form, Balanced Harvest would be feasible, economical, and socially acceptable (e.g. Froese et al. 2016; Pauly et al. 2016; Breen et al. 2016; Burgess et al. 2016). Zhou et al. (2019) noted that Balanced Harvest could in fact be implemented in a number of forms, including ones with a more restricted focus. Zhou et al. (2019) defined Balanced Harvest as:

> *the management strategy and collective fishing activities that impose moderate fishing mortality on each* <u>*utilizable*</u> *ecological group in proportion to its production, to support long-term sustainable yields while minimizing fishing impacts on the relative species, size, and sex composition within an ecosystem* (emphasis added).

Under this definition various implementation strategies could be accommodated including partially balanced harvest focusing on selected ecosystem components (Zhou et al. 2019).

Exploitation of an aquatic community has two components: the overall level of fishing pressure and the degree of selection for species, sizes, and sexes. The effects of these two components are very difficult to disentangle from observations of real fish communities; hence empirical tests of the effects of selective fishing are fraught with difficulty. Given the challenges of empirical tests, several size-based models have been used to test the effects of selective and unselective fisheries on community structure and sustainable yield (e.g. Rochet et al. 2011; Rochet and Benoît 2012; Andersen and Pedersen 2010).

Jacobsen et al. (2014) made a systematic evaluation of this issue using the size-spectrum model of Andersen and Pedersen (2010). The analysis was directed at fish community dynamics and did not consider harvesting of zooplankton or aquatic vertebrates other than fish. In this study, balanced harvesting meant that fishing mortality was scaled in relation to species and size-dependent productivity. "Selective" or "unselective" denoted whether or not juvenile fishes were included in the fishery. In these simulations, unselective balanced harvesting produced a slightly higher total yield than unbalanced and selective fishing (Figure 15.5). Balanced harvesting also had less impact on the overall size structure of the fish community but significantly reduced the mean size of fish in the catch. Jacobsen et al. (2014) further note that under the options that entail harvesting of all size classes[2], most of the catch would not go to direct human consumption but rather would be processed as fish meal for animal food (including aquaculture operations) or other products. The sharply reduced mean size of individuals in the catch may not be acceptable to consumers and the resulting fisheries may not be as profitable. It is widely recognized that further evaluation of the full social and economic consequences and aspects of practical implementation of Balanced Harvest is required (e.g. Jacobsen et al. 2014; Burgess et al. 2016; Zhou et al. 2019). This analysis neatly encapsulates some of the potential benefits and practical difficulties involved in consideration of balanced harvesting.

Partial implementation could focus on taxa that are currently harvested (principally fish and shellfish). Fisheries are collections of fishing gears that are more-or-less selective. A balanced fishery would tune the individual gears and their species allocations to obtain a more balanced harvest across these traditionally harvested communities.

[2] Jacobsen et al. (2014) set a lower size limit at 10 g.

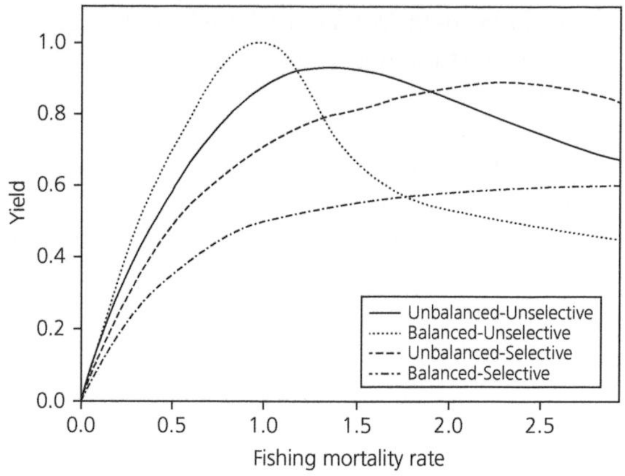

Figure 15.5 Yield from four exploitation patterns representing combinations of size selective harvest [no juvenile harvest (selective); juvenile harvest (unselective)] and balanced harvest [fishing mortality proportional to productivity (balanced); fishing mortality equal over all sizes (unbalanced)]. Adapted from Jacobsen et al. (2014).

Species that are currently overfished would need to be selectively rebuilt before a sustainable balanced harvest could be achieved. Current fishery regulations specify fishing mortality rates in relation to species productivity, accounting for the size selectivity of fishing gears. A balanced harvest would involve lower overall fishing pressure across a wider range of species and sizes than at present (e.g. Fogarty and Murawski 1998). In this sense balanced harvesting is more of a guiding principle than a formulaic prescription to harvest all sizes of all species (see also Zhou et al. 2019). Markets will need to be developed to promote utilization of less-desirable and less-valuable species. It is worth noting that, in fact, the number of species caught in mixed-species fisheries can be substantially larger than the component actually landed for market. In this case, the issue is one of fuller utilization of what is caught (and now discarded) rather than developing new fisheries altogether.

15.4 Defining overfishing in an ecosystem context

The classical definitions of overfishing from a single-species perspective focus on yield, biomass, and demographic structure of the harvested species. The concept of single-species maximum sustainable yield is typically invoked to quantify resource status. Because of the broader remit of EBFM, consideration of impacts on other ecosystem components is necessary to assess the impacts of harvesting on community and ecosystem status. Specification of ecosystem-based reference points to guide management actions remain at an early stage of development, in part because clearly defined and actionable objectives for EBFM have been adopted in few jurisdictions to date.

Community-level status can be framed as a natural extension of approaches aimed at individual species or stocks. However, as we noted in Chapter 12, a direct translation to concepts such as multispecies maximum yield is not possible. The expected yield, population biomass, and demography are now functions of the interactions with other species and differential patterns of exploitation. The yield in general cannot be simultaneously maximized for all species. The problem is particularly acute in mixed-species fisheries driven by technical and biological interactions among species. At the full ecosystem level, of course, a much broader array of factors must be considered.

We have noted the importance of environmental influences, particularly temperature, on production and yield at the individual, population, community, and ecosystem levels. The rapidly accumulating evidence of climate effects on aquatic ecosystems highlights the need to view ecosystem-based reference points in a dynamic context. Timely adjustment of reference points in response to changing production levels will be essential.

Barange et al. (2018) provide a recent comprehensive synthesis of climate effects on global fisheries and aquaculture in marine and freshwater ecosystems. Changes in hydrological cycles and hydrodynamics processes; increases in temperature, stratification, and acidification; decreases in oxygen content; and shifts in species distribution patterns are predicted. The maximum potential catch in global marine ecosystems is projected to decline by 7–12% by mid-century under a business-as-usual scenario. Under a mitigation scenario intended to keep global temperature increase below 2°C, a decline of 2.8–5.3% is predicted, with the strongest negative impacts in low latitude systems (Barange et al. 2018). Freshwater ecosystems have a lower buffering capacity to absorb the effects of climate change and they are subject to greater array of anthropogenic threats (including competing water use demands, habitat destruction and fragmentation, and introduction of non-native species) than marine ecosystems. However, approximately 60% of the freshwater fishery production currently comes from regions projected to experience low to moderate stress related to climate change while stress in other areas already experiencing high impacts due climate change is projected to increase.

15.4.1 Community-level reference points

In Chapter 12 we noted that biological and technical interactions must be considered in developing targets for management at the community and ecosystem levels. In particular, any attempt to remove an overall maximum yield from an assemblage of interacting species places lower-productivity members of the community at risk. We considered two options: (1) adoption of a precautionary management scheme in which fishing pressure is reduced below the level resulting in the maximum yield from the assemblage as a whole with the objective of reducing risk to low-productivity species and (2) defining ranges of fishing mortality rates for component species that provide sustainable yields for each. To illustrate the first option, we considered results from the size-structured model LeMANS applied to the Georges Bank fish community (Worm et al. 2009; see Figure 12.13) and showed that we can readily visualize tradeoffs between overall yield and conservation of biodiversity. In this illustration, reducing fishing mortality from the level resulting in maximum community yield ($F = 0.43$) to $F = 0.2$ substantially reduced the probability of species falling below a safety threshold from nearly 40 per cent to less than 5 per cent. The yield at the lower harvest rate was nearly 90 per cent of the maximum. With a further reduction to $F = 0.1$, no species were driven below the threshold in these deterministic simulations but the yield was less than 60 per cent of the maximum. A significant issue in addressing tradeoffs between yield and biodiversity is evident and management choices depend on the mix of conservation tools available to ameliorate risk to vulnerable species.

The second option considered in Chapter 12 entailed defining ranges of fishing mortality rates for the mix of species harvested to explore the multidimensional space allowing sustainable harvest for all species. Specifying the range of fishing rates that provide "pretty good yield" for all species in the complex has been given particular attention (e.g. Rindorf et al. 2017). In this context, pretty good multispecies yield is defined as the combination of fishing mortality rates that provide a specified percentage of the yield for that stock in a single-species analysis. In most explorations to date, that percentage has been set at $>$ 95 per cent of the single species MSY level and assume that individual stocks can be cleanly targeted. Quite sophisticated analytical methods have been used to determine the "best" options, including applications of game theory which seek to determine Nash equilibria for the species complex. In this context, the Nash equilibrium involves fishing all exploited stocks at rates such that changes in the exploitation rate cannot increase the long-term yield from that stock if exploitation rates on all other stocks are fixed.

Thorpe et al. (2017) extended this approach to consider species/stocks subject to both biological and technical interactions using the LeMANS modeling framework introduced in Chapter 12. It was shown that while modest increases in yield were achievable by fishing at the upper end of the fishing mortality rates giving pretty good yield, risks of population collapse increased considerably relative to choices at the lower end of this exploitation rate spectrum. It was recommended that fishing mortality rates be held to values less than F_{MSY} levels for all stocks.

Thorpe (2019) subsequently explored the performance of a management strategy in which stocks were subject to a common fishing mortality rate and a fixed target biomass level defined as a specified fraction of virgin biomass. This harvest control rule was compared to alternatives, including ones based on determining the Nash equilibrium for the multispecies complex and the use of results of single-species assessments to define the range of fishing mortality rates providing pretty good yield. The strategy employing a common fishing mortality rate for all stocks and the establishment of a biomass target constraint produced results that were comparable with the Nash equilibrium results while substantially reducing the computational burden. Thorpe (2019) concluded that overall management outcomes were less dependent on specifying stock-specific fishing mortality rates than knowing the biomass status of the stock (although income variability of fishers was higher than for the case of stock-specific fishing mortality rates).

15.4.2 Ecosystem-level reference points

In order to establish effective benchmarks for management, a clear definition of what constitutes ecosystem overfishing is necessary. A number of criteria for defining overfishing at an ecosystem level have been put forward. Murawski (2000) suggested that an ecosystem could be considered overfished if one or more of the following conditions were met:

- Biomasses of one or more important species assemblages or components fall below minimum biologically acceptable limits, such that:
    1) recruitment prospects are impaired,
    2) rebuilding times to levels allowing catches near MSY are extended,
    3) prospects for recovery are jeopardized because of species interactions,
    4) any species is threatened with local or biological extinction.
- Diversity of communities or populations declines significantly as a result of sequential "fishing-down" of stocks, selective harvesting of ecosystem components, or other factors associated with harvest rates or species selectivity.
- The pattern of species selection and harvest rates leads to greater year-to-year variation in populations or catches than would result from lower cumulative harvest rates.
- Changes in species composition or population demographics as a result of fishing significantly decrease the resilience or resistance of the ecosystem to perturbations arising from non-biological factors.
- The pattern of harvest rates among interacting species results in lower cumulative net economic or social benefits than would result from a less

intense overall fishing pattern or alternative species selection.

- Harvests of prey species or direct mortalities resulting from fishing operations impair the long-term viability of ecologically important, non-resource species (e.g., marine mammals, turtles, seabirds).

Tudela et al. (2005) connected the basic concept of the amount of primary production appropriated to sustain the fisheries to Murawski's criteria to specify an ecosystem overfishing definition. The primary production required to account for observed catch levels is:

$$PPR = \sum_{i}^{n} \frac{C_i}{c_r} \left(\frac{1}{TE} \right)^{TL_i - 1} \quad (15.8)$$

(Pauly and Christensen 1995) where c_r is a conversion rate from wet weight to carbon (typically assigned a value of 9), C_i is the catch of species i (landings plus discards), TE is the transfer efficiency between trophic levels (here taken to be a constant over all trophic levels), and TL_i is the mean trophic level of species i. In many existing applications, the transfer efficiency is taken to be 10 per cent although, given system-specific information, alternative values can be applied. Estimates of transfer efficiencies by trophic level can be obtained from a valuable repository of Ecopath model results (see www.ecobase.ecopath.org). Extensive compilations of trophic level estimates are available from FishBase (www.fishbase.org) for fish and SeaLifeBase (www.sealifebase.org) for invertebrates. The resulting estimate of PPR expressed as a proportion of the primary production is itself an informative metric (Pauly and Christensen 1995).

Tudela et al. (2005) examined observations for 49 ecosystems to which Murawski's criteria could be applied and for which PPR could be calculated. Multivariate analyses were used to define demarcation points between overfished and sustainably fished systems in relation to PPR and the mean trophic level of the catch. Libralato et al. (2008) extended this approach to estimate the losses in secondary production incurred by removing production at lower trophic levels, thus affecting the energy available to higher trophic levels. The loss index was calculated for each ecosystem and related to ecosystem status according to Murawski's criteria. The probability that an ecosystem would be classified as sustainably fished for a given loss level was then determined. See Libralato et al. (2008) and Coll et al. (2008) for details on derivations and estimation methods.

Link (2005) proposed warning thresholds and limit reference points for a series of indicator types including size structure (mean length and slope of the size spectrum of all species); biomass of species guilds defined by trophic groups or taxonomic affinity; fishery metrics (landings, biomass removed through catch, discards, and by-catch); community and food-web indices (species richness, mean number of interactions per species, number of cycles); and metrics related to scavenger species, gelatinous zooplankton, and hard corals. Fulton et al. (2019) applied these indicators and associated reference points to the Southern and Eastern Scalefish and Shark Fishery and reported good performance of this approach in their management strategy evaluation of this fishery (see Section 15.5).

Link and Watson (2019) proposed a set of production- and biomass-based indicators and associated limit reference points including: (1) the ratio of catches to chlorophyll concentration; (2) the ratio of catches to primary production; and (3) the catch-per-unit-area. Based on theoretical and empirical considerations, Link and Watson set tentative threshold values of the ratio of catches to chlorophyll and catches to primary production not to exceed 1 per mil. Values above this level would be classified as overfished at the ecosystem level. Based on empirical observations, Link and Watson proposed a tentative catch-per-unit-area threshold of 1 t km^{-2}. Values above this level would be designated overfished. Based on these threshold indicators, Link and Watson reported that currently, nearly 40–50 per cent of tropical and temperate large marine ecosystems would be classified as overfished from an ecosystem perspective using these criteria.

15.4.3 System-level yield

The amount of carbon fixed at the base of the food web sets limits on the production at different trophic levels and ultimately the amount of yield that can be extracted, with appropriate constraints, from an aquatic ecosystem. The exact amount is a function of the interplay between bottom-up and top-down controls in different systems. It has long been recognized that total fish yield, size structure, and biomass levels often reflect remarkably conservative properties of aquatic ecosystems and communities (Kerr and Ryder 1988). The apparent greater stability at the ecosystem level reflects overall energetic constraints on system dynamics. In traditional fisheries management, we focus on the dynamics of individual species. Adopting an ecosystem perspective has led to management options that establish a system-wide limit reference point such that the sum of removals from individual elements of the system cannot exceed a specified level. Link (2018) reviewed the conceptual framework for this approach in terms of systems theory and noted its potential benefits in terms of fishery output (yield, value, and stability) and risk management.

A system of this type was implemented by the International Commission for Northwest Atlantic Fisheries (ICNAF) for the Northeast US continental shelf (ICNAF 1974) prior to the implementation of the 200-mile limit. A similar system-level constraint has been in place for the Bering Sea–Aleutian Islands fishery since 1984 (Witherell et al. 2000; Link 2018). Trenkel (2018) proposed a management procedure in which a multispecies total allowable catch (MTAC) would be set. Individual species TACs derived from single- or multispecies models would be established such that their sum would not exceed the MTAC. The overall system catch was set using estimates of multispecies MSY derived from the size spectrum model of Jennings and Collingridge (2015) (see Chapter 13). Trenkel (2018) incorporated potential changes in primary production as a multiplier to the system-level MTAC. In the proposed system, a relative decrease or high variability in primary production would trigger a proportional downward adjustment to the MTAC. This general philosophy could also accommodate other forms of environmental change. An additional precautionary buffer was also recommended. The management procedure is therefore viewed in a dynamic setting that could prove extremely valuable in the face of climate-induced ecosystem changes.

15.5 Management strategy evaluation

The value of testing candidate fishery management strategies in a virtual world has been widely appreciated since Walters (1986) introduced the concept in the context of adaptive management. Hilborn and Walters (1992, p. 237) subsequently brought the concept of simulation testing of model performance and estimation procedures using operating models to a broad audience of fisheries scientists. At about this time, adoption of the related concept of risk assessment was gaining traction in fisheries applications (e.g. Smith et al. 1993). Smith (1994) introduced the term Management Strategy Evaluation (MSE) as a descriptor of these fundamental concepts and their application in assessing the potential efficacy of implementing alternative harvest policies.

While routine simulations are performed by fisheries scientists to test the performance of estimation procedures and harvest control rules, stakeholder engagement at multiple points in the MSE process is needed for the resulting harvest policies to be implemented (Walters 1986; Smith 1994). Punt et al. (2016) identified the key elements of MSE, which we paraphrase here to encompass community- or ecosystem-level MSEs:

- Identify the management objectives in concept and represent them with quantitative performance statistics.
- Identify a broad range of uncertainties (related to biology, the environment, the fishery, and the management system) to which the management strategy should be robust.
- Develop a set of operating models that mathematically represent the fishery system. An operating model must represent the biological/ecological components of the system, the fishery, how data are collected, and how they relate to the modelled system (including measurement error). In addition, an implementation model is required that reflects how management regulations are applied in practice.

- Select the parameters of the operating model(s) and quantify parameter uncertainty (ideally by fitting or "conditioning" the operating model(s) to data from the actual system under consideration).
- Identify candidate management strategies that could realistically be implemented for the fishery system.
- Simulate the application of each management strategy for each operating model.
- Summarize and interpret the performance statistics (e.g. with a decision table) to inform the quantitative tradeoffs among competing goals. (Punt et al. 2016).

A flow diagram of the iterative MSE process is provided in Figure 15.6. Although ecosystem-level MSEs remain less common than evaluation of single-species management strategies, there has been a steady increase in simulation testing of ecosystem-based management strategies. Operating models used for this purpose include LeMANS (Thorpe et al. 2017, Thorpe 2019); EWE (Walters et al. 2005; Mackinson et al. 2018); and Atlantis (Fulton et al. 2019).

One of the earliest MSEs for a fisheries ecosystem was developed for krill (*Euphausia superba*) in the Southern Ocean by the Commission for the Conservation of Antarctic Living Marine Resources (CCAMLR)(De la Mare 1996; Constable 2011). The overarching goal of CCAMLR is to maintain ecological relationships among harvested, dependent, and related species and to restore depleted populations within the convention area. Krill occupy the nexus of the Southern Ocean food web and many fish, mammal, and bird species depend on krill as prey. The selected objective for the krill management procedure is to maintain spawning biomass at three quarters of the unexploited level with a specified probability to ensure adequate food supplies for predators. Spatial management strategies to protect non-target organisms are an integral part of the overall approach. The CCAMLR experience is unique in that ecosystem principles were embedded in its management strategy from the beginning. A review of the implementation and performance of the CCAMLR approach reveals that relatively simple ecosystem-based management procedures can in fact be quite effective (Constable 2011).

In what is perhaps the most comprehensive evaluation of the relative merits of ecosystem-based and multispecies fisheries management relative to conventional single-species management to date, Fulton et al. (2019) made a detailed evaluation of alternative harvesting policies for the Southern and Eastern Scalefish and Shark Fishery (SESSF) in Australia. This analysis involved extensive

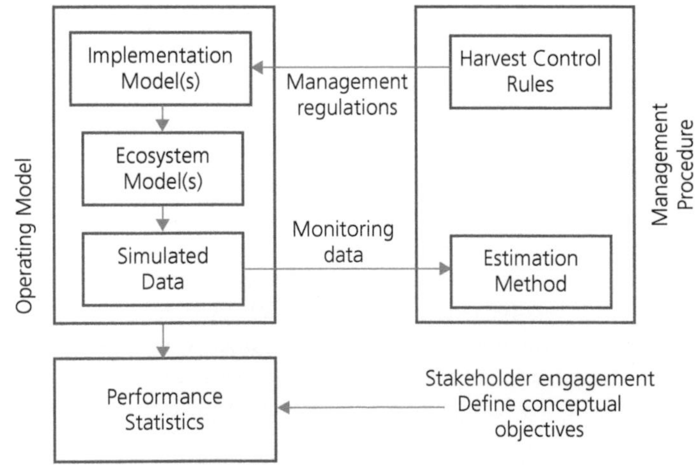

Figure 15.6 Flow diagram for management strategy evaluation (adapted from Punt et al. 2016).

stakeholder involvement and was undertaken using the Atlantis modeling framework. The MSE was tailored to the specific fishery and management history of the scalefish–shark complex with detailed specification of the exploited species, the environmental/oceanographic setting, food-web architecture, and fishery structure and characteristics of this ecosystem. Twenty-seven different management scenarios were simulated with major subdivisions defined by the application of multispecies and EBFM policies versus single-species management and whether these alternative approaches were applied throughout the SESSF domain or in subsections defined by internal jurisdictional boundaries. In these simulations, the EBFM (or integrated management) scenarios employed a mix of Individual Transferable Quotas, limited entry, gear controls, and spatial management. Multispecies strategies were defined as ones in which catches were taken in proportion to productivity within constraints imposed by the mix of existing gears (and their selectivities). The multispecies option was therefore a form of partial implementation of a balanced-harvesting strategy. For detailed description of the component elements within each subdivision, see Fulton et al. (2019). The final overall scenario involved unconstrained fishing.

Fourteen indicators were selected as measures of performance for each of the scenarios tested (see Table 4 of Fulton et al. 2019). Major subdivisions of these metrics included measures of: biomass or abundance; habitat integrity; ratio of total biomass of demersal to pelagic fish; biodiversity; indicators related to catch characteristics and value; employment; and social well-being of individual fishers. Performance of management options throughout the entire domain ranked higher than ones implemented in parts of the region. Within each of these two major subdivisions, the overall performance of the EBFM policy ranked highest, followed by multispecies and then single-species management policies. Examination of the rankings of individual performance indicators reveal distinctions that emerge in comparisons across scenarios. Not surprisingly, the unconstrained fishing scenario had the lowest overall performance score.

Fulton et al. (2019) conclude that the fishery-related and conservation benefits of the integrated approach embodied in EBFM are clear relative to the single-species approach. Fulton et al. (2019) further note that important ecosystem benefits need not come at the expense of reduced catches. The resilience afforded by adopting the ecosystem approach can provide a buffer against the impacts of climate change and expansion of pressures from other anthropogenic stressors. Although this example is of course tailored to a specific real-world fishery, the general conclusions appear to be quite robust and potentially applicable to many other fishery ecosystems.

15.6 Summary

The strategic and tactical elements of EBFM will inevitably be shaped by the objectives chosen by management authorities (preferably determined in direct consultation with stakeholders) and the scientific, fiscal, and administrative resources available for implementation. EBFM will therefore of necessity assume different forms in different places.

Given a set of objectives, we can identify key elements for implementation:

- Objectively define the management units (spatial domain and components of the coupled human-ecological fishery ecosystem).
- Identify the production characteristics and potential yield of the system.
- Specify the types of controls and management options to be employed to meet specified objectives.
- Define reference points for management that relate directly to existing management approaches and adjust in response to climate forcing.
- Evaluate tradeoffs that will inevitably emerge (e.g. tradeoffs between yield and protection of non-target species and habitats).

Maintaining social-ecological resilience lies at the heart of successful EBFM as an overarching principle. If resilience is safeguarded, the flow of ecosystem services (including food from aquatic systems) can be maintained. If we were to identify a single construct critical to maintaining resilience, it would center on preserving diversity at all levels in the social-ecological system. Because diversity can be quantified, it can lead to the specification of clear benchmarks for setting objectives and tracking performance success.

Additional reading

For an overview of fisheries management in an ecosystem context, see Garcia et al. (2003). Link (2010) provides a succinct guide to Ecosystem-based Fisheries Management and the critically important issue of tradeoffs that inevitably arise in multi-objective management. See Fulton and Link (2014) for an overview of models employed in Ecosystem-based Management. Recent overviews of Aquatic Protected Areas are provided for freshwater ecosystems by Finlayson et al. (2018) and for marine ecosystems by Botsford et al. (2019, Chapter 11); see also Jennings et al. (2001, Chapter 17). The ICES Journal of Marine Science Volume 73(6) contains a special theme section on Balanced Harvest. For an early treatment of core concepts of management strategy evaluation see Walters (1986).

Bibliography

Abell, R., Thieme, M. L., Revenga, C., Bryer, M., Kottelat, M., Bogutskaya, N., & Stiassny, M. L. (2008). Freshwater ecoregions of the world: A new map of biogeographic units for freshwater biodiversity conservation. BioScience, 58(5), 403–14.

Abrams, P. A. (1987). On classifying interactions between populations. Oecologia, 73(2), 272–81.

Abrams, P. A. (1994). The fallacies of "ratio dependent" predation. Ecology, 75(6), 1842–50.

Abrams, P. A. (2015). Why ratio dependence is (still) a bad model of predation. Biol. Rev, 90, 794–814. https://doi.org/doi:10.1111/brv.12134

Adams, C. F., Alade, L. A., Legault, C. M., O'Brien, L., Palmer, M. C., Sosebee, K. A., & Traver, M. L. (2018). Relative importance of population size, fishing pressure and temperature on the spatial distribution of nine Northwest Atlantic groundfish stocks. PLoS One, 13(4), e0196583.

Alaska Sea Grant College Program (1999). Ecosystem approaches for fisheries management. University of Alaska, AKSG-99–01, Fairbanks.

Allen, K. R. (1963). Analysis of stock-recruitment relations in Antarctic fin whales. Rapports et Proces-verbaux des Reunions Conseil International pour Exploration de la Mer 164, 132–7.

Allen, K. R. (1971). Relation between production and biomass. Journal of the Fisheries Board of Canada, 28(10), 1573–81.

Allen, K. R. & Kirkwood, G.P. (1988). Marine mammals. In: Gulland, J. A. (Ed.), Fish population dynamics: The implications for management (2nd edition). Wiley, New York.

Allison, E. H., Irvine, K., Thompson, A. B., & Ngatunga, B. P. (1996). Diets and food consumption rates of pelagic fish in Lake Malawi, Africa. Freshwater Biology, 35(3), 489–515.

Alverson, D. L., Freeburg, M. H., Murawski, S. A., & Pope, J. G. (1994). A global assessment of fisheries bycatch and discards. FAO Fisheries Technical Paper 339.

Amoroso, R. O., Pitcher, C. R., Rijnsdorp, A. D., McConnaughey, R. A., Parma, A. M., Suuronen, P., ... Jennings, S. (2018). Bottom trawl fishing footprints on the world's continental shelves. Proceedings of the National Academy of Sciences USA, 115, 10275–82.

Andersen, K. H., Brander, K., & Ravn-Jonsen, L. (2015). Trade-offs between 624 objectives for ecosystem management of fisheries. Ecological Applications, 25, 1390–6.

Andersen, K. H., Jacobsen, N. S., Jansen, T., & Beyer, J. E. (2017). When in life does density dependence occur in fish populations? Fish and Fisheries, 18(4), 656–67.

Andersen, K. H., & Pedersen, M. (2010). Damped trophic cascades driven by fishing in model marine ecosystems. Proceedings of the Royal Society B: Biological Sciences, 277(1682), 795–802.

Andersen, K. P., & Ursin, E. (1977). A multispecies extension to the Beverton and Holt theory of fishing, which accounts for phosphorus circulation and primary production. Meddelelser fra Danmarks Fiskeri-og Havundersoegelser 7, 319–435.

Andersen, R. M., & May, R. M. (1978). Regulation and stability of host-parasite population interactions: I. Regulatory processes. Journal of Animal Ecology, 47, 219–247.

Anderson, C. N. K., Hsieh, C.-h., Sandin, S. A., Hewitt, R., Hollowed, A., Beddington, J., ... Sugihara, G. (2008). Why fishing magnifies fluctuations in fish abundance. Nature, 452, 835–839.

Anon. (1921, March 19). The history of trawling. Fish Trades Gazette, 22–69.

Arditi, R., & Ginzburg, L. R. (1989). Coupling in predator–prey dynamics: Ratio-dependence. Journal of Theoretical Biology, 139(3), 311–26.

Arditi, R., & Ginzburg, L. R. (2012). How species interact: Altering the standard view on trophic ecology. Oxford University Press, New York.

Audzijonyte, A., Pethybridge, H., Porobic, J., Gorton, R., Kaplan, I., & Fulton, E. A. (2019). Atlantis: A spatially explicit end-to-end marine ecosystem model with dynamically integrated physics, ecology and socio-economic modules. Methods in Ecology and Evolution, 10(10), 1814–19.

Fishery Ecosystem Dynamics. Michael J. Fogarty and Jeremy S. Collie, Oxford University Press (2020).
DOI: 10.1093/oso/9780198768937.001.0001

Auster, P. J., & Langton, R. W. (1999). The effects of fishing on fish habitat. American Fisheries Society Symposium, 22, 150–87.

Auster, P. J., & Link, J. S. (2009). Compensation and recovery of feeding guilds in a northwest Atlantic shelf fish community. Marine Ecology Progress Series, 382, 163–72.

Aydin, K. Y. (2004). Age structure or functional response? Reconciling the energetics of surplus production between single-species models and Ecosim. Ecosystem Approaches to Fisheries in the Southern Benguela, 26, 289–301.

Azam, F., Fenchel, T., Field, J. G., Gray, J. S., Meyer-Reil, L. A., & Thingstad, F. (1983). The ecological role of water-column microbes in the sea. Marine Ecology Progress Series, 10, 257–63.

Barange, M., Bahri, T., Beveridge, M. C. M, Cochrane, K. L., Funge-Smith, S., & Poulain, F. (2018). Impacts of climate change on fisheries and aquaculture: synthesis of current knowledge, adaptation and mitigation options. FAO Fisheries and Aquaculture Technical Paper No. 627. FAO, Rome.

Baranov, F. I. (1918). On the question of the biological basis of fisheries. Nauchnge Issledovaniya Ikhtiologicheskii Instituta Izvestiya, 1918(1), 81–128.

Baranov, F. I. (1976). Selected works on fishing gear. Israel Program for Scientific Translations, Jerusalem.

Barbour, C., & Brown, J. (1974). Fish species diversity in lakes. The American Naturalist, 108(962), 473–89.

Barneche, D. R., Burgess, S. C., & Marshall, D. J. (2018). Global environmental drivers of marine fish egg size. Global Ecology and Biogeography, 27(8), 890–8.

Barnett, T. P., Adam, J. C., & Lettenmaier, D. P. (2005). Potential impacts of a warming climate on water availability in snow-dominated regions. Nature, 438, 303–9.

Baskett, M. L. (2006). Prey size refugia and trophic cascades in marine reserves. Marine Ecology Progress Series, 328, 285–93.

Baskett, M. L., Micheli, F., & Levin, S. A. (2007). Designing marine reserves for interacting species: insights from theory. Biological Conservation, 137(2), 163–79.

Basson, M., & Fogarty, M. J. (1997). Harvesting in discrete-time predator–prey systems. Mathematical Biosciences, 141(1), 41–74.

Bauer, B., Horbowy, J., Rahikainen, M., Kulatska, N., Müller-Karulis, B., & Tomczak, M. T. (2019). Model uncertainty and simulated multispecies fisheries management advice in the Baltic Sea. PLoS One, 14(1), e0211320.

Baumgartner, T. R., Soutar, A., & Ferreira-Bartrina, V. (1992). Reconstruction of the Pacific sardine and northern anchovy population over the past two millennia from sediments of the Santa Barbara basin. California Cooperative Fishery Investigations Report (Vol. 33, pp. 24–40).

Beamish, R. J., McFarlane, G. A., & Benson, A. (2006). Longevity overfishing. Progress in Oceanography, 68, 289–302.

Beddington, J. R. (1975). Mutual interference between parasites or predators and its effect on searching efficiency. The Journal of Animal Ecology, 331–340.

Beddington, J. R., Free, C. A., & Lawton, J. H. (1975). Dynamic complexity in predator–prey models framed in difference equations. Nature, 255, 58–60.

Bell, R. J., Fogarty, M. J., & Collie, J. S. (2014a). Stability in marine fish communities. Marine Ecology Progress Series, 504, 221–39.

Bell, R. J., Hare, J. A., Manderson, J. P., & Richardson, D.E. (2014b). Externally driven changes in the abundance of summer and winter flounder. ICES Journal of Marine Science, 9, 2416–28.

Belmaker, J., Ben-Moshe, N., Ziv, Y., & Shashar, N. (2007). Determinants of the steep species area relationship of coral reef fishes. Coral Reefs, 26, 103–12.

Berryman, A. A. (1991). Can economic forces cause ecological chaos? The case of the northern California Dungeness crab fishery. Oikos, 62, 106–9.

Beverton, R. J. H., & Holt, S. J. (1957). On the dynamics of exploited fish populations. London: Chapman & Hall.

Bjørndal, T. (1988). Optimal harvesting of farmed fish. Marine Resource Economics, 5, 139–59.

Blackwood, J. C., Hastings, A., & Mumby, P. J. (2012). The effect of fishing on hysteresis in Caribbean coral reefs. Theoretical Ecology, 5(1), 105–14.

Blackwood, J. C., Okasaki, C., Archer, A., Matt, E. W., Sherman, E., & Montovan, K. (2018). Modeling alternative stable states in Caribbean coral reefs. Natural Resource Modeling, 31(1), e12157.

Blanchard, J. L., Dulvy, N. K., Jennings, S., Ellis, J. R., Pinnegar, J. K., Tidd, A., & Kell, L. T. (2005). Do climate and fishing influence size-based indicators of Celtic Sea fish community structure? ICES Journal of Marine Science, 62, 405–11.

Blanchard, J. L., Jennings, S., Holmes, R., Harle, J., Merino, G., Allen, J. I., & Barange, M. (2012). Potential consequences of climate change for primary production and fish production in large marine ecosystems. Philosophical Transactions of the Royal Society B: Biological Sciences, 367(1605), 2979–89.

Bohaboy, E. C. (2010). Multi-species fish population biomass dynamics of Georges. Thesis, Graduate School of Oceanography, University of Rhode Island, Rhode Island.

Bolker, B. M. (2008). Ecological models and data in R. Princeton: Princeton University Press.

Bolster, W. J. (2008). Putting the ocean in Atlantic history: Maritime communities and marine ecology in the Northwest Atlantic, 1500–1800. The American Historical Review, 113(1), 19–47.

Bordet, C., & Rivest, L. P. (2014). A stochastic Pella Tomlinson model and its maximum sustainable yield. Journal of Theoretical Biology, 360, 46–53.

Botsford, L. W. (1992). Further analysis of Clark's delayed recruitment model. Bulletin of Mathematical Biology, 54(2–3), 275–93.

Botsford, L. W., Castilla, J. C., & Peterson, C. H. (1997). The management of fisheries and marine ecosystems. Science, 277(5325), 509–15.

Botsford, L.W., White, J.W., & Hastings, A. (2019). Population Dynamics for Conservation. Oxford University Press, Oxford.

Bousquet, N., Duchesne, T., & Rivest, L. P. (2008). Redefining the maximum sustainable yield for the Schaefer population model including multiplicative environmental noise. Journal of Theoretical Biology, 254(1), 65–75.

Bowen, W. D., McMillan, J., & Mohn, R. (2003). Sustained exponential population growth of grey seals at Sable Island, Nova Scotia. ICES Journal of Marine Science, 60(6), 1265–74.

Brander, K. M. (1995). The effect of temperature on growth of Atlantic cod (*Gadus morhua* L.). ICES Journal of Marine Science, 52(1), 1–10.

Brander, K. M. (2005). Cod recruitment is strongly affected by climate when stock biomass is low. ICES Journal of Marine Science, 62(3), 339–43.

Brander, K. M. (2010). Reconciling biodiversity conservation and marine capture fisheries production. Current Opinion in Environment, 2, 416–21.

Brasseur, S. M. J. M., Reijnders, P. J. H., Cremer, J., Meesters, E., Kirkwood, R., Jensen, L.F., . . . Aarts, G. (2018). Echoes from the past: Regional variations in recovery within a harbor seal population. PLoS One, 13, 0189674.

Breen, M., Graham, N., Pol, M., He, P., Reid, D., & Suuronen, P. (2016). Selective fishing and balanced harvesting. Fisheries Research, 184, 2–8.

Brewster-Geisz, K. K., & Miller, T. J. (2000). Management of the sandbar shark, *Carcharhinus plumbeus*: Implications of a stage-based model. Fishery Bulletin, 98, 236–49.

Brey, T., Müller-Wiegmann, C., & Zittier, Z. M. C. (2010). Body composition in aquatic organisms—A global data bank of relationships between mass, elemental composition and energy content. Journal of Sea Research 64, 334–40.

Brody, S. (1945). Bioenergetics and growth. Reinhold, New York.

Browman, H. I., & Stergiou, K. I. (2004). Marine Protected Areas as a central element of ecosystem-based management: defining their location, size and number. Marine Ecology Progress Series, 274, 271–2.

Brown, B. E., Brennan, J. A., Grosslein, M. D., Heyerdahl, E. G., & Hennemuth, R. C. (1976). The effect of fishing on the marine finfish biomass in the Northwest Atlantic from the Gulf of Maine to Cape Hatteras. International Commission for the Northwest Atlantic Fisheries Research Bulletin, 12, 49–68.

Brown, B. L., Downing, A. L., & Leibold, M. A. (2016). Compensatory dynamics stabilize aggregate community properties in response to multiple types of perturbations. Ecology, 97(8), 2021–33.

Brown, J. H. (1995). Macroecology. University of Chicago Press, Chicago.

Brown, J. H., Gillooly, J. F., Allen, A. P., Savage, V. M., & West, G. B. (2004). Toward a metabolic theory of ecology. Ecology, 85(7), 1771–89.

Brown, J. H., & Sibly, R. M. (2012). The metabolic theory of ecology and its central equation. In: Sibly, R. M., Brown, J. H., Kodric-Brown, A. (Eds.), Metabolic ecology: A scaling approach (pp 21–33). John Wiley & Sons, Chichester.

Budy, P., & Gaeta, J. W. (2017). Brown trout as an invader: a synthesis of problems and perspectives in North America. In: Lobón-Cerviá, J., Sanz, N. (Eds.), Brown trout: Biology, ecology and management (pp. 523–43). Wiley, Hoboken, NJ.

Bundy, A., Link, J., Miller, T., Moksness, E., & Stergiou, K. (2012). Comparative analysis of marine fisheries production. Marine Ecology Progress Series, 459(157).

Burgess, M. G., Polasky, S., & Tilman, D. (2013). Predicting overfishing and extinction threats in multispecies fisheries. Proceedings of the National Academy of Sciences USA, 110, 15943–8.

Burgess, M.G., Diekert, F. K., Jacobsen, N. S., Andersen, K. H., & Gaines, S. D. (2016). Remaining questions in the case for balanced harvesting. Fish and Fisheries, 17(4), 1216–26.

Burgner, R. L. (1991). Life history of sockeye salmon (*Oncorhynchus gorbuscha*). In Groot, C. & Margolis, L. (Eds.), Pacific Salmon Life Histories. UBC Press, Vancouver.

Cabral, R. B., Gaines, S. D., Lim, M. T., Atrigenio, M. P., Mamauag, S. S., Pedemonte, G. C., & Aliño, P. M. (2016). Siting marine protected areas based on habitat quality and extent provides the greatest benefit to spatially structured metapopulations. Ecosphere, 7(11), e01533.

Caddy, J. F., & Gulland, J. A. (1983). Historical patterns of fish stocks. Marine Policy, 7, 267–78.

Carpenter, S. R., & Kitchell, J. F. (Eds.). (1996). The trophic cascade in lakes. Cambridge University Press, Cambridge.

Carr, M. E., Friedrichs, M. A., Schmeltz, M., Aita, M. N., Antoine, D., Arrigo, K. R., . . . & Bidigare, R. (2006). A comparison of global estimates of marine primary production from ocean color. Deep Sea Research Part II: Topical Studies in Oceanography, 53(5–7), 741–70.

Case, T. (2000). An illustrated guide to theoretical ecology. Oxford University Press, Oxford.

Caswell, H. (2001). Matrix population models: Construction, analysis, and interpretation. (2nd edition.) Sinauer Associates, Sunderland, Mass.

Chang, C. W., Ushio, M., & Hsieh, C.-h. (2017). Empirical dynamic modeling for beginners. Ecological research, 32(6), 785–96.

Chang, Y. J., Sun, C. L., Chen, Y., & Yeh, S. Z. (2012). Modelling the growth of crustacean species. Reviews in Fish Biology and Fisheries, 22, 157–87.

Chapin, F. S., III, Robards, M. D., Huntington, H. P., Johnstone, J. F., Trainor, S. F, Kofinas, G. P., Ruess, R. W., Fresco, N., Natcher, D. C., & Naylor, R. L. (2006). Directional changes in ecological communities and social-ecological systems: A framework for prediction based on Alaskan examples. American Naturalist, 168, 36–49.

Chassot, E., Bonhommeau, S., Dulvy, N. K., Mélin, F., Watson, R., Gascuel, D., & Le Pape, O. (2010). Global marine primary production constrains fisheries catches. Ecology Letters, 13(4), 495–505.

Chavez, F. P., Ryan, J., Lluch-Cota, S. E., & Ñiquen, M. (2003). From anchovies to sardines and back: Multidecadal change in the Pacific Ocean. Science, 299(5604), 217–21.

Checkley, D. M., Raman, S., Mason, K.M., & Maillet, G. L. (1988). Winter storm effects on the spawning and larval drift of a pelagic fish. Nature, 335, 346–8.

Chesson, P. L. (1984). The storage effect in stochastic population models. In: Levin S.A., & Hallam T.G. (Eds.), Mathematical ecology. Lecture notes in biomathematics, vol 54 (pp. 76–89). Springer, Berlin, Heidelberg.

Chotiyaputta, C., Nootmorn, P., & Jirapunpipat, K. (2002). Review of cephalopod fishery production and long term changes in fish communities in the Gulf of Thailand. Bulletin of Marine Science, 71(1), 223–38.

Christensen, V., & Maclean, J. (Eds.) (2011). Ecosystem approaches to fisheries: a global perspective. Cambridge University Press, Cambridge.

Christensen, V., & Pauly, D. (1992). ECOPATH II—a software for balancing steady-state ecosystem models and calculating network characteristics. Ecological Modelling, 61, 169–85.

Christensen, V., Walters, C. J., Ahrens, R., Alder, J., Buszowski, J., Christensen, L. B., ... & Kaschner, K. (2009). Database-driven models of the world's Large Marine Ecosystems. Ecological Modelling, 220(17), 1984–96.

Clark, A. T., Ye, H., Isbell, F., Deyle, E. R., Cowles, J., Tilman, G. D., & Sugihara, G. (2015). Spatial convergent cross mapping to detect causal relationships from short time series. Ecology, 96, 1174–81.

Clark, C. W. (1976). A delayed-recruitment model of population dynamics, with an application to baleen whale populations. Journal of Mathematical Biology, 3(3–4), 381–91.

Clark, C. W. (1982). Concentration profiles and the production and management of marine fisheries. In: Eichhorn W., Henn R., Neumann K., & Shephard R.W. (Eds.), Economic theory of natural resources (pp. 97–112). Physica, Heidelberg.

Clark, C. W. (1985). Bioeconomic modeling and fisheries management. Wiley-Interscience, New York.

Clark, C. W. (1990). Mathematical Bioeconomics. (2nd edition.) John Wiley and Sons, New York.

Clark, C. W. (2010). Mathematical bioeconomics: The mathematics of conservation. (3rd edition.) Wiley, New York.

Clark, S. H., Overholtz, W. J., & Hennemuth, R. C. (1982). Review and assessment of the Georges Bank and Gulf of Maine haddock fishery. Journal of Northwest Atlantic Fisheries Science, 3(1), 1–27.

Clarke, G. L. (1946). Dynamics of production in a marine area. Ecological Monographs, 16, 321–37.

Cohen, A. S., Gergurich, E. L., Kraemer, B. M., McGlue, M. M., McIntyre, P. B., Russell, J. M., & Swarzenski, P. W. (2016). Climate warming reduces fish production and benthic habitat in Lake Tanganyika, one of the most biodiverse freshwater ecosystems. Proceedings of the National Academy of Sciences USA, 113, 9563–8.

Cohen, J. E., Christensen, S. W., & Goodyear, C. P. (1983). A stochastic age-structured population model of striped bass (*Morone saxatilis*) in the Potomac River. Canadian Journal of Fisheries and Aquatic Sciences, 40(12), 2170–83.

Coll, M., Libralato, S., Tudela, S., Palomera, I., & Pranovi, F. (2008). Ecosystem overfishing in the ocean. PLoS One, 3(12).

Colléter, M., Valls, A., Guitton, J., Gascuel, D., Pauly, D., & Christensen, V. (2015). Global overview of the applications of the Ecopath with Ecosim modeling approach using the EcoBase models repository. Ecological Modelling, 302, 42–53.

Collie, J. S., Botsford, L.W., Hastings, A., Kaplan, I. C., Largier, J. L., Livingston, P. A., Plagányi, E., Rose, K. A., Wells, B. K., & Werner, F.E. (2014). Ecosystem models for fisheries management: finding the sweet spot. Fish and Fisheries, 17(1), 101–25.

Collie, J. S., & DeLong, A.K. (1999). Multispecies interactions in the Georges Bank fish community. In: Ecosystem approaches for fisheries management (16th Lowell Wakefield Fisheries Symposium). Alaska Sea Grant College Program, 99–01.

Collie, J. S., & Gislason, H. (2001). Biological reference points for fish stocks in a multispecies context. Canadian Journal of Fisheries and Aquatic Sciences, 58, 2167–76.

Collie, J. S., Gislason, H., & Vinther, M. (2003). Using AMOEBAs to display multispecies, multifleet fisheries advice. ICES Journal of Marine Science, 60, 709–20.

Collie, J. S., Hiddink, J. G., van Kooten, T., Rijnsdorp, A. D., Kaiser, M. J., Jennings, S., & Hilborn, R. (2017). Indirect effects of bottom fishing on the productivity of marine fish. Fish and Fisheries, 18(4), 619–37.

Collie, J. S., Hall, S. J., Kaiser, M. J., & Poiner, I. R. (2000). A quantitative analysis of fishing impacts on shelf-sea benthos. Journal of Animal Ecology, 69(5), 785–98.

Collie, J. S., Hermsen, J. M., Valentine, P. C., & Almeida, F. P. (2005). Effects of fishing on gravel habitats: Assessment and recovery of benthic megafauna on Georges Bank. In: Barnes, P., & Thomas, J. (Eds.), Benthic Habitats and the Effects of Fishing (Vol. 325). American Fisheries Society, Bethesda, MD.

Collie, J. S., Richardson, K., & Steele, J. H. (2004). Regime shifts: can ecological theory illuminate the mechanisms?. Progress in Oceanography, 60(2–4), 281–302.

Collie, J. S. & Spencer, P. D. (1993). Management strategies for fish populations subject to long-term environmental variability and depensatory predation. In: Kruse, G. H., Eggers, D. M., Marasco, R. J., Pautzke, and Quinn, T. J. (Eds.), Proceedings of the International Symposium on Management Strategies for Exploited Fish Populations. Alaska Sea Grant College Program Report 93-02.

Collie, J. S., & Spencer, P. D. (1994). Modeling predator–prey dynamics in a fluctuating environment. Canadian Journal of Fisheries and Aquatic Sciences, 51(12), 2665–72.

Collie, J. S., & Walters, C. J. (1987). Alternate recruitment models of Adams River sockeye salmon, *Oncorhynchus nerka*. Canadian Journal of Fisheries and Aquatic Sciences, 44(9), 1551–61.

Collins, J. W. (1884). A search for mackerel off Block Island, Montauk, and Sandy Hook in November, 1883. Bulletin, IV, for 1884, United States Fish Commission, 49–51.

Connell, J. H. (1961). The influence of interspecific competition and other factors on the distribution of the barnacle *Chthamalus stellatus*. Ecology, 42, 710–23.

Conover, D. O., & Munch, S. B. (2002). Sustaining fisheries yields over evolutionary time scales. Science, 297, 94–6.

Constable, A. J. (2011). Lessons from CCAMLR on the implementation of the ecosystem approach to managing fisheries. Fish and Fisheries, 12(2), 138–51.

Conti, L., & Scardi, M. (2010). Fisheries yield and primary productivity in large marine ecosystems. Marine Ecology Progress Series, 410, 233–44.

Craig, J. F. (Ed.). (2015). Freshwater fisheries ecology. Wiley Blackwell, Oxford.

Craig, J. F., & Kipling, C. (1983). Reproduction effort versus the environment; case histories of Windermere perch, *Perca fluviatilis* L., and pike, *Esox lucius* L. Journal of Fish Biology, 22(6), 713–27.

Cramer, N. F., & May, R. M. (1972). Interspecific competition, predation, and species diversity: A comment. Journal of Theoretical Biology, 34, 289–93.

Crisp, D. J. (1984). Energy flow measurements. IBP Handbook, 16, 284–372.

Crouse, D. T., Crowder, L. B., & Caswell, H. (1987). A stage-based population model for loggerhead sea turtles and implications for conservation. Ecology, 68, 1412–23.

Crowder, L. B., Crouse, D. T., Heppell, S. S., & Martin, T. H. (1994). Predicting the impact of turtle excluder devices on loggerhead sea turtle populations. Ecological Applications, 4(3), 437–45.

Csirke, J. (1988). Small shoaling pelagic fish stocks. In: Gulland, J. A. (Ed.), Fish population dynamics: The implications for management. Wiley, New York.

Cunnane, S. C., & Stewart, K. M. (Eds.). (2010). Human brain evolution: The influence of freshwater and marine food resources. Wiley, Hoboken, NJ.

Curry, G. L., & Feldman, R. M. (1987). Mathematical foundations of population dynamics. The Quarterly Review of Biology, 62(4), 476–7.

Curti, K. L. (2012). Age-structured multispecies model of the Georges Bank fish community. Doctoral dissertation, University of Rhode Island.

Curti, K. L., Collie, J. S., Legault, C., & Link, J. S. (2013). Evaluating the performance of a multispecies statistical catch-at-age model. Canadian Journal of Fisheries and Aquatic Sciences, 70(3), 470–84.

Cushing, D. H. (1982). Climate and fisheries. Academic Press, London.

Cushing, D. H. (1988). The provident sea. Cambridge University Press, Cambridge.

Cushing, D. H. (1995). Population production and regulation in the sea: a fisheries perspective. Cambridge University Press, Cambridge.

D'Ancona, U. (1954). The struggle for existence (Vol. 6). Brill Archive.

Daan, N., & Sissenwine, M. P. (1991). Multispecies models relevant to management of living resources. International Council for Exploration of the Sea Marine Science Symposium 193, Copenhagen.

Dambacher, J., Gaughan, D. J., Rochet, M.-J., Rossignol, P. A., & Trenkel, V.M. (2009). Qualitative modelling and indicators of exploited ecosystems. Fish and Fisheries, 10, 305–22.

Dambacher, J. M., Li, H. W., & Rossignol, P. A. (2003). Qualitative predictions in model ecosystems. Ecological Modelling, 161(1–2), 79–93.

Darimont, C. T., Fox, C. H., Bryan, H. M., & Riemchen, T. E. (2015). The unique ecology of human predators. Science, 349, 858–60.

Daskalov, G. M. (2002). Overfishing drives a trophic cascade in the Black Sea. Marine Ecology Progress Series, 225, 53–63.

Daskalov, G. M., Grishin, A. N., Rodionov, S., & Mihneva, V. (2007). Trophic cascades triggered by overfishing reveal possible mechanisms of ecosystem regime shifts. Proceedings of the National Academy of Sciences USA, 104, 10518–23.

Day, T., & Taylor, P.D. (1997). Von Bertalanffy's growth equation should not be used to model age and size at maturity. The American Naturalist, 149(2), 381–93.

DeAngelis, D. L., Goldstein, R. A., & O'Neill, R. V. (1975). A model for trophic interaction. Ecology, 56, 881–92. https://doi.org/doi:10.2307/1936298

DeAngelis, D. L., Svoboda, L. J., Christensen, S. W., & Vaughan, D. S. (1980). Stability and return times of Leslie matrices with density-dependent survival: Applications to fish populations. Ecological Modeling, 8, 149–63.

Deines, A. M., Bunnell, D. B., Rogers, M. W., Beard, T. D., & Taylor, W. W. (2015). A review of the global relationship among freshwater fish, autotrophic activity, and regional climate. Reviews in Fish Biology and Fisheries, 25, 323–36.

De la Mare, W. K. (1996). Some recent developments in the management of marine living resources. Frontiers of Population Ecology, 599–616.

Denny, M. W. and Gaines, S. (2000). Chance in biology. Princeton University Press, Princeton, NJ.

Deriso, R. B. (1980). Harvesting strategies and parameter estimation for an age-structured model. Canadian Journal of Fisheries and Aquatic Sciences, 37, 268–82.

Deslauriers, D., Chipps, S. R., Breck, J. E., Rice, J. A., & Madenjian, C. P. (2017). Fish Bioenergetics 4.0: An R-based modeling application. Fisheries, 42(11), 586–96.

Desmond, A. J. (1994). Huxley: The devil's disciple (Vol. 1). Michael Joseph, London.

Deyle, E. R., Fogarty, M., Hsieh, C.-h., Kaufman, L., MacCall, A. D., Munch, S. B., ... Sugihara, G. (2013). Predicting climate effects on Pacific sardine. Proceedings of the National Academy of Sciences USA, 110, 6430–5. https://doi.org/10.1073/pnas.1215506110

Deyle, E. R., May, R. M., Munch, S. B., & Sugihara, G. (2016). Tracking and forecasting ecosystem interactions in real time. Proceedings of the Royal Society of London Series B: Biological Sciences, 283, 20152258.

Deyle, E., Schueller, A., Ye, H., Pao, G., & Sugihara, G. (2018). Ecosystem-based forecasts of recruitment in two menhaden species. Fish and Fisheries. 19. 10.1111/faf.12287.

Deyle, E., & Sugihara, G. (2011). Generalized theorems for nonlinear state space reconstruction. PLoS One 6:e18295.

deYoung, B., Heath, M., Werner, F., Chai, F., Megrey, B., & Monfray, P. (2004). Challenges of modeling ocean basin ecosystems. Science, 304, 1463–6.

DFO (Department of Fisheries and Oceans Canada) (2016).Stock Assessment of Northern Cod (NAFO Divs. 2J3KL) in 2016. Canadian Science Advisory Secretariat. Science Advisory Report 2016/026.

Dobson, A. P., & May, R. M. (1987). The effects of parasites on fish populations—Theoretical aspects. International Journal for Parasitology, 17(2), 363–70.

Dominey, W. J., & Blumer, L. S. (1984). Cannibalism of early life stages in fishes. In: Hausfater, G., & Hrdy, S. B. (Eds.), Infanticide: comparative & evolutionary perspectives. Aldine, New York.

Drinkwater, K. F. (2005). The response of Atlantic cod (*Gadus morhua*) to future climate change. ICES Journal of Marine Science, 62(7), 1327–37.

Edwards, A. M., Robinson, J. P., Plank, M. J., Baum, J. K., & Blanchard, J. L. (2017). Testing and recommending methods for fitting size spectra to data. Methods in Ecology and Evolution, 8, 57–67.

Elliot, J. M. (1975). Weight of food and time required to satiate brown trout, *Salmo trutta* L. Freshwater Biology, 5(51), 64.

Elliott, J. M. (1985). The choice of a stock-recruitment model for migratory trout, *Salmo trutta*, in an English Lake District stream. Archiv für Hydrobiologie, 104, 145–68.

Elliot, J. M. (1994). Quantitative ecology and the brown trout. Oxford Series in Ecology and Evolution. Oxford University Press, Oxford, UK.

Essington, T. E., & Hansson, S. (2004). Predator-dependent functional responses and interaction strengths in a natural food web. Canadian Journal of Fisheries and Aquatic Sciences, 61(11), 2215–26.

Essington, T. E., Kitchell, J. F., & Walters, C. J. (2001). The von Bertalanffy growth function, bioenergetics, and the consumption rates of fish. Canadian Journal of Fisheries and Aquatic Sciences, 2001(58), 2129–38.

Fagan, B. M. (2006). Fish on Friday: Feasting, fasting, and the discovery of the New World. Basic Books, New York.

Fagan, B. M. (2017). Fishing: how the sea fed civilization. Yale University Press, New Haven, Conn.

FAO (Food and Agriculture Organization of the United Nations) (2005). Code of conduct for responsible fisheries. FAO, Rome.

FAO (Food and Agriculture Organization of the United Nations) (2018). The State of World Fisheries and Aquaculture 2018—Meeting the sustainable development goals (No. CC BY-NC-SA 3.0 IGO). FAO, Rome.

Fattahpour, H., Zangeneh, H. R., & Wang, H. (2019). Dynamics of coral reef models in the presence of parrotfish. Natural Resource Modeling, 32(2), e12202.

Fenchel, T. (1974). Intrinsic rate of natural increase: The relationship with body size. Oecologia, 14(4), 317–26.

Finlayson, C. M., Arthington, A. H., & Pittock, J. (Eds.). (2018). Freshwater ecosystems in protected areas: Conservation and management. Routledge, Abingdon, UK.

Finney, B. P., Alheit, J., Emeis, K. C., Field, D. B., Gutiérrez, D., & Struck, U. (2010). Paleoecological studies on variability in marine fish populations: A long-term perspective on the impacts of climatic change on marine ecosystems. Journal of Marine Systems, 79(3–4), 316–26.

Fogarty, M. J. (1993a). Recruitment distributions revisited. Canadian Journal of Fisheries and Aquatic Sciences, 50, 2723–8.

Fogarty, M. J. (1993b). Recruitment in randomly varying environments. ICES Journal of Marine Science, 50, 247–50.

Fogarty, M. J. (1998). Implications of larval dispersal and directed migration in American lobster stocks: Spatial structure and resilience. Canadian Special Publication of Fisheries and Aquatic Sciences, 125, 273–83.

Fogarty, M. J. (2014). The art of ecosystem-based fishery management. Canadian Journal of Fisheries and Aquatic Sciences, 71, 479–90.

Fogarty, M. J., & Addison, J. T. (1997). Modeling performance of individual traps: Entry, escapement and soak time. ICES Journal of Marine Science, 54, 193–205.

Fogarty, M. J., Bohnsack, J. A., & Dayton, P. K. (2000). Marine reserves and resource management. In: Sheppard, C., (Ed.) Seas at the Millennium: An Environmental Evaluation, 3, 375–92. Elsevier Science Ltd, Oxford.

Fogarty, M. J., & Botsford, L. W. (2007). Population connectivity and spatial management of marine fisheries. Oceanography, 20(3), 112–23.

Fogarty, M. J., Gamble, R., & Perretti, C. T. (2016a). Dynamic complexity in exploited marine ecosystems. Frontiers in Ecology and Evolution, 4, 68.

Fogarty, M. J., & Idoine, J. S. (1988). Application of a yield and egg production model based on size to an offshore American lobster population. Transactions of the American Fisheries Society, 117, 350–62.

Fogarty, M. J., Incze, L., Hayhoe, K., Mountain, D., & Manning, J. (2008). Potential climate change impacts on Atlantic cod (*Gadus morhua*) off the northeastern United States. Mitigation and Adaptation Strategies for Global Change, 13, 453–66.

Fogarty, M. J., Mayo, R. K., O'Brien, L., & Rosenberg A. A. (1996). Assessing uncertainty & risk in exploited marine populations. Reliability Engineering and System Safety, 54, 183–95.

Fogarty, M. J., & Murawski, S. A. (1998). Large-scale disturbance and the structure of marine systems: Fishery impacts on Georges Bank. Ecological Applications, 8, 6–22.

Fogarty, M. J., & O'Brien, L. (2016). Recruitment in marine fish populations. In: Jakobsen, T., Fogarty, M. J., Megrey, B.A., & Moskness, E. (Eds.), Fish reproductive biology: Implications for assessment and management. (2nd edition.) Blackwell, Oxford.

Fogarty, M. J., Overholtz, W. J., & Link, J. S. (2012). Aggregate surplus production models for demersal fishery resources of the Gulf of Maine. Marine Ecology Progress Series, 459, 247–58.

Fogarty, M. J., Rosenberg, A. A., Cooper, A. B., Dickey-Collas, M., Fulton, E. A., Gutiérrez, N. L., ... Ye, Y. (2016b). Fishery production potential of large marine ecosystems: A prototype analysis. Environmental Development, 17(1), 211–19.

Fogarty, M. J., Rosenberg, A. A., & Sissenwine, M. P. (1992). ES&T Series: Fisheries risk assessment sources of uncertainty: A case study of Georges Bank haddock. Environmental Science & Technology, 26(3), 440–7.

Fogarty, M. J., Sissenwine, M. P., & Cohen, E. B. (1991). Recruitment variability & the dynamics of exploited marine populations. Trends in Ecology and Evolution, 6, 241–6.

Ford, E. J. (1933). An account of the herring investigations conducted at Plymouth during the years from 1924 to 1933. Journal of the Marine Biological Association of the United Kingdom, 19(1), 305–84.

Fowler, C. (2009). Systemic management. Oxford University Press, Oxford.

Fowler, C. W. (1981). Density dependence as related to life history strategy. Ecology, 62(3), 602–10.

Fox Jr, W. W. (1970). An exponential surplus-yield model for optimizing exploited fish populations. Transactions of the American Fisheries Society, 99(1), 80–8.

Frank, K. T., Petrie, B., Choi, J. S., & Leggett, W. C. (2005). Trophic cascades in a formerly cod-dominated ecosystem. Science, 308(5728), 1621–3.

Frank, K. T., Petrie, B., Leggett, W. C., & Boyce, D. G. (2018). Exploitation drives an ontogenetic-like deepening in marine fish. Proceedings of the National Academy of Sciences USA, 115, 6422–7.

Franks, P. J. S., Wroblewski, J. S., & Flierl, G. R. (1986). Behavior of a simple plankton model with food-level acclimation by herbivores. Marine Biology, 91, 121–9.

Fretwell, S. D., & Lucas, H. L. (1970). On territorial behaviour and other factors influencing distribution in birds. Acta Biotheoretica, 19, 16–36.

Friedland, K. D., Stock, C., Drinkwater, K. F., Link, J. S., Leaf, R. T., Shank, B. V., ... & Fogarty, M. J. (2012). Pathways between primary production and fisheries yields of large marine ecosystems. PLoS One, 7(1), e28945.

Froese, R., Walters, C., Pauly, D., Winker, H., Weyl, O. L. F., Demirel, N., Tsikliras, A. C., & Holt, S. J. (2016). A critique of the balanced harvesting approach to fishing. ICES Journal of Marine Science, 73(6), 1640–50.

Fulton, E. A., Bax, N. J., Bustamante, R. H., Dambacher, J. M., Dichmont, C., Dunstan, P. K., ... & Punt, A. E. (2015). Modelling marine protected areas: insights and hurdles. Philosophical Transactions of the Royal Society B: Biological Sciences, 370(1681), 20140278.

Fulton, E. A., & Link, J. S. (2014). Modeling approaches for marine ecosystem-based management. In: Fogarty, M. J. & McCarthy, J. J. (Eds.), The Sea, Vol 16 Marine Ecosystem-based Management (pp. 121–70). Harvard University Press, Cambridge, Mass.

Fulton, E. A., Link, J. S., Kaplan, I. C., Savina-Rolland, M., Johnson, P., Ainsworth, C., ... & Smith, D. C. (2011). Lessons in modelling and management of marine ecosystems: the Atlantis experience. Fish and Fisheries, 12(2), 171–88.

Fulton, E. A., Punt, A. E., Dichmont, C. M., Harvey, C. J., & Gorton, R. (2019). Ecosystems say good management pays off. Fish and Fisheries, 20(1), 66–96.

Gabriel, W. L., Sissenwine, M. P., & Overholtz, W. J. (1989). Analysis of spawning stock biomass per recruit: An example of Georges Bank. North American Journal of Fisheries Management, 9, 383–91.

Gaichas, S., Gamble, R., Fogarty, M., Benoît, H., Essington, T., Fu, C., ... Link. J. (2012). Assembly rules for aggregate-species production models: Simulations in support of management strategy evaluation. Marine Ecology Progress Series, 459, 275–92.

Gaichas, S. K., Fogarty, M., Fay, G., Gamble, R., Lucey, S., & Smith, L. (2017). Combining stock, multispecies, and ecosystem level fishery objectives within an operational management procedure: Simulations to start the conversation. ICES Journal of Marine Science, 74, 552–65. https://doi.org/doi:10.1093/icesjms/fsw119

Garcia, S. M., Kolding, J., Rice, J., Rochet, M. J., Zhou, S., Arimoto, T., ... & Smith, A. D. M. (2012). Reconsidering the consequences of selective fisheries. Science, 335(6072), 1045–7.

Garcia, S. M., Zerbi, A., Aliaume, C., Do Chi, T., & Lasserre, G. (2003). The ecosystem approach to fisheries. Issues, terminology, principles, institutional foundations, implementation and outlook. FAO Fisheries Technical Paper. No. 443. FAO, Rome.

Gause, G. F. (1934). Experimental analysis of Vito Volterra's mathematical theory of the struggle for existence. Science, 79(2036), 16–17.

Gause, G. F. (1935). Experimental demonstration of Volterra's periodic oscillations in the numbers of animals. Journal of Experimental Biology, 12(1), 44–8.

Gause, G. F., & Witt, A. A. (1935). Behavior of mixed populations and the problem of natural selection. The American Naturalist, 69(725), 596–609.

Gellner, G., & McCann, K. (2012). Reconciling the omnivory-stability debate. The American Naturalist, 179(1), 22–37.

Gerking, S. D. (1994). Feeding ecology of fish. Academic Press, Cambridge, Mass.

Getz, W., & Haight, R. (1989). Population harvesting. Princeton University Press, Princeton.

Giller, P. S. (1984). Community Structure and the Niche. Chapman & Hall, London.

Gilpin, M. E., & Ayala, F. J. (1973). Global models of growth and competition. Proceedings of the National Academy of Sciences USA, 70, 3590–3.

Gislason, H. (1999). Single and multispecies reference points for Baltic fish stocks. ICES Journal of Marine Science, 56, 571–83.

Glaser, S. M., Fogarty, M. J., Liu, H., Altman, I., Hsieh, C.-h., Kaufman, L., ... Sugihara, G. (2014). Complex dynamics may limit prediction in marine fisheries. Fish and Fisheries, 15, 616–633. https://doi.org/doi:10.1111/faf.12037

Glaser, S. M., Ye, H., Maunder, M. N., MacCall, A. D., Fogarty, M. J., & Sugihara, G. (2011). Detecting and forecasting complex nonlinear dynamics in spatially structured catch-per-unit- effort time series for North Pacific albacore (*Thunnus alalunga*). Canadian Journal of Fisheries and Aquatic Sciences, 68, 400–12.

Goodale, E., Beauchamp, G., & Ruxton, G. D. (2017). Mixed-species groups of animals: Behavior, community structure, and conservation. Academic Press, Cambridge, Mass.

Gotelli, N. J. (2008). A primer of ecology. (4th edition.) Sinauer, Sunderland, Mass.

Graham, H. W., & Edwards, R. L. (1962). The world biomass of marine fishes. In: Heen, E., & Kreuzer, R. (Eds.), Fish in Nutrition (p. 445). CABI.

Graham, M. (1935). Modern theory of exploiting a fishery, and application to North Sea trawling. ICES Journal of Marine Science, 10(3), 264–74.

Granger, C. W. (1969). Investigating causal relations by econometric models and cross-spectral methods. Econometrica: Journal of the Econometric Society, 424–38.

Grenfell, B. T., Price, O. F., Albon, S. D., & Clutton-Brock, T. H. (1992). Overcompensation and population cycles in an ungulate. Nature, 355(6363), 823.

Guill, C. E., Carmack, E., & Drossel, B. (2014). Exploring cyclic dominance of sockeye salmon with a predator–prey model. Canadian Journal of Fisheries and Aquatic Sciences, 71(959–72).

Gulland, J. A. (1968). The concept of the marginal yield from exploited fish stocks. ICES Journal of Marine Science, 32(2), 256–61.

Gulland, J. A. (1969). Manual of methods of fish stock assessment. Part 1. Fish Population Analysis. FAO Manuals. FAO, Rome.

Gulland, J. A. (1970). The fish resources of the ocean. Fishing News Books, West Byfleet.

Gulland, J. A. (1977). Fish population dynamics. John Wiley and Sons, New York.

Gulland, J. A. (1983). Fish stock assessment: a manual of basic methods (Vol. 223). FAO/Wiley, Chichester.

Gulland, J. A. (1988). Fish population dynamics. (2nd edition.) John Wiley and Sons, New York.

Gulland, J. A., & Boerema, L. K. (1973). Scientific advice on catch levels. Fishery Bulletin, 71, 325–36.

Hall, M. A. (1996). On bycatches. Reviews in Fish Biology and Fisheries, 6, 319–52.

Hall, S. J. (1999). The effects of fisheries on marine ecosystems and communities. Blackwell Science, London.

Hall, S. J., Collie, J. S., Duplisea, D. E., Jennings, S., Bravington, M., & Link, J. (2006). A length-based multispecies model for evaluating community responses to fishing. Canadian Journal of Fisheries and Aquatic Sciences, 63(6), 1344–59.

Hall, S. J., & Mainprize, B. M. (2005). Managing by-catch and discards: How much progress are we making and how can we do better? Fish and Fisheries, 6(2), 134–55.

Halpern, B. S., Lester, S. E., & McLeod, K. L. (2010). Placing marine protected areas onto the ecosystem-based management seascape. Proceedings of the National Academy of Sciences USA, 107(43), 18312–17.

Hammill, M. O., den Heyer, C. E., & Bowen, W. D. (2014). Grey seal population trends in Canadian waters, 1960–2014 (Canadian Science Advisory Secretariat Research Document, Quebec and Maritimes Regions No. 2014/037).

Hannesson, R. (1983). Bioeconomic production function in fisheries: Theoretical and empirical analysis. Canadian Journal of Fisheries and Aquatic Sciences, 40(7), 968–82.

Hansen, G. J., & Carey, C. C. (2015). Fish and phytoplankton exhibit contrasting temporal species abundance patterns in a dynamic north temperate lake. PLoS One, 10(2), 0115414.

Hanski, I. (1982). Dynamics of regional distribution: The core and satellite species hypothesis. Oikos, 38, 210–21.

Hanski, I. (1998). Metapopulation dynamics. Nature, 396(6706), 41.

Hanski, I., & Simberloff, D. (1997). The metapopulation approach, its history, conceptual domain, and application to conservation. In: Hanski, I. & Gilpin, M. E. (Eds.), Metapopulation biology: ecology, genetics, and evolution (pp. 5–26). Academic Press.

Hardin, G. (1968). The tragedy of the commons. Science, 162(3859), 1243–8.

Hardy, A. C. (1924). The herring in relation to its animate environment. Part 1. Fishery Investigations, Series 2, 7, 140 pp.

Hare, J. A., Alexander, M. A., Fogarty, M. J., Williams, E. H., & Scott, J. D. (2010). Forecasting the dynamics of a coastal fishery species using a coupled climate-population model. Ecological Applications, 2(452–64).

Härkönen, T., Dietz, R., Reijnders, P., Teilmann, J., Harding, K., Hall, A., & Rasmussen, T. D. (2006). The 1988 and 2002 phocine distemper virus epidemics in European harbour seals. Diseases of Aquatic Organisms, 68(2), 115–30.

Harley, S. J., Myers, R. A., & Dunn, A. (2001). Is catch-per-unit-effort proportional to abundance? Canadian Journal of Fisheries and Aquatic Sciences, 58(9), 1760–72.

Harris, J. G. K. (1975). The effect of density-dependent mortality on the shape of the stock recruitment curve. Journal du Conseil International Pour Exploration de La Mer, 36, 144–9.

Hartman, K. J., & Brandt, S. B. (1995). Estimating energy density of fish. Transactions of the American Fisheries Society, 124(3), 347–55.

Harvell, C. D., Mitchell, C. E., Ward, J. R., Altizer, S., Dobson, A. P., Ostfeld, R. S., & Samuel, M. D. (2002). Climate warming and disease risks for terrestrial and marine biota. Science, 296(5576), 2158–62.

Hassell, M. P., & Comins, H. N. (1976). Discrete time models for two-species competition. Theoretical Population Biology, 9(2), 202–21.

Hassell, M. P., & Varley, G. C. (1969). New inductive population model for insect parasites and its bearing on biological control. Nature, 223, 1133–7.

Hastings, A. (1997). Population biology: Concepts and models. Springer-Verlag, New York.

Hastings, A., & Botsford, L. W. (1999). Equivalence in yield from marine reserves and traditional fisheries management. Science, 284(5419), 1537–8.

Hastings, A., Hom, C. L., Ellner, S., Turchin, P., & Godfray, H. C. J. (1993). Chaos in ecology—is mother nature a strange attractor? Annual Reviews of Ecology and Systematics, 24, 1–33.

Hastings, A., & Powell, T. (1991). Chaos in a three-species food chain. Ecology, 72(3), 896–903.

Hein, C. L., Roth, B. M., Ives, A. R., & Zanden, M. J. V. (2006). Fish predation and trapping for rusty crayfish (*Orconectes rusticus*) control: A whole-lake experiment. Canadian Journal of Fisheries and Aquatic Sciences, 63(2), 383–93.

Helgason, T., & Gislason, H. (1979). VPA-analysis with species interaction due to predation. ICES CM, 1979/G:52.

Helmuth, B., Harley, C. D., Halpin, P. M., O'Donnell, M., Hofmann, G. E., & Blanchette, C. A. (2002). Climate change and latitudinal patterns of intertidal thermal stress. Science, 298(5595), 1015–17.

Hiddink, J. G., Jennings, S., Sciberras, M., Szostek, C. L., Hughes, K. M., Ellis, N., & Collie, J. S. (2017). Global analysis of depletion and recovery of seabed biota after bottom trawling disturbance. Proceedings of the National Academy of Sciences USA, 114, 8301–6.

Hiddink, J. G., Moranta, J., Balestrini, S., Sciberras, M., Cendrier, M., Bowyer, R., & Hinz, H. (2016). Bottom trawling affects fish condition through changes in the ratio of prey availability to density of competitors. Journal of Applied Ecology, 53(5), 1500–10.

Hilborn, R. (2014). Introduction to marine managed areas. Advances in Marine Biology, 69, 1–13.

Hilborn, R., Stokes, K., Maguire, J. J., Smith, T., Botsford, L. W., Mangel, M., ... & Cochrane, K. L. (2004). When can marine reserves improve fisheries management? Ocean & Coastal Management, 47(3–4), 197–205.

Hilborn, R. & Walters, C. J. (1992). Quantitative fish stock assessment: Choice, dynamics and uncertainty. Chapman-Hall, New York.

Hixon, M. A. (1980). Competitive interactions between California Reef Fishes of the genus Embiotoca. Ecology, 61, 918–931. doi:10.2307/1936761

Hixon, M. A., Johnson, D. W., & Sogard, S. M. (2014). BOFFFFs: On the importance of conserving old-growth age structure in fishery populations. ICES Journal of Marine Science, 71(8), 2171–85.

Hjort, J., Jahn, G., & Ottestad, P. (1933). The optimum catch. Hvalradets Skrifter, 7, 92–127.

Hobday, A. J., Smith, A. D. M., Webb, H., Daley, R., Wayte, S., Bulman, C., ... & Walker, T. (2007). Ecological risk assessment for the effects of fishing: Methodology. Report R04/1072 for the Australian Fisheries Management Authority, Canberra.

Hoenig, N. A., & Hanumara, R. C. (1990). An empirical comparison of seasonal growth models. Fishbyte, 8(1), 32–4.

Hofmann, E., Bushek, D., Ford, S., Guo, X., Haidvogel, D., Hedgecock, D., ... & Zhang, L. (2009). Understanding how disease and environment combine to structure resistance in estuarine bivalve populations. Oceanography, 22(4), 212–31.

Holling, C. S. (1965). The functional response of predators to prey density and its role in mimicry and population regulation. The Memoirs of the Entomological Society of Canada, 97(45), 5–60.

Holling, C. S. (1978). Adaptive environmental assessment and management. John Wiley & Sons, New York.

Hollowed, A. B., Bax, N., Beamish, R., Collie, J., Fogarty, M., Livingston, P., ... & Rice, J. C. (2000). Are multispecies models an improvement on single-species models for measuring fishing impacts on marine ecosystems? ICES Journal of Marine Science, 57(3), 707–19.

Holsman, K. K., & Aydin, K. (2015). Comparative methods for evaluating climate change impacts on the foraging ecology of Alaskan groundfish. Marine Ecology Progress Series, 521, 217–35.

Holt, R. D., & Polis, G.A. (1997). A theoretical framework for intraguild predation. American Naturalist, 149, 745–64.

Hoppensteadt, F. C. (1982). Mathematical methods of population biology (Vol. 4). Cambridge University Press, Cambridge.

Horbowy, J. (2005). The dynamics of Baltic fish stocks based on a multispecies stock production model. Journal of Applied Ichthyology, 21(3), 198–204.

Houde, E. D. (2016). Recruitment variability. In: Jakobsen, T., Fogarty, M. J., Megrey, B. A., & Moksness, E. (Eds.), Fish reproductive biology: Implications for assessment and management. (2nd edition.) Wiley Blackwell, Oxford, UK.

Hsieh, C. H., Anderson, C., & Sugihara, G. (2008). Extending nonlinear analysis to short ecological time series. The American Naturalist, 171(1), 71–80.

Huckins, C. J. F., Osenberg, C. W., & Mittelbach, G. G. (2000). Species introductions and their ecological consequences: An example with congeneric sunfish. Ecological Applications, 10, 612–25.

Hughes, T. P. (1994). Catastrophes, phase shifts, and large-scale degradation of a Caribbean coral reef. Science, 265(5178), 1547–51.

Hutchinson, G. E. (1948). Circular causal systems in ecology. Annals of the New York Academy of Sciences, 50(4), 221–46.

Hutchinson, G. E. (1957). A Treatise on Limnology (Vol. 1). Wiley, New York.

ICNAF (1974). Annual report (volume 24) for the year 1973/74. Dartmouth, Nova Scotia.

Iles, T. C., & Beverton, R. J. H. (1998). Stock, recruitment and moderating processes in flatfish. Journal of Sea Research, 39(1–2), 41–55.

Irigoien, X., & de Roos, A. (2011). The role of intraguild predation in the population dynamics of small pelagic fish. Marine Biology, 158(8), 1683–90.

Irigoien, X., Klevjer, T. A., Røstad, A., Martinez, U., Boyra, G., Acuña, J. L., & Agusti, S. (2014). Large mesopelagic fishes biomass and trophic efficiency in the open ocean. Nature Communications, 5, 3271.

Iverson, R. I. (1990). Control of marine fish production. Limnology and Oceanography, 35, 1593–604.

Ivlev, V. S. (1961). Experimental ecology of the feeding of fishes. Yale University Press, New Haven, Conn.

Ivlev, V. S. (1965). On the quantitative relationship between survival rate of larvae and their food supply. The Bulletin of Mathematical Biophysics, 27(1), 215–22.

Jacobsen, N. S., Gislason, H., & Andersen, K. H. (2014). The consequences of balanced harvesting of fish communities. Proceedings of the Royal Society of London Series B: Biological Sciences, 281(1775), 20132701.

Jacobson, L. D., & MacCall, A. D. (1995). Stock-recruitment models for Pacific sardine (*Sardinops sagax*). Canadian Journal of Fisheries and Aquatic Sciences, 52(3), 566–77.

Jacobson, L. D., & McClatchie, S. (2013). Comment on temperature-dependent stock–recruit modeling for Pacific sardine (*Sardinops sagax*) in Jacobson and MacCall (1995), McClatchie et al. (2010), and Lindegren and Checkley (2013). Canadian Journal of Fisheries and Aquatic Sciences, 70(10), 1566–9.

Jacobson, N. S. (2009). Mixed-stock projections for New England groundfish. M.S. Thesis, School of Marine Science and Technology, University of Massachusetts, New Bedford, Mass.

Jacobson, N. S., & Cadrin, S. (2008). Projecting equilibrium, mixed-species yield of New England groundfish (ICES CM No. 2008:/I:02).

Jakobsen, T., Fogarty, M. J., Megrey, B.A., & Moksness, E. (Eds.) (2016). Fish reproductive biology: implications for assessment and management (2nd edition). Wiley Blackwell, Oxford, UK.

Jenkins, G. P., Young, J. W., & Davis, T. L. (1991). Density dependence of larval growth of a marine fish, the southern bluefin tuna, *Thunnus maccoyii*. Canadian Journal of Fisheries and Aquatic Sciences, 48(8), 1358–63.

Jennings, S. and Blanchard, J. L. (2004). Fish abundance with no fishing. Journal of Animal Ecology, 73, 632–42.

Jennings, S., & Collingridge, K. (2015). Predicting consumer biomass, size-structure, production, catch potential, responses to fishing and associated uncertainties in the world's marine ecosystems. PLoS One, 10(7), 0133794.

Jennings, S., Kaiser, M. J., & Reynolds, J. D. (2001). Marine fisheries ecology. Blackwell Science, Oxford, UK.

Jennings, S., Lee, J., & Hiddink, J. G. (2012). Assessing fishery footprints and the trade-offs between landings value, habitat sensitivity, and fishing impacts to inform marine spatial planning and an ecosystem approach. ICES Journal of Marine Science, 69(6), 1053–63.

Jennings, S., Mélin, F., Blanchard, J. L., Forster, R. M., Dulvy, N. K., & Wilson, R. W. (2008). Global-scale predictions of community and ecosystem properties from simple ecological theory. Proceedings of the Royal Society of London Series B: Biological Sciences, 275(1641), 1375–83.

Jennings, S., Warr, K. J., & Mackinson, S. (2002). Use of size-based production and stable isotope analyses to predict trophic transfer efficiencies and predator–prey body mass ratios in food webs. Marine Ecology Progress Series, 240, 11–20.

Jensen, A. L. (1976). Assessment of the United States lake whitefish (*Coregonus clupeaformis*) fisheries of Lake Superior, Lake Michigan, and Lake Huron. Journal of the Fisheries Board of Canada, 33(4), 747–59.

Jensen, A. L. (1994). Larkin's predation model of lake trout (*Salvelinus namaycush*) extinction with harvesting and sea lamprey (*Petromyzon marinus*) predation: A qualitative analysis. Canadian Journal of Fisheries and Aquatic Sciences, 51(4), 942–5.

Jensen, O. P., Branch, T. A., & Hilborn, R. (2012). Marine fisheries as ecological experiments. Theoretical Ecology, 5(1), 3–22.

Jobling, M. (1994). Fish bioenergetics. Chapman and Hall, London.

Johnson, A. F., Gorelli, G., Jenkins, S. R., Hiddink, J. G., & Hinz, H. (2015). Effects of bottom trawling on fish foraging and feeding. Proceedings of the Royal Society of London Series B: Biological Sciences, 282(1799), 20142336.

Johnson, B. M., Pate, W. M., & Hansen, A. G. (2017). Energy density and dry matter content in fish: New observations and an evaluation of some empirical models. Transactions of the American Fisheries Society, 146(6), 1262–78.

Juffe-Bignoli, D., Harrison, I., Butchart, S. H., Flitcroft, R., Hermoso, V., Jonas, H., ... & Dalton, J. (2016). Achieving Aichi Biodiversity Target 11 to improve the performance of protected areas and conserve freshwater biodiversity. Aquatic Conservation: Marine and Freshwater Ecosystems, 26, 133–51.

Kaiser, M. J., Clarke, K. R., Hinz, H., Austen, M. C. V., & Somerfield, P. J. (2006). Global analysis and prediction of the response of benthic biota to fishing. Marine Ecology Progress Series, 311, 1–14.

Kantz, H., & Schreiber, T. (2003). Nonlinear time series analysis. (2nd edition.) Cambridge University Press, Cambridge, UK.

Kaplan, D. and Glass, L. (1995). Understanding nonlinear dynamics (Vol. 420). Springer-Verlag, New York.

Kareiva, P., & Wennergren, U. (1995). Connecting landscape patterns to ecosystem and population processes. Nature, 373(6512), 299.

Karplus, I. (2014). Symbiosis in fishes: The biology of interspecific partnerships. John Wiley & Sons, Hoboken, NJ.

Kellner, J. B., Litvin, S. Y., Hastings, A., Micheli, F., & Mumby, P. J. (2010). Disentangling trophic interactions inside a Caribbean marine reserve. Ecological Applications, 20(7), 1979–92.

Kellner, J. B., Tetreault, I., Gaines, S. D., & Nisbet, R. M. (2007). Fishing the line near marine reserves in single and multispecies fisheries. Ecological Applications, 17(4), 1039–54.

Kendall, B. E., Fujiwara, M., Diaz-Lopez, J., Schneider, S., Voigt, J., & Wiesner, S. (2019). Persistent problems in the construction of matrix population models. *Ecological Modelling*, *406*, 33–43.

Kerfoot, W. C., & Sih, A. (1987). Predation: Direct and indirect impacts on aquatic communities. University Press of New England, Lebanon, New Hampshire.

Kerr, S. R. (1974). Theory of size distribution in ecological communities. Journal of the Fisheries Board of Canada, 31(12), 1859–62.

Kerr, S. R., & Dickie, L. M. (2001). The biomass spectrum: A predator–prey theory of aquatic production. Columbia University Press, New York.

Kerr, S. R., & Ryder, R. A. (1988). The applicability of fish yield indices in freshwater and marine ecosystems. Limnology and Oceanography, 33, 973–981.

Kilada, R. W., Campana, S. E., & Roddick, D. (2007). Validated age, growth, and mortality estimates of the ocean quahog (*Arctica islandica*) in the western Atlantic. ICES Journal of Marine Science, 64, 31–8.

King, M. (2013). Fisheries biology, assessment and management. (2nd edition.) John Wiley & Sons, New York.

Kingsland, S. E. (1995). Modeling nature. University of Chicago Press, Chicago.

Kitchell, J. F., Koonce, J. F., Magnuson, J. J., O'Neill, R. V., Herman, J., Shugart, H., & Booth, R. S. (1974). Model of fish biomass dynamics. Transactions of the American Fisheries Society, 103(4), 786–98. https://doi.org/10.1577/1548-8659(1974)103>786:MOFBD<2.0.CO;2

Kitchell, J. F., Stewart, D. J., & Weininger, D. (1977). Applications of a bioenergetics model to yellow perch (*Perca flavescens*) and walleye (*Stizostedion vitreum vitreum*). Journal of the Fisheries Board of Canada, 34(10), 1922–35.

Koen-Alonso, M., & Yodzis, P. (2005). Multispecies modelling of some components of the marine community of northern and central Patagonia, Argentina. Canadian Journal of Fisheries and Aquatic Sciences, 62(7), 1490–512.

Kolding, J., van Zweiten, P., Mkumbo, O., Silsbe, G., & Heckey, R. (2008). Are the Lake Victoria fisheries threatened by exploitation or eutrophication? In: Bianchi, G., & Skjoldal, H. R. (Eds.), The ecosystem approach to fisheries (pp. 309–54). Cabi (FAO), Rome.

Kolding, J., Medard, M., Mkumbo, O., & van Zwieten, P. A. M. (2014). Status, trends and management of the Lake Victoria fisheries. In: Inland fisheries evolution and management. Case studies from four continents (No. 579, pp. 49–62). FAO, Rome.

Kostylev, V. E., Todd, B. J., Longva, O., & Valentine, P. C. (2005a). Characterization of benthic habitat on northeastern Georges Bank, Canada. In: Barnes, P. W., & Thomas, J. P. (Eds.), Benthic habitat and the effects of fishing. American Fisheries Society Symposium (Vol. 41), pp.141–152. American Fisheries Society, Bethesda, MD.

Kot, M. (2001). Elements of mathematical ecology. Cambridge University Press, Cambridge, UK.

Kovalenko, K. E., Thomaz, S. M., & Warfe, D. M. (2012). Habitat complexity: approaches and future directions. Hydrobiologia, 685(1), 1–17.

Krebs, C. J. (1999). Ecological Methodology. (2nd edition.) Benjamin/Cummings, San Francisco.

Kricher, J. (2009). The balance of nature: ecology's enduring myth. Princeton University Press, Princeton, NJ.

Kritzer, J. P., & Sale, P. F. (2006). Marine metapopulations. Academic Press, Cambridge, Mass.

Kurlansky, M. (1997). Cod: A biography of the fish that changed the world. Vintage Canada, Toronto.

Lafferty, K. D., Harvell, C. D., Conrad, J. M., Friedman, C. S., Kent, M. L., Kuris, A. M., & Saksida, S. M. (2015). Infectious diseases affect marine fisheries and aquaculture economics. Annual Review of Marine Science, 7, 471–96.

Lafferty, K. D., Porter, J. W., & Ford, S. E. (2004). Are diseases increasing in the ocean? Annual Review of Ecology and Systematics, 35, 31–54.

Langangen, Ø., Edeline, E., Ohlberger, J., Winfield, I. J., Fletcher, J. M., James, J. B., & Vøllestad, L. A. (2011). Six decades of pike and perch population dynamics in Windermere. Fisheries Research, 109(1), 131–9.

Larkin, P. A. (1963). Interspecific competition and exploitation. Journal of the Fisheries Board of Canada, 20(3), 647–78.

Larkin, P. A. (1977). An epitaph for the concept of maximum sustainable yield. Transactions of the American Fisheries Society, 106, 1–11.

Laska, M. S., & Wootton, J. T. (1998). Theoretical concepts and empirical approaches to measuring interaction strength. Ecology, 79(2), 461–76.

Lauck, T., Clark, C. W., Mangel, M., & Munro, G. R. (1998). Implementing the precautionary principle in fisheries management through marine reserves. Ecological Applications, 8(sp1), S72–S78.

Leopold, A. (1949). A Sand County Almanac. Oxford University Press, New York.

Leslie, P. H. (1948). Some further notes on the use of matrices in population mathematics. Biometrika, 35, 213–45.

Leslie, P. H., & Gower, J. C. (1958). The properties of a stochastic model for two competing species. Biometrika, 45(3/4), 316–30.

Lester, N. P., Shuter, B. J., & Abrams, P. A. (2004). Interpreting the von Bertalanffy model of somatic growth in fishes: The cost of reproduction. Proceedings of the Royal Society of London Series B: Biological Sciences, 271(1548), 1625–31.

Lester, S. E., & Halpern, B. S. (2008). Biological responses in marine no-take reserves versus partially protected areas. Marine Ecology Progress Series, 367, 49–56.

Levin, P. S., Essington, T. E., Marshall, K. N., Koehn, L. E., Anderson, L. G., Bundy, A., . . . & Houde, E. (2018). Building effective fishery ecosystem plans. Marine Policy, 92, 48–57.

Levin, P. S., Fogarty, M. J., Matlock, G. C., & Ernst, M. (2008). Integrated ecosystem assessments. US Dept. of Commerce. NOAA Technical Memorandum NMFS-NWFSC, 92.

Levin, P. S., Fogarty, M. J., Murawski, S. A., & Fluharty, D. (2009a). Integrated ecosystem assessments: developing the scientific basis for ecosystem-based management of the ocean. PLoS Biology, 7(1), e1000014.

Levin, P. S., Kaplan, I., Grober-Dunsmore, R., Chittaro, P. M., Oyamada, S., Andrews, K., & Mangel, M. (2009b). A framework for assessing the biodiversity and fishery aspects of marine reserves. Journal of Applied Ecology, 46(4), 735–42.

Levin, S., & Goodyear, C. (1980). Analysis of an age-structured fishery model. Journal of Mathematical Biology, 9, 245–74.

Levins, R. (1968). Evolution in changing environments: Some theoretical explorations (MPB-2). Princeton University Press, Princeton.

Levins, R. (1969). Some demographic and genetic consequences of environmental heterogeneity for biological control. Bulletin of the Entomological Society of America, 15, 237–40.

Levins, R. (1974). Discussion paper: The qualitative analysis of partially specified systems. Annals of the New York Academy of Sciences, 231(1), 123–38.

Levins, R. (1975). Evolution in communities near equilibrium. In: Cody, M. L., McArthur, R. H., & Diamond, J. M. (Eds.), Ecology and evolution of communities (pp. 16–50). Harvard University Press, Cambridge, Mass.

Lewy, P., & Vinther, M. (2004). A stochastic age-length-structured multispecies model applied to North Sea stocks. ICES CM, 2004/FF:20.

Li, H. W., & Moyle, P. B. (1981). Ecological analysis of species introduction into aquatic systems. Transactions of the American Fisheries Society, 110, 772–82.

Libralato, S., Coll, M., Tudela, S., Palomera, I., & Pranovi, F. (2008). Novel index for quantification of ecosystem effects of fishing as removal of secondary production. Marine Ecology Progress Series, 355, 107–29.

Lindeman, R. I. (1942). The trophic-dynamic aspect of ecology. Ecology, (23), 399–418.

Link, J. S. (2002a). Does food web theory work for marine ecosystems? Marine Ecology Progress Series, 230, 1–9.

Link, J. S. (2002b). Ecological considerations in fisheries management: When does it matter? Fisheries, 27(4), 10–17.

Link, J. S. (2005). Translating ecosystem indicators into decision criteria. ICES Journal of Marine Science, 62(3), 569–76.

Link, J. S. (2010). Ecosystem-based fisheries management: confronting tradeoffs. Cambridge University Press, Cambridge, UK.

Link, J. S. (2018). System-level optimal yield: increased value, less risk, improved stability, and better fisheries. Canadian Journal of Fisheries and Aquatic Sciences, 75(1), 1–16.

Link, J. S., & Auster, P. J. (2013). The challenges of evaluating competition among marine fishes: Who cares, when does it matter, and what can one do about it? Bulletin of Marine Science, 89(1), 213–47.

Link, J. S., Lucey, S. M., & Melgey, J. H. (2012). Examining cannibalism in relation to recruitment of silver hake *Merluccius bilinearis* in the US northwest Atlantic. Fisheries Research, 114, 31–41.

Link, J. S., & Watson, R. A. (2019). Global ecosystem overfishing: Clear delineation within real limits to production. Science Advances, 5(6), eaav0474.

Liu, H., Fogarty, M. J., Glaser, S. M., Altman, I., Hsieh, C.-h., Kaufman, L., . . . & Sugihara, G. (2012). Nonlinear dynamic features and co-predictability of the Georges Bank fish community. Marine Ecology Progress Series, 464, 195–207. https://doi.org/10.3354/meps09868

Lloyd, M. (1967). Mean crowding. The Journal of Animal Ecology, 1–30.

Longhurst, A. (2002). Murphy's law revisited: Longevity as a factor in recruitment to fish populations. Fisheries Research, 56, 125–31.

Longhurst, A. (2010). Mismanagement of marine fisheries. Cambridge University Press, New York.

Longhurst, A., Sathyendranath, S., Platt, T., & Caverhill, C. (1995). An estimate of global primary production in the ocean from satellite radiometer data. Journal of Plankton Research, 17(6), 1245–71.

Lorenz, E. N. (1963). Deterministic nonperiodic flow. Journal of Atmospheric Science, 20, 130–41.

Lorenz, M. O. (1905). Methods of measuring the concentration of wealth. Publications of the American Statistical Association, 9(70), 209–19.

Lorenzen, K., & Enberg, K. (2002). Density-dependent growth as a key mechanism in the regulation of fish populations: Evidence from among-population comparisons. Proceedings of the Royal Society of London Series B: Biological Sciences, 269(1486), 49–54.

Lotka, A. J. (1925). Elements of physical biology. Williams and Wilkens, Baltimore, MD.

Lotze, H. K., Lenihan, H. S., Bourque, B. J., Bradbury, R. H., Cooke, R. G., Kay, M. C., . . . & Jackson, J. B. (2006). Depletion, degradation, and recovery potential of estuaries and coastal seas. Science, 312(5781), 1806–9.

Lucey, S. M., Gaichas, S. K., & Aydin, K. Y. (2020). Conducting reproducible ecosystem modeling using the open source mass balance model Rpath. *Ecological Modelling, 427*, 109057.

Ludwig, D., Jones, D. D., & Holling, C. S. (1978). Qualitative analysis of insect outbreak systems: The spruce budworm and forest. Journal of Animal Ecology, 47(1), 315–32.

Ludwig, D., & Walters, C. J. (1985). Are age-structured models appropriate for catch-effort data? Canadian Journal of Fisheries and Aquatic Sciences, 42(6), 1066–72.

Lynam, C. P., & Brierley, A. S. (2007). Enhanced survival of 0-group gadoid fish under jellyfish umbrellas. Marine Biology, 150(6), 1397–401.

MacCall, A. D. (1990). Dynamic geography of marine fish populations. University of Washington Press, Seattle.

MacCall, A. D. (2012). Data-limited management reference points to avoid collapse of stocks dependent on learned migration behaviour. ICES Journal of Marine Science, 69(2), 267–70.

MacCall, A. D., Sydeman, W. J., Davison, P. C., & Thayer, J. A. (2016). Recent collapse of northern anchovy biomass off California. Fisheries Research, 175, 87–94.

Mackinson, S., Platts, M., Garcia, C., & Lynam, C. (2018). Evaluating the fishery and ecological consequences of the proposed North Sea multi-annual plan. PLoS One, 0190015. Retrieved from https://doi.org/10.1371/journal.pone.0190015.

Magnusson, K. G. (1995). An overview of the multispecies VPA—theory and applications. Reviews in Fish Biology and Fisheries, 5, 195–212.

Magurran, A. E. (2004). Measuring Biological Diversity. Blackwell Science, Malden, Mass.

Mallet, J. P., Charles, S., Persat, H., & Auger, P. (1999). Growth modelling in accordance with daily water temperature in European grayling (*Thymallus thymallus* L.). Canadian Journal of Fisheries and Aquatic Sciences, 56(6), 994–1000.

Mangel, M. (1998). No-take areas for sustainability of harvested species and a conservation invariant for marine reserves. Ecology Letters, 1(2), 87–90.

Mangel, M. (2000). Irreducible uncertainties, sustainable fisheries and marine reserves. Evolutionary Ecology Research, 2(4), 547–57.

Mangel, M., & Levin, P. S. (2005). Regime, phase, and paradigm shifts: Making community ecology the basic science for fisheries. Philosophical Transactions of the Royal Society B: Biological Sciences, 360, 95–105.

Margalef, R. (1960). Ideas for a synthetic approach to the ecology of running waters. Internationale Revue der Gesamten Hydrobiologie undHydrographie, 45(1), 133–53.

Marsh, R., Petrie, B., Weidman, C. R., Dickson, R. R., Loder, J. W., Hannah, C. G., . . . Drinkwater, K. (1999). The 1882 tilefish kill—A cold event in shelf waters off the north-eastern United States? Fisheries Oceanography, 8(1), 39–49.

Marshall, C. T. (2016). Implementing information on stock reproductive potential in fisheries management: The motivation, challenges and opportunities. In: Jakobsen, T., Fogarty, M. J., Megrey, B. A., & Moksness, E. (Eds.), Fish reproductive biology: Implications for assessment and management. (2nd edition.) Wiley Blackwell, Oxford, UK.

Marty, G. D., Li, T. J. Q., Carpenter, G., Meyers, T. R., & Willits, N. H. (2003). Role of disease in abundance of a Pacific herring (*Clupea pallasii*) population. Canadian Journal of Fisheries and Aquatic Sciences, 60(10), 1258–65.

May, R. M. (1972). Will a large complex system be stable? Nature, 238, 413–14.

May, R. M. (1973). Stability and complexity in model ecosystems. Princeton University Press, Princeton, NJ.

May, R. M. (1974). Biological populations with non-overlapping generations: Stable points, stable cycles, and chaos. Science, 186–645.

May, R. M. (1975). Biological populations obeying difference equations: Stable points, stable cycles, and chaos. Journal of Theoretical Biology, 51(2), 511–24.

May, R. M. (1976). Simple mathematical models with very complicated dynamics. Nature, 261, 459–67.

May, R. M., & Anderson, R. M. (1983). Epidemiology and genetics in the coevolution of parasites and hosts. Proceedings of the Royal Society of London Series B. Biological Sciences, 219(1216), 281–313.

May, R. M., Beddington, J. R., Clark, C. W., Holt, S. J., & Laws, R. M. (1979). Management of multispecies fisheries. Science, 405, 267–76.

Maynard-Smith, J. M., & Slatkin, M. (1973). The stability of predator–prey systems. Ecology, 54(2), 384–91.

McCallum, H. (2008). Population parameters: Estimation for ecological models (Vol. 3). Blackwell, Oxford.

McCann, K., Hastings, A., & Huxel, G. R. (1998). Weak trophic interactions and the balance of nature. Nature, 395(6704), 794.

McCann, K. S. (2012). Food webs. Princeton University Press Princeton, NJ.

McCann, K., & Yodzis, P. (1994). Biological conditions for chaos in a three-species food chain. Ecology, 75(2), 561–4.

McClatchie, S., Goericke, R., Auad, G., & Hill, K. (2010). Re-assessment of the stock–recruit and temperature–recruit relationships for Pacific sardine (*Sardinops sagax*). Canadian Journal of Fisheries and Aquatic Sciences, 67(11), 1782–90.

McConnell, W. J., Lewis, S., & Olson,.J.E. (1977). Gross photosynthesis as an estimator of potential fish production. Transactions of the American Fisheries Society, 106, 417–23.

McGrady-Steed, J., & Morin, P. J. (2000). Biodiversity, density compensation, and the dynamics of populations and functional groups. Ecology, 81(2), 361–73.

Mcowen, C. J., Cheung, W. W., Rykaczewski, R. R., Watson, R. A., & Wood, L. J. (2015). Is fisheries production within large marine ecosystems determined by bottom-up or top-down forcing? Fish and Fisheries, 16(4), 623–32.

Mercer, M. (Ed.) (1982). Multispecies approaches to fisheries management advice. Canadian Special Publication of Fisheries and Aquatic Sciences 59. Ottowa.

Micheli, F., Amarasekare, P., Bascompte, J., & Gerber, L. R. (2004). Including species interactions in the design and evaluation of marine reserves: some insights from a predator–prey model. Bulletin of Marine Science, 74(3), 653–69.

Miller, C. B. (2004). Biological oceanography. Blackwell, Boston, Mass.

Miller, T. J., Crowder, L. B., Rice, J. A., & Marschall, E. A. (1988). Larval size and recruitment in fishes: Toward a conceptual framework. Canadian Journal of Fisheries and Aquatic Sciences, 45, 1657–70.

Mittelbach, G. G. (2005). Parasites, communities, and ecosystems: conclusions and perspectives. In: Thomas, F., Renaud, F., & Guégan, J.-F. (Eds.), Parasites and ecosystems (pp. 171–6). Oxford University Press, New York.

Mittelbach, G. G. (2012). Community ecology. Sinauer Associates, Sunderland, Mass.

Mittelbach, G. G., Garcia, E. A., & Taniguchi, Y. (2006). Fish reintroductions reveal smooth transitions between lake community states. Ecology, 87(2), 312–18.

Mittelbach, G. G., Turner, A. M., Hall, D. J., Rettig, J. E., & Osenberg, C. W. (1995). Perturbation and resilience: A long-term, whole-lake study of predator extinction and reintroduction. Ecology, 76(8), 2347–60.

Moran, P. A. P. (1950). Some remarks on animal population dynamics. Biometrics, 6(3), 250–8.

Morin, P. J. (1999). Community ecology. Blackwell, Oxford, UK.

Mueter, F. J., & Megrey, B. A. (2006). Using multi-species surplus production models to estimate ecosystem-level maximum sustainable yields. Fisheries Research, 81, 189–201.

Mullon, C., Freon, P., & Cury, P. (2005). The dynamics of collapse in world fisheries. Fish and Fisheries, 6(2), 111–20.

Munch, S. B., Brias, A., Sugihara, G., & Rogers, T. L. (2019). Frequently asked questions about nonlinear dynamics and empirical dynamic modelling. ICES Journal of Marine Science. doi:10.1093/icesjms/fsz209

Munch, S. B., Giron-Nava, A., & Sugihara, G. (2018). Nonlinear dynamics and noise in fisheries recruitment: A global meta-analysis. Fish and Fisheries, 19(6), 964–73.

Murawski, S. (1984). Mixed-species yield-recruitment analyses accounting for technological interactions. Canadian Journal of Fisheries and Aquatic Sciences, 41, 897–916.

Murawski, S. A. (1993). Climate change and marine fish distributions: Forecasting from historical analogy. Transactions of the American Fisheries Society, 122(5), 647–58.

Murawski, S. A. (2000). Definitions of overfishing from an ecosystem perspective. ICES Journal of Marine Science, 57(3), 649–58.

Murawski, S. A., & Finn, A. J. (1988). Biological bases for mixed-species fisheries: Species co-distribution in relation to environmental and biotic variables. Canadian Journal of Fisheries and Aquatic Sciences, 45(10), 1720–35.

Murawski, S. A., & Idoine, J. S. (1992). Multispecies size composition: A conservative property of exploited fishery systems. Journal of Northwest Atlantic Fishery Science, 14, 79–85.

Musick, J. A. (1999). Criteria to define extinction risk in marine fishes. Fisheries, 24(12), 6–14.

Myers, R. A. (1998). When do environment–recruitment correlations work? Reviews in Fish Biology and Fisheries, 8(3), 285–305.

Myers, R. A., Bowen, K. G., & Barrowman, N. J. (1999). Maximum reproductive rate of fish at low population sizes. Canadian Journal of Fisheries and Aquatic Sciences, 56, 2404–19.

Myers, R. A., Mertz, G., Bridson, J. M., & Bradford, M. J. (1998). Simple dynamics underlie sockeye salmon (*Oncorhynchus nerka*) cycles. Canadian Journal of Fisheries and Aquatic Sciences, 55, 2355–64.

NEFSC (Northeast Fisheries Science Center) (2008). Assessment of 19 northeast groundfish stocks through 2007: Report of the 3rd Groundfish Assessment Review Meeting (GARM III) (Northeast Fisheries Science Center Reference Document 08–15; pp. 08–15). Woods Hole, Mass.

Neubert, M. G., & Caswell, H. (2000). Density-dependent vital rates and their population dynamic consequences. Journal of Mathematical Biology, 41(2), 103–21.

Neuheimer, A. B., & Taggart, C. T. (2007). The growing degree-day and fish size-at-age: The overlooked metric. Canadian Journal of Fisheries and Aquatic Sciences, 64(2), 375–85.

Nicolis, G., & Prigogine, I. (1989). Exploring complexity: An introduction. St Martin's Press, New York.

Nilsson, N. A. (1963). Interaction between trout and char in Scandinavia. Transactions of the American Fisheries Society, 92(3), 276–85.

NRC (National Research Council) (2002). Effects of trawling and dredging on seafloor habitat. National Academies Press, Washington, DC.

Nye, J. A., Link, J. S., Hare, J. A., & Overholtz, W. J. (2009). Changing spatial distribution of fish stocks in relation to climate and population size on the Northeast United States continental shelf. Marine Ecology Progress Series, 393, 111–29.

Ogle, D. H. (2016). *Introductory fisheries analyses with R*. Chapman and Hall/CRC, Boca Raton, FL.

Ohlberger, J. (2016). Population coherence and environmental impacts across spatial scales: A case study of chinook salmon. Ecosphere, 7, 01333.

Ohlberger, J., Langangen, Ø., Edeline, E., Claessen, D., Winfield, I. J., Stenseth, N. C., & Vøllestad, L. A. (2011). Stage-specific biomass overcompensation by juveniles in response to increased adult mortality in a wild fish population. Ecology, 92(12), 2175–82.

Oken, K. L., & Essington, T. E. (2015). How detectable is predation in stage-structured populations? Insights from a simulation-testing analysis. Journal of Animal Ecology, 84(1), 60–70.

Paine, R. T. (1969). A note on trophic complexity and community stability. The American Naturalist, 103(929), 91–3.

Paloheimo, J. E. D., & Dickie, L. M. (1965). Food and growth of fishes. I. A growth curve derived from experimental data. Journal of the Fisheries Board of Canada, 22(2), 521–42.

Palomares, M. L. D., & Pauly, D. (1998). Predicting food consumption of fish populations as functions of mortality, food type, morphometrics, temperature and salinity. Marine and Freshwater Research, 49(5), 447–53.

Parrish, J. D., & Saila, S. B. (1970). Interspecific competition, predation and species diversity. Journal of Theoretical Biology, 27(2), 207–20.

Parsons (1967). Contributions of year-classes of blue pike to the commercial fishery of Lake Erie, 1943–59. Journal of the Fisheries Board of Canada, 24(5), 1035–66.

Pascual, M., & Dunne, J. A. (Eds.). (2006). Ecological networks: Linking structure to dynamics in food webs. Oxford University Press, New York.

Patrick, W. S., & Link, J. S. (2015). Hidden in plain sight: using optimum yield as a policy framework to operationalize ecosystem-based fisheries management. Marine Policy, 62, 74–81.

Patrick, W. S., Spencer, P., Link, J., Cope, J., Field, J., Kobayashi, D., & Bigelow, K. (2010). Using productivity and susceptibility indices to assess the vulnerability of United States fish stocks to overfishing. Fishery Bulletin, 108(3), 305–22.

Paulik, G.J. (1973). Studies of the possible form of the stock-recruitment curve. Rapport et Proces-verbaux des Reunions Conseil International pour Exploration de la Mer, 164, 302–15.

Pauly, D. (1986). A simple method of estimating the food consumption of fish populations from growth data and food conversion experiments. Fishery Bulletin, 84, 827–40.

Pauly, D. (1996). One hundred million tons of fish, and fisheries research. Fisheries Research, 25, 25–38.

Pauly, D., & Christensen, V. (1995). Primary production required to sustain global fisheries. Nature, 374(255), 257.

Pauly, D., Christensen, V., & Walters, C. (2000). Ecopath, Ecosim, and Ecospace as tools for evaluating ecosystem impact of fisheries. ICES Journal of Marine Science, 57(3), 697–706.

Pauly, D., Froese, R., & Holt, S. J. (2016). Balanced harvesting: The institutional incompatibilities. Marine Policy, 69, 121–3.

Pella, J. J., & Tomlinson, P. K. (1969). A generalized stock production model. Inter-American Tropical Tuna Commission Bulletin, 13(3), 416–97.

Penn, J. W., Caputi, N., & de Lestang, S. (2015). A review of lobster fishery management: The Western Australian fishery for *Panulirus cygnus*, a case study in the development and implementation of input and output-based management systems. ICES Journal of Marine Science, 72(1), 22–34.

Pennycuik, C. J., Compton, R. M., & Beckingham, L. (1968). A computer model for simulating the growth of a population, or of two interacting populations. Journal of Theoretical Biology, 18, 316–29.

Pérez Roda, A. M., Gilman, E., Huntington, T., Kennelly, S. J., Suuronen, P., Chaloupka, M., & Medley, P. (2019). A third assessment of global marine fisheries discards. FAO Fisheries Technical Paper, 633, 78.

Perretti, C. T., Munch, S. B., Fogarty, M. J., & Sugihara, G. (2015). Global evidence for non-random dynamics in fish recruitment. arXiv preprint arXiv:1509.01434.

Perretti, C. T., Munch, S. B., & Sugihara, G. (2013). Model-free forecasting outperforms the correct mechanistic model for simulated and experimental data. Proceedings of the National Academy of Sciences USA, 110, 5253–7.

Perry, A. L., Low, P. J., Ellis, J. R., & Reynolds, J. D. (2005). Climate change and distribution shifts in marine fishes. Science, 308, 1912–15.

Peters, R. H. (1986). The ecological implications of body size. Cambridge University Press, Cambridge, UK.

Peters, R. H. (1991). A critique for ecology. Cambridge University Press, Cambridge, UK.

Peterson, I., & Wroblewski, J. S. (1984). Mortality rate of fishes in the pelagic ecosystem. Canadian Journal of Fisheries and Aquatic Sciences, 41(7), 1117–20.

Petitgas, P., Secor, D. H., McQuinn, I., Huse, G., & Lo, N. (2010). Stock collapses and their recovery: Mechanisms that establish and maintain life-cycle closure in space and time. ICES Journal of Marine Science, 67(9), 1841–8.

Petraitis, P. (2013). Multiple stable states in natural ecosystems. Oxford University Press, Oxford, UK.

Pielou, E. C. (1969). An introduction to mathematical ecology. Wiley, New York.

Pikitch, E. K. (1987). Use of a mixed-species yield-per-recruit model to explore the consequences of various management policies for the Oregon flatfish fishery. Canadian Journal of Fisheries and Aquatic Sciences, 44(2), 349–59.

Pikitch, E. K., Santora, C., Babcock, E. A., Bakun, A., Bonfil, R., Conover, D. O., . . . & Houde, E. D. (2004). Ecosystem-based fishery management. Science, 305(5682), 346–7.

Pimm, S. L. (1982). Food webs. In: Pimm, S. L., Food webs: Population and community biology, (pp. 1–11). Springer, Dordrecht.

Pimm, S. L., & Lawton, J. H. (1977). Number of trophic levels in ecological communities. Nature, 268(5618), 329.

Pinsky, M. L., Worm, B., Fogarty, M. J., Sarmiento, J. L., & Levin, S. A. (2013). Marine taxa track local climate velocities. Science, 341, 1239–42.

Pitcher, T. J. (2016). Assessment and modelling in freshwater fisheries. In: Craig, J. F. (Ed.). (2016). Freshwater fisheries ecology. John Wiley & Sons.

Pitcher, T., & Hart, P. J. (1982). Fisheries ecology. Croom Helm, London.

Pitcher, T. J., Kalikoski, D., Short, K., Varkey, D., & Pramod, G. (2009). An evaluation of progress in implementing ecosystem-based management of fisheries in 33 countries. Marine Policy, 33(2), 223–32.

Pitcher, T. J., & MacDonald, P. D. M. (1973). Two models for seasonal growth in fishes. Journal of Applied Ecology, 599–606.

Plagányi, É. E. (2007). Models for an ecosystem approach to fisheries. Fisheries Technical Paper 477. FAO, Rome, Italy.

Plagányi, É. E., Punt, A. E., Hillary, R., Morello, E. B., Thébaud, O., Hutton, T., . . . & Smith, F. (2014). Multispecies fisheries management and conservation: tactical applications using models of intermediate complexity. Fish and Fisheries, 15(1), 1–22.

Polis, G. A., Myers, C. A., & Holt, R. D. (1989). The ecology and evolution of intraguild predation: Potential competitors that eat each other. Annual Review of Ecology and Systematics, 20, 297–330.

Polovina, J. J. (1984). Model of a coral reef ecosystem. Coral Reefs, 3, 1–11.

Pope, J. G. (1977). Estimation of fishing mortality, its precision and implications for the management of fisheries. Fisheries Mathematics, 63–76.

Pope, J. G. (1979). Stock assessment in multispecies fisheries, with special reference to the trawl fishery in the Gulf of Thailand. South China Sea Fisheries Development and Coordinating Programme (Vol. 79). FAO, Rome.

Pope, J. G., & Jiming, Y. (1987). Phalanx analysis: An extension of Jones' length cohort analysis to multispecies cohort analysis. In: Length-based Methods in Fisheries Research (ICLARM Conf. Proc., Vol. 13, pp. 177–192). Manila: ICLARM.

Prager, M. H. (1994). A suite of extensions to a nonequilibrium surplus-production model. Fishery Bulletin, 92, 374–89.

Pringle, R. M. (2005). The Nile perch in Lake Victoria: Local responses and adaptations. Bioscience, 9(780–7).

Puccia, C. J., & Levins, R. (1985). Qualitative modeling of complex systems, an introduction to loop analysis and time averaging. Harvard University Press, Cambridge, Mass.

Punt, A. E., Butterworth, D. S., de Moor, C. L., De Oliveira, J. A., & Haddon, M. (2016). Management strategy evaluation: best practices. Fish and Fisheries, 17(2), 303–34.

Quinn, T. J., & Deriso, R. B. (1999). Quantitative Fish Dynamics. Oxford University Press, New York.

Ralston, S., & Polovina, J. J. (1982). A multispecies analysis of the commercial deep-sea handline fishery in Hawaii. Fishery Bulletin, 3(435–48).

Regier, H. A., & Hartman, W. L. (1973). Lake Erie's fish community: 150 years of cultural stresses. Science, 180(4092), 1248–55.

Reznick, D. N., & Ghalambor, C. K. (2005). Can commercial fishing cause evolution? Answers from guppies (*Poecilia reticulata*). Canadian Journal of Fisheries and Aquatic Sciences, 62(4), 791–801.

Rice, J., Daan, N., & and J. Pope, H. G. (2013). Does functional redundancy stabilize fish communities? ICES Journal of Marine Science, 70, 734–42.

Richards, F. J. (1959). A flexible growth function for empirical use. Journal of Experimental Botany, 10, 290–300.

Ricker, W. E. (1946). Production and utilization of fish populations. Ecological Monographs, 16(4), 373–91.

Ricker, W. E. (1954). Stock and recruitment. Journal of the Fisheries Research Board of Canada, 11, 559–623.

Ricker, W. E. (1958). Handbook of computations for biological statistics of fish populations (Vol. 119). Canadian Fisheries Research Board Bulletin.

Ricker, W. E. (1963). Big effects from small causes: Two examples from fish population dynamics. Journal of the Fisheries Board of Canada, 20(2), 257–64.

Ricker, W. E. (1969). Food from the sea. In: Resources and man, a study and recommendation report of the Committee on Resources and Man, US National Academy of Sciences (pp. 87–108). W H Freeman, San Francisco.

Ricker, W. E. (1975). Computation and interpretation of biological statistics of fish populations. Bulletin of the Fisheries Research Board of Canada, 191, 1–382.

Rigler, F. H. (1982a). The relation between fisheries management and limnology. Transactions of the American Fisheries Society, 111(2), 121–32.

Rigler, F. H. (1982b). Recognition of the possible: an advantage of empiricism in ecology. Canadian Journal of Fisheries and Aquatic Sciences, 39(9), 1323–31.

Rindorf, A., Dichmont, C. M., Levin, P. S., Mace, P., Pascoe, S., Prellezo, R., . . . & Vinther, M. (2017). Food for thought: pretty good multispecies yield. ICES Journal of Marine Science, 74(2), 475–86.

Rindorf, A., & Lewy, P. (2012). Estimating the relationship between abundance and distribution. Canadian Journal of Fisheries and Aquatic Sciences, 2(382–97).

Rochet, M. J., & Benoît, E. (2012). Fishing destabilizes the biomass flow in the marine size spectrum. Proceedings of the Royal Society B: Biological Sciences, 279(1727), 284–92.

Rochet, M. J., Collie, J. S., Jennings, S., & Hall, S. J. (2011). Does selective fishing conserve community biodiversity? Predictions from a length-based multispecies model. Canadian Journal of Fisheries and Aquatic Sciences, 68, 469–86.

Rochet, M. J., Collie, J. S., & Trenkel, V. M. (2013). How do fishing and environmental effects propagate among and within functional groups? Bulletin of Marine Science, 89(1), 285–315.

Roos, J. F. (1991). Restoring Fraser River Salmon: A History of the International Pacific Salmon Fisheries Commission 1937–1985. International Pacific Salmon Fisheries Commission, Vancouver.

Rose, G. A. (2004). Reconciling overfishing and climate change with stock dynamics of Atlantic cod (*Gadus morhua*) over 500 years. Canadian Journal of Fisheries and Aquatic Sciences, 61(9), 1553–7.

Rose, K. A., Cowan, J. H., Winemiller, K. O., Myers, R. A., & Hilborn, R. (2001). Compensatory density dependence in fish populations: importance, controversy, understanding and prognosis. Fish and Fisheries, 2(4), 293–327.

Rosenzweig, M. L. (1971). Paradox of enrichment: Destabilization of exploitation ecosystems in ecological time. Science, 171(3969), 385–7.

Rothschild, B. J. (1977). Fishing effort. In: Gulland, J. A. (Ed.), Fish population dynamics (pp. 96–115). John Wiley & Sons, New York.

Rothschild, B. J. (1986). Dynamics of marine fish populations. Harvard University Press, Cambridge, Mass.

Rothschild, B. J., & Fogarty, M. J. (1989). Spawning stock biomass as a source of error in recruitment-stock relationships. Journal Du Conseil International Pour Exploration de La Mer, 45, 131–5.

Rothschild B., J., & Fogarty M., J. (1998). Recruitment and the population dynamics process. The Sea, 10, 293–325.

Roughgarden, J. (1975). A simple model for population dynamics in stochastic environments. The American Naturalist, 109(970), 713–36.

Roughgarden, J. (1998). Primer of ecological theory. Prentice Hall, Upper Saddle River, NJ.

Rouyer, T., Fromentin, J.-M., Ménard, F., Cazelles, B., Briand, K., Pianet, R., Planque, B., & Stenseth, N. C. (2008). Complex interplays among population dynamics, environmental forcing, and exploitation in fisheries. Proceedings of the National Academy of Sciences USA, 105, 5420–5.

Royer, F. and Fromentin, J.M. (2006). Recurrent and density-dependent patterns in long-term fluctuations of Atlantic bluefin tuna trap catches. Marine Ecology Progress Series, 319, 237–49.

Ruggerone, G. T., & Nielsen, J. L. (2004). Evidence for competitive dominance of pink salmon (*Oncorhynchus gorbuscha*) over other salmonids in the North Pacific Ocean. Reviews in Fish Biology and Fisheries, 14, 371–90.

Russell, E. S. (1931). Some theoretical considerations on the "overfishing" problem. ICES Journal of Marine Science, 6(1), 3–20.

Ruzzante, D. E., Wroblewski, J. S., Taggart, C. T., Smedbol, R. K., & and S.V. Goddard, D. C. (2000). Differentiation between Atlantic cod (*Gadus morhua*) from Gilbert Bay, Labrador, Trinity Bay, Newfoundland, and offshore northern cod: Evidence for bay-scale population structure. Journal of Fish Biology, 56, 431–47.

Ryder, R. A. (1965). A method for estimating the potential fish production of north-temperate lakes. Transactions of the American Fisheries Society, 94, 214–18.

Ryder, R. A. (1982). The morphoedaphic index—Use, abuse, and fundamental concepts. Transactions of the American Fisheries Society, 111(2), 154–64.

Ryder, R. A., Kerr, S. R., Loftus, K. H., & Regier, H. A. (1974). The morphoedaphic index, a fish yield estimator—Review and evaluation. Journal of the Fisheries Board of Canada, 31(5), 663–88.

Ryther, J. (1969). Photosynthesis and fish production in the sea. Science, 166, 72–6.

Sahrhage, D., & Lundbeck, J. (1992). A history of fishing. Springer-Verlag, Berlin Heidelberg.

Sainsbury, K. J. (1991). Application of an experimental approach to management of a tropical multispecies fishery with highly uncertain dynamics. ICES Marine Science Symposium 193, 301–320.

Sainsbury, K. J., Campbell, R. A., Lindholm, R., & Whitelaw, A. W. (1997). Experimental management of an Australian multispecies fishery: examining the possibility of trawl-induced habitat modification. In: Global trends: fisheries management (eds. K. Pikitch, D. D. Huppert and M. P. Sissenwine) American Fisheries Society Symposium 20, 107–112.

Schaefer, M. B. (1954). Some aspects of the dynamics of populations important to the management of the commercial marine fisheries. Inter-American Tropical Tuna Commission Bulletin, 1(2), 23–56.

Schaefer, M. B. (1957). Some considerations of population dynamics and economics in relation to the management of the commercial marine fisheries. Journal of the Fisheries Board of Canada, 14(5), 669–81.

Schaefer, M. B. (1965). The potential harvest of the sea. Transactions of the American Fisheries Society, 94(2), 123–8.

Scheffer, M. (2009). Critical transitions in nature and society. Princeton University Press, Princeton, NJ.

Scheld, A.M., & Anderson, C. M. (2016). Selective fishing and shifting production in multispecies fisheries. Canadian Journal of Fisheries and Aquatic Sciences, 74(3), 388–95.

Schnute, J. (1985). A general theory for the analysis of catch and effort data. Canadian Journal of Fisheries and Aquatic Sciences, 42, 414–29.

Scudo, F. M. (1971). Vito Volterra and theoretical ecology. Theoretical Population Biology, 2(1), 1–23.

Semakula, S. N., & Larkin, P. A. (1968). Age, growth, food, and yield of the white sturgeon (*Acipenser transmontanus*) of the Fraser River, British Columbia. Journal of the Fisheries Board of Canada, 25(12), 2589–602.

Sheldon, R. W. and Kerr, S.R. (1972). The population density of monsters in Loch Ness. Limnology and Oceanography, 5(796–8).

Sheldon, R. W., Prakash, A., & Sutcliffe, W. H. (1972). The size distribution of particles in the ocean. Limnology and Oceanography, 3(327–40).

Sheldon, R. W., Sutcliffe, W. H., & Paranjape, M. A. (1977). Structure of pelagic food chain and relationship between plankton and fish production. Journal of the Fisheries Research Board of Canada, 34, 2344–53.

Shepard, M. P., & Withler, F. C. (1958). Spawning stock size and resultant production for Skeena sockeye. Journal of the Fisheries Board of Canada, 15(5), 1007–25.

Shepherd, J. G., & Cushing, D. H. (1980). A mechanism for density-dependent survival of larval fish as the basis of a stock-recruitment relationship. ICES Journal of Marine Science, 39(2), 160–7.

Sherman, K., & Alexander, L. (1986). Variability and management of large marine ecosystems. Westview Press, Boulder, CO.

Sibly, R. M., Brown, J. H., & Kodric-Brown, A. (Eds.). (2012). Metabolic ecology: A scaling approach. John Wiley & Sons, Chichester.

Silliman, R. (1975). Selective and unselective exploitation of experimental populations of *Tilapia mossambica*. Fishery Bulletin, 73, 495–507.

Silliman, R. P., & Gutsell, J. S. (1958). Experimental exploitation of fish populations. Fishery Bulletin, 58, 215–52.

Simonoff, J. S. (1998). Smoothing methods in statistics. Springer-Verlag, New York.

Sindermann, C. J. (1970). Predators and diseases of commercial marine Mollusca of the United States. Annual Report of the American Malacological Union, 35–36.

Sindermann, C. J. (1990). Neoplastic diseases: Principal diseases of marine fish and shellfish. Academic Press, New York.

Sissenwine, M. P. (1978). Is MSY and adequate foundation for optimum yield? Fisheries, 3(6), 22–42.

Sissenwine, M. P., & Shepherd, J. G. (1987). An alternative perspective on recruitment overfishing and biological reference points. Canadian Journal of Fisheries and Aquatic Sciences, 44, 913–18.

Sissenwine, M. P., Fogarty, M. J., & Overholtz, W. J. (1988). Some fisheries management implications of recruitment variability. In: Gulland, J. A. (Ed.), Fish population dynamics (2nd edition), pp. 129–52. Wiley, Chichester, UK.

Skalski, G. T., & Gilliam, J. F. (2001). Functional responses with predator interference: Viable alternatives to the Holling type II model. Ecology, 82(11), 3083–92.

Skern-Mauritzen, M., Ottersen, G., Handegard, N. O., Huse, G., Dingsør, G. E., Stenseth, N. C., & Kjesbu, O. S. (2016). Ecosystem processes are rarely included in tactical fisheries management. Fish and Fisheries, 17(1), 165–75.

Skud, B. E. (1982). Dominance in fishes: The relation between environment and abundance. Science, 216(4542), 144–9.

Slobodkin, L. B. (1961). The growth and regulation of animal numbers. Holt, Reinhardt and Winston, New York.

Slobodkin, L. B. (1972). On the inconstancy of ecological efficiency and the form of ecological theories.

Transactions of the Connecticut Academy of Arts and Sciences, 44, 293–305.

Slobodkin, L. B. (1974). Prudent predation does not require group selection. The American Naturalist, 108(963), 665–78.

Smedbol, R. K., & Wroblewski, J. S. (2002). Metapopulation theory and northern cod population structure: Interdependency of subpopulations in recovery of a groundfish population. Fisheries Research, 55, 161–74.

Smith, A. D. M. (1994). Management strategy evaluation: the light on the hill. Population dynamics for fisheries management, Australian Society for Fish Biology Proceedings, 249–53.

Smith, A. D. M., Brown, C. J., & Bulman, C. M. (2011). Impacts of fishing low trophic level species on marine ecosystems. Science, 333, 1147–50.

Smith, S. J., Hunt, J. J., & Rivard, D. (Eds.). (1993). Risk Evaluation and Biological Reference Points for Fisheries Management Canadian Special Publication of Fisheries and Aquatic Sciences No. 120. NRC Research Press.

Smith, T. D. (1994). Scaling fisheries: The science of measuring the effects of fishing, 1855–1955. Cambridge University Press, Cambridge, UK.

Soetaert, K. & P. M. J. Herman (2009). A practical guide to ecological modelling: using R as a simulation platform. Springer.

Soutar, A., & Isaacs, J. D. (1969). Abundance of pelagic fish during the 19th and 20th centuries as recorded in anaerobic sediment off the Californias. Fishery Bulletin, 72, 257–73.

Spalding, M. D., Fox, H. E., Allen, G. R., Davidson, N., Ferdaña, Z. A., Finlayson, M., . . . & Robertson, J. (2007). Marine ecoregions of the world: A bioregionalization of coastal and shelf areas. BioScience, 57(7), 573–83.

Sparholt, H. and Cook, R. M. (2010). Sustainable exploitation of temperate fish stocks. Biology Letters, 6(1), 124–27. https://doi.org/doi: 10.1098/rsbl.2009.0516

Sparre, P. (1991). Introduction to multispecies virtual population analysis. In: Multispecies models relevant to management of living resources (Vol. 193, pp. 12–21). ICES Marine Science Symposium.

Spencer, P. D., & Collie, J. S. (1997a). Patterns of population variability in marine fish stocks. Fisheries Oceanography, 6, 188–204.

Spencer, P. D., & Collie, J. S. (1997b). The effect of nonlinear predation rates on rebuilding the Georges Bank (*Melanogrammus aeglefinus*) stock. Canadian Journal of Fisheries and Aquatic Sciences 54, 2920–9.

Sprules, W. G. (2008). Ecological change in Great Lakes communities—A matter of perspective. Canadian Journal of Fisheries and Aquatic Sciences, 65, 1–9.

Squires, D. (1987). Fishing effort: Its testing, specification, and internal structure in fisheries economics and management. Journal of Environmental Economics and Management, 14(3), 268–82.

Steele, J. H. (1974). The structure of marine ecosystems. Harvard University Press, Cambridge, Mass.

Steele, J. H. (2001). Network analysis of food webs. In: Steele, J. H., Thorpe, S., Turekian, K., Encyclopedia of Ocean Sciences (pp. 1870–5). Academic Press, San Diego.

Steele, J. H., Collie, J. S., Bisagni, J. J., Gifford, D. J., Fogarty, M. J., Link, J. S., & Durbin, E. G. (2007). Balancing end-to-end budgets of the Georges Bank ecosystem. Progress in Oceanography, 74, 423–48.

Steele, J. H., & Henderson, E. W. (1981). A simple plankton model. American Naturalist, 117, 676–91.

Steele, J. H., & Henderson, E. W. (1984). Modeling long-term fluctuations in fish stocks. Science, 224, 985–7.

Steele, J. H., & Henderson, E. W. (1992). The role of predation in plankton models. Journal of Plankton Research, 14, 157–72.

Steele, J. H., & Schumacher, M. (2000). Ecosystem structure before fishing. Fisheries Research, 44, 201–5.

Stevens, M. H. (2009). A Primer of Ecology with R. Springer-Verlag, New York.

Stock, C. A., John, J. G., Rykaczewski, R. R., Asch, R. G., Cheung, W. W., Dunne, J. P., & Watson, R. A. (2017). Reconciling fisheries catch and ocean productivity. Proceedings of the National Academy of Sciences USA, 114, 1441–9.

Stringer, C., & McKie, R. (1998). African exodus: The origins of modern humanity. Jonathan Cape, London.

Stroud, R. H., & Clepper, H. (1979). Predator–prey systems in fishery management. Sport Fishing Institute, Washington, DC.

Sugihara, G. (1984). Ecosystem dynamics. In May, R. M. (Ed.). Exploitation of Marine Communities: report of the Dahlem Workshop on exploitation of marine communities, Berlin 1984, April 1-6. Springer-Verlag. pp 131–153.

Sugihara, G. (1994). Nonlinear forecasting for the classification of natural time series. Philosophical Transactions of the Royal Society of London Series A, 348, 477–95.

Sugihara, G., & May, R. M. (1990). Nonlinear forecasting as a way of distinguishing chaos from measurement error in time series. Nature, 344, 734–41.

Sugihara, G., May, R., Ye, H., Hsieh, C.-h., Deyle, E., Fogarty, M., & Munch, S. (2012). Detecting causality in complex ecosystems. Science, 338(6106), 496–500.

Sullivan, K. J. (1991). The estimation of parameters of the multispecies production model. In: Multispecies Models Relevant to Management of Living Resources (Vol. 193, pp. 185–93). ICES Marine Science Symposium.

Sutcliffe Jr, W. H., Drinkwater, K., & Muir, B.S. (1977). Correlations of fish catch and environmental factors in the Gulf of Maine. Journal of the Fisheries Board of Canada, 34(1), 19–30.

Sutton, R. T., & Hodson, D. L. (2005). Atlantic Ocean forcing of North American and European summer climate. Science, 309(5731), 115–18.

Swingle, H. S. (1950). Relationships and dynamics of balanced and unbalanced fish populations (Alabama Polytechncal Institute Bulletin No. 274; p. 74). Alabama Agricultural Experiment Station.

Takashina, N., Mougi, A., & Iwasa, Y. (2012). Paradox of marine protected areas: suppression of fishing may cause species loss. Population Ecology, 54(3), 475–85.

Takens, F. (1981). Detecting strange attractors in turbulence. In: Rand, D., & Young, L. S. (Eds.), Dynamical Systems and Turbulence, Warwick 1980. Lecture Notes in Mathematics (Vol. 898, pp. 366–81). Springer, Berlin Heidelberg.

Tanabe, K., & Namba, T. (2005). Omnivory creates chaos in simple food web models. Ecology, 86(12), 3411–14.

Tang, Y. A. (1970). Evaluation of balance between fishes and available fish foods in multispecies fish culture ponds in Taiwan. Transactions of the American Fisheries Society, 99(4), 708–18.

Tanner, J. E., Hughes, T. P., & Connell, J. H. (2009). Community-level density dependence: an example from a shallow coral assemblage. Ecology, 90(2), 506–16.

Taylor, D. L., & Collie, J. S. (2003). Effect of temperature on the functional response and foraging behavior of the sand shrimp *Crangon septemspinosa* preying on juvenile winter flounder *Pseudopleuronectes americanus*. Marine Ecology Progress Series, 263, 217–34.

Temming, A., & Herrmann, J. P. (2009). A generic model to estimate food consumption: Linking von Bertalanffy's growth model with Beverton and Holt's and Ivlev's concepts of net conversion efficiency. Canadian Journal of Fisheries and Aquatic Sciences, 66(4), 683–700.

Thompson, W. F., & Bell, F. H. (1934). Biological statistics of the Pacific halibut fishery. 2. Effects of changes in intensity upon total yield and yield per unit of gear (Report of the International Fisheries Commission No. 8; p. 49). International Fisheries Commission, Seattle, WA.

Thornton, K. W., & Lessem, A. S. (1978). A temperature algorithm for modifying biological rates. Transactions of the American Fisheries Society, 107(2), 284–7.

Thorpe, R. B. (2019). What is multispecies MSY? A worked example from the North Sea. Journal of Fish Biology. https://doi.org/DOI: 10.1111/jfb.13967

Thorpe, R. B., Dolder, P. J., Reeves, S., Robinson, P., & Jennings, S. (2016). Assessing fishery and ecological consequences of alternate management options for multispecies fisheries. ICES Journal of Marine Science, 73(6), 1503–12.

Thorpe, R. B., Jennings, S., & Dolder, P. J. (2017). Risks and benefits of catching pretty good yield in multispecies mixed fisheries. ICES Journal of Marine Science, 74(8), 2097–106.

Tonn, W. M. (1985). Density compensation in *Umbra–Perca* fish assemblages of northern Wisconsin lakes. Ecology, 66(2), 415–29.

Townsend, D. W., Thomas, A. C., Mayer, L. M., Thomas, M. A., & Quinlan, J. A. (2006). Oceanography of the northwest Atlantic continental shelf (1, W). In: Robinson, A. R., & Brink, K. H., The sea (Vol. 14A). Harvard University Press, Cambridge, Mass.

Trenkel, V. M. (2018). How to provide scientific advice or ecosystem-based management. Fish and Fisheries, 19(2), 390–8.

Tsoa, E., Schrank, W. E., & Roy, N. (1985). Generalizing fisheries models: An extension of the Schaefer analysis. Canadian Journal of Fisheries and Aquatic Sciences, 42(1), 44–50.

Tsou, T. S., & Collie, J. S. (2001). Estimating predation mortality in the Georges Bank fish community. Canadian Journal of Fisheries and Aquatic Sciences, 58(5), 908–22.

Tudela, S., Coll, M., & Palomera, I. (2005). Developing an operational reference framework for fisheries management on the basis of a two-dimensional index of ecosystem impact. ICES Journal of Marine Science, 62(3), 585–91.

Turchin, P. (2003). Complex population dynamics: A theoretical/empirical synthesis (Vol. 35). Princeton University Press, Princeton, NJ.

Uchiyama, T., Kruse, G. H., & Mueter, F. J. (2016). A multispecies biomass dynamics model for investigating predator–prey interactions in the Bering Sea groundfish community. Deep Sea Research Part II: Topical Studies in Oceanography, 134, 331–49.

Ursin, E. (1979). Principles of growth in fishes. Symposia of the Zoological Society of London, 44, 63–87.

Ursin, E. (1982). Multispecies fish stock and yield assessment in ICES. Canadian Special Publication of Fisheries and Aquatic Sciences No. 59, 39–47.

Usher, M. B. (1972). Developments in the Leslie matrix model. In: Jeffers, J. N. R. (Ed.), Mathematical models in ecology (pp. 29–60). Symposia of the British Ecological Society, Blackwell, Oxford.

van Wilk, S. J., Taylor, M. I., Creer, S., Dreyer, C., Rodrigues, F. M., Ramnarine, I. W., . . . & Carvalho, G. R. (2013). Experimental harvesting of fish populations drives genetically based shifts in body size and maturation. Frontiers in Ecology and the Environment, 11, 181–7.

Vandermeer, J. & Goldberg, D. (2003). Population ecology: First principles. Princeton University Press, Princeton, NJ.

Vandermeer, J. (1989). The ecology of intercropping. Cambridge University Press, Cambridge.

Vezina, A. F., & Platt, T. (1988). Food web dynamics in the ocean I. Best estimates of flow networks using inverse methods. Marine Ecology Progress Series, 42, 269–87.

Volterra, V. (1926). Fluctuations in the abundance of a species considered mathematically. Nature, 118(2972), 558–60.

Volterra, V. (1928). Variations in the number of individuals in animal species living together. Journal du Conseil International pour Exploration de la Mer, 3, 1–51.

von Bertalanffy, L. (1938). A quantitative theory of organic growth (Inquiries on growth laws II). Human Biology, 10, 181–213.

von Bertalanffy, L. (1957). Quantitative laws in metabolism and growth. The Quarterly Review of Biology, 32(3), 217–31.

Wahle, R. A., & Fogarty, M. J. (2006). Growth and development: Understanding and modelling growth variability in lobsters. In: Phillips, B. F. (Ed.), Lobsters: Biology, management, aquaculture and fisheries (pp. 1–44). Blackwell, Oxford.

Waldie, P. A., Blomberg, S. P., Cheney, K. L., Goldizen, A. W., & Grutter, A. S. (2011). Long-term effects of the cleaner fish *Labroides dimidiatus* on coral reef fish communities. PLoS One, 6(6), 21201.

Walford, L. A. (1946). A new graphic method of describing the growth of animals. The Biological Bulletin, 90(2), 141–7.

Walter, G. G., & Hoagman, W. J. (1971). Mathematical models for estimating changes in fish populations with applications to Green Bay. In Proc. 14th Conference Great Lakes Research (Vol. 170, p. 84).

Walter, G., & Hoagman, W. J. (1975). A method for estimating year class strength from abundance data with application to the fishery of Green Bay, Lake Michigan. Transactions of the American Fisheries Society, 104(2), 245–55.

Walters, C. J. (1986). Adaptive management of renewable resources. Blackburn Press, Caldwell, NJ.

Walters, C. (2000). Impacts of dispersal, ecological interactions, and fishing effort dynamics on efficacy of marine protected areas: how large should protected areas be? Bulletin of Marine Science, 66(3), 745–57.

Walters, C., Pauly, D., & Christensen, V. (1999). Ecospace: prediction of mesoscale spatial patterns in trophic relationships of exploited ecosystems, with emphasis on the impacts of marine protected areas. Ecosystems, 2(6), 539–54.

Walters, C. J., Christensen, V., Martell, S. J., & Kitchell, J. F. (2005). Possible ecosystem impacts of applying MSY policies from single-species assessment. ICES Journal of Marine Science, 62, 558–68.

Walters, C. J., Christensen, V., & Pauly, D. (1997). Structuring dynamic models of exploited ecosystems from trophic mass-balance assessments. Reviews in Fish Biology and Fisheries, 7, 139–72.

Walters, C., & Essington, T. (2010). Recovery of bioenergetics parameters from information on growth: Overview of an approach based on statistical analysis of tagging and size-at-age data. The Open Fish Science Journal, 3(1), 52–68.

Walters, C. J., Hilborn, R., & Christensen, V. (2008). Surplus production dynamics in declining and recovering fish populations. Canadian Journal of Fisheries and Aquatic Sciences, 65(11), 2536–51.

Walters, C., & Kitchell, J. F. (2001). Cultivation/depensation effects on juvenile survival and recruitment: Implications for the theory of fishing. Canadian Journal of Fisheries and Aquatic Sciences, 58(1), 39–50.

Walters, C. J., & Ludwig, D. (1981). Effects of measurement errors on the assessment of stock–recruitment relationships. Canadian Journal of Fisheries and Aquatic Sciences, 38(6), 704–10.

Walters, C. J., & Martell, S. J. (2004). Fisheries ecology and management. Princeton University Press, Princeton, NJ.

Walters, C. J., & Post, J. R. (1993). Density-dependent growth and competitive asymmetries in size-structured fish populations: A theoretical model and recommendations for field experiments. Transactions of the American Fisheries Society, 122(1), 34–45.

Walters, C. J., Stocker, M., Tyler, A. V., & Westrheim, S. J. (1986). Interaction between Pacific cod (*Gadus macrocephalus*) and herring (*Clupea harengus pallasii*) in the Hecate Strait, British Columbia. Canadian Journal of Fisheries and Aquatic Sciences, 43(4), 830–7.

Ware, D. M. (1980). Bioenergetics of stock and recruitment. Canadian Journal of Fisheries and Aquatic Sciences, 37, 1012–24.

Ware, D. M. (2000). Aquatic ecosystems: Properties and models. In: Harrison, P. J. & Parsons, T. R. (Eds.), Fisheries Oceanography: An Integrative Approach to Fisheries and Ecology and Management. Blackwell, Oxford.

Waters, T. F. (1977). The streams and rivers of Minnesota. University of Minnesota Press, Minneapolis.

Waters, T. F. (1999). Long-term trout production dynamics in Valley Creek, Minnesota. Transactions of the American Fisheries Society, 128(6), 1151–62.

Weatherley, A. H. (1972). Growth and ecology of fish populations. Academic Press, San Diego.

Werner, E. E., & Gilliam, J. F. (1984). The ontogenetic niche and species interactions in size-structured populations. Annual Review of Ecology and Systematics, 15(1), 393–425.

Werner, E. E., & Hall, D. J. (1977). Competition and habitat shift in two sunfishes (Centrarchidae). Ecology, 58, 869–976.

Westman, K., Savolainen, R., & Julkunen, M. (2002). Replacement of the native crayfish *Astacus astacus* by the introduced species *Pacifastacus leniusculus* in a small, enclosed Finnish lake: A 30-year study. Ecography, 25(1), 53–73.

Whitney, H. (1936). Differentiable manifolds. Annals of Mathematics, Second Series, Vol. 37(3) (July, 1936 645–80.

Whipple, S. J., Link, J. S., Garrison, L. P., & Fogarty, M. J. (2000). Models of predation and fishing mortality in aquatic ecosystems. Fish and Fisheries, 1(1), 22–40.

Wiff, R., Roa-Ureta, R. H., Borchers, D. L., Milessi, A. C., & Barrientos, M. A. (2015). Estimating consumption to biomass ratio in non-stationary harvested fish populations. PLoS One, 10(11), 0141538.

Wilson, E. O., & Bossert, W. H. (1971). A primer of population biology (Vol. 3). Sinauer Associates, Sunderland, Mass.

Wilson, R., Millero, F. J., Taylor, J. R., Walsh, P. J., Christensen, V., Jennings, S., et al. (2009). Contribution of fish to the marine inorganic carbon cycle. Science, 323, 359–62.

Winberg, G. G. (1960). Rate of metabolism and food requirements of fishes. Fisheries Research Board of Canada. Translations series, no. 194.

Wing, S. R., Botsford, L. W., & Quinn, J. F. (1998). The impact of coastal circulation on the spatial distribution of invertebrate recruitment, with implications for management. Canadian Special Publication of Fisheries and Aquatic Sciences, 125, 285–94.

Wing, S. R., & Jack, L. (2013). Marine reserve networks conserve biodiversity by stabilizing communities and maintaining food web structure. Ecosphere, 4(11), 1–14.

Winship, A. J., Trites, A. W., & Calkins, D. G. (2001). Growth in body size of the Steller sea lion (*Eumetopias jubatus*). Journal of Mammalogy, 82(2), 500–19.

Witherell, D., Pautzke, C., & Fluharty, D. (2000). An ecosystem-based approach for Alaska groundfish fisheries. ICES Journal of Marine Science, 57(3), 771–7.

Wootton, J. T., & Emmerson, M. (2005). Measurement of interaction strength in nature. Annual Review of Ecology and Systematics, 36, 419–44.

Wootton, R. J. (1998). Ecology of teleost fishes (2nd edition.) Springer, Dordrecht.

Worm, B., Hilborn, R., Baum, J., Branch, T., Collie, J. S., Costello, C., ... & Zeller, D. (2009). Rebuilding global fisheries. Science, 325, 578–85.

Ye, H., Beamish, R. J., Glaser, S. M., Grant, S. C. H., Hsieh, C.-h., Richards, L. J., ... & Sugihara, G. (2015). Equation-free mechanistic ecosystem forecasting using empirical dynamic modeling. Proceedings of the National Academy of Sciences USA, 112(13), E1569–76. https://doi.org/10.1073/pnas.1417063112

Ye, H., Clark, A., Deyle, E., & Sugihara, G. (2016). rEDM: an R package for empirical dynamic modeling and convergent cross-mapping.

Ye, H., Deyle, E. R., Gilarranz, L. J., & Sugihara, G. (2015). Distinguishing time-delayed causal interactions using convergent cross mapping. *Scientific reports*, 5, 14750

Yodzis, P. M. (1989). Introduction to theoretical ecology. Harper & Row, New York.

Yodzis, P. M. (1998). Local trophodynamics and the interaction of marine mammals and fisheries in the Benguela ecosystem. Journal of Animal Ecology, 67, 635–58.

Yodzis, P. M. (2000). Diffuse effects in food webs. Ecology, 1, 261–6.

Yurista, P. M., Yule, D. L., Balge, M., Van Alstine, J. D., Thompson, J. A., Gamble, A. E., ... & Vinson, M. R. (2014). A new look at the Lake Superior biomass spectrum. Canadian Journal of Fisheries and Aquatic Sciences, 71, 1324–33.

Zaret, T. M. (1980). Predation and freshwater communities (Vol. 187). Yale University Press, New Haven, Conn.

Zhou, S., Kolding, J., Garcia, S. M., Plank, M. J., Bundy, A., Charles, A., ... & Reid, D. G. (2019). Balanced harvest: concept, policies, evidence, and management implications. Reviews in Fish Biology and Fisheries, 29(3), 711–33.

Zhou, S., Smith, A. D., Punt, A. E., Richardson, A. J., Gibbs, M., Fulton, E. A., ... & Sainsbury, K. (2010). Ecosystem-based fisheries management requires a change to the selective fishing philosophy. Proceedings of the National Academy of Sciences USA, 107(21), 9485–9.

Zug, G. R., Balazs, G. H., Wetherall, J. A., Parker, D. M., & Murakawa, S. K. (2002). Age and growth of Hawaiian seaturtles (*Chelonia mydas*): An analysis based on skeletochronology. Fishery Bulletin, 100(1), 117–27.

Index

Tables, figures, and boxes are indicated by an italic *t*, *f*, and *b* following the page number.

D

Q

R

S